Dance Music Manual

Dance Music Manual

Tools, Toys, and Techniques

Third Edition

Rick Snoman

NEW YORK AND LONDON

First published 2004 by Focal Press
This edition published 2014 by Focal Press
70 Blanchard Rd Suite 402, Burlington, MA 01803

Published in the UK by Focal Press
2 Park Square, Milton Park, Abingdon, Oxon OX14 4RN

Focal Press is an imprint of the Taylor & Francis Group, an informa business

Library of Congress Cataloging in Publication Data
Snoman, Rick.
Dance music manual: tools, toys, and techniques/Rick Snoman.—Third edition.
pages cm
1. Underground dance music—Instruction and study. 2. Underground dance music—Production and direction. I. Title.
MT723.S545 2014
781.49—dc23

2013026006

ISBN: 978-0-415-82564-1 (pbk)
ISBN: 978-0-203-38364-3 (ebk)

Typeset in ITC Giovanni
Project Managed and Typeset by: diacriTech

300

Focal Press
Taylor & Francis Group

This book is dedicated to my wife Lindsay and my children, Neve and Logan.

Table of Contents

Acknowledgements

I would like to personally thank the following for their invaluable help, contributions and support whilst writing this edition:

Ian Shaw
Dakova Dae
Peter Allison-Nichol
Colja Boennemann
Shu Med
John Mitchell
Mike at the Whippin' post
Simon Adams
Everyone on the Dance Music Production Forum
The DJ's who play out our music

I'd particularly like to thank the following:
Amy-Jo for providing the vocals for the audio examples
Lars Pettersson for his tireless efforts in correcting my grammar

Introduction

> You can have all the music plug-ins in the world but that doesn't make you an artist or an EDM producer – it just makes you a collector. The only secret to producing great dance music is through knowing what to use, when to use it and how to use it...
>
> **R. Snoman**

Welcome to the third edition of the *Dance Music Manual*. Since the release of the second edition, technology and techniques have changed in a particularly significant way, and therefore much of this edition had to be totally re-written to reflect these changes.

In particular, I think one of the biggest problems evolving out of the current dance music scene is the attitude banded around by various media that the entire production ethic is down to a few 'secrets', and with nothing more than a laptop and a quick tutorial that reveals these 'secrets', you too can be knocking out hits like Skrillex or David Guetta.

Whilst I wouldn't argue that it is possible to produce a good quality dance track on little more than a laptop and some carefully chosen software, I would argue that good music production amounts to much more than spending a few hours with a magazine or video tutorial!

There are no secrets to producing dance music, and there are certainly no shortcuts either. To be successful as an EDM artist today, you need to be fluent in music theory, composition, arranging, recording, synthesis, sound design, producing, mixing, mastering and promotion, and have an understanding of computers. This takes time to learn and time to practice.

Simply put, producing EDM dance music is far more involved than many first realize. Whilst on face value the music can appear superficially simple, it involves an incredibly intricate and multifaceted process of production ethics. Even something as simple in appearance as the basic house drum loop requires knowledge of synthesis, compression, reverb, delay, filters, limiters and mixing alongside techniques such as syncopation, polymeter, polyrhythm, hyperbeat, compound time and both frequency and pitch modulation. And that's before you even consider composing a bass line. Indeed, the real trick with dance music is making something extremely complex appear incredibly simple. And the simpler it appears, the more complex it usually is to produce.

For this edition, I'm going to discuss the theory and practice of today's modern EDM production ethics. While there have been numerous publications on this subject, some worse than others, the majority have been written by authors who have little experience of the scene, but have simply relied on 'educated guesswork' from listening to the music.

I was first introduced to dance music at the Hacienda in Manchester in 1988 and was a constant regular until 1995. During that time, I began to produce my own dance music using an Atari STE and limited MIDI instruments. I've released numerous white labels, worked as a tape op, engineer and producer, released music under numerous guises from GOD to Phiadra to Red5, officially remixed for large label artists such as Kylie Minogue, Madonna, Christina Aguilera, Lady Gaga and Britney Spears and also met with countless artists in my studios.

I've held seminars across the country on remixing and producing club-based dance music, authored numerous articles and reviews, authored a distance-learning course on digital media for Guy Michelmore at *Music for the Media* and produced a large number of tutorial videos for my site at www.dancemusicproduction.com.

This book is a culmination of the knowledge I've attained over the years, and I hope to offer the reader a shortcut of a good few years through my experience, discussions with other producers and artists and personal observations of the techniques and practices that have been developed over the past 20 years.

However, I cannot stress enough that the purpose of this book is not to act as paint-by-numbers or a cookie-cutting course to your next number one hit. If that's what you're searching for, you won't find it in this book. Despite many companies now jumping in and attempting to cash in on the dance music craze, as I mentioned before, there are no secrets to producing a dance hit.

I can guarantee you that we are not clandestine. We do not meet in undisclosed locations with surreptitious handshakes to share and swap closely guarded production techniques. The techniques are simply discovered through years of observation and analysis of our peer's music and then further developed upon with experimentation.

The aim of this book is simply to demystify the process and save the reader from having to sift through the mine of misinformation born from those looking to make a quick buck by offering useless, impotent information based around guesswork and no practical experience of the music itself.

Perhaps more importantly, though, its purpose is not to say 'This is how it should be done' but rather 'This is how we are currently doing it today, it's up to you to show us how it's done tomorrow'. Dance music will be always based around musicians and producers pushing the technology and ideas in new directions, and simply reading a book or article or watching a tutorial will not turn you into a superstar overnight.

Whilst this book will provide you with the techniques and knowledge required to point you in the right direction, it is your own observations and analysis based around the principles discussed here that will unlock your true potential. Creativity can never be encapsulated in books, pictures, video or the latest sequencer or synthesizer. It is your own creativity and experimentation that will produce the dance records of tomorrow.

Rick Snoman

http://www.dancemusicproduction.com
www.facebook.com/rick.snoman

CHAPTER 1

Music Theory Part 1 – Octaves, Scales and Modes

'Could you put that up an octave just a little?'

Producer

Although perhaps not the most inviting way to start a book on dance music production, some prior knowledge of music theory is vital for any producer looking to create and produce electronic dance music (EDM). While many critics often view this style of music as little more than a series of constant, unchanging repeating cycles that make little use of any musical theory, casual listening belies its intricate complexity.

The fact is keeping a listener entertained for 5 to 10 minutes at a time with what *appears* to be little more than a few repeating patterns requires more than a faith basis that your audience will ingest any number of pharmaceuticals before listening. Indeed, it instead relies on a judicious manipulation of not only production techniques but also basic musical theory. And it's only now – 30 years after its conception that music theorists are beginning to analyse EDM and understand how its manipulation plays such a significant role in its creation.

THE MUSICAL OCTAVE

The first principle in grasping the ideas of music theory is to understand the relationship between notes and intervals in the musical scale, so we should begin by examining a piano keyboard.

As shown in Figure 1.1, regardless of the number of keys spread across the piano keyboard, it consists of the same 12 keys repeated over and over. Each individual key relates to a particular pitch in the musical octave, and to aid with identification, each is named after a letter in the alphabet. Consequently, an octave will always begin and end on the same letter; so if you were to start counting from note C, the octave would end on the following C.

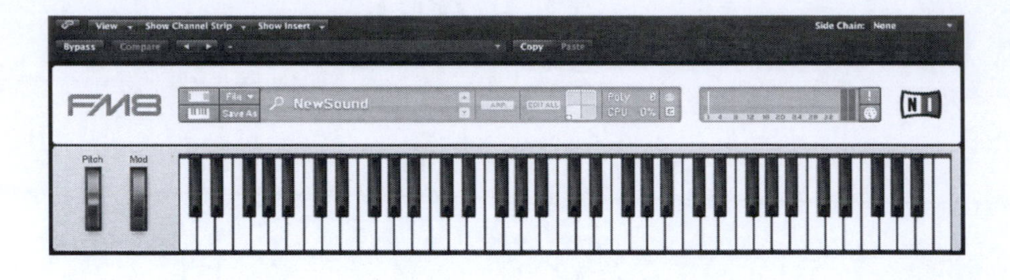

FIGURE 1.1
A piano keyboard

Naturally, a piano keyboard consists of both white and black keys, and it is important to first note that these smaller black notes can be referred to by more than one name. For instance, if we begin on the white, or natural, note of C and we move to the first black note to its right, in effect counting *up* the scale, the black note becomes a *sharp* of the C and is denoted as C#.

Conversely, if we begin at the natural note of D and count *down* the scale to the left, the black note becomes a flat of the D and is written as D*b*. This black note still produces the same principle pitch (i.e. they are enharmonically equivalent); it's simply known by two different names depending on the context of the music. To understand this context, we must examine the theory of intervals.

The distance or spacing between different pitches in the musical scale is termed an *interval*, and in western music, a semitone is the smallest possible interval, or distance, that exists between any two different pitches. For example, the interval between C and C# is one semitone, C# to D is one semitone, D to D# is one semitone and so forth throughout the octave, as shown in Figure 1.2.

Any two semitones can hypothetically be summed together to produce a single *tone*. This means that if you were to play C followed by C#, it would produce two pitches a *semitone* apart, whereas if you were to play C followed by D, it would produce two pitches a *tone* apart. This is because by moving from a C to a D, you're moving by two semitones, that is one semitone plus one semitone is equal to one tone. It's this simple relationship between tones and semitones that forms the fundamental cornerstone of all musical theory.

KEYS AND SCALES

To place all this into a more compositional perspective, an octave can be viewed as the musical equivalent of our alphabet, whereas we select letters from the alphabet to form words, the same can be said about the octave.

In the composition of music, we select a total of seven notes from the octave in which to produce any musical piece, and these same seven notes are used to compose everything from the lead to the chords and the bass line. These notes are not random, however, and are pre-determined by the choice of our musical *key*.

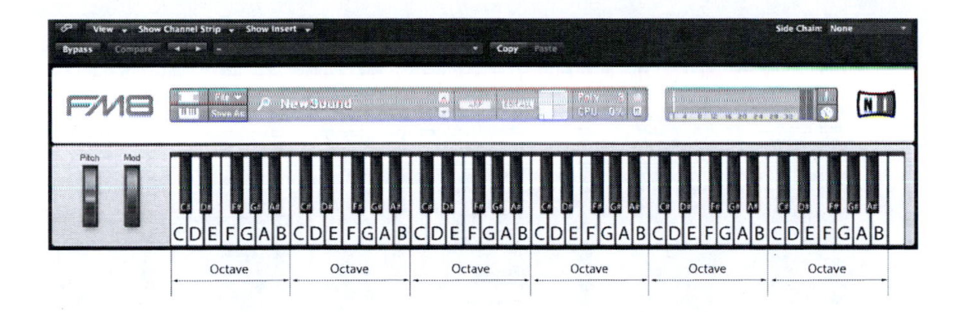

FIGURE 1.2
Intervals on a piano keyboard

The key of music uses a theory to provide the composer with a collective series of notes selected from the octave that will all sound pleasing to our ears when used in a composition.

To better explain this principle, imagine that you wanted to write a piece of music in a major key using a particular note in the octave as the starting, or root note for your composition. Regardless of the starting note you chose, all the major scales follow the pattern laid out below:

Tone – Tone – Semitone – Tone – Tone – Tone – Semitone

To compose any piece of music, we therefore first choose the key we wish to work in and then use the aforementioned formula to calculate the notes that exist in that particular key. So, if you wanted to write in the key of C major, that is, your first note in the scale was C, you would begin counting at C, and by using the formula, it would produce the notes in Figure 1.3.

C – D – E – F – G – A – B – C

Tone – Tone – Semitone – Tone – Tone – Tone – Semitone

Similarly, if you wanted to write in, say, E major, you would begin counting from E, and by using the major formula, this would result in the following notes:

E – F# – G# – A – B – C# – D# – E

In this instance, all the compositional elements of the song from the bass through to the lead and the chords would use a combination of just these notes from the key of the music. Musicians and composers use different keys for their songs because each key can provide a slightly different mood or feel

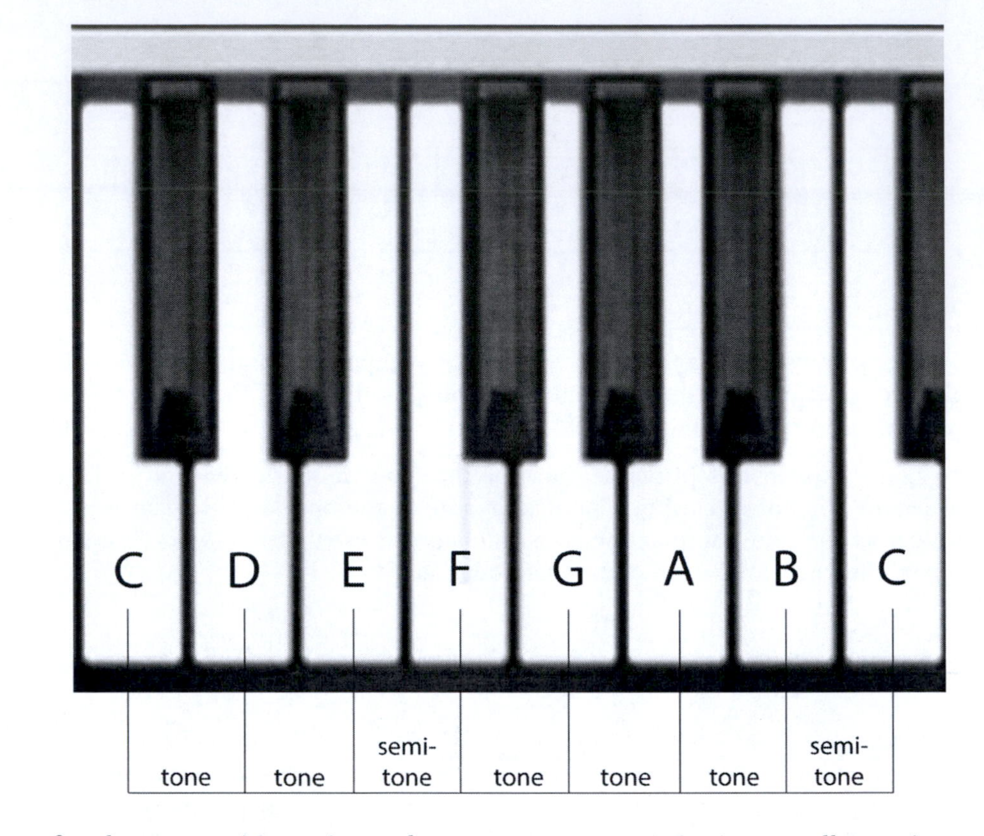

FIGURE 1.3
The arrangement of a major key

for the composition, since when we compose music, it naturally gravitates around the *root note* of the chosen key.

It is generally accepted among many musicians and music theorists that writing in pretty much any major key will produce music that exhibits a happy uplifting vibe, and this is why many pop music records and nursery rhymes are composed and performed in a major key. However, while a good number of dance records are also composed in a major key, it is more common to compose in a minor key.

The minor key is a different musical mode than major and exhibits a much larger shift in the emotional impact and feel of the record than writing in different major keys. Minor keys are sometimes described as sad or somber, but this is an incorrect assumption, and a number of uplifting dance records have also been written in minor keys. Indeed, rather than being described as a somber modal form, minor would probably be better described as sounding more serious or focused than a major key.

As with composing in a major key, a minor key can be written starting from any note in the octave, and the difference lies in the order of the tone and semitone pattern. To create the minor scale, the major tonal pattern is nudged or pushed along by two to the right as shown in Figure 1.4.

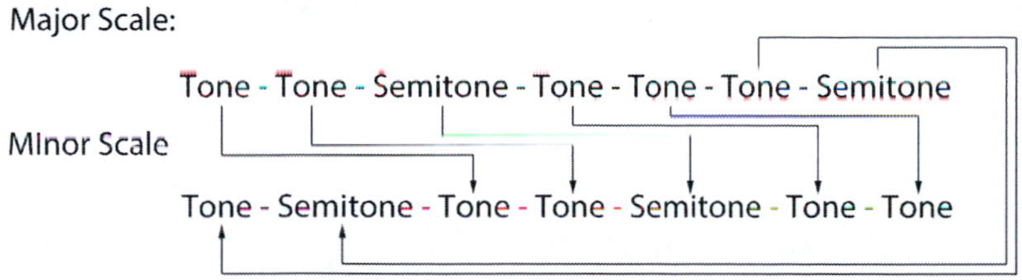

Major Scale:

Tone - Tone - Semitone - Tone - Tone - Tone - Semitone

Minor Scale

Tone - Semitone - Tone - Tone - Semitone - Tone - Tone

FIGURE 1.4
The tone and semitone arrangement of major and minor

Major = Tone – Tone – Semitone – Tone – Tone – Tone – Semitone

Minor = Tone – Semitone – Tone –Tone – Semitone – Tone – Tone

This new minor pattern formula is used to create a number of different keys in the minor scale in the exact same method used to create keys in the major scale. For example, if you were to choose to write in A minor, you would begin counting from A and following the minor pattern, it would produce the following notes:

A – B – C – D – E – F – G – A

Similarly, if you chose to write in the key of C# minor, using the minor formula, the notes would consist of:

C# – D# – E – F# – G# – A – B – C#

Those with some prior musical knowledge may have noticed that in these minor key examples, I didn't randomly choose two minor keys, rather I specifically chose the relative keys to the previous examples in the major key.

All major keys will have a relative minor and vice versa, and this means that a key in both minor and major will contain the exact same notes, just starting with a different root note. For example, the notes in C major consist of:

C – D – E – F – G – A – B – C

Whereas the notes in A minor consist of:

A – B – C – D – E – F – G – A

In these examples, C major and A minor feature the same notes as does the previous example of E major and C# minor. Indeed the only differences between major and minor keys in both these examples are the root notes. Every major key will have a relative minor key that consists of the exact same notes, just having a different root note. This major to minor relationship offers the composer the opportunity to move, or *modulate,* between the two different keys during a composition to add more interest to a musical piece.

You can quickly calculate the relative minor of any major key by taking the sixth note of the major key and playing the minor key starting on that note. In the case of E major, the sixth note is a C#, and therefore, the relative minor root note would be C#. Table 1.1 shows all the relative major and minor keys.

MELODIC AND HARMONIC MINOR

Thus far, I've only discussed the Natural Minor Scale, but unlike the major scale, the minor scale consists of three different scales that were introduced over a period of years due to some inconsistencies that appear with the minor scale.

As previously mentioned, musical keys are designed to provide a series of pleasing intervals between the different notes or pitches in the key. For example, if you play C, C# and D, one after the other, it doesn't sound anywhere near as pleasing as if you were to play C, then D and finally E.

Table 1.1 Relative Major and Minor Keys

Major	Minor
• C major	• A minor
• C sharp/D flat major	• A sharp/B flat minor
• D major	• B minor
• D sharp/E flat major	• C minor
• E major	• C sharp/D flat minor
• F major	• D minor
• F sharp/G flat major	• D sharp/E flat minor
• G major	• E minor
• G sharp/A flat major	• F minor
• A major	• F sharp/G flat minor
• A sharp/B flat major	• G minor
• B/C flat major	• G sharp/A flat minor

However, while the natural minor can provide pleasing intervals between notes, many musicians consider them to sound disjointed or displaced when writing harmonies and melodies, and therefore, both the harmonic and melodic minor scales were introduced.

The harmonic minor scale is very similar to the natural minor scale but with the seventh note increased, or augmented, up the scale by a semitone. Without wanting to go into too much detail until chords are discussed, by raising the seventh note a semitone, it helps to create a more natural and powerful sounding chord progression.

The problem with this approach, however, is that the intervals between the sixth and seventh notes are now three semitones apart, and this is an unusual movement that's not only dissonant to our ears when not used in chords but also incredibly difficult for a vocalist to perform alongside. Indeed, it's due to our enjoyment of singing along with music that makes it very uncommon for any scale to have an interval larger than a tone because it's difficult to move by a tone and a half. Consequently, the melodic minor scale was formed.

In the melodic minor scale, both the sixth and the seventh notes of the natural minor are raised by a semitone each when ascending up the scale, but when descending the scale, it returns to using the natural minor scale. As confusing as this may initially appear, it's generally accepted that if you choose to write in a minor key, it's best to use the notes in melodic minor for all instruments with the exception of any chords or chord progressions whereby it is best to use the harmonic minor for these.

FURTHER MODES

So far I've been discussing the two most popular modes of music, the major and minor scales or as they are often known, the *Ionian* and *Aeolian* modes. However, alongside these most commonly used modes, there are a further six modes available to the musician each of which have their own tonal patterns.

Further modes are created in music much the same way as the major and minor modes discussed previously. For example, in the case of the major, or Ionian mode, we know that C major consists of only white, or natural, notes in the octave, and this provides the familiar tonal pattern of:

Tone – Tone – Semitone – Tone – Tone – Tone – Semitone

This chromatic pattern is then applied across the entire major key range, so regardless of the key you choose to work in, provided it's a major key, the same chromatic pattern is used to determine the notes in that particular key. Similarly, the Aeolian or minor mode is constructed entirely from the natural notes of the octave but rather than starting at C, we start at A to provide the tonal pattern of:

Tone – Semitone – Tone –Tone – Semitone – Tone – Tone

Therefore, it's feasible that we can create a number of further tonal patterns by starting to count at different notes within the octave whilst remaining with only natural notes. For example, if we begin to count from D upwards and ensure we only remain with the natural notes, it results in the tonal pattern of:

Tone – Semitone – Tone – Tone – Tone – Semitone – Tone

This new tonal pattern is known as Dorian mode, and like both Ionian and Aeolian, it allows the composer to write in any number of keys in this mode, each of which will exhibit a very different emotional feel than writing in the same key in either Ionian or Aeolian, and we can create modes from every natural note on the keyboard consisting of:

Phrygian (E):
Semitone – Tone – Tone – Tone – Semitone – Tone – Tone

Lydian (F):
Tone – Tone – Tone – Semitone – Tone – Tone – Semitone

Mixolydian (G):
Tone – Tone – Semitone – Tone - Tone – Semitone – Tone

Locrian (B):
Semitone – Tone – Tone – Semitone – Tone – Tone – Tone

While it's certainly worth experimenting with these modal forms to gain a deeper understanding of how modes and keys can affect the emotional impact of your music, I would nonetheless hesitate in recommending these as the basis of any club-based music track.

From my experience of writing, producing and analysing music in this field, most dance records are written in A minor, D minor or E minor. Indeed, in a recent examination of over a thousand dance records from the past 5 years, more than 42% were written in A minor. A further 21% were in D minor, 16% in E minor, while the rest were a combination of C minor, E_b minor, F# minor, C# minor, G minor, Db minor and their major relatives, in descending order.

While it could be argued that the choice for writing in A minor by many dance musicians is likely due to A minor being the relative of C major; the first scale everyone learns and the only minor scale with no sharps, I believe it has developed into a more carefully considered practical choice rather than one based solely on the simplicity of the scale.

Club DJs employ a technique known as harmonic mixing that consists of matching the musical key of different records. This helps each track flow and ebb into one another to create a single continuous flow of music, a key element of club-based music today. Many dance musicians with a musically trained ear naturally want to produce tracks that can be harmonically mixed easily with other records since it means their track is more likely to be played.

This would also explain the growing popularity of writing in both E minor and D minor. While this allows the artist to be more creative by moving away from the somewhat requisite A minor, both E and D minor keys are harmonically closer to A minor than any other key and therefore prove much less difficult for a DJ to harmonically mix in with the large proportion of A minor records.

Finally, and perhaps more importantly, the most influential and powerful area for dance music's bass reproduction on a club speaker system is 50 to 65 Hz, and the root note of A1 sits at 55 Hz. Since the music will naturally modulate or gravitate around the root note in your chosen key, having the root note of the music at 55 Hz will help to maintain the bass 'groove' energy of the record at a particularly useful frequency.

CONSTRUCTING MELODIES

Understanding how to build a scale in any particular key only forms a small part of the overall picture. Although the majority of club-based music is created in A minor, knowing the key only provides you with the notes, and they still need to be organized effectively to create melodies and motifs.

Over the past 5 years, long elegant melodic lines have slowly been drowned out of EDM and have now become the mainstay of popular music. Although some genres of dance and club-based music still employ fairly complex melodic structures such as *Uplifting Trance* and some *Electro*, almost all genres today rely on numerous motifs interacting poly-rhythmically with one another.

A motif is a short rhythmical idea that provides listeners with a reference point for the track. The opening bars of Eric Prydz *Allein*, the repeating chord patterns in Avicci's *levels*, Calvin Harris' use of rhythmical chords in *Lets Go* and the growing synth rhythm in DeadMau5 *Faxing Berlin* are all excellent examples of motifs. As soon as you hear them, you can immediately identify with the track, and if they are composed well, they can be extremely powerful tools.

Motifs follow similar guidelines to the way that we ask and then answer a question. In a motif, the first musical phrase often applies the question that is then 'answered' by the second phrase. This type of arrangement is known as a 'binary phrase' and plays an important role in almost all club-based tracks.

To demonstrate this, we'll examine the main motif in Avicci's *levels*, a track written in E major (the relative of C# minor). This is a simple motif that follows the theory and became a huge hit amongst clubbers, DJ's, and public alike. Although in the original record the lead consists of a number of synths layered together mixed with side-chain and pitch-slam processing, it can still be broken down into its essence, a simple melodic line as shown in logic sequencers pattern grid in Figure 1.5.

First, the notes in E major consist of:

E – F# – G# – A – B – C# – D# – E

On the COMPANION WEBSITE, you can hear the melody played via a piano and also access the MIDI file.

Examining the motif it can be broken down into a binary phrase, the first phrase consists of:

C#4 – B – G#3 – F#3 – E3 – E3 – E3 – E3 – E3 – E3 – D#3 – D#3 – E3 – E3

And the second phrase consists of:

C#4 – B – G#3 – F#3 – E3 – E3 – E3 – E3 – E3 – E3 – C#3 – C#3 – B2 – B2

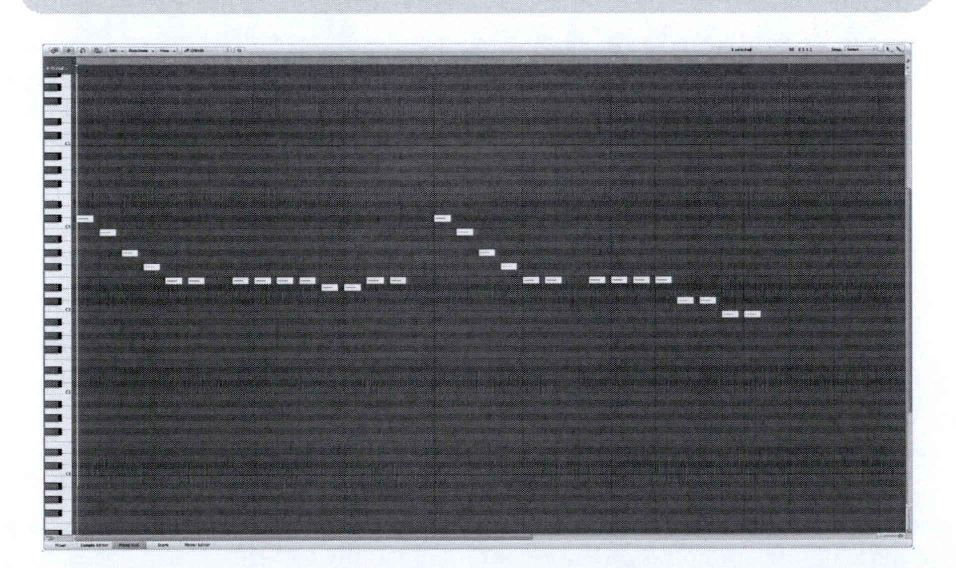

FIGURE 1.5
Avicci levels

Both phrases remain at the same pitch, and it's only at the *end* of each phrase, there is any change between the two. At the end of phrase one, the pitch remains constant with only a slight variation introduced when it drops a semi-tone for two notes at *D#3*, but then rises again back to *E3* to complete the first part of the phrase.

Phrase two repeats the same rhythm and pitch and phrase one with only a slight change applied at the end. Here, the pitch drops further than the previous phrase, moving from the *D#3* in the first phrase to *C#3*, and then rather than returning to *E3*, it moves further down the scale by a complete tone to B2 creating a greater change. Aside from this small amount of pitch variation, both phrases remain incredibly simple and almost identical in terms of rhythm.

This style of repetition in music is often mistaken to be exclusive to EDM, but many other forms of music from classical to pop to rock to metal are often based around the repetition of a single motif. In fact, the principle behind writing any great memorable record is to bombard the listener with the same phrase so that they become embedded in the listener's mind.

In most other genres of music, this is accomplished in such a way as to not to be as noticeable as it is with EDM. For instance, in popular music, it's not uncommon for the vocalist to perform a line that is then backed up by the same vocal line repeated with an instrument, or flourish, to help push the motif into your head. In classical music, the same motif is repeated over and over again but with small flourishes added to each repeat, whether this is in the form of further instruments being introduced to repeat the motif, or small changes made to the motifs rhythmical or pitch characteristics.

From this we can determine that there are three essential elements to creating any motif, and these are simplicity, rhythmical repetition, and small variations. Thus, when it comes to creating a motif, it is generally easiest to compose the rhythmical content first using a single pitch and then move onto introducing pitch changes between phrases at a later stage.

By far the easiest method of accomplishing this is to connect a piano keyboard to a DAW sequencer and keep tapping out rhythms on a keyboard while recording the subsequent MIDI data into the DAW. After 20 minutes, listen back to the MIDI recording and pick out the best rhythms you came up with or mix and match them if required with the sequencers editing tools. Alternatively, you could use a synthesizer's or DAW's arpeggiator to record MIDI patterns and rhythms that you can later edit. The main target at this stage should be aimed at producing a good rhythmical pattern.

A general benchmark to a memorable motif is the ability to hum it. For instance, it's possible to hum the motif on almost all great dance records from the past 20 years (try humming Avicci's levels or Eric Prydz's Pjanoo). If you struggle to hum a motif, it may be too fast or the notes could be too short, or positioned too close together to have any real rhythmical meaning. Once the rhythmical element is down, pitch can then be introduced.

The theory behind melodies and pitch movement is a multifaceted subject that could easily engulf the rest of the book, but very briefly it can be condensed into two simple theories, direction and amount.

Whenever we work with musical notes and pitch, we can move either up or down the scale, and each of these will have a subconscious effect on our emotion. When a subsequent note falls in the scale, it can produce two results, the impression of pulling our emotion downwards, or making the motif sounding more serious. Conversely, if a subsequent note rises in the scale, it appears more uplifting or happier. This is due to frequency content of the notes. When a note increases in pitch its high-frequency content increases, and higher frequencies tend to project a happier, lighter vibe.

This effect is evident when the low-pass filter cut-off of a synthesizer is opened up to allow more high-frequency content through. You hear this effect in countless dance records whereby the music becomes duller, as though it's being played in another room and then gradually the filter begins to open and higher frequencies pass through.

This movement creates a sense of expansion or building as the higher frequency content gradually increases. When in opposite, with the filter closing, higher frequencies are reduced, and this exhibits the feeling of closure, pulling our emotions downwards.

On the COMPANION WEBSITE, you can hear the effects of filters in the creation of builds and drops.

This effect is very similar to the key changes that are often introduced at the end of many pop songs. It's not unusual for a pop song to change to move higher in the octave range during the chorus sections, or for the final chorus in the record to introduce more power and a stronger vibe to help drive the message of the song home.

In direct relation to the pitch moving up and down the scale affecting our emotion or interpretation of the record, the amount the pitch moves *by* can also directly influence our emotion in regard to the energy it conveys. If a motif consists of large pitch jumps from one note to the next and of more than two or three tones at a time, it will make a motif sound more energetic than one that simply steps up or down the scale.

For example, in the key of A minor, playing A – B – C – D in incremental steps doesn't sound anywhere near as energetic or exciting as playing A – D – C – A. However, that's not to suggest a motif should only employ pitch jumps, rather a great motif is born from a mix of both of these practices. In order for a motif to exhibit an energetic feel, it must be carefully mixed with less energetic, stepwise movements in pitch.

By doing so, the listener has a level of comparison to relate with. Action films don't consist of *nothing* but action; there are moments during the movie where it is running smooth, so when the action does begin, it's more exciting. If you were fed with nothing but pure action from beginning to end, you would most probably switch it off, and this same principle applies with motifs. Both steps and leaps should be mixed carefully to produce a motif that exhibits energy.

RHYTHM

Above all, however, with EDM, rhythm remains king. Although pitch movements will inevitably add more energy and interest to any piece, if too much is employed it will begin to sound like a pop record or a record from the eighties.

With the possible exception of uplifting trance, club-based music relies more on its rhythmical energy than it does pitch, and much of this energy is introduced via processing and automated synthesis, not via huge pitch fluctuations throughout the motif.

KISS (Keep It Simple Stupid) applies heavily when creating the rhythmical patterns of motifs for club music, and the complexity today is born from the production aesthetics rather than the musicality. This is something we will discuss in detail later in this book.

CHAPTER 2

Music Theory Part 2 – Chords and Harmony

'A painter paints pictures on canvas. But musicians paint their pictures on silence.'

Leopold Stokowski

To the inexperienced ear, it's easy to assume that harmony is solely the mainstay of a typical chart topping pop music tracks with manufactured divas wailing over the top. Yet, there are many identifiable examples of traditional harmony employed within EDM. Indeed, almost all genres ranging from Euphoric Trance through to Lo-Fi, Progressive House, French House, Chill Out, Electro, Dubstep and Deep House will make selective use of harmony. Moreover, whilst the 'darker' termed genres such as Techno, Tech-House and Minimal are often deemed to be devoid of melody, let alone harmony, they do nonetheless benefit from its concept, because it can be used as a sketchpad for musical ideas and inspiration. For this chapter, I want to examine the basic theory behind chords, structure and the resultant harmony.

Harmony can be described as a series of chords played sequentially to create a chord progression. This progression is used to accompany or support a main melody or motif to create harmony within the music. Used in this way, it can inject emotion and energy and can play such an important role in composition that different harmonies played alongside the same melodies can result in dramatic changes to the mood and energy.

BASIC CHORD STRUCTURES

Any basic chord is constructed from two or more notes that are played simultaneously but traditionally, almost all will consist of three or more notes. The notes chosen to produce the chord must be harmonious with one another, since randomly playing any three notes on a piano/synthesizer keyboard isn't always guaranteed to produce a musically harmonic result. To understand how chords are structured to become harmonious, we first need to re-examine the musical scales discussed in Chapter 1.

There are seven notes in any key of the major and minor scales, and for the purpose of constructing a chord, each note is assigned a number one through seven. For example, C major consists of C – D – E – F – G – A – B, but equally these notes could be named 1 – 2 – 3 – 4 – 5 – 6 –7. Moreover, alongside using numbers to identify notes, they can also be awarded a specific name as shown in Table 2.1.

Both the naming and numbering system applies exactly the same irrespective of the key or scale. This means that regardless of the key or scale, the numbers and names will always correspond to the same note position in the scale. For example, if requested to play a melody in a major key consisting of 5 – 3 – 4 – 2 – 3 – 5, the musician could play the correct notes no matter what the key or scale. Similarly, this same approach could be used using the names of the notes.

The most common chord in the composition of music is the triad. This chord is constructed from just three notes consisting of intervals a third and fifth apart from the first note of the chord. Therefore, if the producer were composing in the key of C major, using C as the root note of the chord, it would consist of the notes C – E – G, since the note of E and G are three and five intervals consecutively above the *root* note of C (Figure 2.1).

Table 2.1 Numbering/Naming the Notes (Using C Major as an Example)

Number	1	2	3	4	5	6	7
C Major	C	D	E	F	G	A	B
Name	Tonic	Supertonic	Mediant	Subdominant	Dominant	Submediant	Leading tone

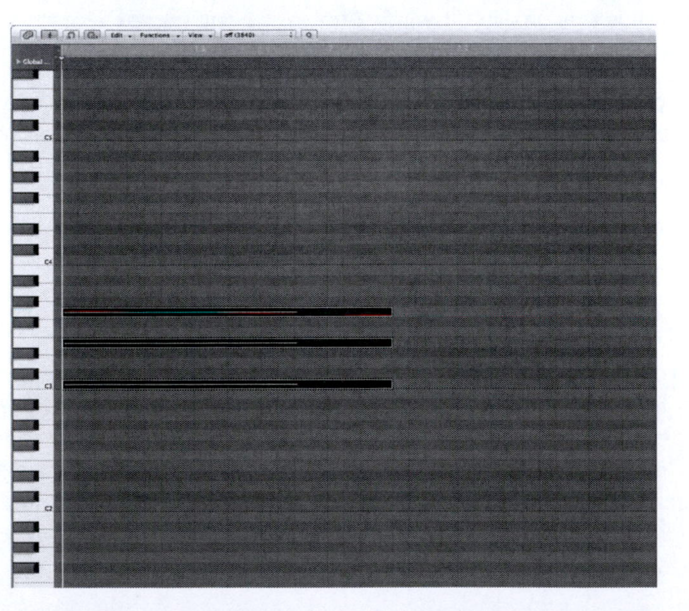

FIGURE 2.1
Logics piano roll editor showing chords CEG

It is possible to construct a chord in any scale using any note within that scale as the root note of the chord. Provided the further notes consist of third and fifth intervals from the root note chosen, the chord will remain harmonious. Consequently, chords consist of not only natural notes but also sharps and flats depending on the key and the scale in use.

Table 2.2 shows that writing in the key of E major with A as the root note of the chord, for example, would create a triad consisting of A – C# – E.

Notably, like the scales they are derived from, chords can be major or minor and understanding the difference between them is essential to the composition of music, since it will greatly influence the message received to the listeners. If the song is to deliver a serious or sombre message then typically the producer would mostly employ minor chords in the progression. On the other hand, if the music is to exhibit more of an uplifting or happy vibe, the producer will generally remain mostly with major chords.

It is possible to quickly determine whether a chord is major or minor by comparing the third note of the chord against a scale that is determined by the root

Table 2.2 **The Notes of the Ionian Scale**

Scale	1	2	3	4	5	6	7
Cb Major	Cb	Db	Eb	Fb	Gb	Ab	Bb
Gb Major	Gb	Ab	Bb	Cb	Db	Eb	F
Db Major	Db	Eb	F	Gb	Ab	Bb	C
Ab Major	Ab	Bb	C	Db	Eb	F	G
Eb Major	Eb	F	G	Ab	Bb	C	D
Bb Major	Bb	C	D	Eb	F	G	A
F Major	F	G	A	Bb	C	D	E
C Major	C	D	E	F	G	A	B
G Major	G	A	B	C	D	E	F#
D Major	D	E	F#	G	A	B	C#
A Major	A	B	C#	D	E	F#	G#
E Major	E	F#	G#	A	B	C#	D#
B Major	B	C#	D#	E	F#	G#	A#
F# Major	F#	G#	A#	B	C#	D#	E#
C# Major	C#	D#	E#	F#	G#	A#	B#

note used for the chord. To better explain this principle, we'll use our previous example of a triad chord in C major with C as the root note, comprising the notes C – E – G.

There are two intervals in this example triad: the third *bottom interval* of E and the fifth *outer interval* of G. To determine whether the chord is major or not, we first consider if the *bottom interval* (E) exits in the scale determined by the root note.

In this particular example, C is the root note of the chord, and therefore, we must check to see if the bottom interval (E) exists in the C major scale. Obviously in this particular example, it does, and therefore, the bottom interval is termed as *major 3rd*. Since this example triad contains a major 3rd, it becomes a major chord.

Remaining in C major, suppose a chord were constructed from the root note of D rather than C, Table 2.4 reveals that this would result in a triad chord consisting of the notes D – F – A.

Whilst we are still composing in the key of C, because the chord is being constructed from the *root note of D*, to determine whether the chord is major or minor, we make comparisons with the notes that exist in the D major scale.

In this instance, the third interval in the chord of D is an F. This note does not exist in the D major scale. Since this note is *not shared* in both the scale, we are writing in and the scale determined by the root note of the chord, the third interval cannot be considered a major 3rd and instead becomes a minor 3rd. Any chord that has a minor 3rd is a minor chord.

Also note that when comparing the D chord to the D major scale, the *outer interval* (fifth) is the note A. The note A is shared by the key we are writing in (C major) and the D major scale, and therefore, this becomes a *perfect fifth*. Perfect fifths have an important relationship with the tonic note of a scale and will be discussed in more detail later in this chapter.

Table 2.3 The Notes in C Major Scale

Scale	1	2	3	4	5	6	7
C Major	C	~~D~~	E	~~F~~	G	~~A~~	~~B~~

Table 2.4 Triad Chord in C Major with D as the Root

Scale	Root	2	3	4	5	6	7
C Major	D	E	F	G	A	B	C
D Major	D	E	F#	G	A	B	C#

Whilst this shared note principle may appear somewhat convoluted, it is important to understand this relationship when constructing chords and chord progressions. All the scales are derived from the same chromatic notes, and therefore, they are all inextricably linked to one another.

For example, if the producer were to write in C major and composed a chord with a root note of E, we would again compare the notes in C major to the notes in the key of E to see if both scales share the same bottom and outer intervals. If they share the same third interval, the chord is a major and if they share the same fifth interval, it is also a perfect fifth.

As long winded as this appears, there are simpler ways to remember these associations. First, the difference between a major and minor triad is the third note and, therefore, by simply adjusting the pitch of the third note by a semitone, the producer can quickly convert between a major and minor triad. Furthermore, roman numerals can be used to describe chords rather than decimal numbers. An example of this is shown in Table 2.5.

In Table 2.5, notice that some of the numerals are in capital whilst others are not. The capital numerals denote major chords whilst the rest denote minor chords. That is, if you were to construct a chord using either the first, fourth or fifth note (I, IV, V) as the root note in any major key, it would create a major chord whereas if you use the second, third, sixth or seventh as the root note (ii, iii, vi, vii), it would result in a minor chord.

There is, however, an important caveat that involves the seventh note in the scale – the leading tone. Once again using C major as the key of the music, if a chord were built using the vii (A) as the root note of the chord, the triad would consist of the notes B – D – F as shown in Table 2.6

Following the previous logic of comparing the root note of the chord to its own particular scale, the third interval is not present in the B major scale that results in the minor chord as suggested by the lower case roman numerals, but there is no perfect fifth either.

Table 2.5 Roman Numerals in Place of Decimal (Major Keys Only)

Note	1	2	3	4	5	6	7
Numeral	I	ii	iii	IV	V	vi	vii

Table 2.6 The Triad Chords in the Ionian Mode

Scale	Root	2	3	4	5	6	7
C Major	B	~~C~~	D	E	F	~~G~~	A
B Major	B	~~C#~~	D#	E	F#	~~G#~~	~~A#~~

In the B major scale, the fifth note is an F# but the note in the B major chord is an F – a semitone lower, or to use the correct term, the note is diminished. This is an important factor to observe when working with chords; when there is a reduction of one semitone on the fifth interval, it results in a *diminished* chord.

We can quickly summarize all of this by simply stating:

1. If the producer constructs a triad chord from the first, fourth or fifth note in any major key, it will result in a major triad.
2. If the producer constructs from the second, third or sixth note in a major key, it results in a minor triad.
3. The producer can quickly change a minor triad into a major triad by moving the bottom interval (third) up by a semitone. Moving a note up by a semitone is known as *augmenting*.
4. If the producer constructs from the seventh note in a major key, it will always result in a diminished triad.

Table 2.7 should clarify what constitutes the different chords.

TRIADS IN MINOR KEYS

Thus far, we've only considered creating triad chords from the major scale, but you can naturally create them from the minor scale too. If you have followed and understood the concept behind the creation of major triad chords, a glance at Table 2.8 will no doubt be self-descriptive.

Notably, in Table 2.8, the dominant (fifth note in the scale) is a major chord even though in theory it would result in a minor chord. This is because

Table 2.7 Major, Minor, Augmented and Diminished Chords

Chord Type:	Mediant (3)	Dominant (5)
Major Triad	Major 3rd	Perfect Fifth
Minor Triad	Minor 3rd	Perfect Fifth
Augmented	Major 3rd	Augmented Fifth
Diminished	Minor 3rd	Diminished Fifth

Table 2.8 Major and Minor Chords (Minor Keys Only)

Number	1	2	3	4	5	6	7
Numeral	i	ii	III	iv	V	VI	VII

in order for harmony to work well, the dominant chord needs to remain as a major chord, and therefore, even if it is a minor chord, the producer will deliberately augment the bottom interval by a semitone. The reasons for this will be explained further in this chapter when discussing progression.

FURTHER CHORDS

The next somewhat logical step upwards from a triad chord is a seventh chord. This is constructed by adding a fourth note to the previous triad examples and sits at the seventh interval above the root note of the chord, resulting in the intervals: 1 – 3 – 5 – 7.

For instance, if constructing a chord in C major, using C as the root note, the chord would consist of the notes C – E – G – B. Similarly, if working in E major with a root note of E, the notes in a seventh chord would consist of E – G# – B – D#.

Seventh chords can be expanded upon yet further by adding another note, this time the ninth note in the scale. For example, continuing in C major and using C as the root note, a ninth chord would result in C – E – G – B – D. This same procedure can continue for further, larger chords, known as eleventh and thirteenth chords. If you have followed the logic behind the creation of triad chords, sevenths and ninths, these further chords shouldn't require further explanation.

Seventh, ninth, eleventh and thirteenth chords can prove useful for thickening up synthesized timbres for large sounding pads and chords. Typically, genres such as Uplifting Trance, Funky House, House and Tech House will employ ninth and eleventh chords in the breakdown since the mix at this point becomes sparse and relies entirely on the chords to carry the music.

NOTATING CHORDS

Regardless of the chord being created, there will come a time when they have to be expressed on paper for another musician or there may be times when you will be required to read them, and therefore, it is important to understand how they are notated.

Chords are always named after the root note and followed with the terminology of what they are. For example, a major chord is written as the root key followed by the word *Major* or more simply *Maj*. Similarly, if the chord is minor, the root note is followed with Minor, if its augmented its followed by *Aug*, if its diminished its followed with *Dim*, and if it's a seventh chord, you simply add a seventh at the end, and so forth. Table 2.9 should clarify this with an example of a chord written in E flat.

Table 2.9 Chord Naming Using Eb as an Example

Chord	Notes	Names
Eb Major	Eb – G – Bb	Eb, EbMaj
Eb Minor	Eb – Gb – Bb	Ebmin, Ebm, Eb-
Eb Diminished	Eb – Gb – A	Ebdim, Eb°
Eb Augmented	Eb – Gb – B	Ebaug, Eb+
Eb Major 7th	Eb – G – Bb – Db	EbMaj 7th

TRADITIONAL HARMONY

The creation of chords is only half of the story, and it's the progression from one chord to the next that creates harmony within music. Indeed, as touched on previously, the choice of chords and how they progress from one to another will determine the energy and mood of the music.

Although from a purely theoretical point of view, there is nothing to stop a musician from creating a number of random chords within a key and playing them one after another to create a progression, this approach will often result in a piece of music that lacks any direction or purpose. Rather, harmony should be composed in such a way that takes the listeners on an interesting journey through music.

In general, to accomplish the effect, harmonies will begin at the key of the music before moving through a number of related chords before eventually returning back to the original key of the music. Obviously, this means that in most instances, the first chords of your harmony are pretty much predetermined since if you're writing in F major then the Tonic chord would be F Major, whereas in the key of G it would be G Major, and so forth.

However, the chords that follow one another to comprise the musical 'journey' should be carefully considered. It's not an ideal situation to simply drop in a few randomly chosen chords and expect it to work well because if the chords do not share any relation to one another, the harmony will appear disjointed.

In order to ensure than harmonies ebb and flow naturally from one chord to the next, there are a number of guidelines that can be employed. For ease of explanation, it is often best to view good harmonies as being made up of a series of paired chords with each pair linked together to produce the full progression. Furthermore, each paired set of chords can either exhibit a feeling of familiarity or alternatively introduce an unexpected twist.

The familiarity, even predictability, of some paired chord progressions comes from the use of what is often termed *natural* or *strong* progression. These types of progressions are created when chords share a note or harmonic relationship with the following or preceding chords notes. One of the most fundamental and most common relationships pertains to that all-important dominant fifth note of the scale.

For example, if the producer were to play C major and G major chords sequentially it would result in a natural sounding progression since the note G is common to both these chords. What's more, both chords also contain the *perfect fifth* discussed earlier. Listeners latch onto this close harmonic relationship between notes in the chords and hear a naturally comfortable progression. The two chords sound as though they should naturally move from one to the other.

Moreover, by reversing this arrangement and moving from V to I can create what's perhaps the most powerful progression of all. Indeed, this progression is so powerful and popular that it finds its way into nearly all music. Understanding why this can prove important for creating progressions, so we first need to re-examine Table 2.1 where each of the note in the scale was named.

The important note to pay attention to is number 7, the leading tone. It's named as such because it leads back onto the tonic, or the tonality, of the music. If a musician were to simply play a few random notes on a piano keyboard and finish on the leading tone of the scale, listeners would be left with an almost instinctual urge that it needs to resolve back to the tonic.

If chords are examined closely, this leading tone exists in the dominant chord, and in fact, because the dominant is so important in music, each note in that particular chord is even awarded its own name as given in Table 2.10.

Since the leading tone is present in the dominant chord, similar to when the scale is played and ends on the leading tone, whenever this particular chord is played we instinctively pick up on the leading tone and feel that instinctual urge for it to resolve back to the tonic. It's for this reason that when working in some keys, whilst the dominant may actually result in a minor chord, the third interval is deliberately augmented to create a major chord. This is to ensure that the all-important feeling of resolve from the dominant to the tonic remains.

Indeed, the emphasis on the tonic of the music plays an incredibly important factor with harmony and should always factor heavily into musical compositions. All great chord progressions are written specifically to provide the feeling of wanting to return to the tonality of the music. It appears as though the music has

Table 2.10 Note Naming of Dominant Chords (C Major as an Example)

Chord Notes:	Chord Name	Key: C Major
First Note	Dominant	G
Third Note	Leading Tone	B
Fifth Note	Supertonic	D

an indescribable desire to move back to the tonic and once it does, it results in a cadence, the feeling of resolution.

At the same time, as the relationship of the I – V and V – I chord progression is incredibly powerful, and some records have used just this alone, it does provide a somewhat short progression that can be lengthened with the introduction of further chords.

The first and most obvious choice for the next chord is the subdominant since this is a major chord and, as its name suggests, is a 'sub' of the dominant and, therefore, naturally leads into the dominant chord. In fact, so powerful is the IV –V – I progression that literally thousands on thousands of records have been composed with just those three chords.

Whilst a huge amount of records have used just these three chords, and many have become lasting hits in pop music, remaining with just these three chords can often result in music sounding stale, or worse still, like chart pop music. Consequently, it's worthwhile employing the minor chords to add some more interest to a progression.

As discussed, the V – I should generally be considered the end of the harmony, since it leads or returns to the tonic of the music, and therefore, additional chords are best introduced before the V.

Knowing what chords to use to continue to create a strong progression borrows from the previously discussed tonal relationship of the leading tone. Simply put, the producer can create a number of strong sounding progressions using any chords provided they ensure the root notes of chords are four notes up or five notes down from the leading tone, or third note of the preceding chord.

Using this formula, by counting five notes down the current working key of C major from the leading tone of the V chord whereby we arrive at the note D. Taking account of our fourth/fifth guideline, the note D must form the root note of the chord and as you can see from Table 2.11, D forms the tonic of the ii chord. Therefore, our progression could now consist of (Figure 2.2):

ii – V – I

Table 2.11 The Triad Chords that Naturally Occur in C Major

Name:	Tonic	Supertonic	Mediant	Subdominant	Dominant	Submediant	Leading
Numeral:	I	ii	iii	IV	V	Vi	vii
Notes:	C E G	D F A	E G B	F A C	G B D	A C E	B D F
Chord:	Major	Minor	Minor	Major	Major	Minor	Minor

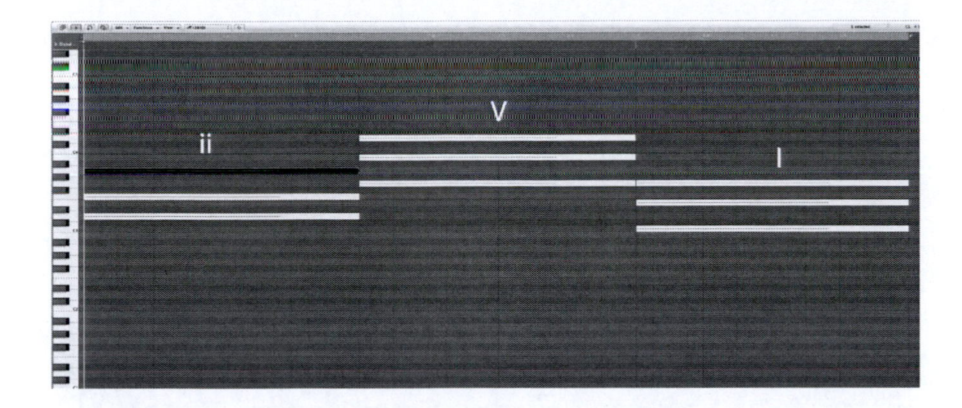

FIGURE 2.2
The ii – V – I chord progression in piano roll

This progression can be further augmented by counting down five intervals from the root note of the ii chord, and this would result in the vi chord; therefore, the progression could now consist of:

vi – ii – V – I

This technique has now provided two pairs of chords that using the formula delivers a natural sounding progression (Figure 2.3). This can be heard on Chapter 2 of the COMPANION WEBSITE.

A Natural Chord Structure

The producer could continue to construct a chord progression using this method for as long as required, but whilst it is certainly a very useful formula for constructing natural sounding progressions, if does tend to sound predictable and hence, boring. Indeed, any chord progression that is constructed entirely of strong progressions will be predictable and if repeated throughout the length of a piece of music, it can become tedious.

Whilst chord progressions should remain fairly predictable, the producer should throw in the odd unexpected chord to catch the listeners off guard and inject some spice into the music. An unexpected or weaker chord partnership is part of a progression that doesn't follow the previously discussed formula.

It should be noted here that the chord itself isn't weak, there are no weak or strong chords; instead it simply refers to the relationship of notes when chords are paired together as part of a progression. Weak chords are introduced into progressions in order to make the music that little more interesting by catching the listeners off guard.

Generally speaking, the larger the numbers of weak chord pairs that are introduced into the progression the more unique the music will sound. However, similar to the overuse of strong chords, if too many weak chords are used, the

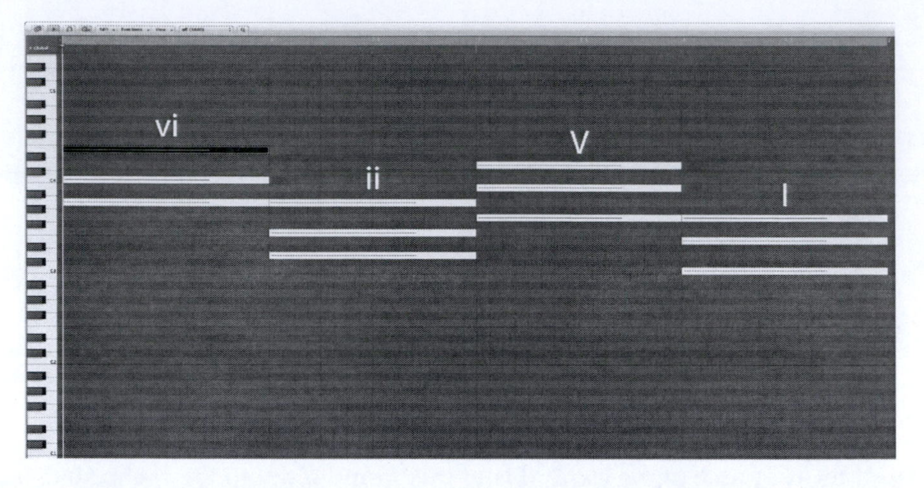

FIGURE 2.3
Chord structure in a piano roll

music will move to the opposite end of the scale. It will become difficult for the listener to determine the key of the music and as a result can appear disjointed or incomplete.

Even as there are no definitive rules in constructing a progression, they can be likened to taking a car journey from home to work and back again. The driver begins at home and progresses through the journey to the destination before returning back home again. The same as the driver wouldn't start the car and suddenly appear at work, neither should chords.

The producer will often begin at the tonic (or home key) of the music and then progress to a chord that is close to home. This could, for example, be the V chord since this is the dominant chord and shares the largest relationship with the I chord. The progression could then move onto the subdominant chord (IV) since this has a relationship with the V chord, but is harmonically 'further away' from the I chord. This could then be followed with an unnatural or weaker chord pairing. In relation to the car journey from home to work, this weaker chord pairing could be compared to a diversion related to the car journey, which adds interest to the journey.

By moving from a strong chord pairing to a weaker pairing it can sometimes appear misplaced or ill fitting for the song. If this occurs, a generally accepted technique is to lead into a weak chord pairing but immediately follow that with a strong chord. That is, the producer will move from a strong chord into a weak chord by choosing a chord that doesn't share any notes or the chord formula with its predecessor. This weak insertion can make the progression appear disjointed but the subsequent chord then shares a note with the preceding weak chord and resolves the disjointed feel almost immediately. The result is a *Strong-Weak-Strong* progression, a technique that reaffirms the harmony and prevents it from sounding too disjointed in the music. Finally, the progression would then move to a stronger progression and to the dominant that loops round back to the start of the journey again, creating the V – I chord cadence.

A simple chord progression.

Of course, this approach shouldn't be taken literally otherwise all chord structures would sound the same (even though most actually do sound the same). Instead, it should be treated more as a guideline for building progressions that will result in the musical journey making more 'sense' to the listeners.

CHORD APPROACH IN EDM

Although the theory of progression above is typical of most music, including EDM, some genres of dance music will twist this approach. A typical example would be trip-hop, a genre of EDM that relies on melancholy progressions that are often created through relying on a large number of weak progression pairs throughout the verses but resolves to stronger chord progressions during the chorus sections. This approach can often help to create the emotion of the song, making the music appear particularly anxious during the verse but more confident during the chorus. *Portishead* have often moved from unnatural progressions to natural in a few of their releases, and *massive attack* have released some records where the chords start unnatural or weak but move to natural and strong as the song progresses.

MINOR PROGRESSIONS

Thus far, we have only considered creating chord progressions in a major key and this is simply due to the fact that many producers will learn the major keys first, and hence, it is easier to relate music theory to major. However, since most EDM is produced in a minor key, we must consider the differences between progressions in a major and minor key.

First, when producing a progression in a minor key, the most commonly used chords happen to be the I – iv and VI, compared to I – IV and V of the major scale. Second, when a producer composes a progression in a minor key, there is an almost instinctual urge to move to a major key. As a consequence it isn't unusual to move from a minor progression to major during specific parts of a production.

Apart from these exceptions, writing in a minor key follows the same guidelines as composing in a major key with both strong and weak progressions. However, solidly adhering to these guidelines to construct progressions will limit your musical palette. Whilst it cannot be denied that a proportionate amount of music has been written using just these guidelines, the producer can create more interesting harmonies by adapting on these guidelines and thinking outside of the box.

FURTHER PROGRESSIONS

One of the easiest ways in which to add more excitement to a progression is to create a chord inversion. Inversions can be particularly useful if the producer wishes to introduce some change but remain in close proximity to the preceding chord. This is particularly useful in breaking up a strong progression whilst maintain some familiarity or if the producer is relying on a 'pedal tone' bass. This is where the bass instrument is held at the same pitch through a number of chord changes. In some dance music styles such as French House, this is typically in the bridge and when this occurs, it's not unusual to use inversions of the same chord whilst the bass remains at either the tonic or dominant note of the chord.

To explain the principle of inversion, we must return to earlier in this chapter, where we discussed how a triad chord is built upon the first, third and fifth note of a scale. For example, from the exercises previous in this chapter, C major contains the notes C – E – G, but the position of these can be altered to create an inversion.

A producer can create a first inversion by playing the chord but ensuring the root note of the chord is played an octave higher. This results in E now being the root note of the chord, presenting the chord E – G – C as shown in Figure 2.4.

The producer can then modify this chord further by taking the root note of E from the first inversion and moving this to an octave higher resulting in the G note becoming the root of the chord, thus creating a second inversion (Figure 2.5).

Due to inversions sharing similarities with the original chord, they're most often used to introduce some variety and are commonly introduced in the middle of a pair of chords that are creating a strong progression. By inserting

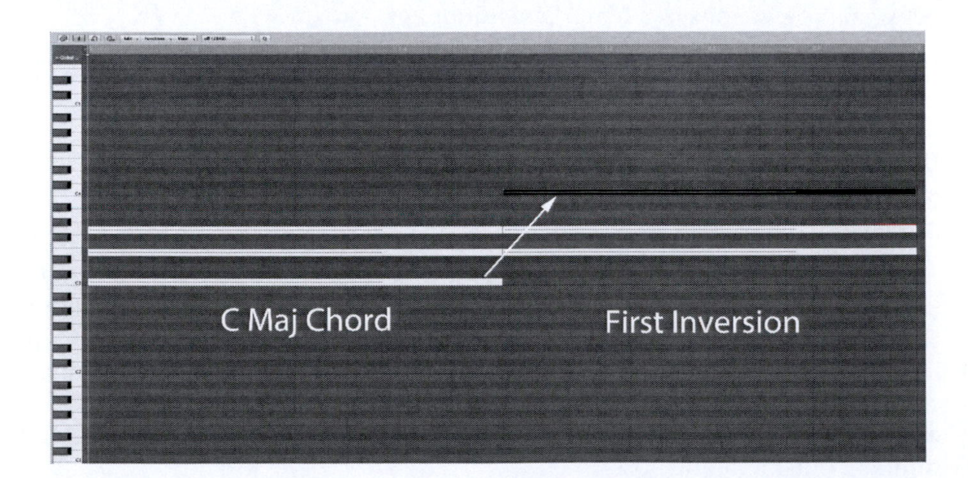

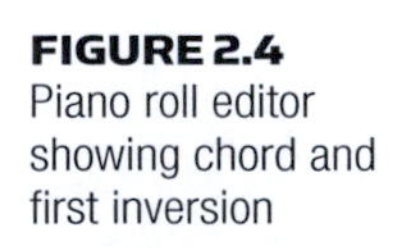

FIGURE 2.4
Piano roll editor showing chord and first inversion

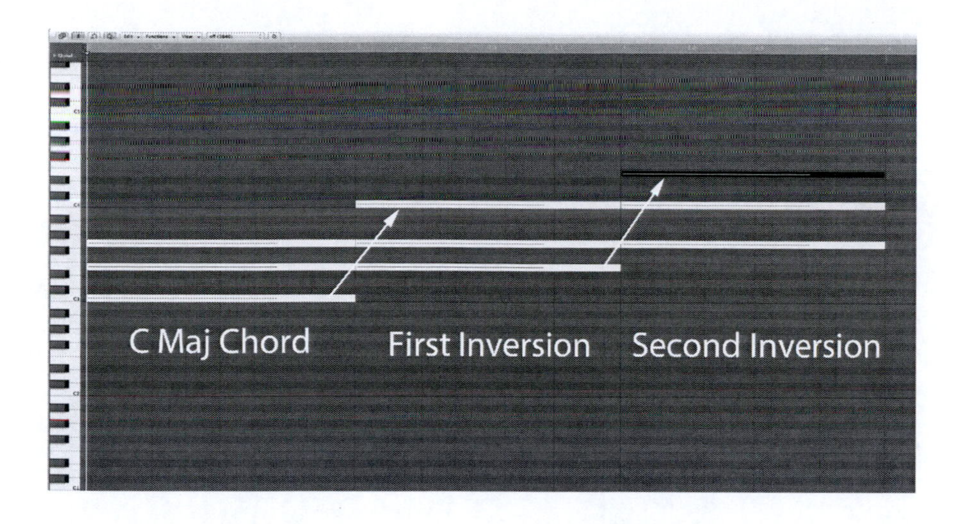

FIGURE 2.5
Piano roll editor showing chord, first and second inversion

a first or second inversion between these, the inversion becomes known as a *passing chord*, since it provides the feeling of *passing over* to the next note in the progression. However, inversions should be used sparingly and with some forethought to the harmonic journey of the music.

Inversions will often introduce a progressively unnatural or weaker sounding progression with movement from the original chord to the second inversion being the weaker of the two. This style of inversion and progression is not as weak as other chords with less of a relationship, but they nevertheless do come across as unstable, and therefore, they shouldn't be used to end a progression.

SEVENTHS

Another approach to add interest to a progression is to modify some of the chords by creating sevenths. Typically, the I, IV and V often benefit from the addition of a fourth note to create the seventh chord but of these the most popular approach is to only modify the V chord. The creation of a seventh here creates a *dominant function* since it serves to increase the feeling that the music wants to return to the tonic, or key of the music. This can often be used to great effect to lead the listener into believing the progression is about to return to the tonic but instead the progression then moves through a number of different chords before then returning to the tonic (Figure 2.6).

If the producer chooses to add a major seventh to the V chord and then lead further into the progression rather than return to the tonic, they must exercise some care so as not to 'loose' the listener. As touched upon, a progression should be viewed as a journey, and therefore, it has to make some 'sense'. Just as you wouldn't walk down your garden path towards your front door and suddenly find yourself in the garage, the progression must follow this same principle.

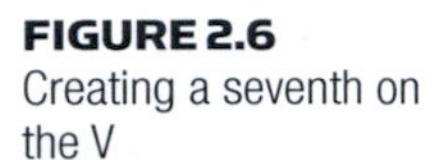

FIGURE 2.6
Creating a seventh on the V

If the producer adds a major seventh to the V chord and doesn't return to the tonic, its advisable to move to the vi, then IV, then V and finally to I. This could be compared to the walk down your path towards your door and then just popping over to check the garage door is locked before entering your house.

SUSPENSION

Another alternative to create interest to chords and progressions is to suspend the chord. Chords can be suspended by simply augmenting the bottom interval (third interval) by a semitone. For instance to create a suspended I chord in C major, the producer would augment the E to F and to suspend the V chord, the producer would augment the B to a C.

It's not by accident that I have chosen both the I and V chords as examples here since these are the most common chords to be suspended. Of course, this is only music theory so naturally any chord is open to this modification, but it generally produces the best results on the I and V chords.

Irrespective of what chord the producer chooses to suspend, it should generally be resolved immediately afterwards by returning to a standard chord. Resolve is incredibly important for a *suspended third*, more commonly known as a *sus4* (as in the bottom interval is augmented by a semitone), since it creates a large amount of tension in the chord, giving the listeners the feeling that it desperately needs to resolve.

It should be noted that sus4 is not the only suspension available and there are others, but the most commonly used suspension after sus4 is *sus9*. Similar to the creation of the sus4, raising the tonic, or root note, of the chord by a tone, creates a sus9. Generally speaking, this doesn't create as much tension as a sus4, and therefore, it isn't as important to resolve, although it is dependant on the chord itself.

SECONDARY DOMINANTS

Yet another method to spice up a chord progression is to introduce secondary dominant chords. This type of chord creates the feeling that a note that isn't the tonic sound like it is the tonic and thus creates more interest to the music. To understand what these are and how they are employed, we need to revisit the dominant chord.

As discussed previously, the relationship between the dominant and the tonic has a vital role in music. This is because the dominant chord contains the leading tone, a central note that is one semitone below the root note of the tonic chord. As a consequence, when we hear the dominant chord, we feel the instinctual urge for it to be resolved by the tonic. The producer can emulate this type of behaviour using two different chords and changing the arrangement of the notes within the chord.

For example, consider the first minor chord in the C major scale – the ii chord – consisting of the notes D – F – A. This is a minor chord in the C major scale because the F is not shared in the D major scale, instead D major's scale contains the note F# and therefore when in C major, it produces a minor chord.

If the producer chose to augment the bottom interval of the ii chord by a semitone, it would result in a chord comprising D – F# – A. By doing so, this would accomplish two things; it would be introducing the note F# that doesn't exist in the C major scale resulting in an accidental note for the scale, but more importantly, it would change the minor chord into a major chord. This could be viewed as *borrowing* the I chord from the D major scale even though in theory the producer is working with the second chord in the major scale, and this is always a minor chord.

Nevertheless, if the producer were to play a chord progression in C major and during the course of the progression introduce the second augmented chord as a major instead of its usual minor, it will immediately grab the attention of the listeners, even if the listener is pretty much tone deaf. We have all grown surrounded by music that follows specific patterns and even if we only hear three or four chords, we instinctively know what the home key of the music is. We may not be able to identify or name the key or notes but we naturally know that the second chord in the scale is a minor and when it isn't, it catches us unexpectedly.

Despite the chord being unexpected, however, listeners have the eerie ability to recognize why its been changed to a major and just as they often expect a resolve from V to I, they anticipate the chord following this new major to resolve in much the same way. Indeed, the D chord now shares similar characteristics to the dominant chord in that it now has a leading tone that needs resolving and this resolution occurs with a secondary chord whose root note is either a fifth above or fourth below. This is why they're known as secondary dominants, it is not necessarily a dominant chord but it acts like one.

This technique can be applied to any minor chord, but when creating a major chord in this way it must be expressed properly. In the example of converting the D into major, it should not be notated as II since this will only result in confusion with musicians, and instead, it is notated as V/V.

By and large, when a producer introduces a secondary dominant, it is commonplace to resolve it with a chord whose root note is either a fifth up or fourth down, emulating the effect of moving from a dominant to a tonic. This adds a resolve and prevents the note from sounding out of place in the progression but isn't always necessary. It is possible to move to other chords and the producer's ears should be the final judge of whether it sounds correct or not.

KEY MODULATION

Changing key part way through a song is a popular choice for many of the more chart-orientated dance records to prevent them from sounding repetitious. However, this approach also has its place in the more melodic driven genres of EDM such as Trance, Trip hop, and French/Disco house. It can be used to provide additional energy and anticipation after breakdowns but at the same time must to be applied cautiously; otherwise, it can appear particularly cliché.

Generally speaking, within EDM, key modulation occurs in one of two places, when moving between different sections of an arrangement or at the main breakdown, just as the build up begins and the whole track is revealed. Key changes in these two situations are often applied delicately or abruptly depending on the track and the producer's own personal preference.

If applied abruptly, it commonly occurs directly after the main breakdown, a drum roll, break, or pause in the music and generally at the point where the music is moving from the dominant to the tonic. When employed at this point, the dominant is played in the original key and the tonic enters in the new key.

If the transition needs to be more gradual then typically the composition will work up to a dominant seventh before then moving to a triad in the new key. By doing so, when in the dominant seventh, there are four notes available and these can be re-arranged, as previously discussed, to help smooth the transition to a triad chord in the new key by sharing as many notes as possible with one another. If at all possible, a smooth a natural progression could be achieved by ensuring the leading tone of the dominant is a semitone below the tonic of the new key.

Another option to smooth over the transition from one key to another is to either introduce a secondary dominant chord as previously touched upon or employ a 'modulating' chord. A modulating or 'pivot' chord is a chord that is shared by the current key and the key the producer wishes to move too.

For instance, if working in the key of D with the idea of moving to the key of A, it can be accomplished by using the I chord since the D Chords tonic (I) contains the same notes as the subdominant (IV) in A, so the producer could play D –E7 – A. Note here that a seventh is introduced to help ease the transition.

Whenever modulating to a new key the producer should aim to employ it near the end of the music and move upwards in tone rather than downwards. Employed here, it produces the final energetic punch of the music, driving the message home and can be particularly effective if the vocals or rhythm section increases with this modulation. Whilst it is possible to move downwards in key, this type of movement tends to remove the energy of the song, and whilst it can be useful if your drawing the track to a close, it's a generally accepted principle that you move upwards, not downwards.

Above all, changing key must be approached carefully and even more so if a vocalist is performing with the music. Apart from key modulation appearing overtly cliché if used too much, modulating to new keys can create problems for the vocalist since they may not be able to perform in the new key. Consequently, it's generally only employed in dance music if it features no vocals or a vocoder is being used on the vocals, since the producer can increase the pitch of a vocoder more easily than with a live performer.

Naturally, these should all be considered simple examples and are by no means an exhaustive concept to harmony. Rather, they are intended as examples and starting points to producing harmonies, but above all any great harmony will incorporate both predictability and a few small, unexpected twists. These can be introduced with just some lateral thinking and innovative planning on your part.

For instance, borrowing a chord from the relative minor or major of the key you are currently writing in can produce interesting results. For example, in C major, the producer could borrow a chord from A minor, the relative of C major. However, taking this approach, the producer must be careful not to over use it since it can result in the song appearing unfinished or disorganized.

Alternatively, with a careful selection of chords, it is possible to make the music appear as though it is moving from one key to another without actually moving. For example, the chords F and G appear in both A minor and C major, so a simple chord progression of Am/F/G could make it appear as though the key is in A minor that moves into major and then back to minor again.

What's more, chords can often be replaced with different ones, a technique known as substitution. For example, the I chord contains the same two notes as in the vi chord so the producer could substitute the I for a vi, or perhaps the IV with an ii since these both contain two of the same notes.

HARMONIC RHYTHM

Whilst re-arranging the notes within a chord or being innovative with the chords in a progression can add some spice to music, it's equally important that the progressions feature symmetry in order to make the harmony coherent. Without this, even if the chords all form a natural or strong progression, the music can still appear incoherent or wrong so we must also examine the principle of harmonic rhythm.

Harmonic rhythm is the name given to the length, pacing and rhythmical structure of a chord progression. All music is constructed from a series of interlinked passages or phrases that tie together to produce the final result and these must all work together to produce a rhythm that augments the surrounding melody and bass. This is accomplished through pairing chords together to create strong or weak progressions and then arranging them to occur symmetrically and rhythmically.

This borrows from a principle discussed earlier in this chapter of pairing chords together to produce strong or weak progressions and then joining these together to create the complete harmony.

For example, the producer may choose to pair chords one and two, three and four, five and six and finally seven, returning to chord one that would result in a type of 'crossover pair'. Figure 2.7 shows how paired chords could work together.

In Figure 2.7, the seventh chord is the dominant that naturally wants to resolve to the tonic (I) at the beginning of the progression. These two chords create the paired crossover progression, since the tonic is not only paired with the preceding dominant (V) at the end of the previous bar but also paired with the following subdominant chord (IV).

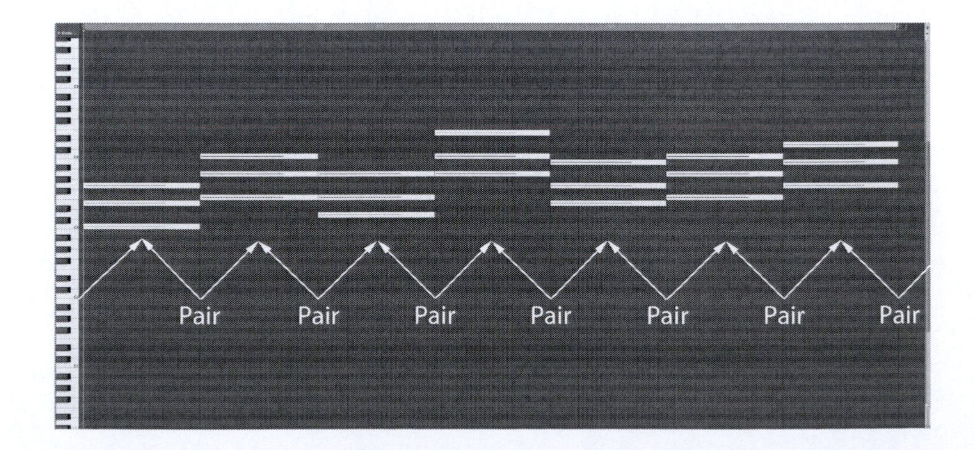

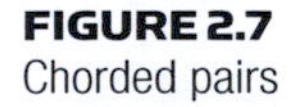

FIGURE 2.7
Chorded pairs

Each paired 'set' of chords are generally strongly harmonious with one another, but may perhaps sound 'weaker' when moving from one strong pair to another strong pair in the progression. This style of motion is shown in Figure 2.8:

The choice on where to place weaker progressions is down to the producer's creativity but as discussed previously, chords should be arranged similar to a physical journey otherwise the music may appear disjointed to the listener. In general, music should be largely predictable with the odd unexpected event thrown in to maintain interest. Additionally, although there is no limit to how many pairs of chords can be used to create a progression, it shouldn't be overtly long.

Also, harmony is perhaps best viewed as a series of interconnecting links in that each paired set of chords produces its own musical phrase but all these chords together produce another musical phrase. This latter musical phrase will designate the start and end point of a much larger music phrase within the music, resulting in the loop. When the chords start again, the melody and bass should generally both loop round and start again, so the progression should be designed with this movement in mind. With the progression length determined, the producer can then formulate a rhythm for the harmony.

Naturally, all music features rhythm but a common misconception is that with EDM, the rhythm is generated entirely by the percussive drum elements. This approach often results in music that sounds tepid or boring since all the individual aspects of music, harmony, bass, and melody should feature independent rhythms. With harmony, this is accomplished through the pacing and placement of the chords over a series of bars.

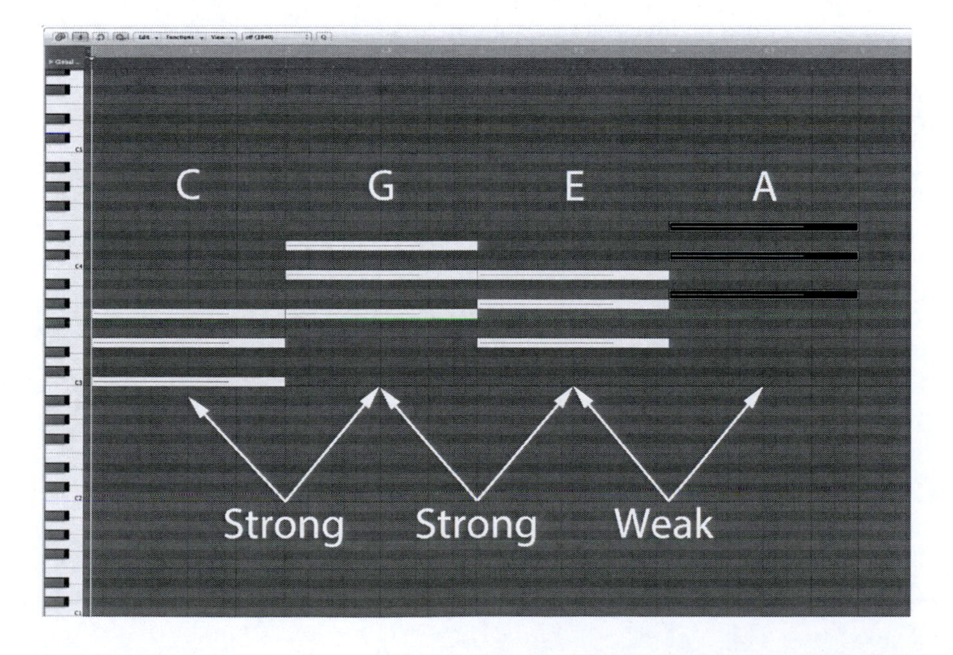

FIGURE 2.8
C to G = strong, G to E = strong, E to A = weak

Just as the producer must carefully consider the chords used in a progression to introduce a good amount of predictability to the listener, they must do the same with the placement and speed of the chords in order to create harmonic rhythm. Here, however, the aim is to make the rhythm entirely predictable to the listener. So although the producer may inject the odd surprise in the chords themselves, the harmonic rhythm remains entirely predictable.

For a harmonic rhythm to remain predictable a chord progression should remain symmetrical. Within music, this can be accomplished by equally spacing the chords over a given number of bars. For example, the producer may have a progression running over four bars and to attain some symmetry they may position two chords in the first bar, one in the second, two in the third and, finally, one in the fourth as in Figure 2.9.

Alternatively, the producer could employ one chord per bar for all four bars, two chords per bar for all four bars, or one chord in the first bar, two in the second, three in a third, two in the fourth, three in the fifth and, finally, one in the sixth.

Provided the producer aims for some kind of symmetry to the progression, it will appear predictable and the listeners will be able to predict when the next chord change will occur. This prediction on behalf of the listeners can be incredibly important because if they can't the music can sound disjointed or incomplete. Chapter 2 on the COMPANION WEBSITE features an example of an asymmetrical progression followed by a symmetrical progression. Whilst they are both *theoretically* correct, the second example appears more coherent and complete.

Alongside the symmetry, the producer must also consider the speed or frequency the chord changes occur since this can have a large influence of the feel of the music. For example, a common method to add excitement to music is to begin with a slow harmonic rhythm and gradually increase the speed of

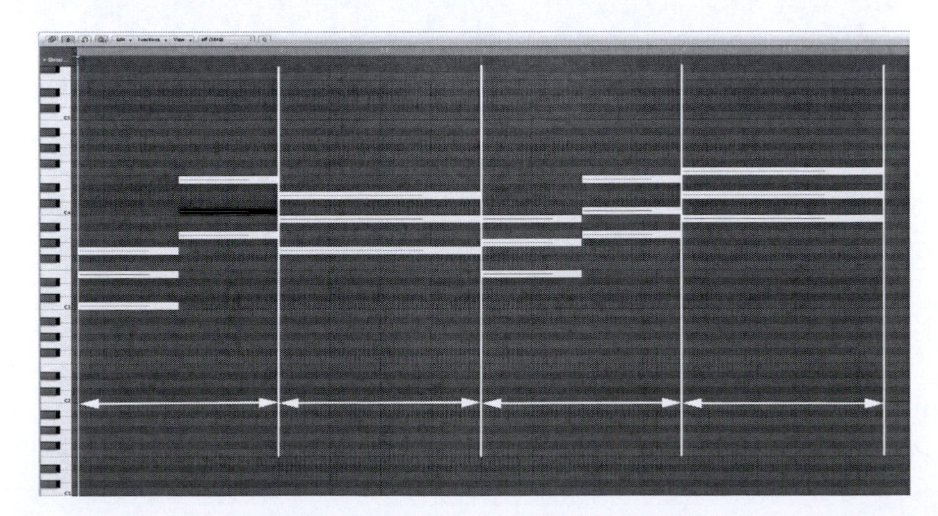

FIGURE 2.9
2 chords – 1 chord – 2 chords – 1 chord

this rhythm as the music progresses. This is a popular technique employed in some genres of Trance, the intro and body section will use a chord progression with perhaps one chord per bar whilst in the reprise (the uplifting section), the chord progression changes to two chords per bar.

On the other end of the scale, genres such as Lo-Fi, Chill Out and Trip Hop often employ a slow harmonic rhythm with a slow rhythm to create the laid back feel of the music, whilst genres such as disco house and French house will use a slow harmonic rhythm mixed with a faster bass or melody line. This latter approach has come to form one of the most important compositional and arrangement aspects of EDM. Termed a *canonic progression*, the chord progression will complete in, say eight bars, whilst the melody will complete in four. This will be investigated in more detail in the Arrangement chapter later in this book.

CADENCE

It would be imprudent to close a chapter dedicated to harmony without first touching upon cadence. For the production of dance music, it isn't necessary to fully comprehend cadence and to discuss it thoroughly would involve an exhaustive examination of harmonic analysis and musical form that would easily require the rest of this book. Consequently, this should be considered an overview of the subject and certainly enough for the production of dance music.

As touched upon numerous times in this chapter, music can be viewed as a story in that it takes the listener on a journey. A cadence could be considered as a way of bringing the story to a close or to bring the end of a particular musical phrase or loop. In many respects, cadence also occurs in films and TV dramas. Although these are often full of unexpected twists and turns, the viewer more or less knows what to generally expect in the way of scenes and the film in general.

The viewer will intuitively feel each scene working towards to the next and ultimately, the final scene with the build up to where the villains receive everything they deserve and the protagonist overcomes all. If there were no tension or build to this final scene, or if the protagonist lost, the story wouldn't feel properly resolved to the viewer and it would result in disappointed.

Whilst receiving something that isn't expected can be rewarding, more often we have a feeling of greater reward if we receive what we expect. The way in which a director builds up into the next scene and constructs the grand finale, building on the viewer's expectation, is what separates a great film from an average one.

When cadence is employed in music it is to achieve this same goal. The music is building upon the listeners expectations, guiding them through the music and building up to signify that they are arriving at the end of a musical phrase or passage. The way in which the producers controls this cadence and how it is employed to build upon the listener's expectations is often what separates a great piece of music from a run of the mill.

Notably, cadence occurs in all the elements of music and not just the harmony. The rhythm section with its drum rolls, the harmony, the bass and even the musical structure will all utilize cadence to lead the listener into the next stage of the song. Of all these, however, it is the harmony that is the most important signifying aspect of cadence and this is why returning to the tonic, or home 'key' of the music is so important.

There is much more to cadence than simply dropping in a tonic chord, though. Whilst it isn't entirely necessary to end every passage of music on the tonic chord music must eventually return there to complete the story, it's inadvisable to simply drop in a tonic and hope for the best.

As with the film analogy, it could be compared to the protagonist walking into a room with the villain and suddenly walking out victorious. It wouldn't feel like an appropriate ending, a huge chunk would be missing and it is the same with harmony. Music must lead into the tonic to make it feel like it has to inevitably occur. Therefore, when the term cadence is used, it doesn't just denote the tonic note but also the build up to it beforehand. Obviously, cadence has been discussed in this chapter in regards to harmony but it is important to recognize the terminology used since if working with a musician they may use terms such as perfect cadence, plagal cadence, half cadence or deceptive cadence.

A perfect cadence has already been discussed in detail in this chapter and simply means moving from the dominant (V) chord to the tonic (I) chord. A plagal cadence, however, consists of moving from the subdominant (IV) to the tonic (I). This is sometimes used in music to denote the end of a short passage such as a verse, but it is generally followed with a chorus or similar that uses a perfect cadence to give the music the completed feel. Finally, a half cadence consists of moving from any chord to the V chord, whilst the deceptive cadence is moving from a dominant (V) chord to any other chord than the tonic (I) chord.

CHAPTER 3

Music Theory Part 3 – Fundamentals of EDM Rhythm

'All you need is a feel for the music. There are people that have been to college to study music and they can't make a simple rhythm track, let alone a hit record...'

Farley 'Jackmaster' Funk

Whilst an understanding of musical scales, melodies and harmonies is understandably significant for anyone aiming to produce music, a much more substantial concept for the electronic dance music producer is that of rhythm. Indeed, it's the manipulation of rhythm that creates the very essence of all electronic dance music but this is a concept that is far more complex than it may initially appear.

For this chapter, I'll discuss the theory of both tempo and time signature and, more importantly, examine the various techniques that EDM producers employ to create more interesting rhythmical interactions so that the cyclic nature of the music can sustain repeated listening.

TEMPO AND TIME SIGNATURES

Tempo and time signature are inseparably linked in music. Tempo is fundamental to all music and is a time-based measurement based on the number of beats that occur every minute (aka Beats Per Minute). Whether this beat is perceptible as it is with EDM or imperceptible – as with classical – the main purpose is to aid musicians or listeners to keep in time with the flow or pace of the music.

Time signature is used to inform the musician how many beats there are per bar and how long each of these beats should be. Displayed similar to a fraction with one number sat on top of another, the top number (the *numerator*) denotes how many beats are contained within a measure or bar (measure and bar are interchangeable terms), whilst the lower number (the *denominator*) represents what size of note is equal to a beat. To further understand this concept, we need to examine the subdivision of a musical bar and note.

A musical bar can be viewed as a single unit capable of holding one whole musical note (semibreve) or any divisions thereof. This means that rather than play a whole note for the duration of a bar, you could divide the semibreve in half and instead play two half notes (minims) for the duration of the bar.

Alternatively, you could divide a whole note in half but then divide just *one* of the resulting minims in half again to produce a total of two-quarter notes (crotchets) and a minim. Taking this approach even further, you could then subdivide just one of these crotchets to produce two 1/8th notes (quavers) to then play two quarter notes, a crotchet and a minim for the bar in any order you choose. This style of subdivision can continue as far as required and using a mix of both rests and notes it becomes possible to produce the numerous rhythmical patterns that music depends upon.

Figure 3.1 shows this same concept of subdivision but in a more familiar setting for many EDM producers, a sequencers piano roll editor set to the most common time signature of 4/4. With this time signature, each bar of music is divided equally into four individual *beats* and each beat is exactly one-quarter note in length. These subdivisions of the bar are clearly displayed in the sequencers piano roll grid.

For obvious reasons, the same as it is possible to subdivide a musical beat (crotchet in 4/4) into any number of smaller notes it is equally possible to subdivide the sequencers grid into smaller divisions or pulses. However, when doing so within a sequencer, this subdivision of the grid is not necessarily to

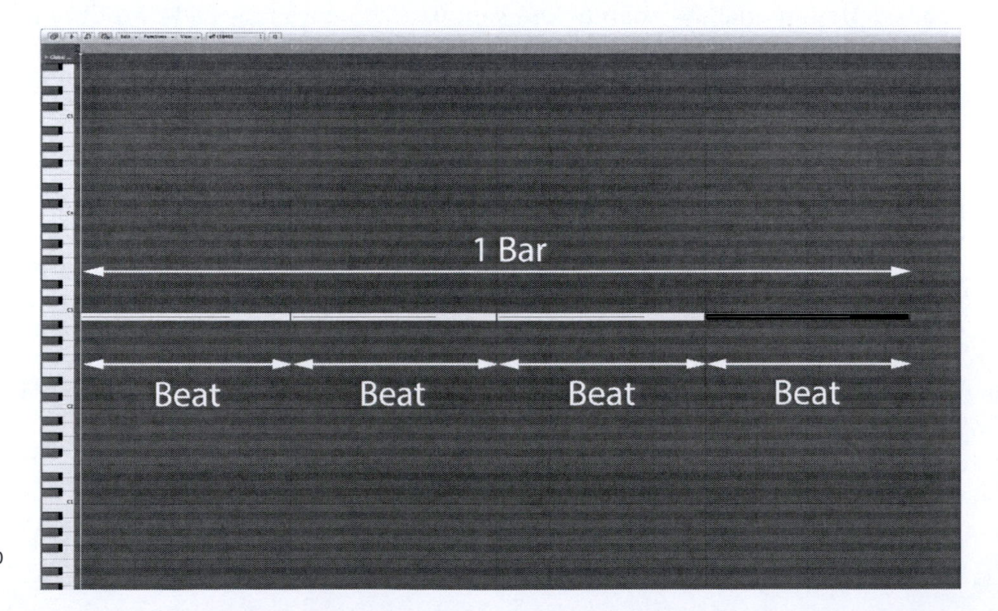

FIGURE 3.1
Logics piano roll set to 4/4 time signature

determine the *size* of the notes, as is the case with musical notation, but rather to determine the *positions* that are available for which to place a note.

For example, if a beat in the sequencers grid were to be split in half – a technique known in music theory as compounding the beat – both of these newly created pulses could then be used for the placement of a note, that is, you have the opportunity to place a note to start either at the beginning or in the middle of the beat. If we were to subdivide this further, we could create yet more pulses within the beat resulting in more positions to place musical notes.

Figure 3.2 shows Logic Pro's piano roll and its resultant grid after compounding the beat a number of times. The resulting grid from this subdivision of beats and bars is referred to as the quantization grid and many computer-based sequencers will permit a quantization up to a maximum of 192. Often referred to as *192 PPQN* or **P**ulses **P**er **Q**uarter **N**ote, it offers the theoretical possibility to subdivide any single musical beat in half continuously up to a maximum number of 192 times.

Obviously, at this maximum setting, it would produce a quantization grid so unnaturally small that you would require a microscope to view it, thus many sequencers limit the grid to show a maximum of 64 evenly spaced pulses and any movement further than this is accomplished by simply nudging a note backwards and forwards to positions in between the physically displayed grid.

Notably, this same process of compounding a beat can also work in reverse. Rather than subdivide a beat into two, it is possible to reduce the quantization so that two beats are amalgamated to produce the note equivalent or

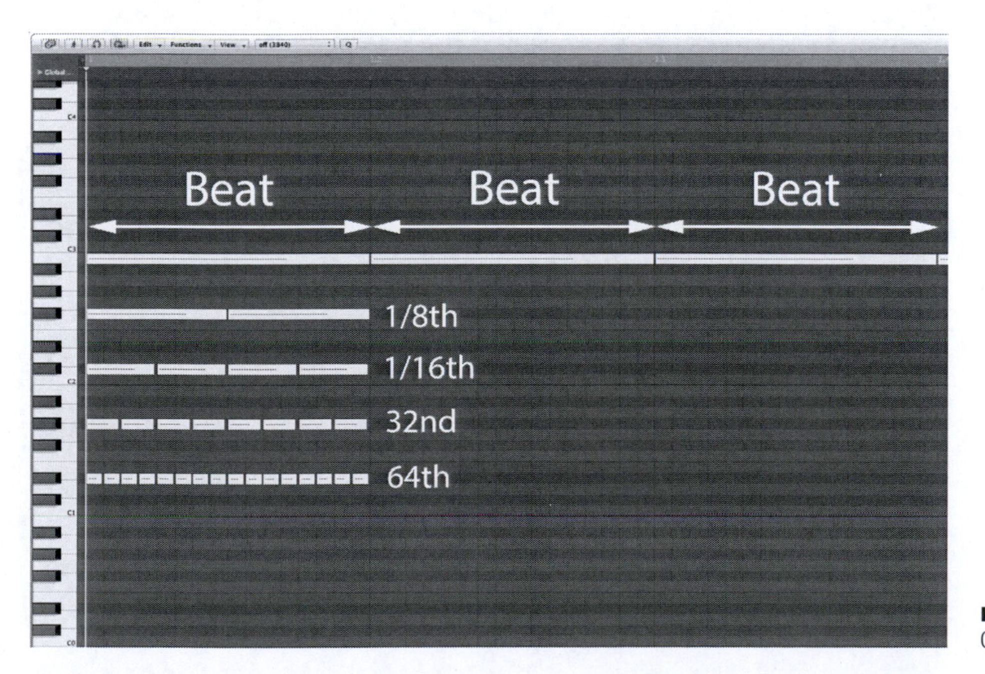

FIGURE 3.2
Compounding the beat

positioning of a minim or half note. Here, there would only be two pulses available within the bar and therefore only two positions with which to place a note within the bar. This process of turning two beats into one is known as hyperbeat, since the producer is effectively *crossing* over the beat.

It's this continual process of compounding or creating hyperbeat by subdividing the quantization grid into larger or smaller pulses that is perhaps the most important concept for the production of electronic dance music.

Almost all EDM is based around a series of constant repeating rhythmical structures rather than extravagant melodic lines and vocals. As a consequence, the dance producer must employ a number of varied techniques in order to create more complex rhythmical structures. The hypothesis is that by creating interlacing rhythmic textures of both chromatic and non-chromatic instruments, the mind struggles disseminating the information and therefore despite being repetitious in nature the music can withstand sustained listening.

The key to obtaining this effect is to divide both beat and bars by both equal and unequal amounts to produce a series of asymmetrical pulses. Unfortunately, however, many sequencers are limited in this regard and therefore the producer must carefully consider the placement of notes.

For the first example, we'll consider the structure of a typical EDM drum rhythm whereby it's usual to compound each beat into four pulses. With the almost requisite time signature of 4/4, this would result in 16 even pulses across the length of the bar (four pulses per beat × four beats per bar).

Figure 3.3 shows the quantization grid split into the 16 pulses and also shows positioning of a basic, yet typical, EDM drum loop. The kick is positioned to

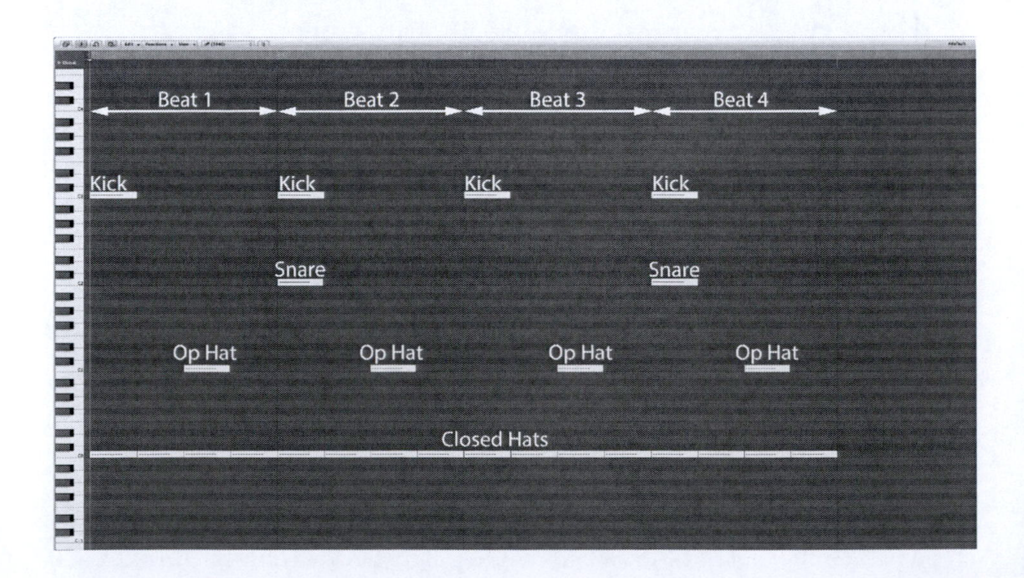

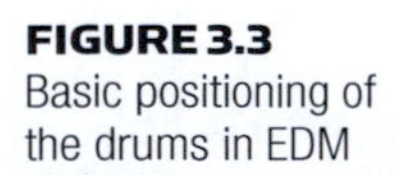
FIGURE 3.3
Basic positioning of the drums in EDM

occur on the down beats (the first and third beat) and the off beats (the second and fourth beats). In a sequencers grid, these positions are at 1, 5, 9 and 13.

A snare backs up the weaker off beats at positions 5 and 13 whilst an open hi-hat is positioned on 3, 7, 10 and 15. Finally a closed hat occurs on all of the 16 pulse positions. This produces the 'floor to the floor' drum rhythm that is present in many forms of dance music.

This, however, is only considered the very initial starting point of any EDM drum rhythm since this alone would be unable to maintain sustained listening for any extended period of time. In order to permit sustainability for the entirety of the track, further techniques are then introduced on top of this basic rhythm.

TRIPLETS

The first and perhaps most common technique that is employed in almost all genres of EDM is an effect known as *hemiola*. This is the result of creating a triplet rhythm that, when played alongside the aforementioned even quantization grid, creates a form of cross meter modulation. To accomplish this effect, the producer changes the sequencers quantization grid from an even value into an odd one.

Some sequencers that are aimed specifically towards dance musicians, such as Ableton Live, have a triplet grid setting that is accessed by right clicking in the MIDI window of the piano roll editor whilst others such as Logic require the quantization value for the bar to be set at 1/12 rather than its more common 1/16 value. Using this quantization value, the length of the bar is subdivided unequally.

Figure 3.4 shows the resulting grid from setting a triplet pattern. Rather than the usual four equal subdivisions of the bar, there are three equal

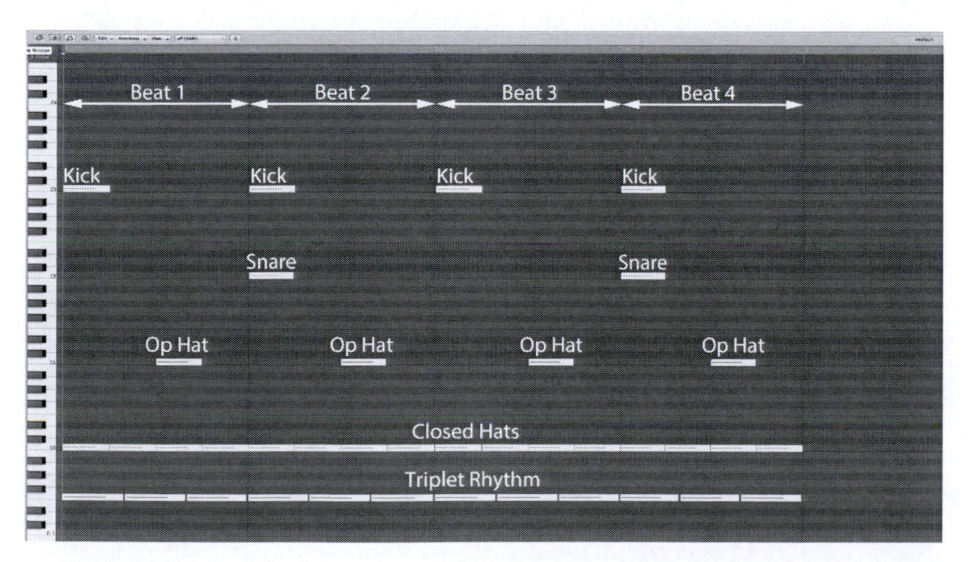

FIGURE 3.4
Subdividing the bar into a triplet pattern

subdivisions instead. By using an odd division of the bar, the pulses are offset and it is now possible to place notes in positions that were not available when subdividing equally.

When notes are placed to start in these three positions, the effect is a rhythm that is offset against the evenly distributed kick and snare and this produces a complex rhythmical interpolation that the mind struggles to disseminate.

Just this single and somewhat simple effect can change a constant repeating structured pattern into a fascinating complex musical structure that can withstand sustained listening and features on just about every EDM track in one respect or another.

> An example of the triplet effect is in Chapter 3 of the COMPANION WEBSITE; here the triplet effect has been panned to the left so you can hear it.

Naturally, this can be used to produce complex polyrhythmic combinations between instruments. A typical example of this is 3:2, popular in genres such as techno and tech house, one rhythm of three notes moves over another rhythm composed of two notes.

As useful as this polyrhythmic technique is for the dance producer it can only be employed once at any one time. If numerous polyrhythms occur simultaneously alongside the standard note pulses, the music will appear confused and loose coherence. Consequently, EDM musicians will commonly develop on a single polyrhythm by introducing asymmetric phrasing.

Thus far, we have considered subdividing the bar unequally to produce effects such as triplets. In this example, the individual events (notes) do not sound concurrently between rhythms but the actual phrase length will always remain the same. For example, in the basic drum example discussed earlier in this chapter, a 4/4 time signature was employed resulting in four quarter notes per bar. A kick drum occurs on all four beats of the bar and this results in an evenly spaced distribution of kicks.

As shown in Figure 3.4, with hemiola introduced it resulted in an uneven subdivision of the bar and where three notes occur to every four of the basic beat. That is, the bar is divided into three equal positions and a note is positioned on each of these equal positions.

Whilst this polyrhythmic technique certainly adds interest to the rhythm it is nevertheless unchanging. Due to the equal spacing, the same triplet rhythm occurs in the exact same position in every bar for the duration that it's employed for and therefore listeners begin to draw expectations that this will continue to occur throughout.

However, a common technique for EDM is to maintain this triplet pattern over a number of bars and then after a fixed number of bars, reduce the triplet pattern to two or one note for a single bar. This creation of asymmetrical phrasing throughout a series of bars creates a denial of expectation in the listener and results in a more interesting mix.

An example of the asymmetrical triplet effect (panned left so you can hear the effect).

This effect is not limited to polyrhythm through bars and can equally be employed in single bars by employing notes of differing lengths. For example, if the bar is subdivided by a 16 quantize grid, notes that are 3 + 3 + 3 + 3 + 4 could be employed. When played alongside a standard 4/4 kick drum, it introduces an asymmetrical resolution to the melody. This style of approach is shown in Logic Pro's sequencer in Figure 3.6 and can be heard on the COMPANION WEBSITE.

An example of the asymmetrical notes in the bar (panned left so you can hear the effect).

Whilst these techniques certainly help to maintain interest, in order to prolong sustained and repeated listening they are commonly combined with further techniques such as polymeter, syncopation and swing.

Of these techniques, polymeter is perhaps one of the most evident and striking, involving one or more simultaneous conflicting rhythms extending *beyond* a bar and resulting in a rhythmical movement in and out of synchronization between different instrument channels.

This effect is accomplished by the producer first creating a pattern or rhythm in a 4/4 time signature and then switching to a different time signature for

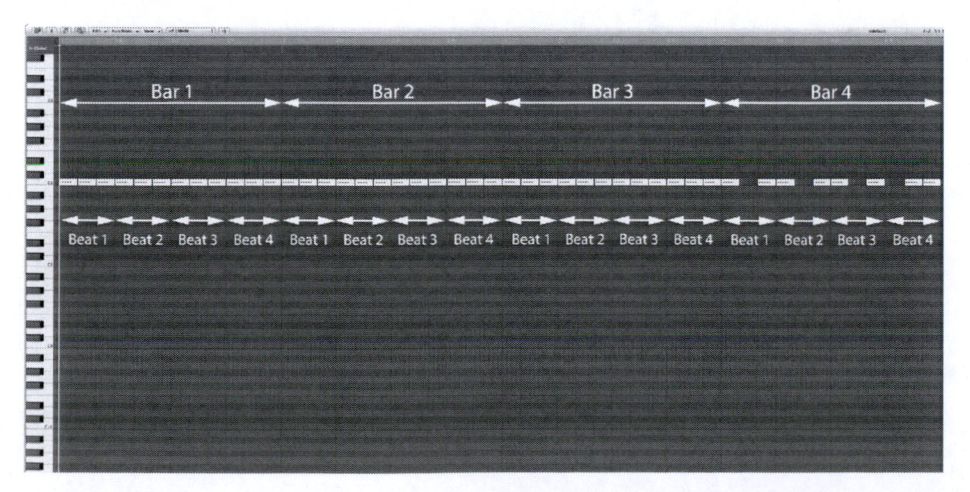

FIGURE 3.5 Asymmetrical subdivision over bars with triplet

the creation of the next rhythmical pattern. The alternative signature can be anything from 5/4 to 9/4 but by employing a metrical difference, it results in two rhythms of differing bar lengths.

For example, if 4/4 were used for the creation of one pattern and then 5/4 for a secondary pattern, the two patterns would be of a different length and consequently move in and out of sync with one another over a number of bars. By employing this technique, the listener is offered two different rhythm signatures that they can 'lock' onto that will permit repeated listening. This technique is shown in Figure 3.7.

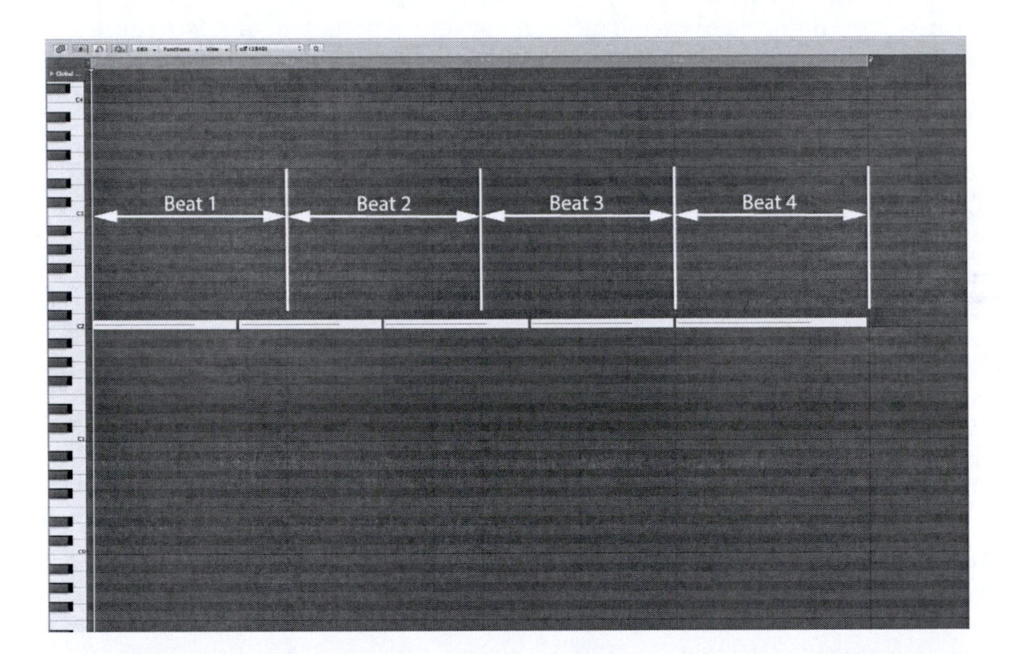

FIGURE 3.6
Asymmetrical subdivision of a bar

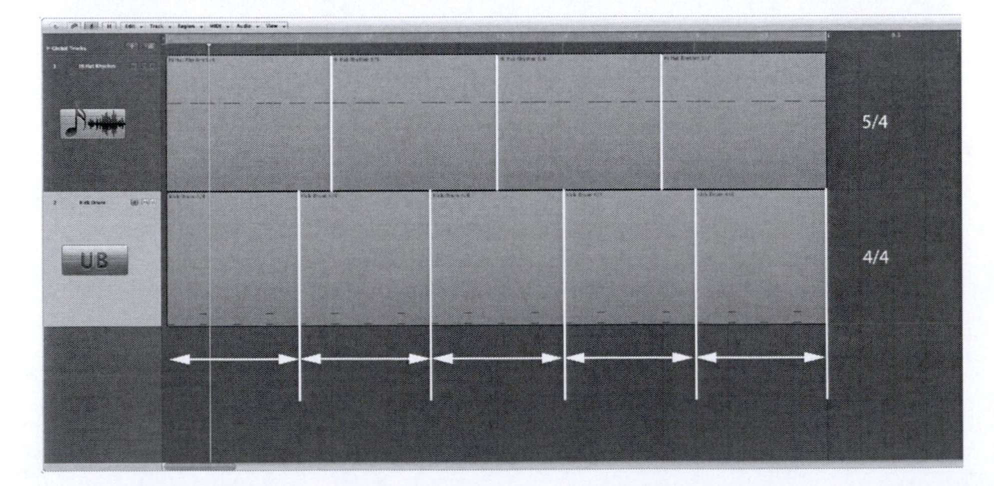

FIGURE 3.7
Polymeter

SYNCOPATION

Further interest can be added to repeating patterns with the use of syncopation. This is where the stress of the music occurs off the beat rather than on it. All musical forms are based upon patterns of strong and weak working together. In classical music, it is the constant variation in these strong/weak dynamics that produces emotion and rhythm. Within electronic dance music, these dynamic variations are often accentuated by the rhythmic elements.

As touched upon previously, in 4/4 time signature the first and third beats of the bar are referred to as the on beats and this is because they are the likeliest place that a musician would choose to change chords or perhaps introduce a new instrument. As a consequence, the first and third beats will often receive the accent of the music, which is notes that occur on these two beats will often be performed a little louder than notes occurring at any other position in the bar. Typically, in most forms of music the first beat, also known as a downbeat, will receive the strongest accent whilst the third will receive a slightly weaker accent.

However, if the accent is used on a different note that occurs off beats 3 and 4, it results in syncopation and this is a particularly powerful effect to employ. With little more than emphasizing notes that occur off beat using a sequencers velocity commands the producer can simulate the effect of hitting some notes harder than others, thereby creating different levels of brightness and energy in the sound. By changing the syncopation every few bars on rhythmical elements such as hi-hats or on chromatic instruments, it can produce significant changes in a simple repeating rhythm.

Indeed, syncopation happens to be one of the most important, yet most overlooked principles in the production of electronic dance music. The dynamic variation in timbres created by accenting notes that occur either on or off the beat is one of the main contributing factors to creating groove and therefore it's a theory that shouldn't be overlooked.

There are, of course, further contributing factors towards creating groove in a record. Alongside syncopation, timing differences, however slight, can also contribute heavily. By adjusting the timing and dynamics of each instrument, the producer can inject groove into a recording.

For instance, if the kick drum occurs on all four beats and a snare occurs on the second and fourth, moving the snare's timing forward or backward by just a few milliseconds can make a significant difference to the feel of the record and the groove in general. Programming parts to occur slightly behind the beat creates a laid-back feel whilst positioning them to occur just ahead of the beat front of the beat results in a more intense surging feel. These are the basic principles behind introducing swing into music and happen to one of the key factors to producing techno, minimal, tech house and house music.

The use of swing in electronic dance music is of vital importance and can be traced back to the early drum machines such as the Emu SP1200 and Akai MPC 60. Since these machines worked on a strict quantize grid basis, the resultant rhythms appeared particularly strict and therefore swing quantize could be employed to emulate the natural timing discrepancies of human players. By increasing the swing percentage, the drum hits are randomly moved by increasing amounts off the quantize grid. The higher the swing value, the more the randomness occurs and the further off the grid the instruments are positioned.

All digital audio workstations today including Logic, Ableton, Reason, Cakewalk, Cubase, Pro Tools and Studio One feature a wealth of swing and quantize options to achieve this human feel but it should be noted that each sequencer applies it differently and thus produces differing results. A swing quantize applied in Cubase, for example, may sound very different with the same settings applied in Studio One. To many of the older dance musicians, the AKAI MPC units have always featured the best swing although in my experience of both these machines, I feel this to be based around bias and myth rather than fact.

Nevertheless, despite all DAWs featuring swing quantize, it's inadvisable to leave the injection of groove to a machine. Many EDM producers will employ swing to get them in the general ballpark but then spend many, many hours adjusting each individual note position to create the effect they want to achieve.

CHAPTER 4

Basic Electronics and Acoustic Science

'If you want to find the secrets of the universe, think in terms of energy, frequency and vibration.'

Nikola Tesla

With an understanding of music theory, we can turn attention towards electronics and acoustic sciences. As random and conflicting to the production of music as this may appear, its usefulness is evident in the name: *electronic* dance music.

This is a style of music that has developed from the use and abuse of electronic equipment, and whether this equipment is hardware or virtual renditions within an audio workstation, it relies on the producer having an understanding of the science behind it. After all, if you don't comprehend the basic concepts your equipment is built around, you'll be limited to little more than blind button pushing, and second-guessing.

To discuss how both these can play such a significant role in the production of electronic dance music, we first need to investigate some basic electronics along with the science of acoustics.

When any object vibrates, the surrounding air molecules will be forced into vibration and begin to move spherically outwards in all directions. This spherical outwards momentum of air molecules could be visualized through considering the reaction when an object is dropped into a pool of water. The moment the object strikes the water, the reaction is a series of small waves spreading spherically outwards. Each individual motion of the wave, whether this occurs through water or through air, follows the same principle and consists of any number of movements between compression and rarefaction.

With air molecules, if a tuning fork is struck, the forks first move closer towards one another and this compresses the air molecules between the forks. The moment the forks move in the opposite direction, air molecules collect again in between the forks to fill in the free space. These are then consequently compressed when the forks return on the next cycle, and this motion continues until the forks come to rest. The series of alternating compressions and rarefactions pass through the air and force the cycles to continue spreading further afield, just like dropping an object into water.

The numbers of rarefactions and compressions, or 'cycles', that are completed every second are termed the *operating frequency* and measured in Hertz. This is named after the German physicist Heinrich Rudolf Hertz who first described and documented the propagation of sound through various media. From this, we can determine that if the compression and rarefaction cycle completes 300 cycles per second, it would have a resultant frequency of 300 Hz; whilst if there were 3,000 cycles per second, it would have a frequency of 3 kHz.

The frequency also determines the perceived pitch, and the faster the frequency is, the higher we determine the pitch to be. Thus, we can determine that the faster an object vibrates, or 'oscillates', the shorter the cycle between compression and rarefaction are and the higher the pitch becomes. An example of this action is shown in Figure 4.2.

As an object vibrates, it must repeatedly pass through the same position as it moves back and forth through its cycle. Any particular point during this cycle is termed the *phase* and is measured in degrees, similar to the measurement of a geometric circle. As shown in Figure 4.3, each cycle starts at position zero, passes back through this position known as the 'zero crossing' and eventually returns to zero. Consequently, if two objects were vibrating at different frequencies and both were mixed together, both waveforms would begin at the same zero point but the higher frequency waveform would overtake the phase of the lower frequency. Provided that these waveforms continued to oscillate at their respective frequencies, they will eventually catch up with one other before repeating the process all over again. This produces an effect termed *beating*.

The speed at which waveforms beat together is determined by the frequency differential between the two, but it's the deliberate manipulation of this beating effect that forms the most fundamental element of modern synthesis. By mixing any number of simple waveforms together and adjusting the pitch of each, it is possible to create a vast number of complex waveforms.

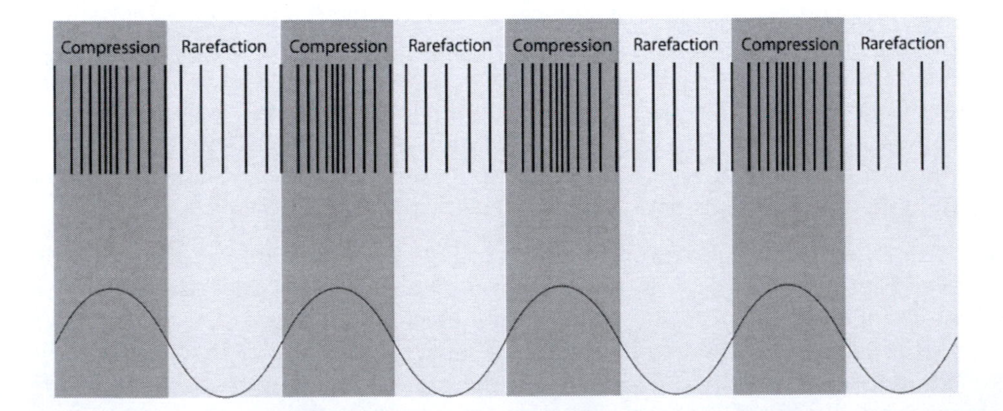

FIGURE 4.1
Compression and rarefaction of air molecules

Whilst any basic sinusoidal wave will only consist of a fundamental frequency – a single frequency that determines the pitch – when it is combined with further sinusoidal waves each detuned from one another, it creates additional overtones known as harmonic. It is these additional harmonics that contribute to any sound hear.

Indeed, this doesn't just occur within synthesis but occurs with all everyday objects. Dropping a cup, for example, would result in the cup creating

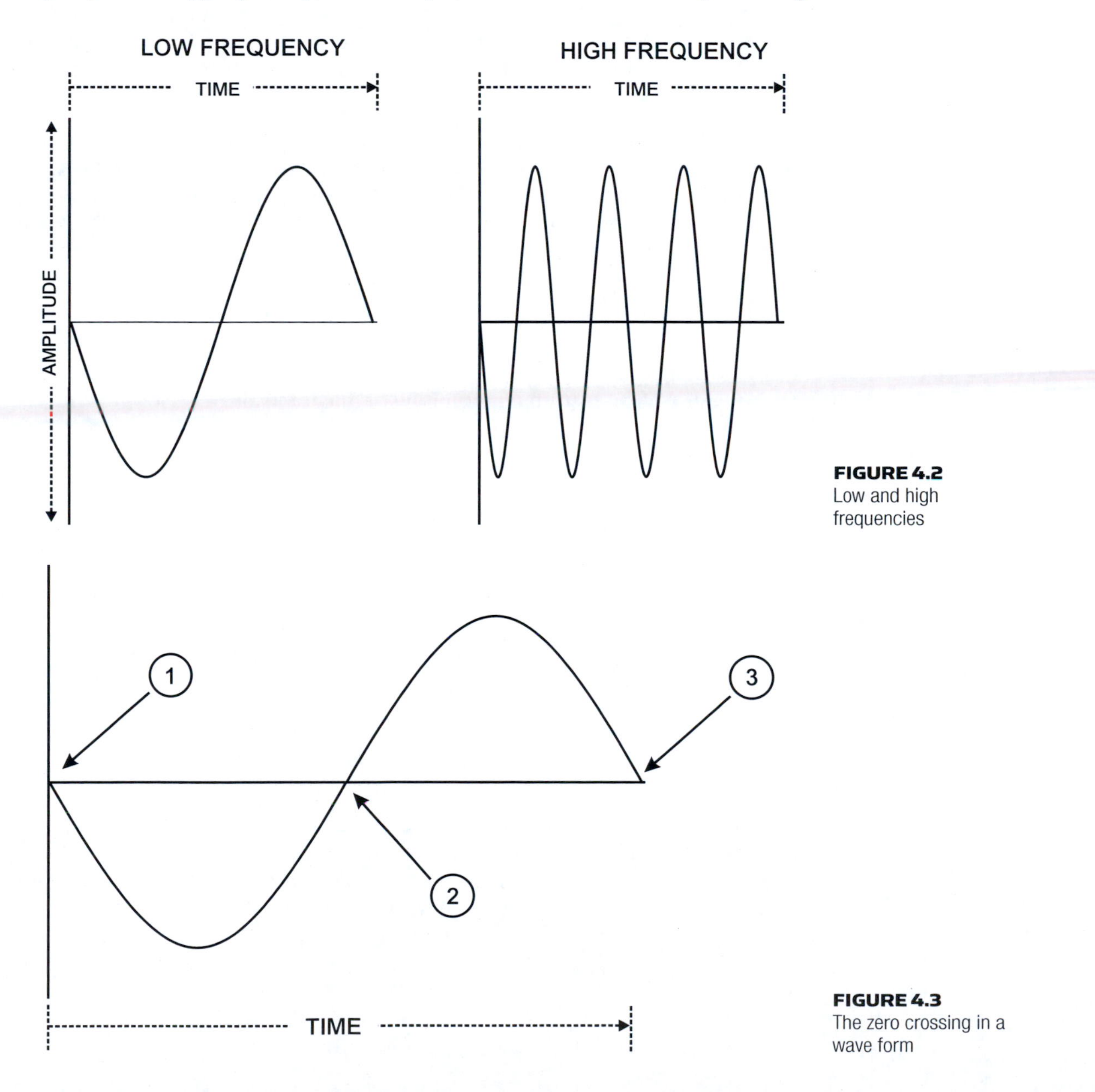

FIGURE 4.2 Low and high frequencies

FIGURE 4.3 The zero crossing in a wave form

its fundamental frequency as it strikes the floor, but further fundamental frequencies would occur from air molecules forcing a host of other objects into motion too. The different parts of the cup as it smashes and objects close by will all begin to resonate sympathetically and produce fundamental frequencies of their own. As these combine, numerous harmonic overtones are generated, but since these are all random frequencies, it rarely results in a pleasing tone.

Conversely, musical instruments are tuned to produce harmonics that occur at specific integers of the fundamental to produce a pleasant sound. For example, the wires in a piano instrument are adjusted so that each oscillates at a precise frequency.

When a piano note is struck, it sets into motion a mallet that strikes the corresponding wire forcing it to oscillate. As the piano wire oscillates, it produces the fundamental frequency or pitch of the struck note, but the subsequent vibrations of air molecules also set a number of other note wires in the piano into motion too, creating a series of harmonic overtones. These overtones combine to produce the piano's distinctive sound.

The mathematical relationship with one another within an instrument was an occurrence that was first realized by Pythagoras (the same Greek philosopher of math triangle geometry, $a^2 + b^2 = c^2$). Pythagoras noted that the harmonics generated by an instrument occur at specific intervals of the original fundamental frequency.

The first frequency would occur at an octave above the fundamental, whereas the next would be at 3:1, followed by 4:1, then 5:1 and so forth. Consequently if the note *A* were struck on a piano, the initial wire would oscillate at 440 Hz, but this would set more wires into motion, the next being at twice the original frequency (or an octave higher) at 880 Hz, followed by a third harmonic at 1,320 Hz ad infinitum.

The problem many scientists faced before the introduction of computers was how to conceptualize this relationship, and the harmonic relationships of different sounds onto paper. Since the harmonic content or 'timbre' (the French word for colour) of a sound can be particularly complex with harmonics interacting with one another in numerous ways, recreating the resulting waveform of any sound without the aid of a computer proved impossible.

Subsequently, Jean Fourier, a French scientist, theorized that no matter how complex any sound is, it could be broken down into a series of frequency components, and by using only a given set of harmonics, it would be possible to reproduce it in a simple form. To quote him directly:

> Every periodic wave can be seen as the sum of sine waves with certain lengths and amplitudes, the wave lengths of which have harmonic relations.

Jean Fourier – 1827.

The mathematical theory is particularly complex, but it is fundamentally based on the concept that any sound is determined by the volume relationship of the fundamental frequency to its associated harmonics and their evolution over time. And it's from this Fourier theorem that the complex audio waveforms displayed in any digital audio workstation are derived alongside the basic waveshapes that adorn many synthesizers. This evolution is shown in Figures 4.4 and 4.5.

Addition of sine waves to create a square wave

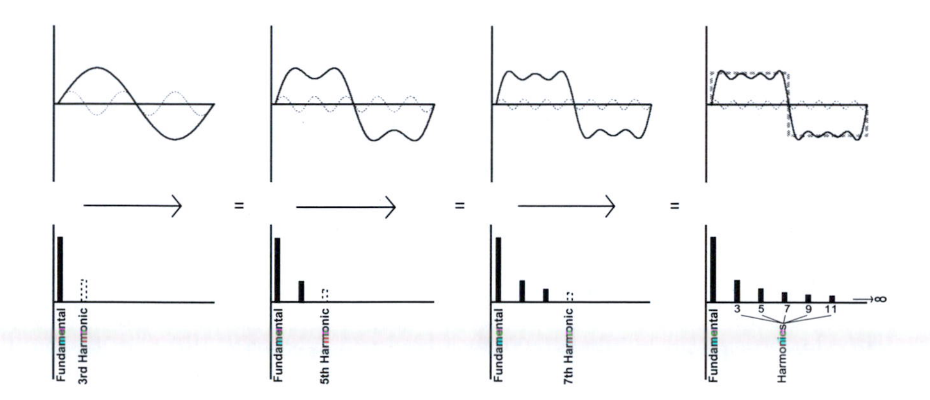

Addition of sine waves to create a sawtooth wave

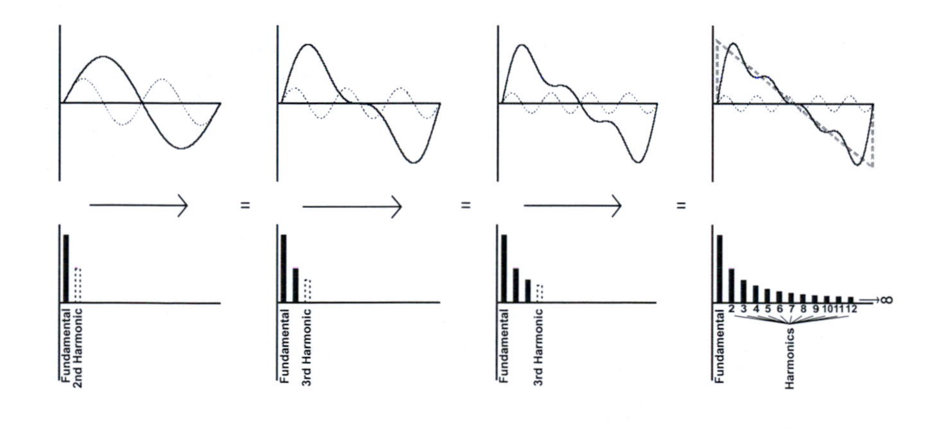

Legend
Current wave form
Next wave to be added
Eventual wave form

FIGURE 4.4
How multiple sound waves create harmonics (1)

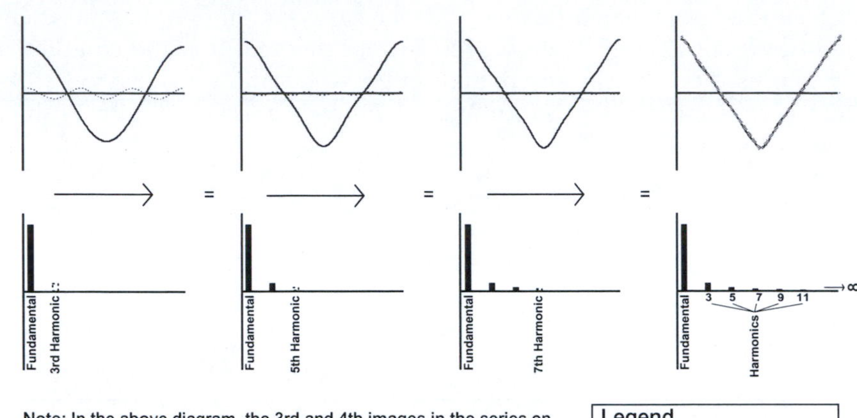

Note: In the above diagram, the 3rd and 4th images in the series on the top row appear to have none, or virtually no wave being added. This is because the odd harmonics decrease in level exponentially. For example, the 3rd harmonic is 3^2 of the level (1/9), the 5th harmonicis 5^2 of the level (1/25), and so forth.

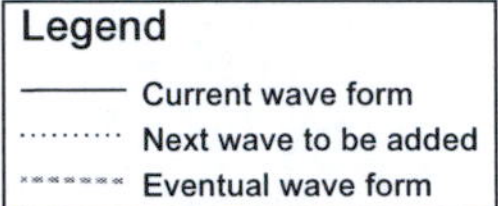

FIGURE 4.5
How multiple sound waves create harmonics (2)

VOLUME AND GAIN

Thus far, we've considered the creation of both pitch and timbre, but the final characteristic we must consider is volume. Changes in volume are a result of the amount of air molecules an oscillating object displaces. The more air an object displaces, the louder the perceived sound will become. This volume, termed *amplitude*, is measured by the degree of motion of the air molecules within the sound waves that correspond to the extent of rarefaction and compression that accompanies a wave.

In many musical instruments, the amount of air displaced is too small for the sound to be audible, and therefore, they often employ different forms of amplification. In acoustic instruments, this is accomplished through the principle of forced vibration. Returning to the previous example of a piano, when the wire is struck by the mallet, its vibrations not only set other wires in motion but also a large wooden board located beneath. This sounding board is forced into motion via the displaced air molecules, and since this board moves a greater amount of air, the sound is amplified.

Naturally, neither of these methods of amplification offers any physical control over the amplitude. If the level of amplification can be adjusted, then the ratio between the original and the newly adjusted amplitude is termed the *gain*.

Gain is measured in decibels, a unit of measurement first conceived by Bell Laboratories when they needed to express how much power was being lost whilst sending voltages over vast lengths of telephone cable. Originally, this loss of power was measured in *Bells*, but being such a large measurement when it is applied to audio, it is subdivided into tenths of a bell, resulting in the term *decibel*.

DECIBELS

Decibels are most suited to measuring audio because they more accurately represent the non-linearity curve of human hearing. Natural evolution has resulted in us developing both frequency-dependent hearing and an incredibly large audible range. This makes measuring the complete auditory range difficult without expressing incomprehensible numbers.

For example, if we were to measure our hearing using a measurement of micro-pascals, the difference between normal conversation levels and an airliner taking off would be a good few trillion. This is a number so fantastically huge that it is not only difficult to write down but also difficult to enter into a calculator with any degree of certainty. Decibels, however, are based on the 10 base logarithmic scale since this quantifies human hearing more accurately.

Understanding decibels requires an understanding of logarithmic scales and in particular working with base 10. An example of this is shown in Figure 4.6. Here, the lower numbers represent the sound intensity, and each respective number is to the power of 10. For example, 10 to the power of 10 is 100 ($10 \times 10 = 100$), whilst 100 to the power of 10 is 1,000 ($100 \times 10 = 1{,}000$) and 1,000 to the power of 10 is 10,000 ($1{,}000 \times 10 - 10{,}000$) and so forth through the scale.

Above each of base 10 number is the equivalent measurement in decibels, and using this chart, it can be determined that a movement from 10 to 20 dB would be equivalent to a 10-fold increase in intensity (to the power of 10). Similarly, 30 dB would be 100 times more intense than 10 dB and 40 dB would be 1,000 times more intense. This type of scale offers a very good representation of how our hearing behaves and is the reason decibels are used.

An important concept to understanding the use of decibels in audio engineering is that they are not a direct measurement and are instead used as a way to express the ratio or difference between two different levels – a reference level and the current level. This is why any decibel measurement is (or should be) followed by a suffix. The suffix letters that follow the decibel will determine the reference level being used. For example, if the producer were measuring noise levels at a concert, then sound pressure level (dB-SPL) would be used as the reference level. Using this reference, 0 dB is configured to the lowest threshold of human hearing range, and therefore, anything above this measurement would be considered a fairly accurate representation of its loudness. I state fairly here because loudness, unlike gain, is difficult to quantify since it depends on a number of factors such as the listener's age and the frequency.

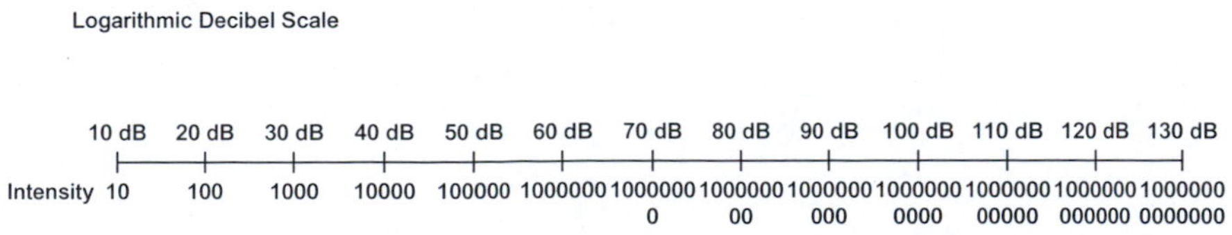

Logarithmic Decibel Scale

	10 dB	20 dB	30 dB	40 dB	50 dB	60 dB	70 dB	80 dB	90 dB	100 dB	110 dB	120 dB	130 dB
Intensity	10	100	1000	10000	100000	1000000	10000000	100000000	1000000000	10000000000	100000000000	1000000000000	10000000000000

FIGURE 4.6
The logarithmic scale

Generally speaking, the human ear can detect frequencies from as low as 20 Hz up to 20 kHz; however, this depends on a number of factors. Indeed, whilst most of us are capable of hearing frequencies as low as 20 Hz, the perception of higher frequencies changes with age. Most teenagers are capable of hearing frequencies as high as 18 kHz, while the middle-aged tend not to hear frequencies above 14 kHz. A person's level of hearing may also have been damaged, for example, by overexposure to loud noise or music.

In addition, human hearing has developed in such a way that we will perceive some frequencies to be much louder than others even if they're all played at the same gain. This inaccurate response is part of our survival mechanism and developed to permit us to single out the human voice amongst a series of other sounds or situations. For example, atop a very windy hill, we would need to be able to decipher another human shouting that there is a hungry looking bear approaching above the sound of the wind and trees rustling.

Since the human voice is centred at 3 to 4 kHz, this is where our ears have become most attuned. This means that at conversation level, our ears are most sensitive to sounds occupying the midrange, and any frequencies higher or lower than this must be physically louder in order for us to perceive them to be at the same volume. Indeed, at normal conversation levels, it's 64 times more difficult to hear bass frequencies and 18 times more difficult to perceive the high range. However, this perceived relationship changes as the gain increased.

If the volume is increased beyond normal conversation level (approximately 70 dB), the lower and higher frequencies gradually become perceivably louder than the midrange. In the 1930s, two researchers Fletcher and Munson from

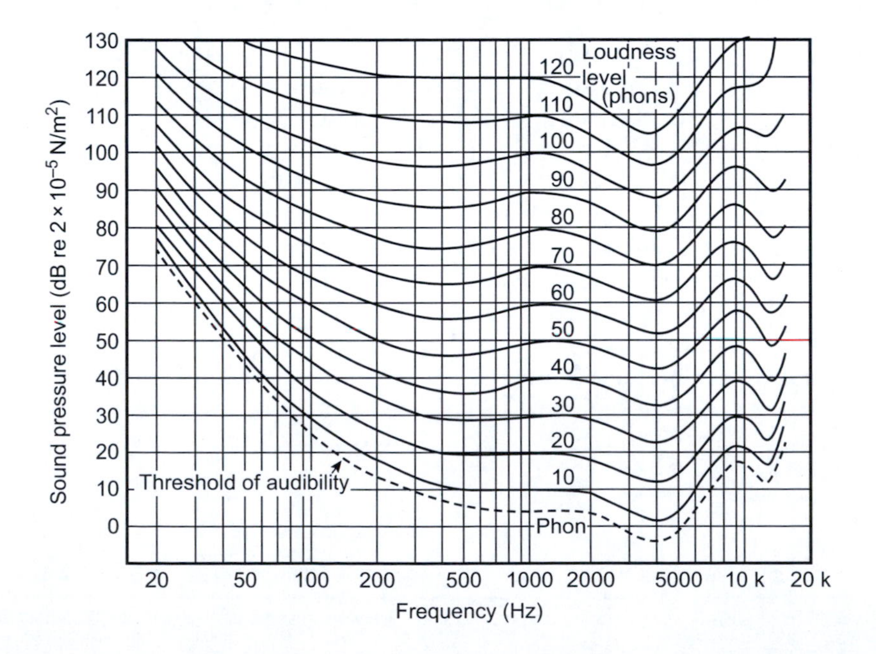

FIGURE 4.7
The Fletcher Munson contour control

Bell laboratories were the first to experiment and accurately measure this uneven hearing response of the ear, and thus, it is often referred to as the *Fletcher Munson Contour Control*.

To put this into some perspective, if you were to measure a 60 Hz sine wave at 30 dB with an SPL meter, despite the measuring equipment stating that there is a 30 dB sound present, it would be physically inaudible to humans, and hence – from a hearing perspective – the measurement would be useless. To counter this, the International Electrotechnical Commission (IEC) introduced the principle of weighted filters.

Weighting filters refer to a number of different styles of filters that are designed to reduce or enhance specific frequencies along the sonic spectrum. These consist of A, B, C, D, Z and the newly introduced K system, but in most audio applications, *A-weighting* proves the most common since this is believed to accurately represent human hearing at low levels. With this filter, the filter slowly opens allowing more frequencies through until it reaches its apex at approximately 2 to 3 kHz where it then begins to close again. Figure 4.8 shows this weighting filter in action.

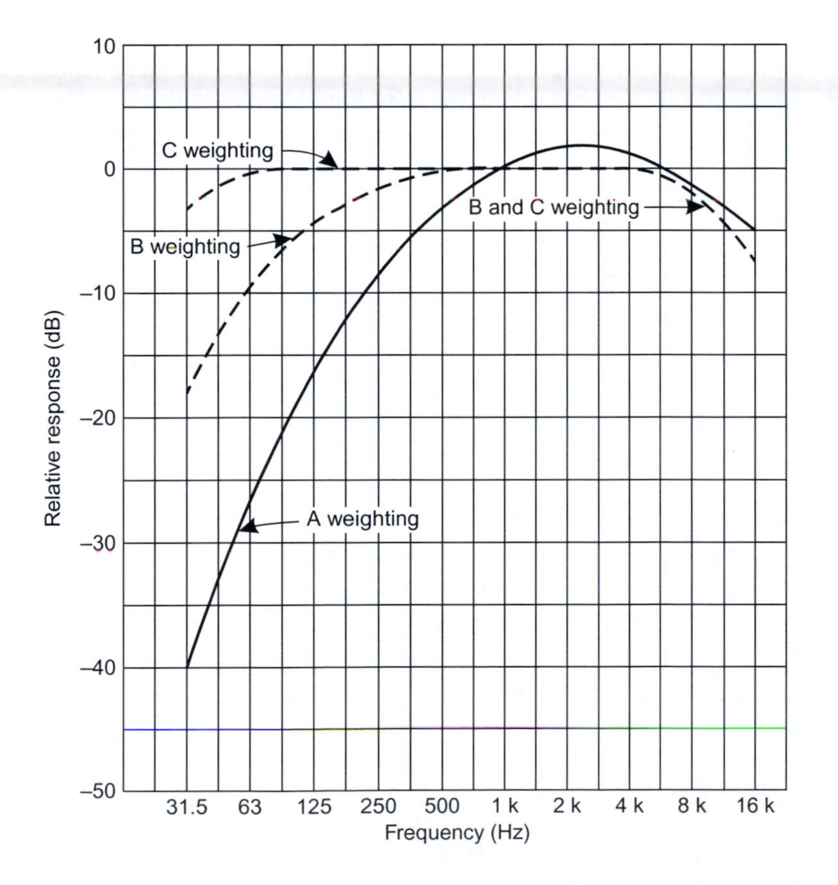

FIGURE 4.8
Weighted filters

When filters are employed, the filter type is added to the end of the decibel measurement. For example, many microphone pre-amplifiers and processors may rate their signal-to-noise ratio (SNR) at 120 dBA. This relates to the ratio between the noise created by the device and the level of the desired signal.

In fact, this SNR is one of the most important properties to factor in when employing any production hardware. Often quoted as the SNR, this measurement is used by many manufacturers to quantify the amount of noise that will be generated by a hardware device processing a signal. And this applies to all electrical devices, including audio interfaces, speaker monitors and computers. Indeed, the noise created by the equipment and any cables is an unfortunate, yet unavoidable aspect that all producers must deal with at one time or another. This alongside any EDM producer having to rely on electronic equipment means that some basic knowledge of electronics can be essential for production work, whether that's choosing a particular pre-amp or microphone or simply choosing an audio interface to work with. Consequently, we'll quickly examine some basic electronics principles and develop further on it in later chapters.

As many reading this are no doubt already aware, all matter is composed of atoms and each atom consists of a central nucleus surrounded by a series of orbiting electrons. The central nucleus consists of positively charged protons and neutrally charged neutrons. The orbiting electrons are negatively charged and, much like the principle of magnetism, as the electrons are negatively charged they repel one another but are equally attracted to the positive protons in the nucleus.

It was discovered that if energy were applied, it was possible to knock electrons away from one nucleus and onto the next, and it is this movement of electrons that can create electricity. The energy applied is voltage, and the number of electrons that are put into motion by the application of voltage is referred to as the current, which is measured in ampere.

Diverse materials offer different types of resistance to the flow of the electrons. While a material such as wood will resist the flow of electrons, other materials such as copper or steel will readily permit the flow. However, even though copper and steel permit the free flow of electrons, it is not completely free and they do still offer some resistance to the flow. This is especially the case in music equipment and computers since not only do the circuit boards offer some resistance, each and every element fitted to that board, from the transistors to capacitors and switches, will also resist the flow.

This resistance to the flow of electrons results in heat, and this is often termed by engineers as Johnson or Nyquist noise but more typically manufacturers will state it as *thermal noise* in their publications and specifications of equipment. This is why all manufacturers will state the SNR of equipment using a weighted A filter.

A further problem manifests itself with electromagnetism. As current flows through a conductive device, such as a piece of wire, it results in the creation of a circular magnetic field surrounding the wire. The direction of the magnetic circular motion is determined by the direction of the current but the larger the current, the larger the resultant electromagnetic force becomes. To further compound the situation, the higher the resistance of a cable, the higher the flow of the current must be in order to fight the resistance, which in turn increases the amount of electromagnetic interference.

Since resistance will increase as the cable length increases or cable thickness decreases, it poses problems for many studios since the cable runs are often quite long, and they're commonly bound together to run down a single conduit. The result is that electromagnetic interference increases as it travels around instruments and through to earth, manifesting itself as hiss, loss of high-frequency detail, and more commonly a low-frequency hum termed *ground hum*.

Both electromagnetic interference and thermal noise obviously result in a series of potential problems for any EDM producer, particularly those who look to expand their studios beyond just a computer. However, before we look further into how to prevent these problems, we should first investigate the hub of any studio, the Digital Audio Workstation.

CHAPTER 5

MIDI, Audio and the Digital Audio Workstation

'Now, lets hear what that looks like...'

AES salesperson demonstrating a DAW

With an understanding of basic music theory and some acoustic science, we can begin to examine the theory and tools pertaining to the production of electronic dance music and where better to start than with the centrepiece of many dance musicians studio; the digital audio workstation.

Provided the host computer is powerful enough, a digital audio workstation is capable of performing all of the functions that would have required a complete studio set-up only 10 years ago. Within DAW software it's possible to record and edit audio, apply effects and processing onto audio, employ software representations of classic synthesizers as well as mix and master the resulting music. In fact, the DAW has become such a powerful application that some artists create their entire tracks on little more than a laptop.

A digital audio workstation, however, is as individual as the music that can be created on one and it would be misleading to recommend one specific workstation over another. With their constant upgrade patterns that consistently introduce new features, it is also impossible to discuss the pros and cons of each individual workstation and therefore this chapter will not discuss the various sequencer-specific options but rather discuss the most commonly shared features to offer the reader a good grounding to the working principles employed.

Most DAWs divide their editing and operational functions over a number of different windows and the first, and perhaps most important of these, is the arrange window. This is the page you're most commonly greeted with upon opening any DAW interface and consists of an overview of the current music project.

The arrangement page consists of a number of tracks that can be created by the producer and are commonly listed down the left hand side of the window.

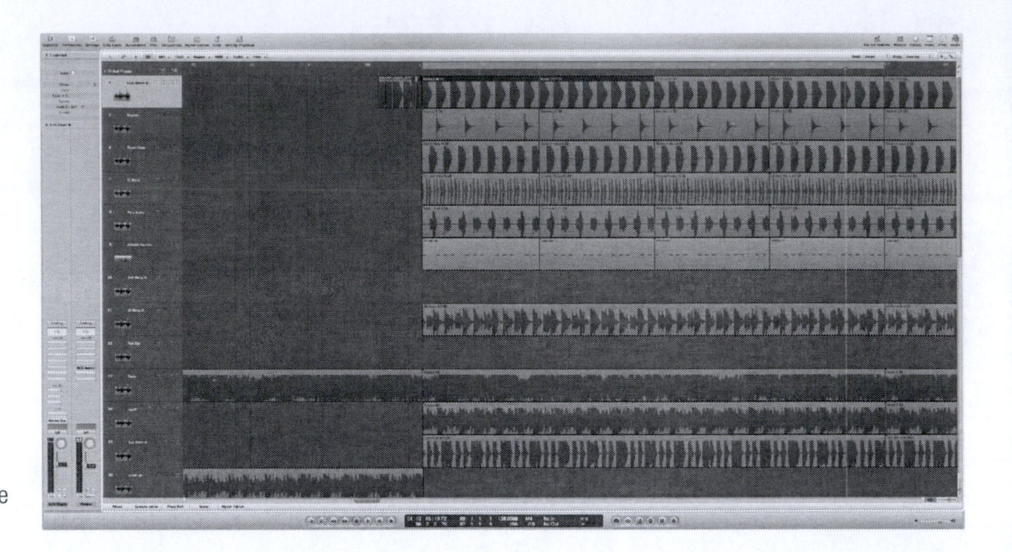

FIGURE 5.1
The arrangement page in logic pro

When playback is initiated (commonly by pressing the space bar, or clicking play on the transport control) a timeline moves from left to right on the arrange page, playing back all these tracks simultaneously.

Within this arrangement page it is possible to perform most of the basic editing and arrangement functions, from adjusting volumes and panning of a selected track, inserting effects or processors through to slicing a tracks contents into smaller segments (often called events or regions) with a scissor tool, or to copy, repeat and move these events across the window to produce an arrangement. For example, if there is a four-bar drum loop in the project, this 'event' could be copied continually so that it plays for any number of bars throughout the arrangement. The same could then be performed with the bass and lead. The different types of tracks that can be employed in the arrange window can be sequencer dependent but all will permit audio, MIDI and virtual instrument tracks.

Audio tracks are employed to contain audio that has been recorded by the producer direct into the sequencer or taken direct from a sample CD or mp3 file. These can all be dragged and dropped into the workstations arrange window where many workstations will copy the audio file to a specific location and then insert an audio 'event' or 'region' into the arrange page.

This means that the actual audio file is *never* actually placed into workstations arrange page and instead the newly created event or region is nothing more than a graphic representation that signifies the starting and ending point of the audio file held on the hard disc. By employing this technique if the event or region is, say, two-bar long and is repeated a hundred times in the arrange page, the workstation simply repeatedly plays the *same* single two-bar audio file a hundred times. This reduces the amount of hard disc space required since it

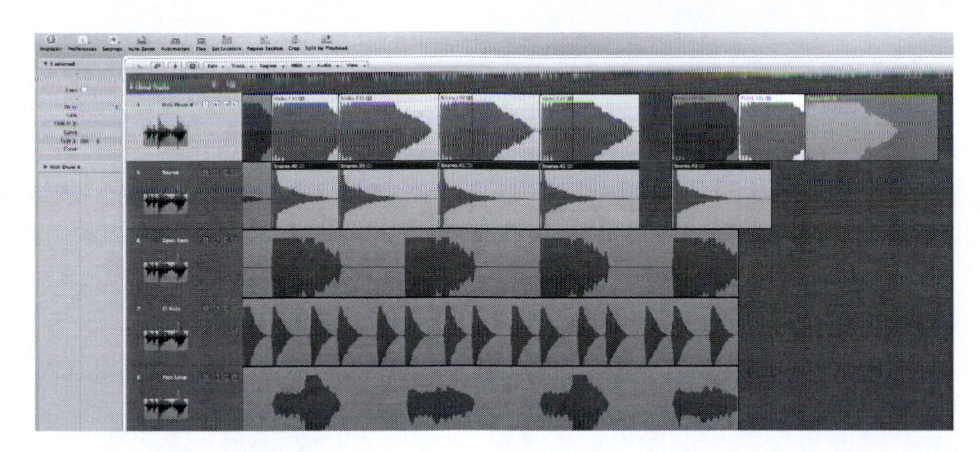

FIGURE 5.2
Audio 'regions' sliced and edited in logic pro

doesn't have to store multiple copies of the same audio file and also speeds up the workstation's process.

Moreover, any edits made to the event or region become non-destructive. In other words, if the producer slices an event into numerous parts and arranges these in a different order across the arrange page, it does not affect the original audio file stored on the hard disc and therefore any changes to an event can be undone easily. There are numerous edits that can be applied to an audio event or region on the arrange page from moving, copying and crossfading through to fading the audio in or/and out. The commands and features available will depend on the workstation in question.

MIDI TRACKS

MIDI tracks and virtual instruments tracks are fundamentally the same in that these both contain MIDI events. MIDI is a protocol that was established to permit digital musical instruments to communicate with one another. The event data transmitted in the protocol consist of simple commands such as when to play a note, the length of the note, and how hard the note was struck (velocity). It can, however, also be used to transmit many more complex commands from changing parameters, such as modulation and pitch through to commands informing any attached devices to dump their memory.

Within a workstation, a standard MIDI track is designed to send MIDI note data outside of the DAW and onto any attached hardware MIDI devices whilst on a virtual instrument track, the MIDI remains internally routed to a software emulation of an instrument that is graphically represented within the confines of a workstation.

Regardless of whether the track is a virtual instrument or MIDI track, only very basic editing is permitted within the arrange window. Here the producer can simply insert, delete, slice and move events around the arrange page, much in the same way as audio. Further in depth MIDI editing must be accomplished with a specific piano roll editor. This could be viewed as an advanced musical conductor, sending MIDI event data instructions to any number of external hardware or internal virtual synthesizers.

As shown in Figure 5.3, down the left hand side is a piano keyboard that denotes the pitch. The higher up towards the top of the screen the notes are, the higher the pitch becomes. To the right is the quantization grid and when playback begins any notes in this field are played from left to right at a speed determined by the tempo of the piece.

Inserting notes into the quantize grid is as simple as choosing a 'Pencil' Tool from the DAW's toolbox, clicking in the grid and holding down the mouse button whilst you draw in the length of the note required. If numerous notes drawn in at the same position in time, the notes will be chorded as the synthesizer plays back all the notes simultaneously. Alongside physically entering notes the producer can also record these data live. Through connecting a MIDI capable keyboard to the DAW through MIDI or USB, event data such as notation and velocity can be recorded live into the piano roll. Once a series of notes have been recorded (or drawn in), the timing can be modified and notes can be lengthened, shortened or deleted to perfect the performance.

As discussed in the previous chapter on rhythm, when inserting notes into this grid the current quantize value must be considered. As previously discussed, the piano roll editor is split into a series of quantized grids and the current quantize setting will dictate where any newly inserted notes may be positioned. All DAWs (and hardware MIDI step sequencers) will specify this quantization resolution in terms of pulses per quarter note (PPQN), and the higher this is, the more note positions become available within the bar.

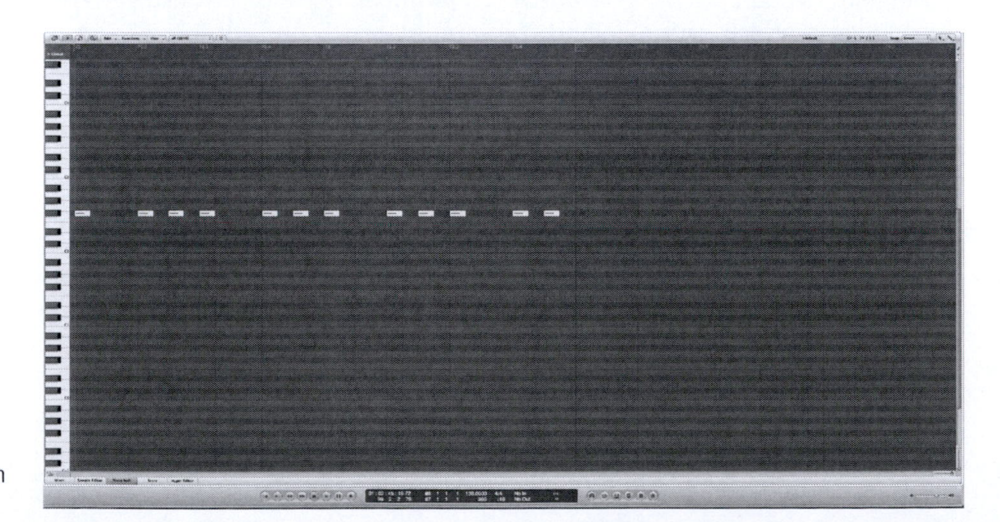

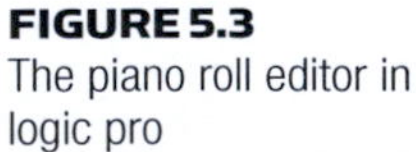

FIGURE 5.3
The piano roll editor in logic pro

For example, if the producer is employing a workstation with a PPQN of 4 it would be impossible to insert any notes that are less than a 16th in length because there isn't a high enough resolution to yield an integer number of clock pulses. Thus, not only would the producer be unable to create notes smaller than a 16th, but also it would also be impossible for any note timing to be adjusted to less than this either. This resolution is typical of the early analogue drum machines that were used to create dance rhythms and are responsible for the somewhat metronomic behavior and sound of early EDM drum patterns.

In the early 1990s, Roland came to the conclusion that a PPQN of 96 would be suitable for capturing most human nuances, while a PPQN of 192 would be able to capture the most delicate human 'feel'. Subsequently, most step sequencers will employ this resolution whilst many software DAWs have adopted a much higher 480 PPQN and above.

Alongside the standard note event data, sequencers and workstations will also record and transmit controller event data. These are often the most overlooked yet influential parts of event data since the textural motion of a sound plays a key role in the production of electronic dance music. No sound in EDM remains constant and it's the evolution of its texture or pitch over the length of a note – no matter how short – that provides much of the interest within dance music.

Controller event data are often termed *CC data* or *Control Change* message information and is transmitted alongside the note data to inform the virtual or hardware synthesizer of any further modifying values such as velocity, after-touch, filter cut-off, modulation or panning. This information can be recorded live through the producer moving any number of controllers on a hardware MIDI interface (assuming the interface sends the CC to be recorded) or entered manually into the sequencer.

Many workstations will permit the producer to enter control event data through a MIDI event page or more commonly, they can be drawn in direct through a pencil into a designated area in the piano roll editor window. Of the two, drawing in events is much simpler and a more tactile approach that requires less programming knowledge but the event page can be useful for making very accurate fine tuning to the previously drawn-in events. The most common approach, however, is to record movements live through an attached keyboards modulation wheel and then assign these recorded movements to the parameter you want to automate on the synthesizer.

Re-assigning controller event data depend on the workstation being employed but in many this is typically implemented in the form of a drop down box. The producer clicks on the drop down box to choose a different parameter and the recorded controller automation will automatically be re-assigned to the new parameter. It is important to note here, however, that not all virtual or hardware instruments will understand these assignments.

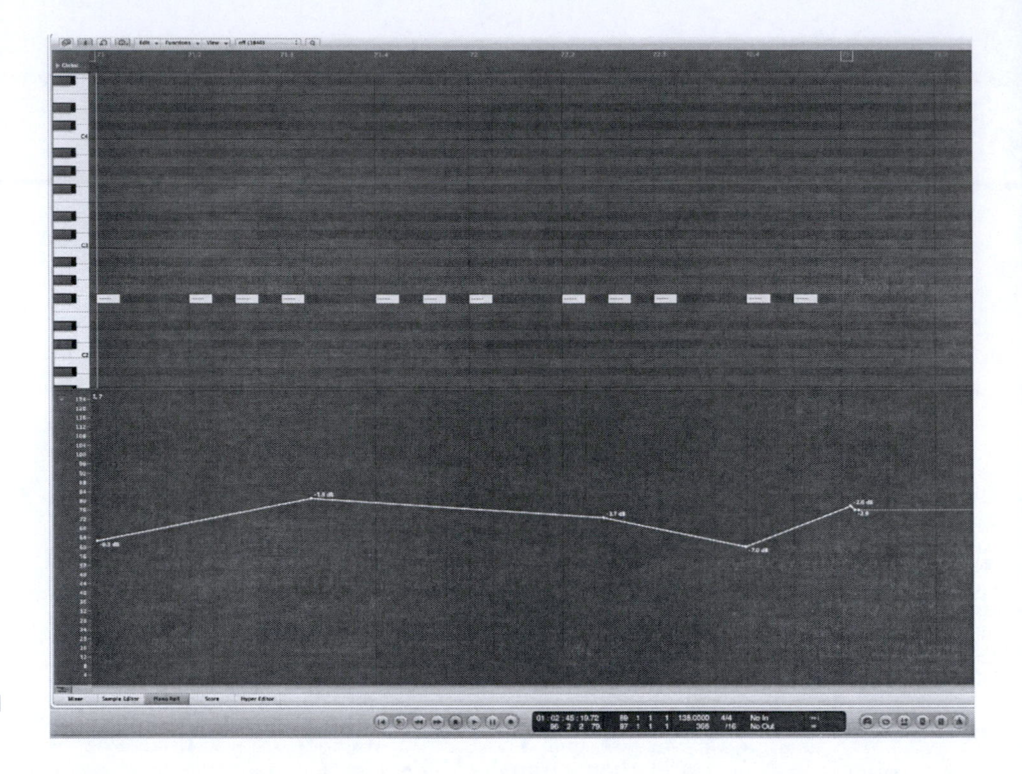

FIGURE 5.4
Modulation movements recorded in logic pro

Controller event data are recorded and transmitted as two main values, the first value determines the parameter to adjust whilst the second determines by how much this parameter is to be modified. In order for this to perform reliably, there must be some form of uniformity between different synthesizers so that they will all respond consistently to the incoming event data. Because of this, the General MIDI standard was developed.

General MIDI is a generalized list of requirements that any synthesizer must adhere to for it to feature the GM symbol. Amongst determining what sounds are assigned to certain numbers and how percussion sounds should be mapped across a keyboard it also deals with the MIDI controller event messages that should be recognized.

There are 128 possible controller event messages on any MIDI capable device and these are numbered from 0 to 127. In the GM standard, many of these are hard wired to control particular functions on the synthesizer so provided that the CC number is encompassed by the GM standard and the synthesizer is GM approved, the producer is guaranteed that it will control the correct parameter no matter what synthesizer the messages are transmitted too. For example, the modulation wheel is assigned to CC1 and panning is assigned to CC10 in the GM standard and therefore any GM-compatible synthesizer will understand both of these commands.

As touched upon previously, each of these controller events must have a secondary variable associated with it to permit the producer to set how much the specified controller should be moved by. Indeed, every controller event message must be followed by this second variable that informs the receiving device how much that particular CC controller is to be adjusted.

Some controller events will offer up to 128 variables (0–127) while others will feature only two variables (0 or 1) and these will perform as on/off switches. What's more, some messages will have both positive and negative values. A typical example of this is the pan controller, which takes the value 64 to indicate central position. Hence, when panning a sound, any values lower than this will send the sound to the left of the spectrum while values greater than this will send the sound to the right of the stereo spectrum. A full list of the GM controller event specifications are listed in the appendix of this book but most notably, pitch bend is *not* amongst them.

This is because pitch bend is a very particular MIDI parameter and requires more values than the typical 0–127 offered by controller events. Since pitch bend sends and receives values between –/+ 8190, it is not included in the GM CC list but almost all synthesizers, whether plug-in or hardware, alongside all sequencer will remain within specifics in order to offer compatibility. Indeed, even though many plug-in instruments will not feature a GM compatible logo since they do not completely adhere to the GM standard in terms of tones, they will nevertheless follow the CC commands laid down in the standard. This is because all sequencers will transmit these values as standard.

On the subject of these plug-in synthesizers, perhaps the largest development to the modern digital audio workstation was the development and introduction of these virtual instruments and effects.

First introduced by Steinberg with Virtual Studio Technology (VST) in 1996, it permitted sequencers to integrate virtual instruments and effects processors into a workstations environment. These virtual applications could be modelled on rare or vintage analogue technology and installed into a studio system at a fraction of the cost of the original hardware units. What's more, since they are inserted direct into the workstations environment, all the parameters can be controlled with a mouse and can also be fully automated. What's more, as the audio output from the plug-ins is routed internally it can be directed through any number of channels in the workstations mixer negating the requirement from external cabling and complex audio interfaces.

Plug-ins are available in a number of formats including Audio Units, Universal Binary, Direct X and VST 2 & VST 3. Audio Units and Universal Binary are Mac only interfaces for use within their OSX operating systems, whilst Direct X is a Windows only application. VST 2 & 3, however, are available for both platforms but are workstation specific. Logic Pro 9 on the Macintosh, for example, will only accept AU and UB plug-ins and currently offers no support for VST.

FURTHER EDITING WINDOWS

Most producers will spend most of their time in the workstation switching between the arrange page and piano roll since these are the most commonly used during the creation of a track but further windows are available including sample editing windows and mixer windows.

Whereas basic audio editing such as slicing, arranging and crossfading can be accomplished in the arrange window, more complex audio editing becomes available in the sample edit window. The features included here vary greatly depending on the workstation being used but typically they will feature options such as normalizing, sample reversing, time stretching, dynamics, slicing and fade in and out.

Whilst some of these functions are available in the arrange page it is important to differentiate the difference between applying edits in the arrange page and in the sample editor. Whereas in the arrange page edits are only applied to an event or region that point to the audio file on the hard disc, any edits applied in the sample editor will *directly alter* the sample stored on the hard disc. Thus, if the producer makes an edit in the sample editor such as reversing the sample, every event or region contained in the arrange page will reflect this change.

Therefore, if the producer wishes to make an edit to single region in the arrange page such as reversing the sample in the sample editor, it is wise to first use the workstations features to 'duplicate' or 'convert' the currently selected event into a secondary audio file. By doing so any changes made to this audio file will not affect the rest of the events or regions in the arrange page that point to the original file.

Beyond the sample-editing window, a final window that will feature in many audio workstations is the mixer. This window displays the mixer and its

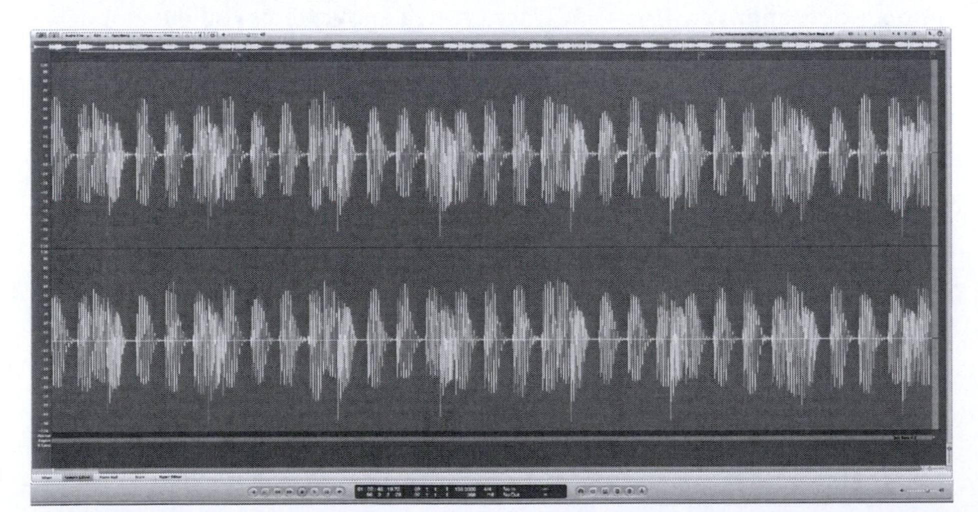

FIGURE 5.5
Sample editing page in logic pro

associated features such as inserts, sends, aux channels, group channels and so forth. Typically, in many workstations as soon as an audio, MIDI or Instruments channel is created in the arrange window, a new channel will be introduced into the mixer to permit the producer to add effects, panning and change the overall volume of the channel. The mixer will be discussed in more detail in a later chapter.

Depending on the sequencer, yet further editing windows may be available such as notation or environments but often these become more sequencer specific and therefore are beyond the scope of this book. However, before drawing this chapter to a close, it would be improper not to briefly discuss hardware sequencing.

Whilst many dance musicians will rely on a digital audio workstation for their production, a good number of artists will employ hardware step sequencers for the majority of their productions and only employ a digital audio workstation at the end of the creative process to mix and master the music.

Although wanting to avoid the argument of digital audio workstations versus hardware step sequencing, many artists feel that the timing of hardware sequencers is unparalleled and it's only possible to create strict rhythms with the use of a step sequencer. To test this theory, I produced a drum loop on a Rhizome Step Sequencer and in Logic Pro – you can hear the results on Chapter 5 of the companion website.

Rhythmical timing differences from a DAW and hardware sequencer.

Whether there is a noticeable difference is a matter of personal opinion but one thing that is certain is working on a step sequencer involves a very different approach than in a workstation. Both require their own production ethics and the difference in approach means that different creative avenues can

FIGURE 5.6
Mixer page in logic pro

FIGURE 5.7
The feeltone Rhizome

be explored depending on whether the artist is employing a step sequencer or a digital audio workstation. Indeed, the growing popularity of the Elektron MachineDrum, Analog4 and Rhizome is testament to how hardware sequencing hasn't been completely replaced by the DAW. It's therefore recommended that rather than simply settling on an audio workstation it is equally worth investigating hardware step sequencers to help boost your creativity.

CHAPTER 6

Synthesizers

'Play the music, not the instrument'.

Anonymous

Despite EDM having developed significantly over the past 20 or so years by encompassing audio manipulation and production techniques only made possible with the latest audio workstations and plug-ins, the humble analogue subtractive synthesizer still remains as the single strongest musical component in the production of EDM.

Much of the analogue synthesizer's popularity can be attributed to this form of synthesis being the most immediately accessible but with the introduction of virtual instruments running completely inside the DAW, the more complex forms of synthesis such as frequency modulation, additive and granular are beginning to resurface as viable alternatives due to the larger screen estate. However, in order to employ any of these synthesizers beyond simply preset surfing and basic tweaking, the producer first requires a good grounding of their operational theory.

All synthesizers are formed around the principle of combining a number of basic sounds or oscillators together to create a timbre that is rich in harmonics. Once this is accomplished, the sound is then further sculpted through the use of filters and time-based movement, and modulation is then applied through a series of available modifiers to create motion in the sound. How this is accomplished depends on the synthesizer in question but an understanding of analogue subtractive synthesis goes a long way to comprehending all other forms of synthesis, so we'll begin by investigating analogue subtractive synthesis.

A subtractive analogue synthesizer fundamentally consists of the following three components:

- Any number of oscillators that can be combined and mixed together to produce a timbre that is rich in harmonics.
- A filter to sculpt the resulting harmonics.

FIGURE 6.1
A plug-in instrument modelled on subtractive analogue synthesis

- An amplifier to define the overall level of the sound over time.
- Modifiers to add time-variant modulation to the timbre.

OSCILLATORS

The oscillator should be considered the heart of *any* synthesizer whether additive, analogue or frequency modulated since these produce the basic tones of the instrument.

In original hardware analogue synthesizers, the oscillators were triggered through different voltages that were transmitted by a note press on the keyboard. This *control voltage* is unique to each key on the piano keyboard allowing the oscillator to determine the pitch it should reproduce. Obviously, these control voltages must be precise in order to prevent the oscillators pitch from drifting, and in many instances, the synthesizers had to be regularly serviced to ensure that they remained in tune, but for some aficionados, this drifting was part of their analogue charm.

A similar style of operation applies today with synthesizers but rather than employ control voltages, each note is awarded a number and from the number transmitted, the synthesizers CPU can determine the note that has been depressed and trigger the correct pitch from its oscillator/s.

In many of the early synthesizers, there were only three types of oscillator waveforms: square, sawtooth and triangle waveforms, but over the years further waveforms were introduced including sine, noise and pulse waves. The most basic of all these waveforms is the sine wave.

1. The Sine Wave
   A sine wave is the simplest wave shape based on the mathematical sinusoidal function. A sine wave consists of only a fundamental frequency and does not contain any harmonics; therefore it is audible as a single fundamental tone, comparable to a whistle. They are not ideally suitable for subtractive synthesis, and this is the reason as to why they were not introduced in the first analogue synthesizers. If the fundamental is removed, there is no sound. Consequently, the sine wave is often used independently to create sub-basses or whistling timbres, but it is more commonly mixed with other waveforms to introduce body or power to a sound.
2. The Square Wave
   A square wave is the simplest waveform that can be produced by an electrical circuit since it can only exist in two states: high and low. This wave produces a large number of odd harmonics that results in a sound that could be described as mellow, hollow or woody. This makes the square particularly suitable for emulating wind instruments adding width to strings and pads, or for the creation of deep, wide bass sounds or leads.
3. The Pulse Wave
   Pulse waves are often confused with square waves, but unlike a square wave, a pulse wave allows the width of the high and low states to be adjusted thereby varying the harmonic content of the sound. It is unusual to see both square and pulse waves featured in a synthesizer, and instead they often just employ a pulse wave, because if not adjusted it produces the sound of a square wave. These waveforms permit you to produce thin reed-like timbres along with the wide hollow sounds created by a square wave. A popular technique is to employ cyclic modulation on the width of the pulse wave

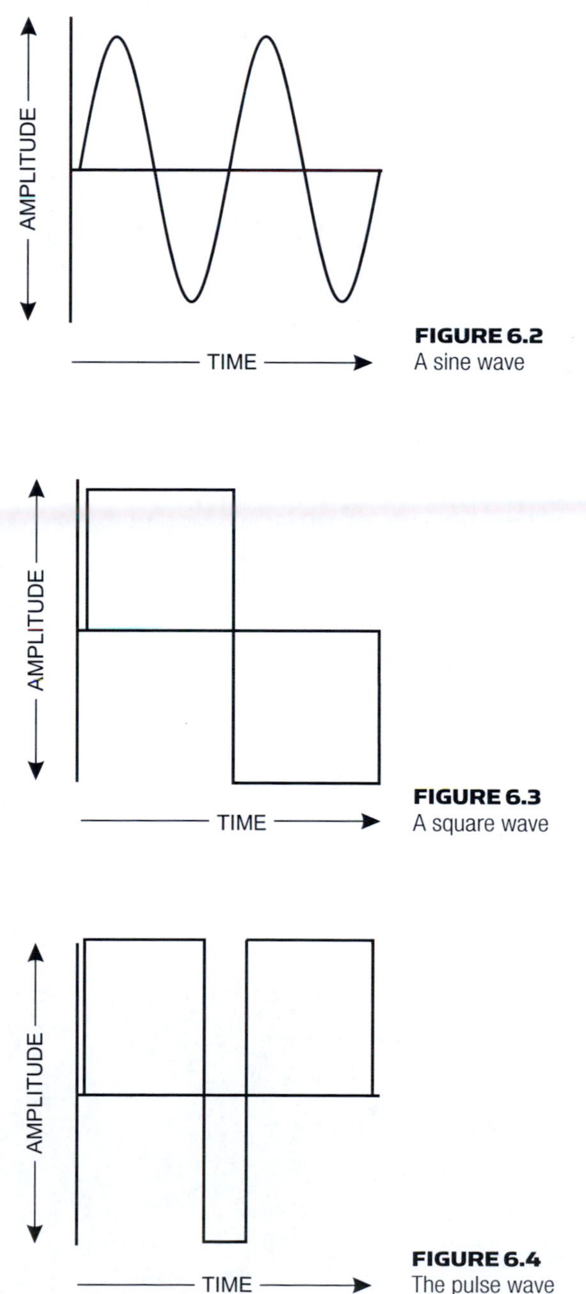

FIGURE 6.2 A sine wave

FIGURE 6.3 A square wave

FIGURE 6.4 The pulse wave

so it moves from its different states over time, creating a complex yet interesting timbre.

4. The Triangle Wave
The triangle wave shape features two linear slopes and is similar in respect to the square wave in that it only consists of odd harmonics. As with the square wave, this produces a timbre that could be described as thin and partially hollow. Typically this type of waveform is often mixed with a sine, square or pulse wave to add a sparkling or bright effect to a sound and is often employed on pads to give them a glittery feel.

5. The Sawtooth Wave
The sawtooth wave is the most popular and most used waveform oscillator in most synthesizers. It produces a proportionate number of even and odd harmonics and therefore produces a timbre with the strongest harmonic content. This large harmonic contents means it is particularly well suited for sweeping and sculpting with filters. Typically, a saw wave is used in the construction of any timbre that requires a complex, full sound.

6. The Noise Wave
Noise waveforms are unlike the other five waveforms because they create a random mixture of all frequencies rather than actual tones. Noise waveforms are commonly 'pink' or 'white' depending on the energy of the mixed frequencies they contain. White noise contains the same amount of energy throughout the frequency range and is comparable to radio static, while pink noise contains differing amounts of energy at different frequencies and therefore we perceive it to produce a heavier, deeper hiss. Indeed, the darker the color of noise, the duller the frequencies it will appear to exhibit.
Noise is most useful for generating percussive sounds and was commonly used in early drum machines to create snares and handclaps but it has more recently been employed in both bass and lead timbres to add higher frequency energy for them to cut through a mix.

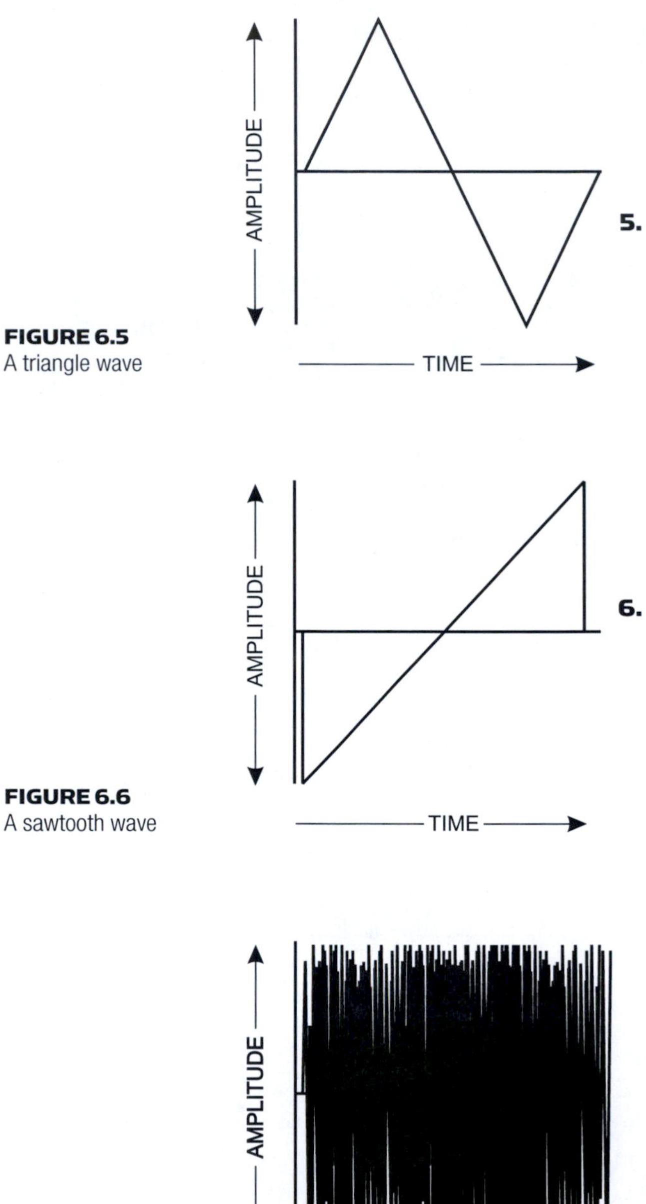

FIGURE 6.5
A triangle wave

FIGURE 6.6
A sawtooth wave

FIGURE 6.7
Noise waveform

FURTHER WAVEFORMS

Today, with computer processing reaching huge speeds, the number and style of oscillators have increased substantially from the early days of synthesis. Many plug-in synthesizers now offer numerous waveforms ranging from tri-saw, pulse, variable pulse, tri-pulse to numerous differing wave shapes that are sampled from a multitude of different sources.

Many of these 'newer' waveforms are based around the three basic analogue oscillators and are employed to prevent having to mix numerous basic wave-forms together, a task that may reduce the number of available oscillators. For example, a tri-saw wave is commonly a sample-based oscillator consisting of three sawtooth oscillators blended together to create a sound that is rich in harmonics. Wave shapes also fall into this sampled category but are commonly samples of real word sounds and instruments. For example, some may consist of the attack stage of a piano repeated over whilst other may consist of a series of noises. With these forms of oscillators, a control is usually offered to set the start point of the sample permitting the producer to create a wide variety of timbres through nothing more than starting the sample part way through its cycle.

Whether a synthesizer features only three or four oscillators or has a wealth available, the foundations for programming using them remain the same. The producer mixes together the available oscillators, detunes them from one another to introduce beating and hence further harmonics and then modifies or modulates the harmonics with the available modulation parameters on the synthesizer.

Detuning is accomplished using a 'detune' parameter that is commonly featured next to each oscillator. Typically, detuning is configured in octaves, tones and cents (100ths of a tone). During programming, the producer will often work in odd rather than even detuning between similar oscillators because detuning by even amounts will introduce harmonic content that mirror the harmonics already provided by the first oscillator. The result is the harmonics that are already present are amplified in gain rather than the process introducing new harmonics. This is not to suggest detuning by even amounts should not occur, however, since even tuning can aid in solidifying the current timbre rather than making it appear busier.

When detuning oscillators from one another, there is obviously a limit whereby the oscillators will separate and create two distinct individual pitches. This typically occurs if the detuning exceeds 20 Hz but can be less depending on the manufacturer's choice of detuning algorithms.

Detuning so that both oscillators feature independent pitches (each has its own audible distinctive fundamental frequency) can be used to good effect if the ratio of detuning remains at integers. For example, detuning by an octave will result in the second harmonic increasing in gain due to the fundamental of the detuned oscillator matching the 2:1 harmonic of the first. This has the effect of thickening up the sound and can be particularly useful for adding weight to the timbre. Further detuning of 3:1, 4:1 and 5:1 can also produce harmonically pleasing results.

Depending on the synthesizer, further control may be offered in the oscillator section for syncing oscillators together. This permits two oscillator cycles to be synced to one another, and most synthesizers will automatically sync secondary waveform to the first oscillator's cycle. This means that no matter where in the cycle any other oscillators are, when the first starts its cycle again, they are all forced to begin their cycle too.

For example, if two square oscillators were detuned from each other by an octave, every time the first oscillator restarts its cycle so too will the second, regardless of the position in its own cycle. This tends to produce a timbre with a constantly changing harmonic structure in a particularly powerful programming strategy.

Moreover, if the second oscillators pitch is modulated through an external source or the synthesizers own internal engine, as the pitch changes the timbre changes dramatically creating a strong distorted style lead. In the 1990s, a popular approach was to use a filter envelope to control the pitch movement over time. This way as the filter changes so does the sync, producing a timbre typical of early rave.

Once the signal leaves the oscillators, it is commonly routed into a mixer section. Here, the output gain of each individual oscillator can be adjusted. The principle here is to adjust the volume of each oscillator to help mix and combine them together to produce a cohesive whole.

In the mix section, some synthesizers may also feature additional parameters such as ring modulation. Employing this, the frequencies of two oscillators (commonly just oscillator one and oscillator two in a multi-oscillator synth) are entered into the modulator where the sum and difference of the two frequencies are output.

As an example, if one oscillator produces a signal frequency of 440 Hz (A4 on a keyboard) and the second produces a frequency of 660 Hz (E5 on a keyboard), the frequency of the first oscillator is subtracted from the second.

$$660\text{ Hz} - 440\text{ Hz} = 220\text{ Hz A3}$$

Then the first oscillator's frequency is added to that of the second.

$$660\text{ Hz} + 440\text{ Hz} = 1{,}100\text{ Hz C\#6}$$

Based on this example, the difference of 220 Hz provides the fundamental frequency while the sum of the two signals, 1,100 Hz, results in a fifth harmonic overtone. Both of these frequencies would be mixed together and then output from the ring modulator.

As simple as this may initially appear unless both oscillators entering the ring modulation were sine waves, every harmonic generated from each oscillator would be subject to the ring modulator resulting in incredibly complex timbres. Ring-modulated timbres are often described as sounding metallic or bell like, both of which provide a fairly accurate description.

The option to add noise may also be included in the oscillator's mix section to introduce additional harmonics, making the signal leaving the oscillator/mix section full of frequencies that can then be shaped further using the options available.

FILTERS

After the mix section, the signal will often be routed into the filter stage of the synthesizer. If the oscillator's signal is thought of as piece of wood that is yet to be carved, the filters are the hammer and chisels that are used to shape it. Filters are used to chip away pieces of the original signal until a rough image of the required sound remains.

This makes the filters the most vital element of subtractive synthesis because if the available filters are of poor quality, then few sound sculpting options will be available, and it will be impossible to create the sound you require. Indeed, the choice of filters combined with the oscillators' waveforms is often the reason why specific synthesizers must be used to recreate certain 'classic' dance timbres.

The most common filter used in basic subtractive synthesizers is a low-pass filter. This is used to remove frequencies above a defined cut-off point. The effect of any filter is progressive, meaning that more frequencies are removed from a sound the further the control is reduced, starting with the higher harmonics and gradually moving to the lowest. This means that if the filter cut-off point is reduced far enough, all harmonics above the fundamental can be removed, leaving just the fundamental frequency. And in some instances, depending on the filter, it is also possible to remove the fundamental too.

While it may appear initially ridiculous to work at creating a harmonically rich sound by detuning oscillators only to then remove them with a filter, by doing so it allows the producer to determine the precise color or timbre of the sound. What's more, if the filter is rhythmically modified or shaped through a synthesizer modulator, the real-time movement can make the timbre much more interesting to the ear.

Indeed, rhythmical or real-time manipulation is an absolutely fundamental aspect of sound design because we naturally expect dynamic movement throughout the length of a note. Using our previous example of a piano string being struck the initial sound is very bright, becoming duller as it dies away. This effect can be simulated by opening the filter as the note starts then gradually sweeping the cut-off frequency down to create the effect of the note dying away.

It should be noted that when using this style of effect, frequencies that lie above the cut-off point are not attenuated at right angles to the cut-off frequency and the rate at which they die away will depend on the transition period of the filter. This is why different filters that essentially perform the same function can create beautiful sweeps, whilst others can produce quite uneventful results.

When a cut-off point is designated, small quantities of the harmonics that lie above this point are not removed completely and are instead attenuated by a certain degree. The degree of attenuation is dependent on the transition band of the filter being used.

The gradient of this transition is important because it defines the sound of any one particular filter. If the slope is steep, the filter is said to be 'sharp' and if the slope is more gradual, the filter is said to be 'soft'.

To fully understand the action of this transition requires some in-depth electrical knowledge of resistor–capacitor circuits that is beyond the purpose of this book but the result is that filters are measured in dB transitions and typically, the smallest is 6 dB. This is known as a single-pole filter, since when plotted on a graph, it looks like a tent draped over a single pole and results in a 6 dB attenuation of frequencies per octave. Typically, single-pole filters are the mainstay of EQ's due to the transition period and synthesizers will often tend to employ 12 or 24 dB filters, otherwise known as two-pole and four-pole filters. Two-pole filters will attenuate 12 dB per octave and four-pole filters provide 24 dB per octave attenuation.

Because four-pole filters attenuate 24 dB per octave, they make substantial changes to a sound and so tend to sound more artificial than sounds created by a two-pole filter. Therefore, it's important to decide which transition period is best suited to the sound. For example, if a 24 dB filter is used to sweep a pad it will result in strong attenuation throughout the sweep, while a 12 dB will create a more natural flowing movement.

If there is more than one filter available in the synthesizers, you may be able to combine them in series or parallel. This means that two 12 dB filters could be summed together to produce a 24 dB transition, or one 24 dB filter could be used in isolation for aggressive tonal adjustments with the following 12 dB filter used to perform a real-time filter sweep.

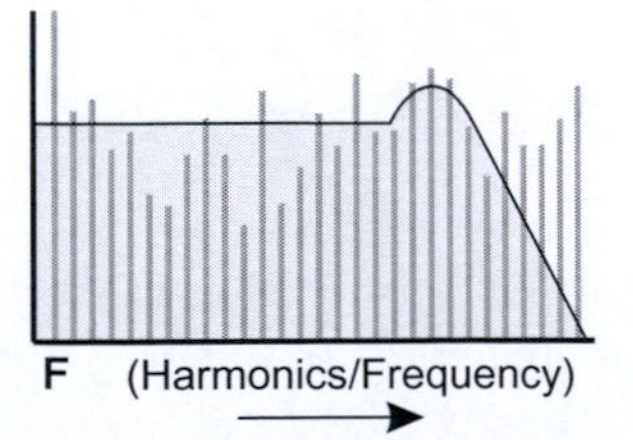

FIGURE 6.8
The action of a low-pass filter

Although low-pass filters are the most commonly used type, there are numerous variations including high pass, band pass and notch and comb. These utilize the same transition periods as the low-pass filter, but each has a wildly different effect on the sound.

A high-pass filter has the opposite effect to a low-pass filter, first removing the low frequencies from the sound gradually moving towards the highest. This is less useful than the low-pass filter because it effectively removes the fundamental frequency of the sound, leaving only the fizzy harmonic overtones.

Because of this, high-pass filters are rarely used in the creation of instruments and are predominantly used to create effervescent sound effects or bright timbres

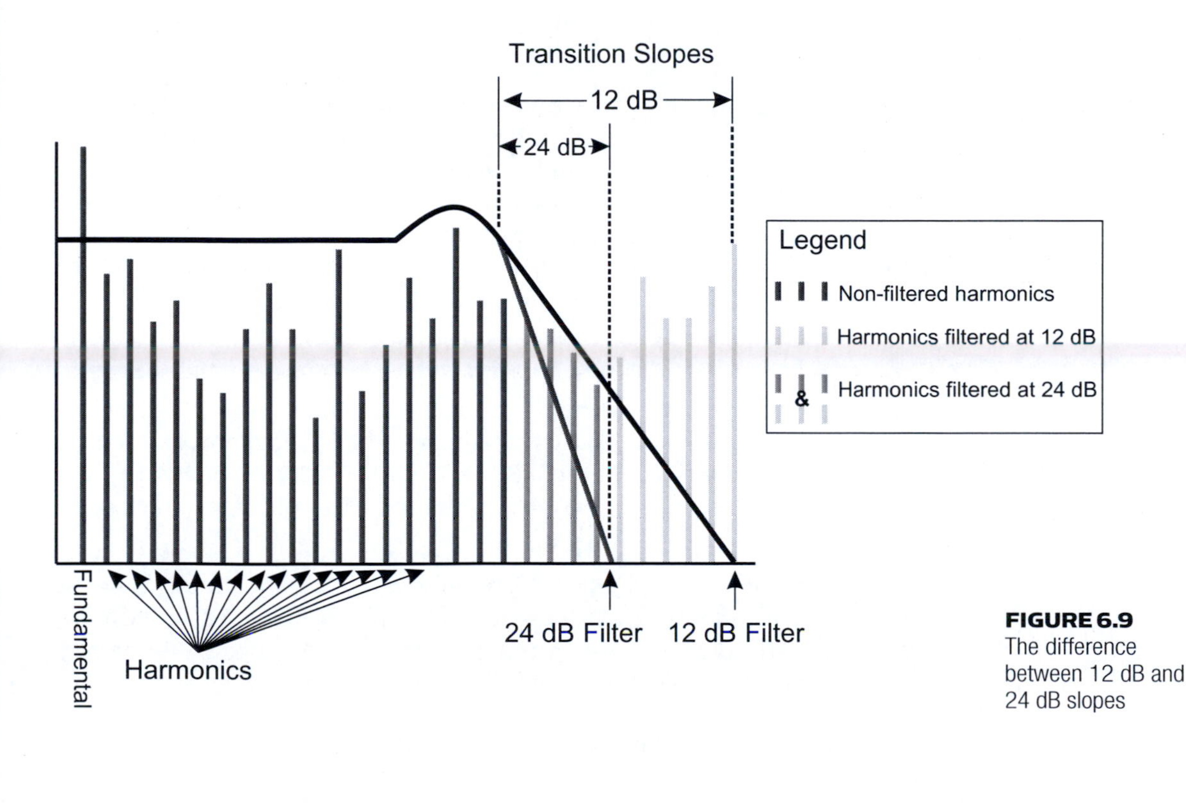

FIGURE 6.9
The difference between 12 dB and 24 dB slopes

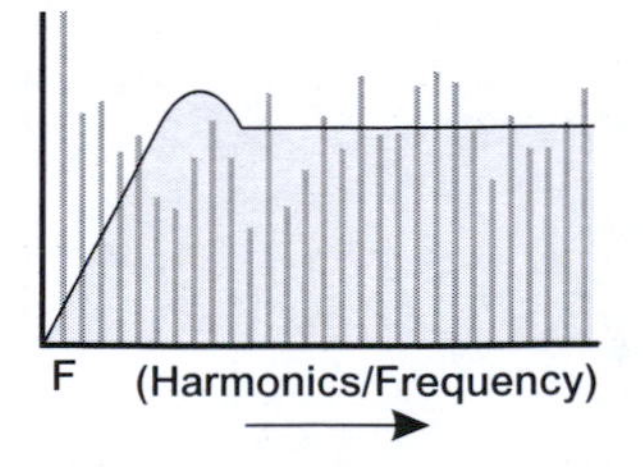

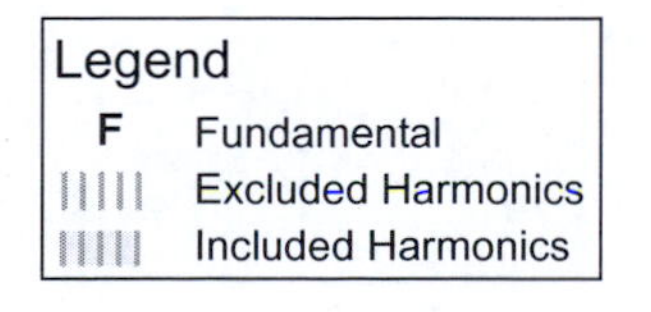

FIGURE 6.10
The action of a high-pass filter

that can be laid over the top of another low-pass sound to increase the harmonic content.

The typical euphoric trance leads are good example of this as they are often created from a tone with the fundamental overlaid with numerous other tones that have been created using a high-pass filter. This prevents the timbre from becoming too muddy as a consequence of stacking together fundamental frequencies.

In both remixing and dance music it's commonplace to run a high-pass filter over an entire mix to eliminate the lower frequencies, creating an effect similar to a transistor radio or telephone. By reducing the cut-off control, gradually or immediately, the track morphs from a thin sound to a fatter one, which can produce a dramatic effect in the right context. More on this will be discussed in a later chapter.

If both high- and low-pass filters are connected in series, then it's possible to create a band-pass, or band-select, filter. These permit a set of frequencies to pass unaltered through the filter while the frequencies on either side of the two filters are attenuated. The frequencies that pass through unaltered are known as the 'bandwidth' or the 'band pass' of the filter, and clearly, if the low pass is set to attenuate a range of frequencies that are above the current high-pass setting, no frequencies will pass through and no sound is produced.

Band-pass filters, like high-pass filters, are often used to create timbres consisting of fizzy harmonics. They can also be used to determine the frequency content of a waveform since sweeping through the frequencies each individual harmonic can often be heard. Because this type of filter frequently removes the fundamental, it is often used as the basis of sound effects or lo-fi and trip hop timbres or to create very thin sounds that will form the basis of sound effects.

Although band-pass filters can be used to thin a sound, they should not be confused with band-reject filters, which can be used for a similar purpose. Band-reject filters, often referred to as notch filters, attenuate a selected range of frequencies effectively creating a notch in the sound – hence the name – and usually leave the fundamental unaffected. This type of filter is handy for scooping out frequencies, thinning out a sound while leaving the fundamental intact, making them useful for creating timbres that contain a discernable pitch but do not have a high level of harmonic content.

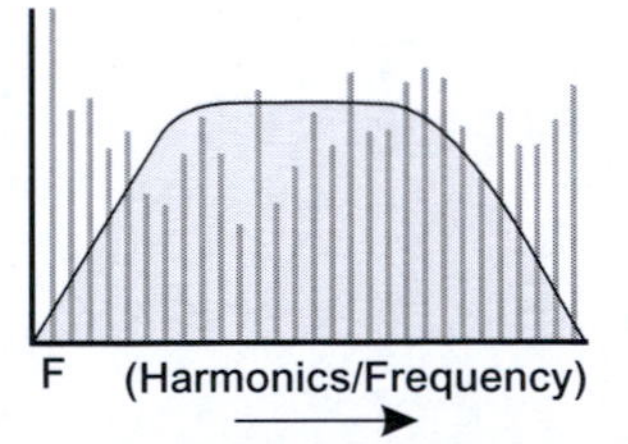

FIGURE 6.11
The action of a band-pass filter

One final form of filter is the comb filter. With these, some of the samples entering the filter are delayed in time and the output is then fed back into the filters input to be reprocessed to produce the results. This style of filter effectively creates a comb appearance – hence the name. Using this method, sounds can be tuned to amplify or reduce specific harmonics based on the length of the delay and the sample rate. This makes it useful for creating complex sounding timbres that cannot be accomplished any other way.

As an example, if a 1 kHz signal is put through the filter with a 1 ms delay, the signal will result in phase because 1 ms is coincident with the inputted signal, equaling one. However, if a 500 Hz signal with a 1 ms delay were used instead, it would be half of the period length and so would be shifted out of phase by 180°, resulting in a zero. It's this constructive and deconstructive period that creates the continual bump then dip in harmonics, resulting in a comb-like appearance when represented graphically, as in Figure 6.13.

This method applies to all frequencies, with integer multiples of 1 kHz producing ones and odd multiples of 500 Hz (1.5 kHz, 2.5 kHz, 3.5 kHz, etc.) producing zeros. The effect of using this filter can at best be described as highly resonant and its use is limited to extreme sound design rather than the more basic sound sculpting. Similarly, because of the way they operate, it is rare to find these styles of filter featured on a synthesizer, and with the possible exception of Native Instruments Massive Synthesizer, they usually only appear as a third-party effect.

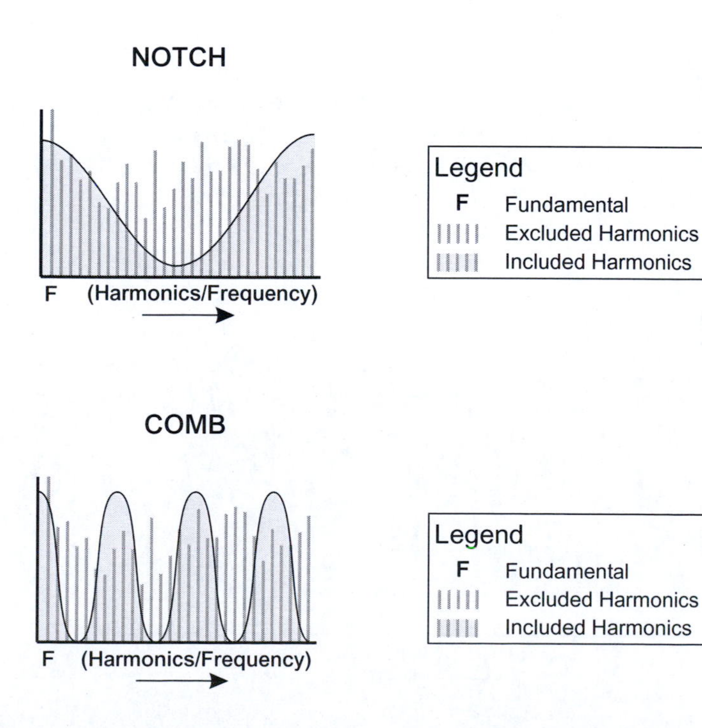

FIGURE 6.12
The action of a notch filter

FIGURE 6.13
The action of the comb filter

One final element of sound manipulation in a synthesizers filter section is the resonance control. Also referred to as Peak or Q, this refers to the amount of output of the filter that is fed back directly into the input. The result of this is that it emphasizes any frequencies that are situated around the cut-off frequency.

This has a similar effect to employing a band-pass filter at the cut-off point, effectively creating a peak. Although this also affects the filter's transition period, it is more noticeable at the actual cut-off frequency than anywhere else. Indeed, as you sweep through the cut-off range the resonance follows the curve, continually peaking at the cut-off point. In terms of the timbre, increasing the resonance makes the filter sound more dramatic and is particularly effective when used in conjunction with low-pass filter sweeps.

On many analogue and DSP analogue-modelled synthesizers, if the resonance is turned up high enough it will feed back on itself. As more and more of the signal feeds back, the signal is exaggerated until the filter breaks into self-oscillation.

Self-oscillation is where a filter creates its own sine wave with a frequency equal to that of the set cut-off point. By then adjusting the filter's cut-off, it is possible to tune the sine wave to any frequency. In the early days of synthesis, this provided a solution for the synthesizers that didn't feature a sine wave. Today, self-oscillation isn't as important since sine waves are now common, but it can still be useful for creating sound effects. For example, pushing a filter into self-oscillation and then modulating the filters cut-off frequency can produce complex effects and sweeping bass timbres.

The keyboard's pitch can also be closely related to the action of the filters, using a method known as pitch tracking, keyboard scaling, or more frequently 'key follow'. On many synthesizers, the depth of this parameter is adjustable allowing you to determine how much or how little the filter should follow the pitch.

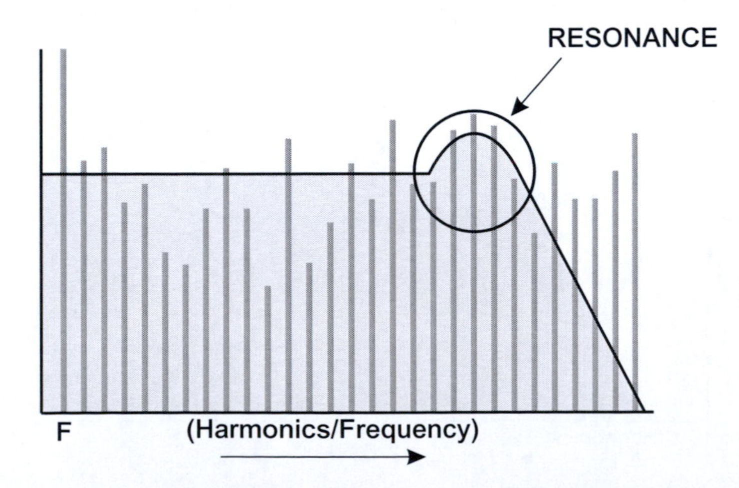

FIGURE 6.14
The effect of resonance

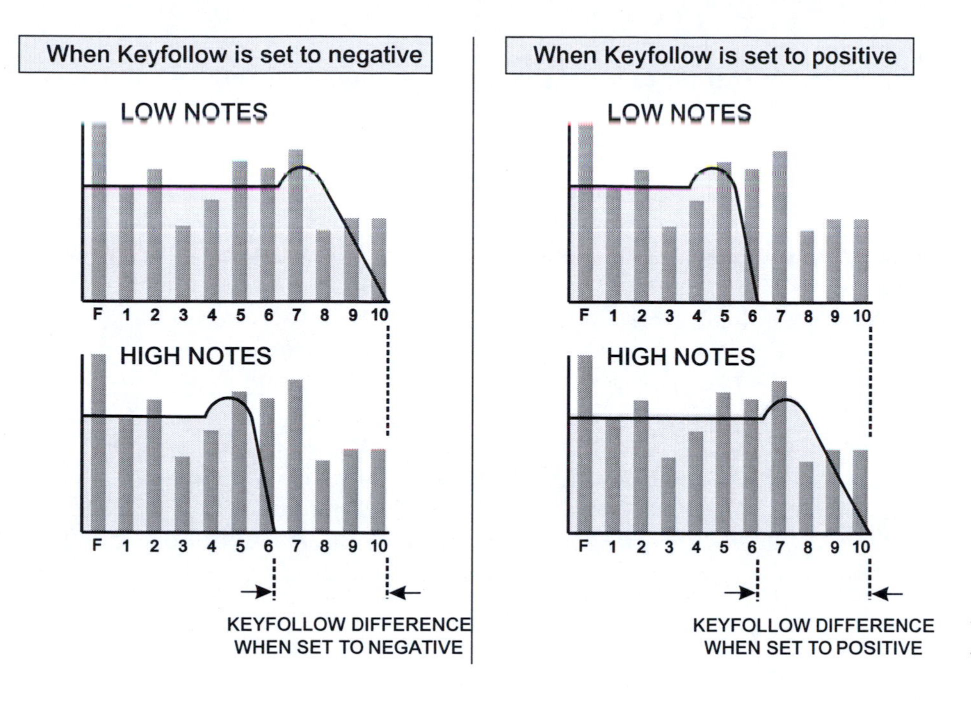

FIGURE 6.15
The effect of filter key follow

When this parameter is set to its neutral state (neither negative nor positive) as a note is played on the keyboard, the cut-off frequency tracks the pitch and each note is subjected to the same level of filtering. If this is used on a low-pass filter, for example, the filter setting remains fixed so as progressively higher notes are played fewer and fewer harmonics will be present in the sound, making the timbre of the higher notes mellower that of the lower notes. On the one hand, if the key follow parameter is set to positive, the higher notes will have a higher cut-off frequency and the high notes will remain bright. If, on the other hand, the key follow parameter is set to negative, the higher notes will lower the cut-off frequency, making the high notes even mellower than when key follow is set to its neutral state. Key follow is useful for recreating real instruments such as brass, where the higher notes are often mellower than the lower notes and is also useful on complex bass lines that jump over an octave, adding further variation to a rhythm.

Moreover, some filters may also feature a saturation parameter, which offers a method to overdrive the filters. If applied heavily, this can be used to create distortion effects but more often it's used to thicken out timbres and add even more harmonics and partials to the signal to create rich sounding leads or basses.

THE AMPLIFIER

Few, if any, acoustic instruments start and stop immediately. It takes a finite amount of time for the sound to reach its amplitude and then decay away to silence again; thus the 'envelope generator, (an EG)' – a feature of all synthesizers – can be used to shape the volume of a timbre with respect to time.

This permits the producer to control whether a sound starts instantly the moment a key is pressed or builds up gradually and how the sound dies away (quickly or slowly) when the key is released. These controls usually comprise four sections, named Attack, Decay, Sustain and Release (ADSR), each of which determine the shaping that occurs at certain points during the length of a note. An example of this is shown in Figure 6.16.

Attack

The attack parameter determines how the note starts from the point when the key is depressed and the period of time it takes for the sound to go from silence to full volume. If the period set is quite long, the sound will 'fade in', as if you are slowly increasing the gain. If the period set is short, the sound will start the instant as a key is pressed. Most instruments utilize a very short attack time.

Decay

Immediately after a note has begun, it may initially decay in volume. For instance, a piano note starts with a very loud, percussive part but then drops quickly to a lower volume while the note sustains as the key is held down. The time the note takes to fade from the initial peak at the attack stage to the sustain level is known as the 'decay time'.

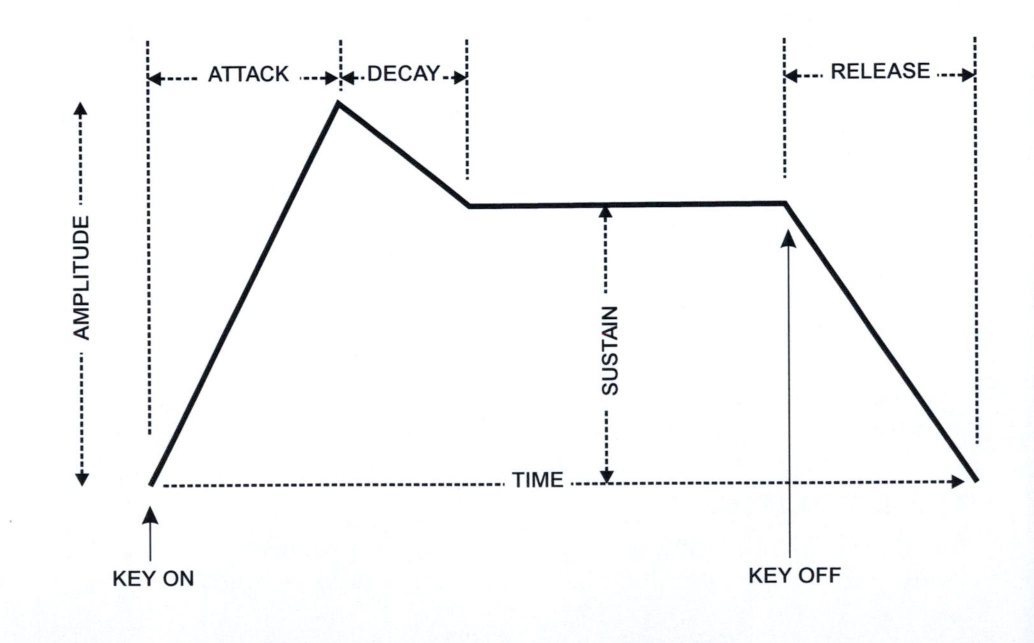

FIGURE 6.16
The ADSR envelope

Sustain

The sustain period occurs after the initial attack and decay periods and determines the volume of the note while the key is held down. This means that if the sustain level is set to maximum, any decay period will be ineffective, because at the attack stage the volume is at maximum so there is no level to decay down to. Conversely, if the sustain levels were set to zero, the sound peaks following the attack period will fade to nothing even if you continue to hold down the key. In this instance, the decay time determines how quickly the sound decays down to silence.

Release

The release period is the time it takes for the sound to fade from the sustain level to silence after the key has been released. If this is set to zero, the sound will stop the instant the key is released, while if a high value is set the note will continue to sound, fading away, as the key is released.

Although ADSR envelopes are the most common, there are some subtle variations, such as Attack-Release (AR), Time-Attack-Delay-Sustain-Release (TADSR) and Attack-Delay-Sustain-Time-Release (ADSTR). Because there are no decay or sustain elements contained in most drum timbres, AR envelopes are often used on drum synthesizers. They can also appear on more economical synthesizers simply because the AR parameters are regarding as having the most significant effect on a sound, making them a basic requirement. Both TADSR and ADSTR envelopes are only usually found on more expensive synthesizers. With the additional period, T (Time), in TADSR, for instance, is possible to set the amount of time that passes before the Attack stage is reached.

It's also important to note that not all envelopes offer linear transitions, meaning that the attack, decay and release stages will not necessarily consist entirely

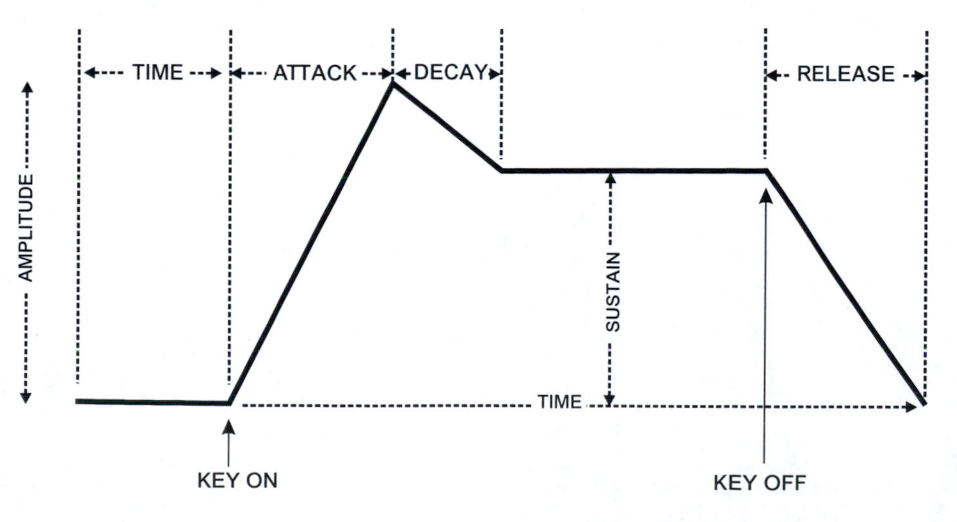

FIGURE 6.17
The TADSR envelope

of a straight line as shown previously. On some synthesizers these stages may be concave or convex, while other synthesizers may allow you to control whether the envelope stages should be linear, concave or convex. The differences between the linear and exponential envelopes are shown in Figure 6.18.

Alongside an envelope to control the amplitude of the final sound, many synthesizers will also feature an envelope that is hard wired to the filter. Using this, it is possible to employ tonal changes to the note while it plays. A typical example of an envelope-controlled filter is the plucky style timbres that appear in Progressive House. By employing a zero attack, short decay, and zero sustain level on the EG, a sound that starts with the filter wide open before quickly sweeping down to fully closed is produced. As the track progresses, the producer commonly automates the decay parameter to gradually increase throughout the length of the track.

Most synthesizers also offer additional tools for manipulating sound in the form of modulation sources and destinations, often termed a *modulation matrix*. Using this it's possible to use the movement of one parameter to modify another independent parameter. The number of modifiers available along with the destinations they can affect is entirely dependent on the synthesizer. Many synthesizers feature a number of EGs alongside the filter and amplitude to permit the producer to control other parameters with an envelope. Alternatively, the synthesizer may permit the filter or amplitude envelope to also control further parameters.

In addition to envelopes, synthesizers will also offer a further oscillator that is only used to modulate another parameter. These low-frequency oscillators (LFOs) produce output frequencies in much the same way a VCO. The difference is that whereas a typical oscillator produces an audible frequency (within the 20 Hz to 20 kHz range), a low frequency produces a signal with a relatively low frequency that is inaudible to the human ear (in the range 1 Hz to 10 Hz).

These oscillators are used to modulate other parameters known as 'destination', to introduce additional movement into a sound. For instance, if an

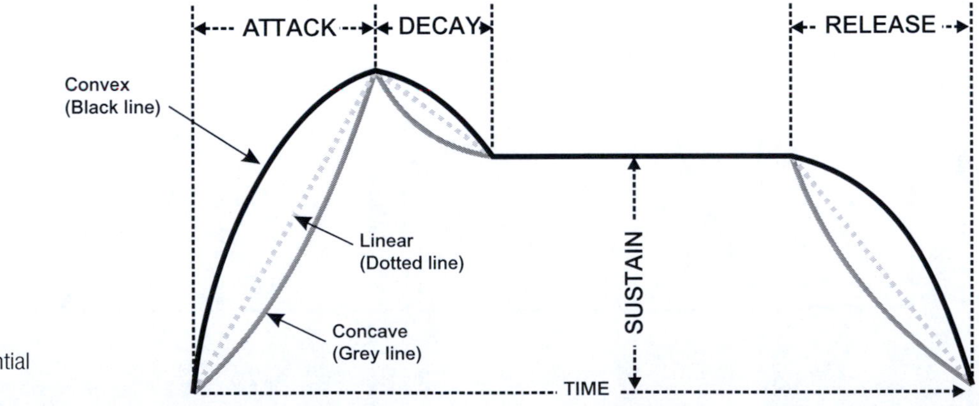

FIGURE 6.18
Linear and exponential envelopes

LFO is set to a relatively high frequency, say 5 Hz, to modulate the pitch of a normal oscillator, the pitch of the oscillator will rise and fall according to the speed and shape of the LFO waveform and an effect similar to vibrato is generated.

If a sine wave is used for the LFO, then it will essentially create an effect similar to a wailing police siren. Alternatively, if this same LFO is used to modulate the filter cut-off, then the filter will open and close at a speed determined by the LFO, while if it were used to modulate an oscillator's volume, it would rise and fall in volume recreating a tremolo effect.

The waveforms an LFO features depend entirely upon the synthesizer in question, but they commonly employ sine, saw, triangle, square, and sample/hold waveforms. The latter is usually constructed with a randomly generated noise waveform that momentarily freezes every few samples before beginning again.

Alongside a choice a of waveform, an LFO must also offer controls over how much it will augment the destination (usually termed amount on a synthesizer), a rate control to control the speed of the LFOs waveform cycles, and some may also feature a fade-in control. The fade-in control adjusts how quickly the LFO begins to affect the waveform after a key has been depressed. An example of this is shown in Figure 6.19, and this is often used to emulate wind instruments since vibrato commonly appears after the initial onset of a tone. In regards to electronic dance music, this is often employed on long strings with the LFO modulating either pitch or filter to introduce fluctuations as it continues.

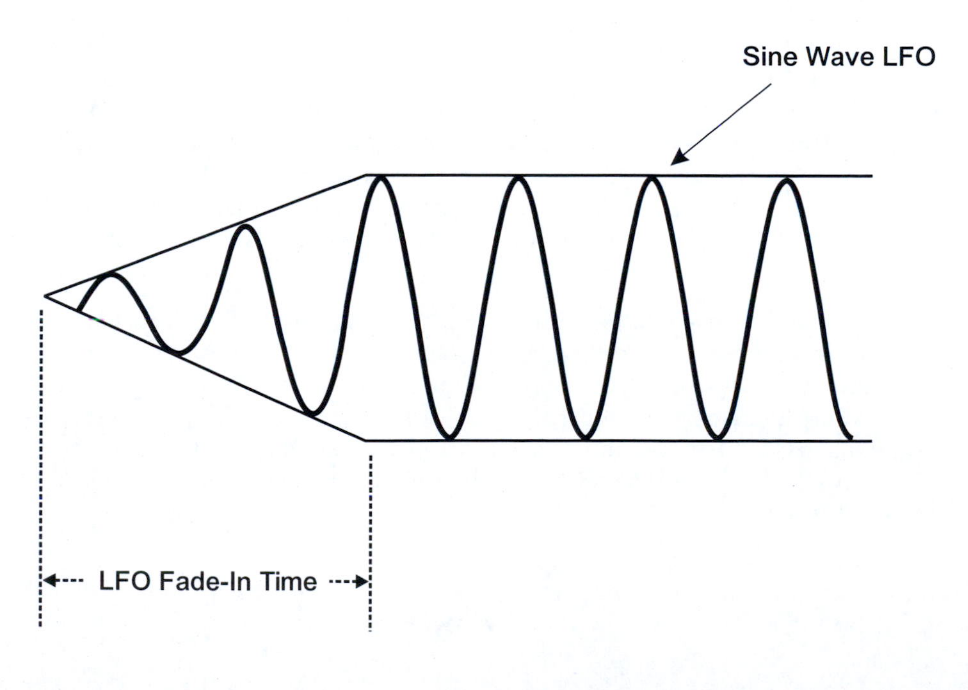

FIGURE 6.19
LFO fade-in

The LFOs on more capable synthesizers may even have access to their own envelopes. This offers control of the LFOs performance over a specified time period permitting it to fade-in after a key has been pressed and employs decay, sustain and release. This is rare, however, and typically most will only employ a fade in and fade out parameters (an Attack/Release envelope).

The destinations an LFO can modulate are entirely dependent on the synthesizer being used. Some synthesizers may only allow LFOs to modulate the oscillator's pitch and the filter, while others may offer multiple destinations and more LFOs. Obviously, the more LFOs and destinations that are available, the more creative options you will have at your disposal.

PRACTICAL APPLICATION

Understanding the different parameters in a synthesizer and actually using them to knowledgably program a sound are a world apart, and there is no substitute for practice and hearing each part of a synthesizer in action. With this in mind, it is sensible to experiment with a short example to aid in the understanding of the components and the effect that they can impart on a sound.

Using the synthesizer of your choice, clear all the current settings so you are starting from a basic patch. On many synthesizers, this is known as 'initializing' so it may feature a button labelled 'init' or 'init patch'.

Begin by pressing and holding C3 on your synthesizer, or alternatively if you are controlling the synthesizer through MIDI – program in a continual note. The purpose of this exercise is to hear how the sound develops as you begin to modify the controls of the synthesizer so the note needs to play continually.

Select a sawtooth wave for each of the two oscillators, if there is a third oscillator that you cannot turn off, set this to a triangle oscillator. Now, detune one sawtooth from the other until the timbre begins to thicken. This is a tutorial to grasp the concept of synthesis so continue to detune until you hear the oscillators separate from one another and then step back again until they just begin to merge back together into a continuous tone rather than two individual tones. If you are using a triangle wave, detune this against the two saws and listen to the results.

Locate the VCA envelope and start experimenting with the parameters. You will need to release C3 and then press it again so you can hear the effect that the envelope is having on the timbre. Experiment with these envelopes until you have a good grasp on how they can adjust the shape of a timbre, once you're happy you have an understanding apply a fast attack with a short decay, medium sustain and a long release. As before, for this next step, you will need to keep C3 depressed.

In the filter section, experiment with the filter settings. Start by using a high-pass filter with the resonance set around midway and slowly reduce the filter cut-off control. Note how the filter sweeps through the sound, removing the lower frequencies first, slowly progressing to the higher frequencies. Also experiment with the resonance and note how this affects the timbre. Do the same with the notch and band-pass before finally moving to the low-pass. Set the low-pass filter quite low, along with a low resonance setting – you should now have a static 'buzzing' style timbre.

The timbre is quite monotonous so use the filter envelope to inject some life into the sound. This envelope works on the exact same principles as the VCA with the exception that it will control the filters movement. Set the filters envelope to a long attack and decay but use a short release and no sustain and set the filter envelope to maximum positive modulation. If the synthesizer has a filter key-follow, this can be used so the filter will track the pitch of the note being played and adjust itself. Now try depressing C3 to hear how the filter envelope controls the filter, essentially sweeping through the frequencies as the note plays.

Finally, to add some more excitement to the timbre, find the LFO section. Generally, the LFO will have a rotary control to adjust the rate (speed), a selector switch to choose the LFO waveform, a depth control and a modulation destination. Choose a triangle wave for the LFO waveform, hold down C3 on the synthesizer's keyboard, turn the LFO depth control up to maximum and set the LFO destination to pitch. As before, hold down the C3 key and slowly rotate the LFO rate (speed) to hear the results. If you have access to a second LFO, try modulating the filter cut-off with a square wave LFO, set the LFO depth to maximum and experiment with the LFO rate again.

Just this simple exercise should result in you becoming more accustomed to the synthesizer and the effect that each parameter can impart onto the basic sound. The companion website also contains a small quick-time movie of this exercise.

All forms of synthesis, regardless of what they are, follow a similar pattern of synthesis beginning with the formation of complex timbres followed by the use of tonal shaping with envelopes and tonal modulation through the use of modifiers. Indeed, the only significant difference appears in the form of the initial sound generation engine. Analogue synthesis will use oscillators, frequency modulation, employs operators, granular synthesis utilizes grains, sample and synthesis relies on samples and additive relies on combining simple waveforms.

Frequency Modulation is a form of synthesizer developed in the early 1970s by Dr John Chowning of Stanford University that was later developed upon by Yamaha, leading to the release of the now legendary DX7 and TX81Z synthesizers: a popular source of bass sounds for many dance musicians for over 20 years.

Unlike analogue, FM synthesis produces the basic sound using operators rather than oscillators. Despite the name, however, an operator is little more than an oscillator that features its own amplitude envelope and can only produce sine waves. Sounds using this method of synthesis are generated through using the output of the first operator to modulate the pitch of the second, thereby introducing additional harmonics. Like an analogue synthesizers, this means that each FM voice requires an absolute minimum of two oscillators in order to create a very basic sound.

Within FM, the operator that is used to modulate is termed as a 'modulator' and the operator being modulated is termed as a 'carrier'. In early FM synthesizers, there were a total of six operators available, and these could be connected and configured in any number of ways. For example, the producer could use a modulator to modulate a carrier, which in turn modulates another carrier before modulating a modulator that modulates a carrier to produce a harmonically rich timbre. This could then be modulated further with an LFO and also augmented with a filter and amplitude envelope in exactly the same method as analogue subtractive synthesis.

Even with only six operators, there are numerous ways the operators can be connected together but not all produce very musical results; so in order to simplify the operation, many synthesizers employed a series of fixed algorithms. These contained preset combinations of modulator and carrier routings and provided the best starting points for programming an FM synthesizer. A typical algorithm from the DX-7 is shown in Figure 6.20.

Notably, it is possible to emulate this same FM characteristic in analogue synthesis using two oscillators. In this situation, the first oscillator acts as a modulator and the second acts as a carrier. If the first oscillator's output is routed

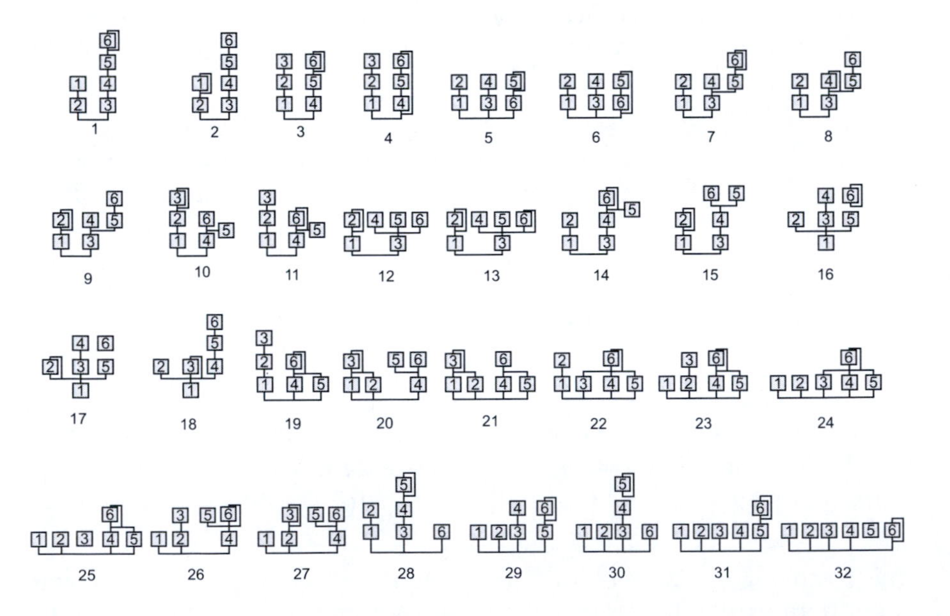

FIGURE 6.20
FM algorithms

into the modulation input of the second oscillator and notes are played on the keyboard, both oscillators play their respective notes, but the pitch of the second oscillator will change in time with the frequency of the first resulting in a somewhat rudimentary form of FM synthesis. Although many manufacturers will term this possibility FM synthesis in their manuals, more accurately it is simply 'cross modulation'.

Nonetheless, due to the nature of frequency modulation in both FM and analogue, the resulting timbres typically tend to sound metallic and digital in character, particularly when compared to the warmth generated by the drifting of analogue oscillators. Also due to the digital nature of FM synthesizer, the hardware synthesizers fascia panel generally contains very few real-time controllers. Instead numerous buttons adorn the front panel forcing you to navigate and adjust any parameters through a small LCD display. The intervention of virtual instruments has changed this approach due to a much larger screen estate but only a few manufacturers have really taken to producing FM-specific synthesizers.

The most immediate example is Native Instruments FM8, a particularly powerful FM synthesizer that can not only accept preset files and sysex from the original DX7, DX100 and DX27 synthesizers but also features a more complex engine allowing it to produce analogue style timbres. *Skrillex* is known to employ the FM8 synthesizer for a good number of his sounds.

SAMPLE AND SYNTHESIS

Sample and synthesis is another form of synthesis that employs samples in place of the oscillators. Typically, these samples do not solely consist of the entire timbre of an instrument but instead are samples of the various stages of an instrument's sound. For instance, a typical sample-based synthesizer may contain five different samples of the attack stage of a piano, along with a sample of the decay, sustain and release portions of the sound. This means that it is possible to mix the attack of one sound with the release of another to produce a complex timbre.

In many of the hardware synthesizers, this approach was commonly a result of four individual tones that could be mixed together to produce a timbre and each of these individual tones could have access to numerous modifiers including LFO's, filters and envelopes. This obviously opens up a whole host of possibilities not only for emulating real instruments but also for creating complex sounds. This method of synthesis has become the de-facto standard for any synthesizer producing realistic instruments. By combining both samples of real-world sounds with all the editing features and functionality of analogue synthesizer, they can offer a huge scope for creating both realistic and synthesized sounds.

GRANULAR SYNTHESIS

Granular synthesis is rarely employed in hardware synthesizers due to its complexity, but there are a few software synthesizers that do employ it such as Spectrasonics Omnisphere. This method of synthesis operates by building up sounds from a series of short segments of sounds called 'grains'. This is best compared to the way that a film projector operates: where are series of still images each slightly different from the last are played sequentially at a rate of around 25 pictures per second. This continual movement fools the eyes and brain into believing there is a smooth continual movement.

Granular synthesizer operates in this same manner through using tiny fragments of sound rather than still images. By joining a number of these grains together, an overall tone is produced that develops over a period of time. To do this, each grain must be less than 30 ms in length as the human ear is unable to determine a single sound if they are less than 30 to 40 ms apart. This phenomenon is known as the Haas effect; a psychoacoustic theory that states any sound that occurs within 40 ms or less from receipt of the direct signal will not be perceived as a distinct event.

In order to employ granular synthesis to any extent, a certain amount of control must be offered over each grain, and this can make granular synthesis difficult to produce since any one sound can be constructed from as many as a thousand grains. If employed carefully, however, it is possible and a granular synthesizer will generally offer five parameters:

- Grain length
  This can be used to alter the length of each individual grain. As previously touched upon, the human ear can differentiate between two grains if they are more than 30 to 50 ms apart but many granular synthesizers usually go above this range, covering from 20 to 100 ms. By setting this length to a longer value, it's possible to create a pulsing effect.
- Density
  This is the percentage of grains that are created by the synthesizer. Generally it can be said that the more grains that are created, the more complex a sound will be, a factor that is also dependent on the grain shape.
- Grain shape
  Commonly, this offers a number between 0 and 200 and represents the curve of the envelopes. Grains are normally enveloped so that they start and finish at zero amplitude, helping the individual grains mix together coherently to produce the overall sound. By setting a longer envelope (a higher number), two individual grains will mix together, which can create too many harmonics and often results in the sound exhibiting lots of clicks as it fades from one grain to the other.
- Grain pan
  This is used to specify the location within the stereo image that each grain is created. This is particularly useful for creating timbres that inhabit both speakers.

- Spacing
  This is used to alter the period of time between each grain. If the time is set to a negative value, the preceding grain will continue through the next created grain. This means that setting a positive value inserts space between each grain, however, if this space is less than 30 ms in length, it will be inaudible.
  The sound produced with granular synthesizer depends on the synthesizer in question. Usually each grain consists of single frequencies with specific waveforms or occasionally they are formed from segments of samples or noise that have been filtered with a band-pass filter. Thus, the constant change of grains can produce sounds that are both bright and incredibly complex, resulting in a timbre that's best described as glistening. After creating this sound by combing the grains, the sound can be sculpted using the modifiers discussed previously the envelopes, filters and LFOs.

ADDITIVE SYNTHESIS

Additive synthesis is a method used to create sounds that emulate the way they occur in nature. As mentioned in Chapter 4 on acoustic science, any sound that occurs in the real world is little more than a combination of sine waves. The fundamental frequency from a sine wave will put further objects into motion, each of which generates their own fundamental frequency. All these mix together to produce the complex sounds that we can hear.

This same method is employed in additive synthesis. Commonly the synthesizer will only feature sine waves and the producer has the opportunity to adjust the frequency and amplitude of each. As more and more sine waves are mixed together, the timbre becomes more complex. After a complex sound has been generated, it can then be modulated and sculpted with the usual modifiers.

Additive synthesis is a particularly powerful form of synthesis but due to the sheer number of oscillators required to produce even a simple sound, and the control required over each, it isn't the most popular format, and there are few instruments that feature it. The most popular at the time of writing is Camel Audio's Alchemy and Native Instruments Razor.

VECTOR SYNTHESIS/WAVETABLE SYNTHESIS

Wavetable or vector synthesis has already been discussed earlier in this chapter. The principle is that the oscillators are all formed via a series of very short static samples and producing timbres from these consists of the producer morphing between different wavetable samples for the length of the timbre. This form of synthesis proved popular in the early 1990s but soon faded from existence and has now become an addition to the more common oscillators. The Access Virus Ti synthesizer features a number of wavetables that can be used as one, or both, of its main oscillators.

CHAPTER 7
Samplers

'There are £250,000 worth of uncleared samples on my latest album. Get listening!'

Norman Cook (Fatboy Slim)

The introduction of the sampler and sampling in the 1980s became responsible for one of the biggest technological and creative shifts in the history of music. However, the audio workstation and in particular digital audio editing have overshadowed what was originally viewed as a marvel of modern technology.

Since it is now possible to cut, slice, move, stretch, compress and edit audio directly in the workstations editor, what was once limited to a sampler has become commonplace within a workstation. What's more, with all sample CD's now in WAV and AIFF format, audio can be imported direct into an arrange page and sample-based players such as Native Instruments Kontakt catering for a vast range of exotic multi-sampled instruments, practical sampling has taken something of a backseat.

However, the humble sampler still has a lot more to offer than the basic editing functions that can be performed in a workstation. And with a growing number of artists beginning to rely on step sequencing samplers such as the Rhizome and the Swedish Elektron Octatrack, an understanding of samplers, how they operate and some of their creative uses is still as paramount today as it's ever been.

A sampler can be viewed as the digital equivalent of an analogue tape recorder but rather than recording the audio signal onto a magnetic tape, it is recorded digitally into random access memory (RAM) or directly onto the computer's hard disk drive. After any audio signal has been recorded, it can then be manipulated using a series of editing parameters similar to those on synthesizers. This includes amplitude and filter envelopes, LFOs and in some units, the ability to mix synthesizer style oscillator with the sample. In addition, any sampled sounds can also be played back at varying pitches. To accomplish this,

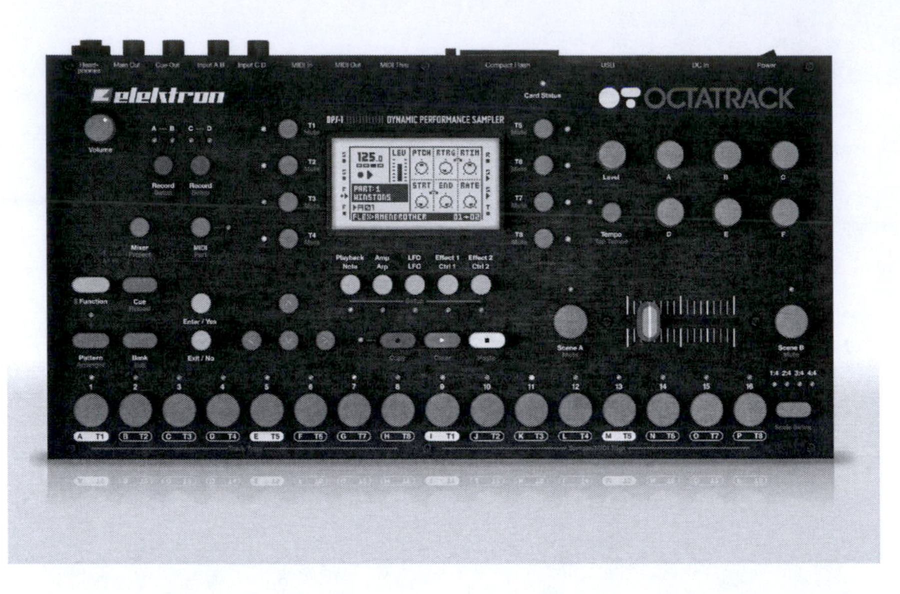

FIGURE 7.1
The elektron octatrack

the sampler artificially increases or decreases the original sample frequency in relation to the note that is depressed.

This means that if the sample's speed is increased or decreased by a significant amount it will no longer sound anything like the original source. For instance, if a single key strike from a piano is sampled at C3 and is played back at this same pitch from a controller keyboard, the sample will play back perfectly. If, however, this same sample is played back at C4, the sampler would increase the frequency of the original sound by 12 semitones (from 130.81 Hz to 523.25 Hz), and the result would sound more like two spoons knocking together than the original recording.

These extreme pitch adjustments are not always a bad thing, particularly for a creative user, but to recreate a real instrument within a sampler, it must be sampled at every few notes. Indeed, most samplers only reproduce an acceptable instrument sound four or five keys from the original root key so if an original instrument is needed throughout the key-range samples should be taken every few notes of the original source instrument. For example, with a piano it is prudent to sample the keys at C0, E0, G0 and B0, then C1, E1, G1, B1 and so forth until the entire range has been sampled. This technique is known as 'multi-sampling' and has occurred on every sample-based instrument available.

Naturally, recording sounds in this way will equate to a large number of samples, particularly if they are also recorded at a number of different velocities and settings. This means that the more accurate the reproduction, the more samples that are required and the more that are taken, the larger the file becomes and the more memory the sampler must have available to play these back.

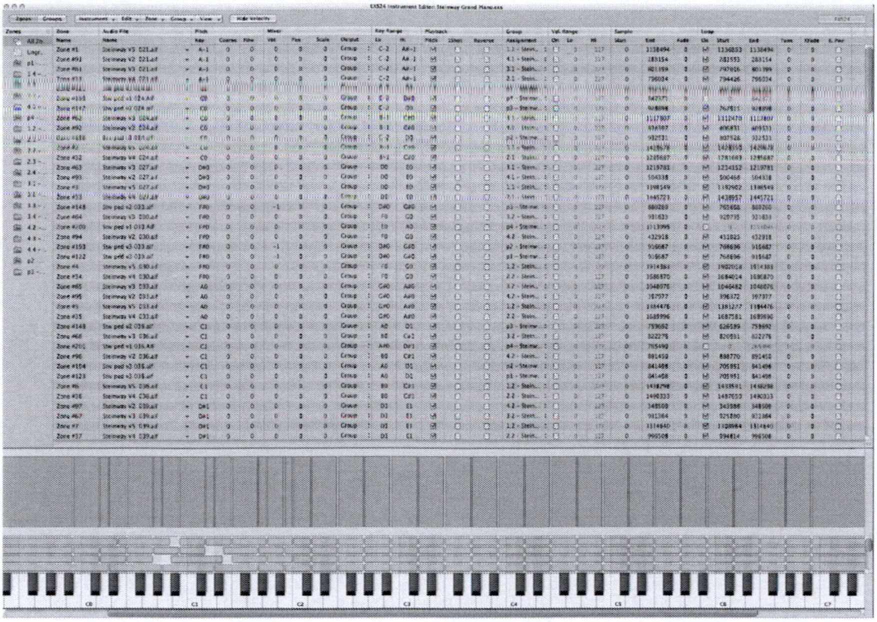

FIGURE 7.2
A multi-sampled instrument in logics EXS

Memory is rarely a problem today inside a computer since most sample-based instruments will stream direct from a hard disc but in many of the step-based sequencers; memory can still be a concern. Because these samplers hold the sounds in their onboard RAM, the maximum sampling time is limited by the amount of available memory.

At full audio bandwidth, one minute of mono recording will use approximately 5 megabytes (MB) of RAM. In the case of sampling a keyboard instrument in its most basic form, this could equate to 80 MB of memory, and you can double this if you want it in stereo. Consequently, various techniques have been adopted to make the most of the available memory, the first of which is to loop the samples.

As the overall length of a sample determines the amount of RAM that is required, reducing the sample's length means more samples will fit into the memory space. Because most sounds have a distinctive attack and decay period but the sustain element remains consistent, the sustain portion can be continually looped for as long as the key is held, moving to the release part after the key is released. This means that only a short burst of the sustain period must be sampled, helping to conserve memory.

This is more difficult than it may appear, however, and the difficulty arises from what may appear to be a consistent sustain period of a sound is rarely static due to slight yet continual changes in the harmonic structure. If only a small segment of this harmonic movement is looped, the results would sound unnatural.

Conversely, if too long a section is looped in an effort to capture the harmonic movements, the decay or some of the release period may also be captured, and again, when looped, the final sound will still be unusual. In addition, any looping points must start and end at the same phase and level during the waveform. If not, the difference in phase or volume could result in an audible click as the waveform reaches the end of its looped section and jumps back to the beginning.

Some samplers offer a work around to this and automatically locate the nearest zero crossing to the position you choose. While this increases the likelihood that a smoother crossover is achieved, if the waveform's level is different at the two loop points, there will still be a glitch. These glitches can, however, be avoided with crossfading.

Using this, the operator can fade out the end of the looped section and overlap it with a fade in at the start of the loop. This creates a smooth crossover between the two looping points, reducing the possibility that glitches are introduced. Although this goes some way to resolve the problems associated with looped samples, this is not always the best solution because if the start and end of the looped points are at different frequencies there will be an apparent change in the overall timbre during the crossfade. Unfortunately, there is no quick fix for avoiding the pitfalls associated with successfully creating a looped segment, so success can only be accredited to patience, experimentation, and experience.

SAMPLE CD'S

Although sampling an instrument can be a rewarding experience, it can nonetheless be incredibly time consuming, particularly in terms of setting up, connections and levels, so in many instances it is easier to simply purchase a sample CD. Although I use the term sample CD here, typically it is only available in electronic download format direct to your hard disc, and there is no actual physical medium involved, but nonetheless the term sample CD is still used by everyone.

Almost all 'sample CDs' available today are in WAV or AIFF format. These are the most common file formats that are recognized by all audio workstations and hardware platforms. Both of these formats can be 'dragged and dropped' from the hard drive (or copied digitally from a CD if there is physical media involved) direct into the workstations arrange page, or into compatible software sampler where they can then be key mapped and/or edited. The disadvantage, however, is that if these sounds happen to be multi-samples of a complete instrument once they are placed into a sampler, they must be named, loops created if needed, crossfades created and key ranges defined. This can be incredibly time consuming, taking a good few hours to set up a multi-sampled instrument depending on the sampler's interface.

Consequently, if the 'CD' consists of multi-sampled instruments, it will more likely be supplied in EXS24 (Logic Pro's EXtreme Sampler) or Kontakt (Native Instruments) formats. Here, the audio samples are stored alongside a specific data format that contains information about sustain loops, key and velocity ranges. With this, the producers only need to copy the folder to their computer and then use the relevant sampler to open the file whereby all the samples are loaded and key-mapped automatically.

Many software samplers today will also be compatible with the older hardware sampler standards of AKAI and E-MU. Although it is very rare to come across any newly released sample CDs in these formats, they were the most popular formats available during the 20-year reign of hardware sampling, and therefore there is a huge back catalogue of these available CDs. Note, however, that sometimes importing an AKAI format CD into a software sampler may not always work properly, and some software samplers will interpret the data differently. This commonly results in differences in the timbre and texture of the sample, but in more severe cases the key-mapping may not come out quite as expected, the sustain samples may not loop properly or the sampler may just crash altogether.

FIGURE 7.3
Logics Pro's **EX**treme **S**ampler

FIGURE 7.4
The author's AKAI S5000

If the sample CD consists of loops rather than multi-sampled instruments, then the formats may once again be different and can consist of ACID, APPLE and REX. Here, both ACID and APPLE loops are again audio files, but they contain additional tempo information so that when inserted into Sony's Acid Music Studio or Apple's Logic Pro, the loops will automatically time stretch or compress to suit the current working tempo. Moreover, if the tempo of the project is adjusted, the samples will automatically change to suit the new tempo.

A similar principle applies with REX and REX2 files. This is a proprietary format conceptualized by Propellerhead software for their ReCycle software. With this software, musical loops (most typically drum loops) could be imported whereby they would be analyzed for their transients and any number of 'slices' would be applied to cut the loop into individual rhythmic components. Once this is accomplished, the tempo or pitch of each can be freely adjusted. Moreover, this file can then be exported from ReCycle as a REX (a mono file) or REX2 (a stereo file) and imported into any sequencer for the same editing procedures.

Notably, many sequencers feature this style of automated transient detection and loop slicing as a standard now, although different sequencers will use different terminology. Apple's Logic Pro refers to this as Flex Time editing whilst Cubase calls it Vari-Audio or Audio Warp.

BASIC SAMPLING PRACTICES

Many of the standard sample CDs will consist of loops or single hits rather than multi-sampled instruments. Indeed, almost all multi-sampled instruments now appear in the form of a virtual instrument, and most commonly employ the Kontakt engine. By doing so, the producer simply installs the instrument and doesn't have to worry about mapping the keys across a range.

However, if there is only one sampled note available, the producers have little choice to insert it into their sampler and physically key map it. To accomplish this, the 'root' key must first be set in the sampler. This is the key that the sampler will use as a reference point for pitching the subsequent notes up or down along the length of the keyboard. For instance, bass samples are typically set up with the root key at C1, meaning that if this note is depressed the sample will play back at its original frequency. This is consequently stretched across the keyboard's range as much as required. In most software or hardware samplers, the root key is set in the key-range or key-zone page along with the lowest and highest possible notes that will be available for that particular sample.

Using this information the sampler automatically spreads the root note across the defined range and the bass can be played within the specified range of notes. To gain the best results from the subsequent pitch adjustments on each key, it is preferable that the sampler's key range is set to six notes above and below the root note. This allows the sample to be played over the octave, if required, and will produce more natural results than if the root note were to be stretched a complete 12 semitones up or down.

When setting a key zone for bass sounds, it may be possible to set the range much lower than six semitones as pitch is determined by the fundamental frequency and the lower this is, the more difficult it is to perceive pitch. Thus, for bass samples, it may possible to set the lowest key of the key zone to 12 semitones (an octave) below the root without introducing any unwanted artifacts. In fact, pretty much anything sounds good if it's pitched down low enough.

Most competent step sequencers and samplers will also allow a number of key zones to be set across the keyboard. This makes it possible to have, say, a bass sample occupying F0 to F1 and a lead sample occupying F2 to F3. If set up in this way, both the bass and lead can be played simultaneously from an attached controller keyboard. In taking this approach, it is worth checking that each key zone can access different parameters of the sampler's modulation engine. If this is not possible, settings applied to the bass will also apply to the lead.

After the samples are arranged into key zones, you can set how the sampler will behave depending on how hard the key is hit on a controller keyboard. This uses 'velocity modulation' and can be useful for recreating the movement within sounds. This can be a vital application since it can be used to add expression to melodies and prevents the static feel that is often a result of using a series of sampled sounds. A typical application for this is velocity to low-pass filter so that the harder the key is hit the more the filter opens.

'Velocity crossfading' and 'switching', if available, may also worth experimenting with. Switching involves taking two samples and using velocity values to determine which one should play. The two samples are imported into the same key range and hitting the key harder (or softer) results in switching between the two different samples. Velocity crossfading employs this same principle but morphs the two samples together creating a (hopefully) seamless blend rather than an immediate switch between one and the other.

SAMPLE LOOPS

Alongside single synthesizer hits, the majority of sample CDs feature a somewhat ubiquitous number of pre-programmed or pre-recorded drum loops to suit the genre of the CD. These are typically dragged direct into the sequencers arrange page, sliced, diced, re-arranged and played alongside the music but a much more creative avenue is available if these loops are imported into a sampler and triggered through MIDI.

By dropping a loop into a sampler, the loop will start over from the beginning every time a key is pressed. Thus, if a key on a controller keyboard is tapped continually, the loop will start repeatedly, producing a stuttering effect. This technique can be used to great effect to create breakdowns in dance music.

Alternatively, the sampler can be set to 'one-shot trigger mode', so that a quick tap on a controller keyboard plays the sample in its entirety even if the key is released before the sample has finished playback. This is useful if you want the original sample to play through to its normal conclusion while triggering the same sample again to play over the top. This technique can be used to create complex drum loops and is a staple of producing techno, a few instances of the same loop are playing simultaneously, each out of time and pitched differently.

This form of sample triggered looping shouldn't be confused with the term 'phrase sampling' as this consists of sampling a short musical phrase that is only used a couple of times throughout a song. This method is commonly employed in dance music for short vocal phrase or hooks that are triggered occasionally to fit with the music.

Nonetheless, phrase sampling can also be developed further to test new musical ideas. By slicing your own song into a series of four-bar sections and assigning each loop to a particular key on the keyboard, it becomes possible to trigger the loops in different orders so that you can determine the order that works best. This is one of the fundamental approaches behind step sequencing and produces more immediate results and ideas that would not normally be considered when arranging in a workstation.

TIME COMPRESSION

Thus far, we've assumed that the phrase or loop that has been sampled is at the same tempo and/or pitch as the rest of the mix, but this may not be the case. Accordingly, all workstations, samplers and step sequencers will provide some form of pitch shifting and time stretching functions so that the pitch and tempo of the phrase or loop can be adjusted to fit with the music. Both these functions are self-explanatory: the pitch shift function changes the pitch of notes or an entire riff without changing the tempo and time stretching adjusts the tempo without affecting the pitch.

These are both useful tools to have but it's important to note that the quality of the results is proportionate to how far they are pushed. While most pitch shifting algorithms remain musical when moving notes up or down the range by a few semitones, if the pitch is shifted by more than five or six semitones the sound may begin to exhibit an unnatural quality. Similarly, while it should be possible to adjust the tempo by 25 BPM without introducing any digital artifacts, any adjustments above this may introduce noise or frequency crunching that will compromise the audio.

This can sometimes be a wanted side effect, such as in the production of drum and bass, but typically the producer will want to avoid it. In this instance, it is advisable to place the loop direct into the workstation and employ its time variant slicing features to slice the loop into constituent parts.

Once sliced, the loop can be adjusted in tempo, exported from the workstation and re-imported back into the sampler.

It should be noted that this form of time variant beat slicing is generally only useful on transient material such as drum loop since the automated process scans the audio for transients. If the audio is a vocal loop or similar, it is less likely to feature transients and therefore the only option is to physically cut the audio into separate words, and sometimes even syllables, and then time stretch each to complete the loop.

Of course, not all loops are taken from sample CDs and in many cases they are taken from other records. Although I cannot condone lifting samples from another artist's record without prior consent from said artist, it would nonetheless be naive to suggest that it doesn't occur. Much of the Hip-Hop musical movement and many house records have their foundations formed from another record. Indeed, although there are plenty of sample CDs dedicated to both these genres, they are often avoided since everyone will have access to the very same CDs and instead producers search through the more obscure or most unlikely resources. For example, Roger Sanchez's massive house anthem 'Another Chance' was a sample taken direct from the beginning of Toto's 'I Won't Hold You Back' and Eric Prydz 'Call on Me' sampled Steve Winwood's Valerie. In the interests of accuracy, Steve Winwood's track Valerie was not actually sampled, the track was reconstructed and Steve Winwood was so impressed with the results when presented with it he re-recorded the vocals for Eric Prydz.

Nonetheless, with many dance records, the more obscure the original record is the better it generally is as it's unlikely any other producers will have access to the same records and if required, you'll be able to acquire copyright permission easily. These records can be sourced in the majority second-hand and charity shops.

Sampling shouldn't just be restricted to other records, however, and it can be equally advantageous to sample some of your own synthesizer's. For example, sampling a mono-timbral synthesizer and layering that sample across a number of keys in the sampler can be used to create a unison effect. Alternatively, if employing a hardware plug-in instrument such as the Virus Ti or M-Audio Venom, the units DSP can be saved for other tasks by multi-sampling the instrument and employing it in a software sampler. Indeed, multi-sampling the synthesizer at every key would recreate the synthesizer in the sampler and in some instances the sampler may offer more synthesis parameters than are available on the synthesizer enabling you to, say, synchronize the LFO to the tempo, use numerous LFOs, or access numerous different filter types.

Similarly, sampling from films can often open a number of creative avenues. Christopher Bertke, AKA *Pogo*, has made a large number of records using nothing more than small samples taken from a specific film or scene and sequencing them together. His 'Upular' remix consists of samples taken only from Disney/Pixar's 'UP' and is currently reaching towards 8 million views on YouTube alone. Alternatively, with some creative thought, a recording of a ping pong ball being thrown at a garage door can become a tom drum, hitting a rolled up newspaper against a wall can make an excellent house drum kick, and the hissing from a compressed air freshener can produce great hi-hats.

Whilst many musicians simply rely on the sampler to record a drum loop (which can be accomplished with any sequencer), they are not using them to their full extent and experimentation is the real magic to any sampler. Therefore, what follows are some general ideas to get you started in experimental sampling.

Creative Sampling

Although cheap microphones are worth avoiding if you plan to record vocals, they do have other creative uses. Connecting one to an audio interface (or step sequencer) and striking the top of the mix with a newspaper of your hand can be used as the starting point of kick drums or scratching the top of the microphone can be used as the starting point to Guiro's.

You should also listen out for sounds in the real world to sample and contort – FSOL have made most of their music using only samples of the real world. Todd Terry acquired snare samples by bouncing a golf ball off a wall and Mark Moore used an aerosol as an open hi-hat sound.

Hitting a plastic bin with a wet newspaper can be used as a thick slurring kick drum and scraping a key down the strings of a piano or guitar can be used as the basis for whirring string effects (it worked for Dr. Who's TARDIS anyway). Once sampled these sounds can be pitched up or down, or effected as you see fit. Even subtle pitch changes can produce effective results. For example, pitching up an analogue snare by just a few semitones results in the snare used on drum 'n' Bass, while pitching it down gives you the snare typical of Lo-Fi.

Sample Reversing

Probably the most immediate effect to try but while it is simple to implement it can produce great results. The simplest use of this is to sample a cymbal hit and reverse it in the sampler. If played at a much lower frequency, it can be used as an uplift sound effect in an arrangement whilst if played higher it can be employed as an incidental sound effect. More creative options appear when you consider that reversing any sound with a fast attack but long decay or release will create a sound with long attack and an immediate release.

For example, a guitar pluck can be reversed and mixed with a lead that isn't. The two attack stages will meet if they're placed together, and these can be cross-faded together. Alternatively the attack of the guitar could be removed so that the timbre begins with a reverse guitar that moves into the sharp attack phase of a lead sound or pad. On the other hand, if a timbre is recorded in stereo, the left channel could be reversed while the right channel remains as is to produce a mix of the two timbres. You could then sum these to mono and pitch them up and down the key range.

Pitch Shifting

A popular method used by dance producers is to pitch shift a vocal phrase, pad or lead sound by a fifth as this can often create impressive harmonies. If pads or leads are shifted further than this and mixed in with the original it can introduce a pleasant phasing effect.

Perceptual Encoding

Any perceptual encoding devices (such as minidisk) can be used to compress and mangle loops further. Fundamentally, these work by analyzing the incoming data and removing anything that the device deems irrelevant. In other words, data representing sounds that are considered to be inaudible in the presence of the other elements are removed, and this can sometimes be beneficial on loops.

Physical Flanging

Most flanger effects are designed to have a low noise floor that isn't much use if you need a dirty flanging effect for use in some genres of dance. This can, however, be created if you own an old analogue cassette recorder. If you record the sound to cassette and applying a small amount of pressure on the drive spool the sound will begin to flange in a dirty uncontrollable manner, this can then be re-recorded into the workstation and sampler to be played at different pitches.

Transient Slicing

Although also possible within a workstation arrange page, a popular technique for House tracks is to design a pad with a slow resonant filter sweep and then sample the results. Once in a sampler, the slow attack phase of the pad is cut off so that the sound begins suddenly and sharply. As the transient is the most important part of the timbre, this creates an interesting side effect that can be particularly striking when placed in a mix.

Creative use of samplers comes from a mix of experience, experimentation and serendipity. What's more, all of the preceding examples are just that – examples to start you on the path towards creative sampling and you should be willing to push the envelope much further. The more time you set aside to experiment, the more you'll learn and the better results you'll achieve.

SAMPLES AND CLEARANCE

I couldn't possibly end a chapter on sampling without at least discussing some copyright law. Ever since the introduction of sampling, single hits, melodies, vocal hooks, drum loops, basses and even entire sections of music from countless records have been sampled, manipulated and otherwise mangled in the name of art. Hits by James Brown, Chicago, Donna Summer, Chic, Chaka Khan, Sylvester, Lolita Holloway, Locksmith and Toto, amongst innumerable others have come under the dance musicians' sample knife and been re-modeled for the dance floor.

In the earlier years of dance, artists managed to get away with releasing records without clearing the samples; this was because it was a new trend and neither the original artists nor the record companies were aware of a way that they could prevent it. This changed in the early 1990s, when the sampling of original records proved it was more than just a passing fad. Today, companies know exactly what to do if they hear that one of their artists' hits has been sampled and in most cases they come down particularly hard on those responsible. This means that if you plan to release a record that contains a sample of another artist's motif, drum, vocals or entire verse/chorus without first applying for clearance you may end up with a lawsuit.

To help clear up some of the myths that still circulate – I spoke to John Mitchell, a music solicitor who has successfully cleared innumerable samples for me and many other artists.

What is copyright?

'Copyright exists in numerous forms and if I were to explain the entire copyright law it would probably use up most of your publication. To summarize, it belongs to the creator and is protected from the moment of conception, exists for the lifetime of the creator and another 70 years after their death. Once out of copyright the work becomes "public domain" and any new versions of that work can be copyrighted again'.

So would it be okay to sample Beethoven or another classical composer?

'In theory, yes, but most probably not. The question arises as to where did you sample the recording from because it certainly wouldn't have been from the original composer. Although copyright belongs to the creator, the performance is also protected through copyright. When a company releases a compilation of classical records they will have been re-recorded and the company will own the copyright to that particular performance. If you sample it you're breaching the performance copyright'.

What if you've transcribed it and recorded your own performance?

'This is a legal wrangle that is far too complex to discuss here since it depends on the original composer and the piece of music. While the original copyright may

have expired, a company may have purchased the copyright and own the transcript. My advice is to check first before you attempt to sample or transcribe anything'.

So what can you get away with when sampling copyright music or speech?

'Nothing, there was a rumor circulating that if the sample is less than 30 seconds in length, you don't need any clearance but this isn't true. If the sample is only a second in length and its copyright protected, you are in breach if you commercially release your music containing it'.

What if the sample were heavily disguised with effects and EQ?

'You do stand a better chance of getting way with it but that's not to say you will. Many artists spend a surmountable amount of time creating their sounds and melodies and it isn't too difficult to spot them if they're used on other tracks. It really isn't worth the risk!'

Does this same copyright law apply to sampling snippets of vocals from TV, radio, DVD and so on?

'Yes, sampling vocal snippets or music from films breaches a number of copyrights, including the script writer (the creator) and the actor who performed the words (the performer)'.

What action can be taken if clearance hasn't been obtained?

'It all depends on the artist and the record company who own the recording. If the record is selling well then they may approach you and work a deal, but be forewarned that the record company and artist have the upper hand when negotiating monies. Alternatively, they can take out an injunction to have all copies of the record destroyed. They also have the right to sue the samplist'.

Does it make any difference how many pressings are being produced for retail with the uncleared samples on them?

'No, if just one record were released commercially containing an illegal sample, the original artist has the right to prevent any more pressings being released and also has the rights to sue'.

What if the records were released to DJs only?

'It makes no difference. The recording is being aired to a public audience'.

How much does it cost to get sample clearance?

'It depends on a number of factors. How well known the original artist is, the sampled record's previous chart history, how much of the sample you've used in your track and how many pressings you're planning to make. Some artists are all too willing to give sample clearance because it earns them royalties while other will request ridiculous sums of money'.

How do you clear a sample?

'Due to the popularity of sampling from other records, there are companies appearing every week who will work on your behalf to clear a sample. However, my personal advice would be to talk to the Mechanical Copyright Protection Society. These guys handle the licensing for the recording of musical works and started a Sample Clearance Department in 1994. Although they will not personally handle the clearance itself – for that you need someone like myself – they can certainly put you in touch with the artist or record company involved'.

CHAPTER 8

Compressors

'We have become the tools of our tools'.

Henry David Thoreau

Compressors are perhaps the most essential processor in the dance musician's toolbox. Originally designed to do little more than control peaks in audio its outright abuse has come to pretty much definite part of the sound of electronic dance music. However, to be able to abuse the processor, the producer first has to understand their original purpose and use.

Originally, compressors were introduced in order to control signal levels and prevent clipping or overloading of signals that would result in distortion in the mixing desk or recording device. For example, if the producer were recording a vocal track that requires the performer to whisper in the verses but shout in the chorus, this sudden shift in dynamics from quiet to loud makes it difficult to maintain an acceptable recording level.

If the producer were to set the recording levels to capture the quiet sections, when the performer performs louder at the chorus, it will overload the recording device and result in distortion. However, if the recording levels were set to capture the louder chorus parts without distortion, in the verses where the performer whispers there will be a very low dynamic ratio between the vocals and the thermal noise generated by the equipment. This means the thermal noise would be evident and ruin the overall quality of the recording.

The obvious solution to this dilemma would be to *ride the gain* during recording. The producer or engineer would sit by the gain control on the recording device and increase or decrease it whenever the source becomes too quiet or too loud. And in fact, this is how the very early radio engineers used to control the volume levels to prevent clipping the broadcast antennas.

Gain riding isn't the perfect solution, however, since it means that the producer not only has to hover over the recording device preparing to capture signals, it also requires some advance warning of upcoming changes in gain.

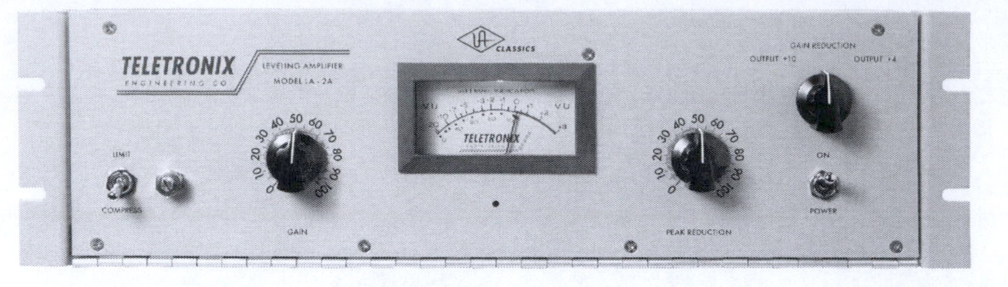

FIGURE 8.1
The first commercial studio compressor – the Teletronix LA-2A Levelling Amplifier

Either that or the producer had to have lightening fast reflexes. It was because of this a radio engineer named Jim Lawrence came up with a solution he referred to as a 'levelling amplifier'.

The LA1 as it becomes known, employed a photo resistor that monitored the incoming signal. As the signal became louder, the photo resistor would grow brighter, and therefore its impedance increased resulting in a reduction in gain of the signal at the output. As the input level decreased, the photo resistor dimmed and the impedance reduces resulting in an increase in gain level at the output. Jim Lawrence went on to form the Teletronix Company to mass-produce the now classic and highly coveted LA-2A levelling amplifier that, internally, works on exactly the same principle as the LA1.

Its main purpose, like any other compressor today was to control the signal levels that entered the unit. By inserting a levelling amplifier or compressor between the signal source and the recording device or broadcast output antenna, it will reduce the gain of any signal that exceeds a certain threshold, therefore preventing any clipping of the signal. Although this has remained as the main use for a compressor, its purpose of controlling dynamics in audio presents a larger number of uses. For example, by managing the dynamics in a recording it can make mix down easier for the producer.

If the amplitude of a recording doesn't remain constant, the fluctuations in volume can result in the quieter sections disappearing behind other instruments in the mix, but if the producer increases the volume in the quiet sections, when the recording becomes louder, it could be too prominent in the mix. Whilst this could quite easily be repaired today with the use of mix automation, a more reliable and easier practice is to simply compress the recording to even out the dynamics.

By compressing a source signal, any part of the signal that exceeds the compressor threshold will automatically be reduced in gain. However, the parts that do not exceed the threshold will remain untouched. This means that the dynamic range of the loudest and quietest sections is reduced, and hence, it consists of a more even volume throughout. This dynamic restriction is shown in Figure 8.2.

The effect of this style of compression is two-fold. First, it will level out any volume fluctuations in the signal and thus provide an average signal level

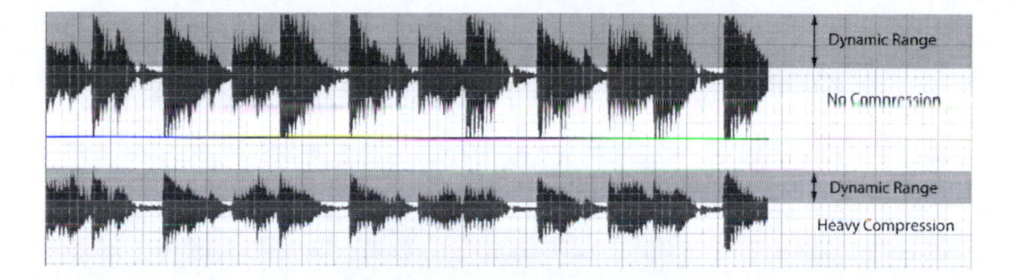

FIGURE 8.2
Before and after compression

that allows the producer to further increase the amplitude without fear of any audio spikes resulting in clipping and distortion. More importantly, however, when the dynamic range of a sound is reduced, we will perceive it as being louder.

Even though we may not introduce any physical increase in gain through a compressor, human hearing has developed in such a way that sustained sounds appear much louder than shorter transient hits, even if the transient happens to be at much higher amplitude than the sustained sound.

This is the reason why a TV program can often appear quieter than the advertisements that follow it. The advertisements have a more restricted dynamic range than the TV program, and therefore, we perceive them as louder. This change in dynamics and perceived volume between a program and advertisement has long been a recognized problem amongst the industry, and recently, a new K weighting filter system has been introduced to prevent it (the EBU R128 standard).

In addition to levelling dynamics of an entire recording, a compressor can be used to change and contort the overall dynamic character and response of any individual sound. This means it can be used to make sounds not only appear louder but also warmer, more rounded, smoother and if applied correctly to a mix can produce an overall sound with more power and attitude. Of particular note to electronic dance producers, though, it's the use of a compressor's side-chain that has defined the sound we now come to associate with 21st century dance music; the all powerful kick that appears to punch a hole in all the rest of the mix every time it strikes.

Before we go further into these techniques, however, the producer first requires a fundamental understanding of compression, the theory, and how each parameter on the compressor can be employed affect a sounds dynamic envelope. To accomplish this, we can examine how a compressors action behaves on a 1 kHz sine wave.

Figure 8.3 shows a sine wave that remains at a constant gain before suddenly increasing in gain and then later decreasing again. In this example, the sine wave has theoretically been recorded and played back on a 16-bit device and, therefore, offers a –96 dB dynamic range. We cannot increase the gain of the loudest part of the sine wave any further since if we do, it will increase beyond

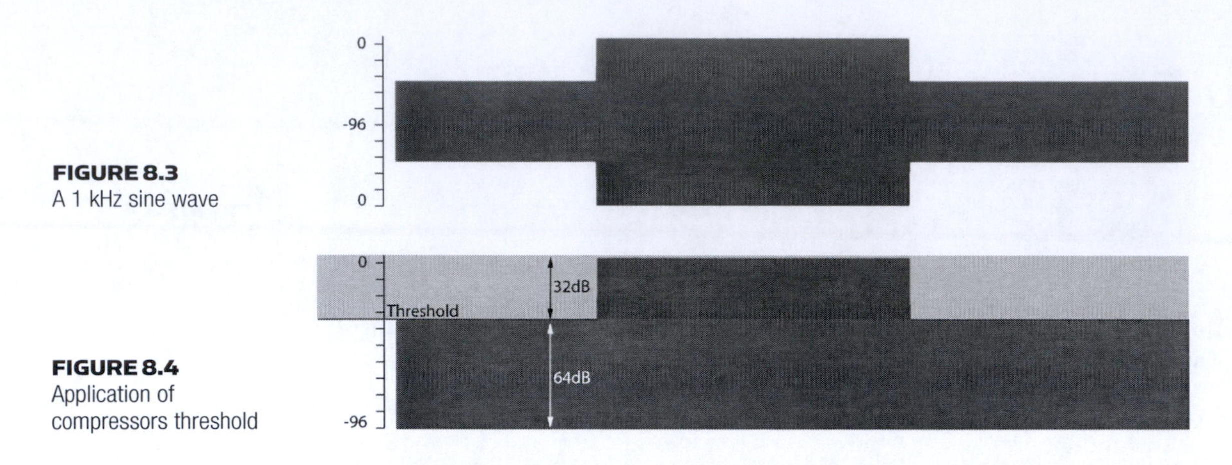

FIGURE 8.3
A 1 kHz sine wave

FIGURE 8.4
Application of compressors threshold

96 dB and above 0 dB, the digital limit. This would clip the outputs of the workstation and result in distortion.

The first parameter on a compressor is the threshold. This is used to determine whereabouts in the signals amplitude range the compressor should begin to act. Once the example sine wave breaches this threshold, the compressor will introduce gain reduction.

The threshold parameter on almost all compressors is calibrated in decibels and ideally the producer should set the threshold so that the average signal level always lies *just below* the threshold. By doing so, if there are any spurious spikes or sudden changes in gain that results in the signal exceeding this threshold, the compressor will activate and gain reduction is applied.

For this particular example, the threshold is set to –32 dB. By doing so, it is best to consider this as creating two independent dynamic ranges in the audio. There is the dynamic range of the compressor that has become –32 dB and the dynamic range that's leftover of 64 dB (32 dB + 64 dB = 96 dB). This behaviour is shown in Figure 8.4

With this application, when the sine wave exceeds the threshold, it will trigger the compressor into action. How much compression is applied to this exceeded signal is dictated by the ratio parameter.

The ratio parameter determines the difference in signal level between the audio entering the compressor to the signal level that leaves the compressor. For example, if the ratio parameter were set to 4:1, for every 4 dB that the signal exceeds the threshold, it will be reduced to only 1 dB at the output of the compressor.

To put this into practice in the current example, if the producer were to set a particularly high ratio to 8:1, for every 8 dB the sine wave exceeds the threshold, the compressor will reduce it to just 1 dB. Since the sine wave breaches the compressors threshold by 32 dB, the compressor will reduce every 8 dB in this compressors dynamic range to just 1 dB.

It is possible to calculate the amount of gain reduction applied in this situation with simple math by dividing 8 by 32 dB. Since there are 4 × 8's in 32, we could speculate that the compressor will 'squash' the 32 dB of the sine wave that exceed the threshold into just 4 dB. As Figure 8.5 shows, this modifies the dynamic envelope beyond its original shape.

The resulting sound from this form of heavy-handed compression can either be beneficial or detrimental depending on what the producer wishes to accomplish. For example, here the sine wave signal that exceeded the threshold has been compressed, and thus its original peak has been massively reduced in gain. This reduction in the overall dynamic range brings it closer in amplitude to the rest of the sine wave. The result of this is that the overall gain of this wave could not be increased further since it would not clip the workstation.

Although this type of compression is not ideally suited towards natural instruments and vocals since it can steal the sound of its realistic nature, it could nevertheless be employed in sound design. For example, the kick in dance music has grown to be one of the powerful elements in the mix and requires plenty of body to create the punch. Thus, if a kick is lacking in body by applying this style of compression, the dynamic range between the transient and body will be reduced and this effectively results in increasing gain of the kicks body, producing a wider thicker sounding kick drum.

Similarly, using the same form of heavy compression on a snare drum would result in a flatter, stronger sounding snare. The compressor would clamp down on its attack stage diminishing the transient, in effect reducing the dynamic range between the transient and body. The end result is a snare drum typical of house and trance.

Whilst this heavy compression has some obvious sound design benefits, it should also be used with caution. We gain a significant amount of information about sounds from the initial attack transients and any immediate and heavy gain reduction will reduce high-frequency content and, as already seen, will severely modify the dynamic envelope. In the case of real world instruments and vocals, we know how they should naturally sound and, therefore, suddenly crushing the dynamic envelope in this way can result in a very obvious and lack lustre result.

Sudden gain changes can also introduce further problems. For example, instruments that contain sustained low-frequency waveforms, such as bass, can

FIGURE 8.5
8:1 Ratio compression

also constantly trigger the compressor. Since low-frequency waveforms are particularly long, if the compressor reacts immediately, it can become confused over the positive and negative states of the waveform and treat them as individual signals. This can result in constant gain changes being applied throughout the waveforms cycle.

ATTACK TIME

In order to avoid the problems associated with immediate gain changes, many compressors feature an attack parameter. Commonly configured in milliseconds, this parameter is used to determine how slowly the gain reduction will be applied once the threshold is breached. This parameter is often misquoted and mistaken for controlling how long the compressor *waits* before compression is applied but *compression will always begin as soon as the threshold is breached* otherwise the compressors action would be immediately evident. Instead, the attack parameter denotes the amount of time it will take for the compressor to reach its maximum gain reduction.

In regard to the dynamic envelope of a timbre, this parameter can seriously affect the tonal content and character. Assuming the threshold remains at –32 dB and the ratio set to 8:1, if the attack parameter were set to 100 ms, the compression would slowly increase gain reduction over a period of 100 ms. This can result in the compression initially bypassing the beginning transient of a sound and gradually compressing only the latter part.

As shown in Figure 8.6, this parameter could be used to seriously modify the dynamic envelope, and therefore, it should not be viewed as an option to apply heavy compression in a transparent way. Whilst it can be used to allow the initial transient through fairly unmolested, the constant increase towards a heavy-gain reduction results in a strong redesign of the dynamic envelope.

If the producer requires transparent results, a much higher threshold and lower ratio can be applied (or since it's the relationship between threshold and ratio that defines the amount of gain reduction, a lower threshold and higher ratio) combined with a fast to medium attack time. This would permit the initial transient to pass through largely unmolested so as not to reduce the high-frequency content, after which gain reduction would gradually increase to bring the remainder of the signal under control.

FIGURE 8.6
The effects of attack parameter

KNEE

If a high-gain reduction is necessary but the results still need to remain transparent, a compressors soft knee can sometimes be employed. Many compressors within the plug-in domain, alongside many hardware compressors, feature a push button or switch to permit the producer to alternate between soft and hard knee.

The knee refers to how the compressor will react as the signal begins to approach the threshold setting. On a hard knee, as soon as the threshold is breached, gain reduction is applied. In a soft-knee application, however, the compressor continually monitors the signal, and gain reduction is gradually applied as the signal approaches the threshold.

Typically, with a soft knee, the gain reduction begins when the signal is within 3 dB to 14 dB of the current threshold (dependent on the compressor), and as the signal grows ever closer to the threshold, gain reduction is increased until the threshold is exceeded whereby full-gain reduction is applied.

This permits the compressor's action to be less evident and is particularly suitable for use on real instruments and vocals whereby the producer may need to apply heavy-gain reduction but also requires for the compressors action to remain transparent. The action of different knee settings is shown in Figure 8.7.

If the compressor offers no option to adjust the knee and is pre-configured to a hard knee, it will likely feature a push button or switch to configure between RMS (root mean square) and peak mode. Although these options aren't just the mainstay of hard-knee compressors and do appear on other styles, they are aimed more towards compressors that react immediately on a threshold breach.

Any compressor that utilizes a hard knee will attempt to control transients as they breach the threshold, however, even with the fastest possible attack setting, it is possible for some short transient sounds such as hi-hats, kick drums, snares and toms to breach the threshold for such a short period of time that

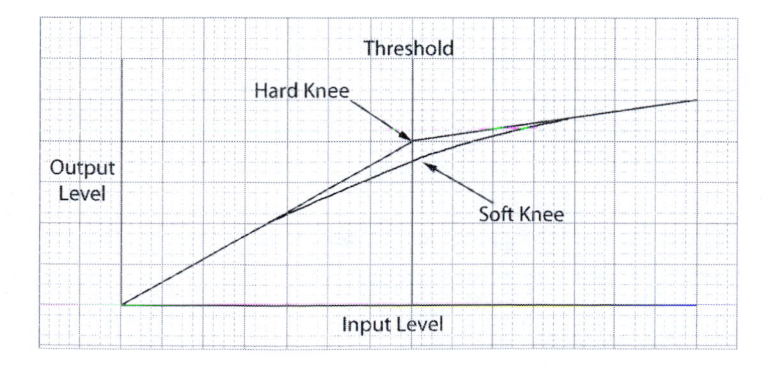

FIGURE 8.7
The action of soft and hard knee

by the time the compressor has detected a threshold breach, the sound has finished, and it has been completely missed by the compressor.

Consequently, when applying compression to short transient sounds such as drum hits, it's often wise to employ the peak mode detection. With this engaged the compressor becomes hyper sensitive to short sharp peaks and will apply gain reduction as soon as signals draw close to the threshold rather than wait until after the sound has breached. By doing so, any transient peaks are controlled before they overshoot the threshold.

If the producer wishes to maintain transparent compression, peak mode should only be engaged when working with short transient hits. If it's applied to other instruments with a more sustained signal, the compressor will commonly clamp down too quickly resulting in a loss of high-frequency detail and the instrument appearing less defined.

In these instances, RMS proves a more sensible option since in this mode, the compressor will relax and only detect and control signals with an average signal level rather than short sharp peaks. Typically these features will only appear on hardware compressors since many software plug-in compressors employ a 'look ahead' system to monitor the audio channels signals before they even reach the compressor and this renders Peak and RMS detection circuitry redundant.

RELEASE TIME

In order to maintain transparent compression, a further control is required to prevent sudden gain changes as the compressor stops applying its gain reduction. As with the attack, the release parameter is configured in milliseconds and allows the user to control the speed at which the compressor relaxes from gain reduction.

Contrary to many articles written about compression, the release parameter does *not* only activate when the signal drops below the compressors current threshold setting. Indeed, many compressors do not monitor the threshold in this way and often the release activates after the attack has completed and the output level begins to fall below the input level. This means that both attack and release can occur multiple times *during* gain reduction and, therefore, both combined can strongly affect the dynamic envelope and tonal content of any sounds if not set thoughtfully.

Figure 8.8 shows the effects the release setting can have on the dynamic envelope. With a release time of 1,000 ms, the timbre spends longer in the compression cycle than if no release was employed and consequently modifies the dynamic envelope. Although this may not appear too significant, our ears can pick up on small details, and if compression is applied heavily, it can be very noticeable.

FIGURE 8.8
With long release settings, the timbre remains in the compression cycle for longer

Many of the plug-in compressors today employ auto-release functions whereby the compressor continually monitors the signal and applies different release times depending on the incoming program material. These commonly provide the best solution when you don't want the compression to be evident but, as always, your ears should be the final judge as to whether auto-release or manual provides the sonic results.

HOLD

Whether auto release or manual release is used, since the release parameter can activate during the gain reduction when a drop in gain reduction is sensed, there can be occasions where the release will activate too soon. This is particularly the case when applying compression to low-frequency waveforms as the different states activate and deactivate gain reduction. To prevent this from occurring many compressors will feature a hold parameter. Again, configured in milliseconds this allows the producer to determine a finite period of time that the compressor should wait before initiating the release phase.

MAKE-UP GAIN

Since a compressor reduces the gain of any peaks and the dynamics in a signal, it will equally reduce the overall amplitude of a signal. Consequently, after compressing a signal, it will be necessary to increase the gain of the compressed signal back up to its original amplitude. Caution must be applied when increasing this gain, though, and it should be set so that the compressors output is the same level as its input because louder will invariably always sound better. This is because we hear more of the room's response to the frequencies and the sound therefore appears to surround us.

As touched upon in previous chapter, at low-amplitude levels, human hearing becomes more sensitive to frequencies between 3 KHz and 4 KHz, as this is where the frequency of the human voice is centred. As volume increases, both the low and high frequencies become more prominent resulting in a feel of increased power from the bass and more clarity from the highs.

Consequently, even if the compressor has been configured poorly and results in unwanted dynamic restructuring of the audio, if the make-up gain is adjusted so that the compressors output is effectively louder than the

FIGURE 8.9
The softube CL1B compressor

uncompressed signal, the compressed signal will generally still sound better regardless of how poor the compression applied.

In this situation, the compressor output level must be set by 'ear' and not by relying on signal levels or metres. Since a compressor reduces the dynamic ratio of audio, it produces a stronger average signal level than the input and our hearing has developed itself in such a way that a sustained sound will appear much louder than a short sharp transient hit even if this transient happens to be at a much higher amplitude than the sustained sound.

PRACTICAL COMPRESSION

Typically, the first use of compression will be for its original design application; to control the dynamics of an audio file during the recording stages or to level out the average signal level to prevent the audio disappearing behind other instruments in a mix. In both these instances, transparent compression would commonly be the aim but every compressor, whether digital plug-in or hardware, will impart its own character regardless of how carefully they are set-up and so the choice of compressor is the first hurdle to overcome.

The first obstacle is to choose between either solid state or valve topology. In the past 20 years, *analogue* and *valve* have become the buzzwords of electronic dance music production. Of the popular compressor plug-ins released over the past five years, close to 80% have been either digital emulations of older analogue or retro valve gear or programmed to at least emulate the 'sound' of valves and analogue. This shows the obvious trend towards using valve style compression and is due to the introduction of second order harmonic distortion to a signal.

Second-order harmonic distortion – in the original hardware – was a result of the random movement of electrons in the valve design that occurred at exactly twice the frequency of the compressed signal. The result of this coloration is, subjectively, a pleasant warmth or character applied to any audio that is compressed. As the amount of compression or amplitude is increased, the distortion increases, respectively, resulting in a warmer tone the more heavily the compression is applied.

Solid-state circuitry doesn't behave in this same manner. By replacing the valves with transistors, the second harmonic distortion is removed and the compressor maintains a particularly transparent sound. Solid state *does* still colour the sound but to many producers not in a subjectively nice way. In addition, if solid-state circuitry is driven too hard, it results in direct distortion and not the pleasant type.

This is not to suggest that producers only use valve equipment and avoid solid state altogether. Any great production is a careful mix of both solid state and valve to produce the best results. Many producers will no doubt agree that solid-state circuitry reacts much faster than valve and produces a more defined sound than valves. This lack of audio 'smothering' from solid state results in many producer choosing to employ solid state during recording stages and then applying valve afterwards. After all, there is no "undo" in real-life recording situations.

In addition to valves and solid state, another deciding factor as to the choice of compressor is based on the detection system it employs since this too will contribute towards its sonic colour. Indeed, the detection differences and variations between compressors are the reasons why many electronic dance producers don't just own one compressor and often have a collection of different styles in their plug-in folder or hardware racks.

The different styles of compression can roughly be categorized into variable MU, field effect, optical, VCA, and digital.

Variable MU

The variable MU could be viewed as the first ever mass-produced compressors available on the market. These compressors use valves for the gain-control circuitry and do not commonly have a ratio control. Instead, the ratio is increased in proportion to the amount of incoming signal that exceeds the threshold. That is, the more the level overshoots the threshold, the more the ratio increases.

While these compressors do offer attack and release stages, they're not ideally suited towards material with fast transients even with their fastest attack settings. Also, due to the valve design, the valves run out of dynamic range relatively quickly so it's unusual to acquire more than 15 dB to 20 dB of gain reduction before the compressor runs out of energy.

Despite this, MU compressors are renowned for their distinctive lush character and are often used to add character to vocals, bass, guitars and keyboard instruments. Both UAD and Waves produce perhaps the best software emulations of MU compressors with the acclaimed Fairchild 680 and in hardware the most notorious is the Vari-MU by Manley.

FET

FET compressors use a field effect transistor to vary the gain. These were the first transistors to emulate the actions of valves and as a result provide fast attack and release stages. This makes the FETs particularly suited for use on drum hits and can serve as an excellent starting point for compressing kicks and snares to beef them up.

Similar to MU, FET suffers from a limited dynamic range but tend to sound better the harder they are driven. They can pump very musically and are well suited for gain pumping a mix. UAD and waves produce software emulations of FET compressors with the 1186LN compressors and soft-tube produce the excellent FET compressor. In hardware, reproduction versions of the early FETs, such as the UREI 1186LN Peak Limiter and the LA Audio Classic II, are a worthwhile alternative.

Optical

Optical compressors employ a light bulb that reacts to the incoming audio by glowing brighter or dimmer depending on the intensity of the incoming signal. A phototransistor tracks the level of illumination from the bulb and adjusts the gain. Since the phototransistor must monitor the light bulb before it takes any action, some latency is created in the compressor's response so the more heavily the compression is applied the longer the envelope times tend to be. Consequently, most optical compressors utilize soft-knee compression.

This soft-knee response aids to create a more natural sounding and transparent compression but does mean these compressors are not ideal for compressing short transients such as drums. Despite this, optical compressors are often the choice for compressing vocals, basses, electric guitars, and drum loops, providing that a limiter follows the compression.

Plug-in giants UADs LA3A and the Joe Meek SC2 Pro Tools plug-in are popular choices for Opto compressors but one of the most accomplished is the soft-tube CL1B compressor. In hardware, the choice is more limited to the ADL 1500, the UREI LA3 and UREI Teletronix LA-2A and the Joe Meek C2.

VCA

VCA compressors are perhaps the fastest compressors available that also offer the highest possible amounts of gain reduction coupled with the best transparency. This makes the VCA one of the best compressors to use for general duties where transparency is of paramount importance but they do distort heavily if pushed hard and the less qualified VCA units will often remove high-frequency detail on compression.

It's unusual to use this style of compressor on a dance mix due to a distinct lack of colouration but they are sometimes used to pump a mix due to the fast reaction times of the envelopes. Many plug-in manufactures produce VCA compressors and emulations. UAD produce the infamous Drawmer dB × 160 (the first commercially available VCA compressor) whilst Brainworx produce the excellent VSC-2 quad discrete compressor.

Possibly the most recognized hardware VCA compressor is the Empirical Labs Stereo Distressor, this is a digitally controlled analogue compressor that allows the user to switch among VCA, solid-state and op-amps at the flick of a switch. Two versions of the Distressor are available to date: the standard version and the British version. Of the two, the British version produces a much more natural, warm tone (I'm not being patriotic) and is the preferred choice of many producers.

Digital Compressors

Computer-based digital is the generic name for any software compressor that hasn't been programmed to emulate any other method of compression but has simply been designed to act as a standard compressor, of which there aren't many available. Since it is possible to emulate classic compressors in the digital realm many developers prefer to release compressors based on valves since these are guaranteed to shift more units, and therefore, the standard no-nonsense digital compressors are often lacking.

Nonetheless, whilst these compressors do not feature any real coloration to attract producers, the digital-based units are often the most precise in terms of all functions, from threshold through ratio, attack and release parameters. Often they will employ a look-ahead function to monitor the audio channel before it reaches the compressor and this allows them to predict and apply compression without any chance of the audio being missed. This results in digital being the most transparent of compressors and although transparency is not always a selling point for compression, it can be useful to maintain during recording stages.

INITIAL SETTINGS

Once the producer has made the decision on what compressor to use for the job at hand, the parameters requires setting up. For most normal compression duties such as recording or dynamic limiting for mixing, the common purpose is program the compressor to remain fairly transparent.

To accomplish this, start playback of the audio through the compressor and set the ratio at 4:1. With the audio playing, gradually reduce the threshold parameter until the loudest parts of the signal read between −8 dB and −10 dB on the gain reduction metre. Set the attack parameter to 0 ms or the fastest possible option and the release should be set around 500 ms. Using these as preliminary settings, you can then adjust them further to suit any particular sound.

The higher the dynamics of the instrument that you're compressing are, the higher the ratio and lower the threshold should be. This will help to keep any massively varying dynamics under control, but it is important to listen carefully to the results as you adjust the parameters and in particular pay attention to the initial transient of the audio.

Begin to slowly increase the attack whilst listening and watching the gain reduction metre. The attack parameter should be configured so that the transient doesn't appear to be dulled by the compressors action whilst ensuring that not too much of the signal is creeping past the compressor unaffected. During this adjustment, the compressor should be bypassed often to compare the compressed and uncompressed version.

Once the attack is set correctly, the release should be configured so that the action of the compressor isn't immediately noticeable but also so that the compressor has time to recover when the signal has left the compression cycle. This is best accomplished by reducing the release until you can hear the action of the compressor and then begin to increase to the point just beyond being noticeable.

It is important to ensure that the release is set short enough that gain reduction stops in between notes by checking the gain reduction metre drops to 0 dB during any passages that you do not want in the compression cycle. For example, if you were compressing a hi-hat, so that the gain reduction metre reads −8 dB on each hat strike, the release should be set so that the gain reduction metres drops to 0 dB between each hat.

For example, if the gain reduction were to drop to say −2 dB between each hit rather than 0 dB, it's safe to assume that it would only require 6 dB of gain reduction rather than 8 dB, or alternatively the release parameter should be reduced. If not, the compressor will permanently apply 2 dB of gain reduction throughout since its unable to recover before the next hat strike. This inability to recover can result in the compressor distorting the transient that follows the silence and remodels the dynamic envelope. This means the producer must also take into account insert effects chains. If a compressor is being applied for dynamic control rather than sound design, placing a compressor *after* a delay effect would result in the delays also being compressed, and therefore, recovery time may not be possible.

If the compressor being employed offers a choice of soft- and hard-knee options, you must also consider the structure of the incoming dynamic envelope. If the signal features a sharp short transient stage then a hard-knee would possibly produce the best results provided the attack is set appropriately to allow the initial transient through unmolested. If the sound has a slower attack stage or you wish to maintain a very natural sound, then soft knee would possibly provide the better option.

Finally, by ear, you should increase the gain of the compressed audio signal, so it's roughly the same level as the signal entering the compressor. Possibly the best way to accomplish this is to consistently bypass the compressor to hear the signal before and after compression and even out the make-up gain so that they appear to be at roughly the same audio level.

EDM COMPRESSION

Whilst compression should remain transparent during recording or dynamic levelling purposes, deliberately making a compressors action evident forms the cornerstone for the sound of dance music.

As touched upon previously, compression can be used as a sound design tool to modify the dynamic envelope of individual timbres. For example, by employing heavy compression (a low threshold and high ratio) on an isolated snare drum, the snare's attack avoids compression but the decay is squashed. This decreases the transient to body dynamic ratio producing a stronger 'thwack' style timbre common in many genres of EDM.

Similarly, employing a similar approach on bass or lead timbres can result in a more powerful, energetic sound or, if applied heavily, it can result in the body and release becoming louder that the attack. This produces a sound that appears to 'suck' upwards as it progresses and is a common approach in some genre of EDM.

Another approach that has gained significant use in more recent years is the abuse of side-chains within compressors. Many hardware and software compressors feature a side-chain function. With hardware this appears as a pair of jack inputs whilst in software models, it usually appears as a drop-down menu somewhere on the compressors interface. Using these, the producer can insert a secondary signal from a hardware device or another audio channel within the digital workstation that can control how the compressor behaves on a signal entering at its main inputs.

A good example of side-chain compression is the effect of when a radio DJ speaks, the music lowers in volume and when they stop the volume of music increases again. This is accomplished by having the music signal input fed into the main inputs of the compressor or in the case of a sequencer, placing a plug-in compressor on the track containing music. The DJ's microphone is connected to the side-chain input, or in software, a secondary-audio channel in the sequencer can be configured to act as a side-chain input to the compressor. Using this configuration, the threshold of the compressor is controlled by the side-chain signal, so whenever the side chain signal is present the compressor activates.

This type of side-chain application has a number of practical uses during mixing by 'ducking' instruments to allow others to sit over the top more easily. For example, in a typical mix, a lead guitar or instrument will often occupy the same frequencies as the human voice. This results in a mix that can appear cluttered making it difficult to distinguish the voice or hear the words clearly. Placing a compressor onto the guitar track and using the vocals as a side-chain signal can avoid this. By setting the compressors ratio, attack and release parameters appropriately every time, the vocal channel plays the compressor will activate and lower the gain of the lead guitar permitting the vocals to be heard clearly.

This same technique has become requisite in dance music production, although rather than the vocals acting upon the main signal; a kick drum is used instead. This effect referred to as gain pumping is accomplished by running the entire mix through a compressor and inserting a kick into the side-chain inputs. By doing so, every time a kick occurs the mix ducks and produces the effect of a kick-drum punching holes in the mix.

This can easily be accomplished in digital audio workstations by first duplicating the kick-drum channel to produce two channels. The second channel is set to no output (so it produces no sound at the mixers outputs) and a compressor is placed across the entire mix – this can be a bounced down mix on a single channel or all the current mix channels sent to a single buss (more on this in a Chapter 11). The second-kick channel with no output is then used as the side-chain insert.

If you don't have a mix available, the companion website contains two tracks – a complete mix track and a kick-drum track that you can use to practice. Set your workstations tempo to 134 BPM and insert both audio files from Chapter 8 companion website into individual channels in your digital audio workstation. Insert a compressor with a side-chain option on the mix track and set the compressors ratio to 4:1 with a fast attack and release. Begin playback and then slowly reduce the threshold;, as you do so, the mix will begin to pump with every kick. Finally, experiment with different attack, release and threshold settings until it produces the effect you prefer.

The COMPANION WEBSITE contains a video file example of pumping a mix.

This effect is not the mainstay of pumping an entire mix and is often employed in many genres to produce bass-lines, melodic leads and rhythmic pads. Employing the exact same technique but placing the compressor on a sustained bass note can produce a pumping bass, and this is a technique that is becoming increasingly popular in most genres of EDM.

The COMPANION WEBSITE contains a video file example of pumping a bass.

This technique is also used in genres that use heavy and loud bass elements that play consecutively with the kick drum. Since a heavy kick and heavy bass playing together can merge together creating a muddy sounding feel, by pumping the bass, it can be avoided. With a compressor inserted onto the bass track and a kick used as a side-chained signal, using a ratio of 3:1, a fast attack and medium release (depending on the sound) every time the kick occurs the bass will drop in volume, creating free space for the kick thereby preventing conflicts.

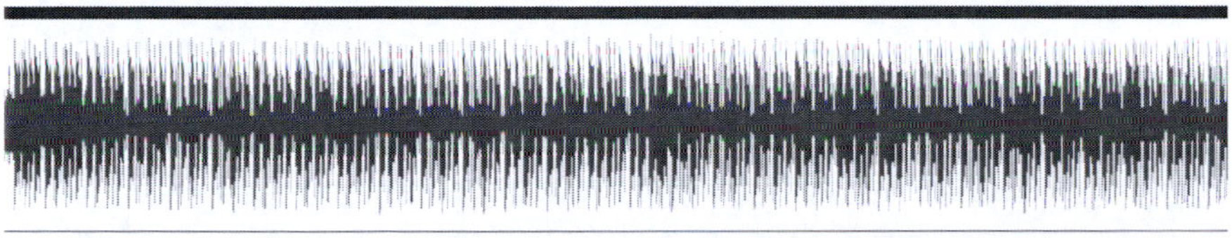

FIGURE 8.10
The waveform of a mix

Some genres of EDM, such as chill out, don't require such a heavy-handed pumping method, and here it's generally accepted to simply gain pump the mix without side chaining the kick. This requires a more careful and considered approach, however, since it's possible to completely loose any excursion and end up with a flat sounding mix.

Figure 8.10 shows a typical out mix in the waveform editor of an audio workstation. It's clear from looking at the waveform that the greatest energy, that is the loudest part of the loop, is derived from the kick drum. If a compressor is inserted across this particular mix, and the threshold of the compressor is set just below the peak level of the loudest part, each consecutive kick will activate the compressor. If the ratio, attack and release are then set carefully, it is possible to create a gentler form of gain pumping that can appear more cohesive to the mix.

A good starting point for this is to use a ratio of 3:1, mixed with a fast attack and release time. Slowly reduce the threshold until it's only activating on each kick in the example mix. An easy way to accomplish this is to watch the gain reduction metre on the compressor and ensure it only illuminates every time a kick occurs.

Finally, experiment with the release time of the compressor and note how as the release is lengthened, the pumping appears to be less dramatic. Similarly, note that as the release time is shortened, the gain pumping becomes more evident. This is a result of the compressor activating on the kick before rapidly releasing when a fall in gain reduction is detected. The rapid change in volume produces a similar effect as side chain pumping but is less obvious and produces a gentler feel to the mix.

The key element to this effect lies with the timing of the release parameter, and this depends on the tempo of the drum loop. It must obviously be short enough for the compressor to recover before the next kick but also long enough for the effect to sound natural so experimentation and careful listening are crucial.

CHAPTER 9
Processors

'Don't do what you've heard, do what you think you gonna hear'"

Dennis Ferrer

Although compression has certainly earned its place as a requisite processor for attaining the 'sound' of electronic dance music today, it's certainly not the only processor employed in its production. Indeed, a number of further processors play a vital role for both the sound design process and the requisite feel of the music. Therefore, this chapter concentrates on the behaviours of the different processors that are widely used in the design and production of dance music.

LIMITERS

A limiter is a dynamic processor that can be compared with the action of a compressor but rather than compress a signal by a specified ratio, a limiter prevents the signal from ever exceeding its threshold. Therefore, no matter how loud the input signal may be the limiter will compress it so that it never violates the current threshold setting.

This style of action is often termed 'brick wall' limiting, but some limiters will permit a very slight increase in level above the threshold. The amount of limiting overshoot permitted depends on the manufacturer or programmer but generally soft limiting very rarely exceeds a few decibels and is only employed to permit the odd transient to overshoot. This action permits for a more natural or open sound from the limiter and is the preferred choice of many producers.

Since a limiters action is restricted to control any peaks that may create distortion, many limiters only require three controls: an input level, a threshold and an output gain. An input is employed to configure the overall signal level entering the limiter, whereas the threshold, like a compressor, is employed to configure the level at which the limiter will begin attenuation of the signal.

Any signals that exceed this threshold are immediately compressed to prevent clipping or distortion of the outputs, and as a result, an output control can then be used to increase the average signal level.

As discussed in previous chapters, since we determine a sustained timbre to appear louder than a transient, by employing a limiter, it is possible to increase the perception of loudness in the mix or timbre. However, as with compression, a limiter can very quickly and quite severely modify the dynamic envelope of any sound, and therefore, it must be used with some caution. Judicious application of limiting with little regard to the dynamics can result in a timbre that appears louder and thus more appealing, despite completely destroying the timbres natural energy and presence.

In order to maintain a more natural uncoloured sound, some limiters do feature an additional release parameter. Using this, the producer can determine the time it takes for the limiter to recover after a signal has been limited. As with compression envelopes, this too must be applied cautiously to ensure that the limiter has enough time to recover before the next threshold breach. If the limiter cannot recover in time, it will distort the transients that follow it and transform their dynamic envelopes.

Since the main purpose of any limiter is to prevent transients from overshooting the threshold and creating distortion at the outputs, there is technically no requirement for an attack parameter since as soon as a breach is detected, it should be limited. However, since many software plug-ins have the advantage to 'look ahead' at the incoming audio channel they may employ a 'fixed' attack parameter.

This attack isn't commonly user definable and a soft or hard setting will be provided that operates in a similar respect to the knees on a compressor. That is, a hard attack activates the limiter as soon as a peak is close to overshooting whilst a soft attack has a smoother curve with 10 ms or 20 ms. This approach can often reduces the likelihood of clicks or snaps that result from the limiter acting too quickly on the incoming signal and also creates a more natural form of limiting. These software look-ahead limiters are sometimes referred to as ultra-maximizers (Figure 9.1).

The types of signals that require limiting are commonly those with an initial sharp transient peak, and therefore, limiters are generally employed to control the 'crack' from snare drums, kick drums, and any other wayward peaks in audio that may contribute to clipping and distortion. However, like a compressor, a limiter will reduce the dynamic ratio between the loudest and quietest parts of any signal, and therefore, they are often employed on a complete mix to increase its volume.

By reducing the dynamic ratio of a piece of music, its average signal level can be increased, and thus it can be made louder. This has contributed to what is now termed the 'loudness war' as natural dynamics suffer in the effort to produce the loudest mix. Whilst this argument rages on, the dance producer must

FIGURE 9.1
An ultramaximizer

nevertheless exercise caution when limiting in order to ensure there is a careful balance between loudness and the kicks excursion.

Within EDM, the kick drum provides the main rhythmical element of the music that thumps the listeners in the chest and that energy is a direct result of the degree it moves the monitor speaker's cones. The more the speaker cone is physically moved the greater the punch of the kick is. This speaker cone 'excursion' is directly related to the size of the kick's peak in relation to the rest of the music's waveform.

If the difference between the peak of the kick and the main body of the music is reduced too much through severe limiting, as the average signal level increases the kick reduces in energy. This is because as the dynamic range is restricted more and more, more of the music will move the cone by the same amount. The effect of this strong dynamic restriction is a mix that although sounding loud, lacks the energetic punch of the kick. The effects of this application are shown in Figure 9.2 and can be heard on the companion website.

The effects of strong dynamic restriction.

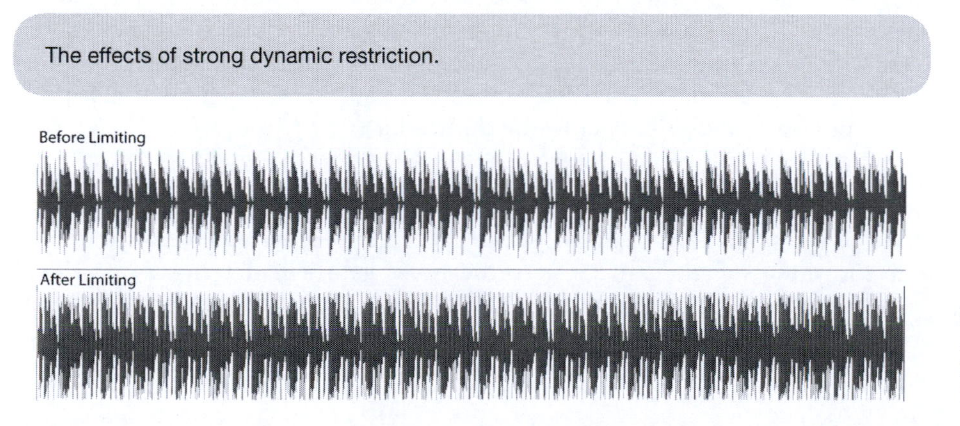

FIGURE 9.2
A mix before and after heavy application of limiting

On average for a dance mix, approximately 3 dB to 6 dB is a reasonable amount of limiting without the kick suffering in response but the exact figure depends entirely on the mix. If the mix has already been quite heavily compressed it would be wise to avoid any more than 3 dB at the limiting stage, otherwise the dynamics may be removed altogether.

NOISE GATES

A noise gate is a dynamic processor that attenuates signals *below* a given threshold. Its purpose is to remove any low-level noise that may be present during what should be a silent passage.

For example, during a recording session with a vocalist it's common practice to leave the recording device running throughout the length of the track in order to capture all the nuances of a single performance. Consequently, there will be times where the vocalist rests for a moment or two between bars of the song. During these passages, it's possible for the microphone to pick up extraneous noises such as the vocalist tapping their foot, resting, breathing heavily, clearing their throat, or adjusting their stance. By employing a noise gate and setting the threshold parameter below the performers' performance level, every time the performer stops, the signal moves below the threshold and the noise gate will activate resulting in complete silence.

Whilst in theory a noise gate could simply consist of a threshold parameter, the practice is a little more complicated. First, we must consider that not all sounds will start and stop abruptly and some instruments will progressively increase in gain. Violins, strings and pads, for example, can often gradually increase in gain over a period of time and if a noise gate only featured a threshold parameter these sounds would suddenly appear once they exceeded the noise gates threshold. Similarly, if they also were to gradually reduce in gain, they would be abruptly silenced as they fall below the threshold again. Therefore, all noise gates will feature attack and release parameters.

Both attack and release parameters are configured in milliseconds and are similar in many respects to a compressors envelope, permitting the producer to determine the attack and release times of the gates action. The attack parameter will determine how quickly the gate opens after the threshold is breached whilst the release parameter determines how quickly the gate will close after the input signal has fallen below the threshold.

As with compression both of these must be configured carefully since an incorrect adjustment could result in both envelopes modifying the dynamic envelope of the signal. For example, if the gates attack is set to a slower rate than the incoming signals gain increase the noise gate will determine the signals gain increase. Similarly, if the release were too short, the dynamic envelope would be further manipulated. The possible dynamic modifications of incorrect adjustments are shown in Figure 9.3.

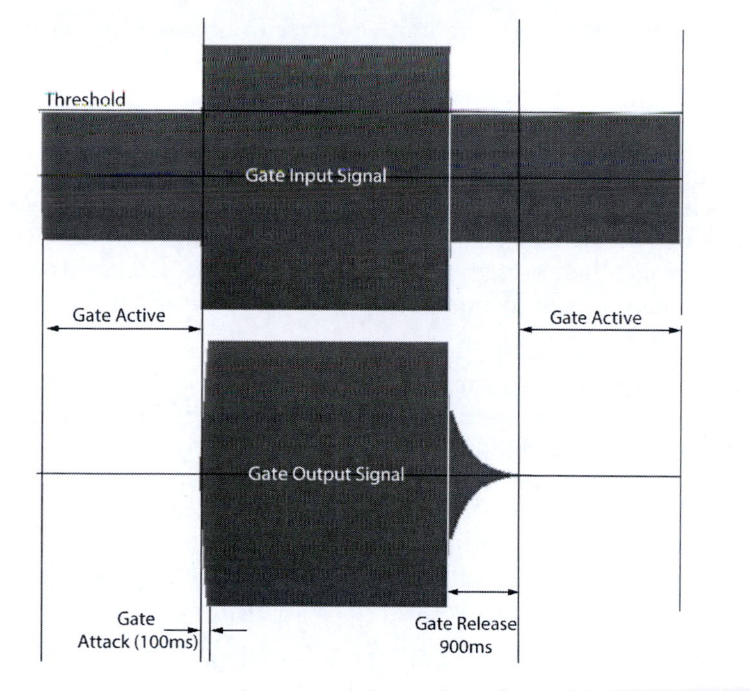

FIGURE 9.3
A noise gate modifying the dynamic envelope

Another consideration with the operation of a noise gate is that not all signals will remain at a constant volume throughout their period. When a performer holds a note, for example, it will result in a varying degree of amplitude and if this occurs close to the threshold of the gate, the signal may consistently hover above and below this threshold. With a short attack and release, the result is a constant activation and deactivation of the gate producing an effect referred to as 'gate chatter'. Alternatively, with a longer attack and release it would produce severe volume fluctuations throughout a performance.

To prevent these problems from occurring, gates will feature an automated or user definable hold time. With this, the producer can determine a specific amount of time that the gate will pause before activating its release after a signal has fallen below the threshold. By doing so, if a signal falls below the threshold for only a short period before returning above it, the hold parameter will prevent the gate from chattering.

Notably, the hold function is often confused with a similar gate process termed hysteresis, but the two have a substantial difference. Whereas a hold function will force the gate to wait for a pre-defined amount of time before closing, hysteresis adjusts the thresholds tolerance independently for both opening and closing the gate. For example, with hysteresis if the gates threshold were set at −12 dB, the audio signal must breach this before the gate will open. However, the same signal must fall a few extra decibels below −12 dB before the gate will close again. This action produces results very similar to a hold function but often produces more natural sounding results that a hold control and, therefore, is the preferred option for real world instruments and vocals.

In addition to hold or hysteresis, many gates may feature a range control, and this can be employed to control the amount of attenuation for signals that fall below the threshold. In many instances, a typical gate will close completely when a signal falls below its threshold in order to produce total silence. However, there are occasions whereby the producer doesn't require complete silence and wishes to maintain some background noise.

Examples of this may be when recording drums or acoustics guitars. Drum strikes will heavily resonate and acoustic guitars suffer from fret squeal as the performers fingers move up and down the fret. These aid in producing a sound we expect to hear, and if this is removed via a gate or other means, the resultant sound appears false or deemed to be the missing character. A similar effect can also occur with vocalists in regards to breaths. Typically, a vocalist will take a large intake of breath before beginning the next line of a verse or chorus, if a noise gate removes this breath, it can sound particularly unnatural, the listeners instinctively know that vocalists have to breathe!

A range control is calibrated in decibels and permits the producer to define by how much any signals that fall below the gates threshold are attenuated. The more the range is increased, the more the signals gain is reduced, until at its maximum setting, the gate will silence the signal altogether. Typically, this parameter is configured so that any signals that do fall below the threshold are only just audible when placed within the mix in order to provide a more realistic nature to the sound.

Many gates, and in particular software emulations, will feature side-chain inputs that operate in the exact same manner as a compressors side chain. With gates, these are often termed "key" inputs, but the principle is the same and permits a workstations audio channel to be input into the gates key inputs that in turn can control the threshold. This action has various creative uses but perhaps the most common use is for the producer to place the noise gate on an audio track with a sustained timbre, such as a pad or string section and then program a hi-hat rhythm on another track. This hi-hat rhythm is then employed as the key input of the gate so that each time a hat occurs the gate opens and the sustained pad timbre is heard. Alternatively, if the workstation permits it, the producer may input a virtual instrument into the gates key inputs, so that each time a note is struck on the instrument, the gate opens allowing the pad sound through.

Obviously, both these actions will supersedes the gates threshold setting but the attack, release, range, hold, hysteresis and range controls are often still available allowing you to control the reaction of the gate on the audio signal.

TRANSIENT DESIGNERS

Transient designers are yet another form of dynamic processor but rather than the previous examples that are employed to control dynamics, a transient designer is employed to deliberately modify the dynamic envelope. This deliberate modification of a dynamic envelope is a significant and important sound design tool for the creation of electronic dance music, and therefore, a transient designer is one of the most fundamental tools to own (Figure 9.4).

FIGURE 9.4
The SPL transient designer

Most transient designers are simple and only feature two controls: attack and sustain. These permit the producer to modify the attack and sustain characteristics of a pre-recorded audio file the same as when using a synthesizer. Since we determine a significant amount of information about a sound through its attack stage modifying this can change the appearance of any timbre. For example, by reducing the attack stage of a timbre, it's possible to move it further back into a mix whilst increasing the attack will bring it to the forefront. They also prove to be indispensible when constructing drum loops since modification of a snare or hi-hat attack and release stage can affect the groove of the loop.

EQ

EQ is principally a frequency-specific volume control that permits the producer to attenuate or increase the gain of a specific band of frequencies. This application can be in the context of an individual timbre or more often in the context of a complete mix to provide each instrument its own space to breathe and be heard. As basic as this premise is, however, EQ is an incredibly difficult processor to master and despite the relatively small amount of parameters offered the theory behind its application can be quite complex.

The most commonly used EQ system is a parametric and paragraphic (Figure 9.5). Both these systems are the same with the exception of the display. A parametric EQ will only feature the parameters whilst a paragraphic with feature both parameters and a graphic display to show how the EQ is being modelled.

Regardless of which is used, both typically consist of three main parameters to offer the producer control over frequency, bandwidth and gain. The frequency

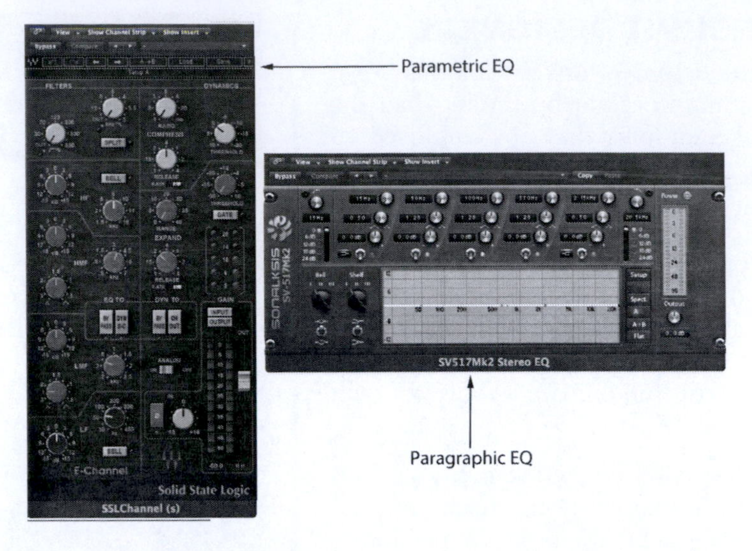

FIGURE 9.5
A parametric and paragraphic EQ

pot is used to select the central frequency that the producer wishes to either enhance or remove whilst the bandwidth pot determines how much of the frequency range either side of this central frequency should be affected. After this, the gain permits the producer to increase or attenuate the gain of these selected frequencies.

For example, suppose the producer wished to EQ a lead timbre that contained frequencies from 600 Hz to 9 kHz and the frequencies between 1 kHz and 7 kHz required a cut. Using the frequency parameter, the producer would home in on the centre frequency (4 kHz) and then set the bandwidth wide enough to affect 3 kHz either side of the centre frequency. By then reducing the gain of this centre frequency, the frequencies 3 kHz either side would be attenuated too. The size of the bandwidth is often referred to as the Q (Quality) and the smaller the Q number, the larger the 'width' of the bandwidth.

As simple as this premise may appear it's a little more complicated in theory since Q is not constant and will change depending on the amount of boost or cut applied. For example, if a particularly heavy boost or cut is applied, the bandwidth will begin particularly wide and become progressively finer as it carves further into the frequency range. This action is shown in Figure 9.6.

This non-constant Q action makes exact measurements difficult to accomplish therefore the actual value of the Q is determined by measuring 3 dB into the boost or cut being applied. This is shown in Figure 9.7.

Whilst taking a measurement 3 dB into a cut or boost will provide a reliable method for determining the current width value of the Q, it is inadvisable to refer to a Q value as a specific range of frequencies and instead it should be referred to in octave values. This is because octaves are exponential, and therefore, a Q frequency value effects will be very different depending on the octave it is being applied too.

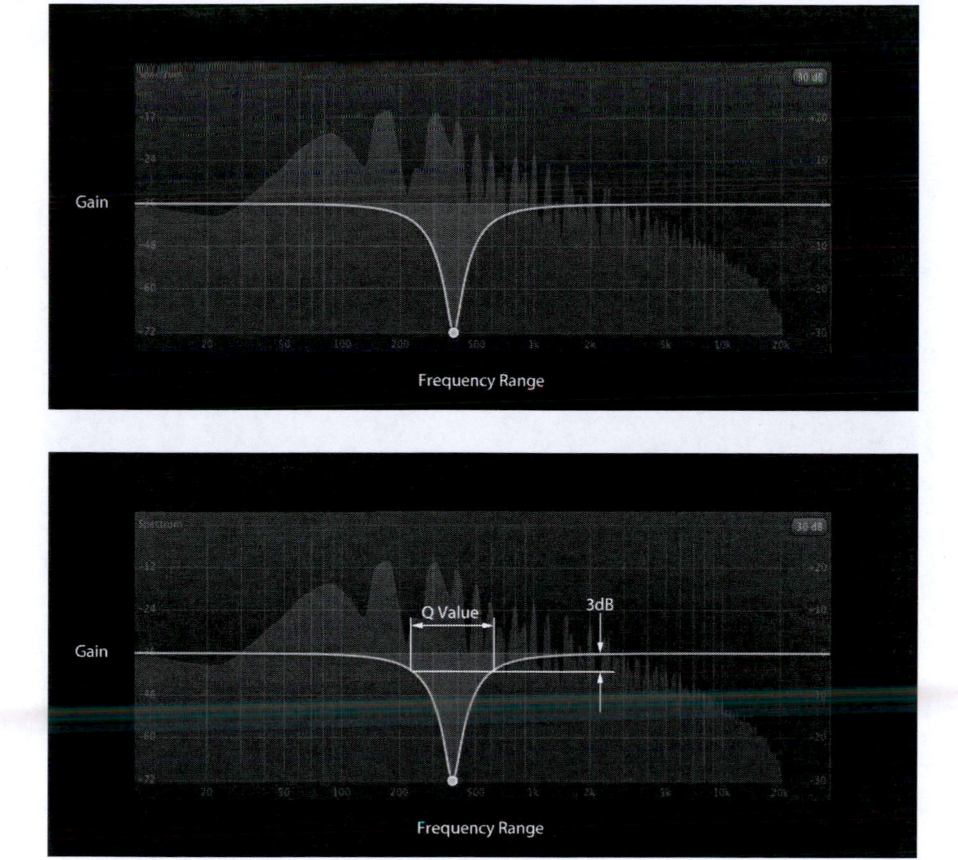

FIGURE 9.6 The action of non-constant Q

FIGURE 9.7 The measurement of Q

As discussed in a previous chapter on acoustic science, Pythagoras discovered that the relationship of pleasing harmonics was strongly related to mathematics and that exponential increases in frequencies were responsible for harmonics that the human ear found pleasing. From this, he also determined that doubling a frequency was equal moving up by an octave in the musical scale.

This means that if we consider that A1 on a musical scale is 110 Hz, by doubling that frequency it would produce the tone A2 occurring a 220 Hz. Moreover, if the frequency of A2 were doubled to 440 Hz, it produces a tone at A3, and if that again were doubled to 880 Hz it produces the tone A4. This exponential relationship of frequencies and the octave is shown in Figure 9.8.

With this in mind, if a producer decided to apply an EQ cut of 6 dB at 500 Hz, it would occur between A3 (440 Hz) and A4 (880 Hz) on the musical scale. If the Q value here were to carve out 232 Hz between these two octaves it would leave 208 Hz unaffected in this particular octave. Thus, a frequency cut with a Q of 232 Hz removes a large proportion of frequencies in this octave (Figure 9.9) (440 Hz between A3 and A4, removing 232 Hz from that = 440 Hz – 232 Hz = 208 Hz).

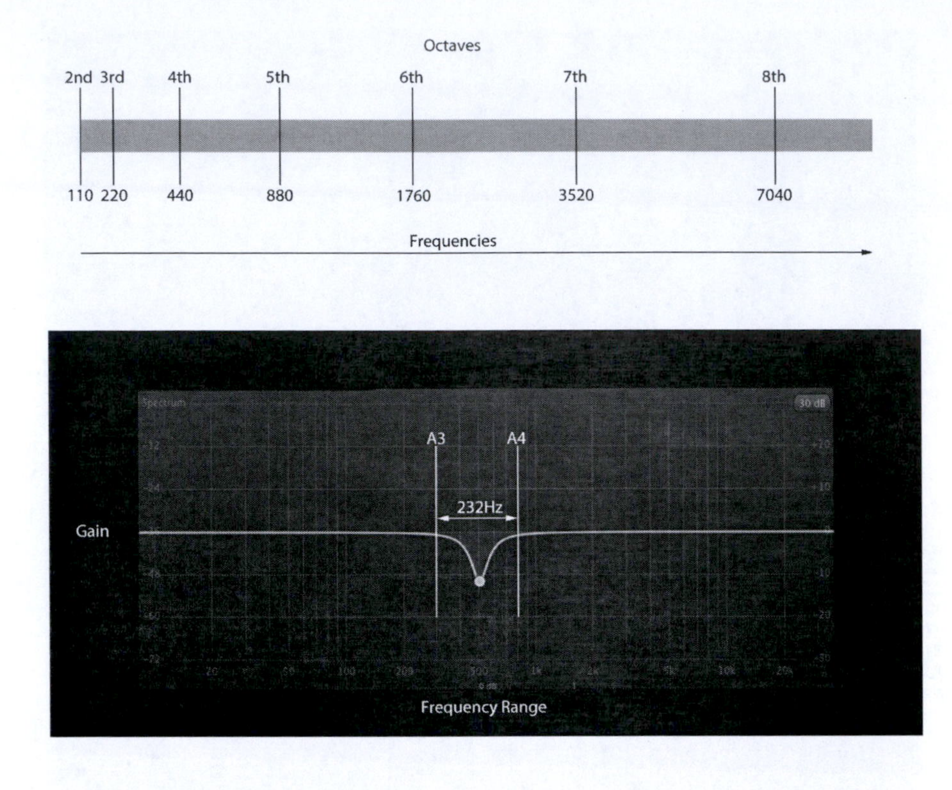

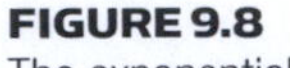
FIGURE 9.8
The exponential increase of frequencies related to octaves

FIGURE 9.9
232 Hz cut applied between octave A3 and A4

Compare this to if the producer were to attempt to apply that same Q of 232 Hz between A6 and A7 on the scale (Figure 9.10). Here, A6 is 1,760 Hz and A7 is 3,250 Hz, a difference of 1,760 Hz. If the cut were placed at 2 kHz with a Q of 232 Hz it would leave 1,528 Hz unaffected in this octave range. Comparatively this is an insignificant amount not even removing a quarter of the frequencies in this octave range compared to over half the frequencies removed when applied between A3 and A4.

As a consequence producers will refer to Q in terms of octave measurements rather than frequencies. By referring to a Q's value in octaves rather than frequencies, the Q will carve out an exponential number of frequencies depending on the octave, and therefore, we perceive it to affect the same amount of frequencies regardless of the octave. With this in mind, engineers and producers tend to rely on a number of basic Q values for production work. These are

- **Q of 0.7 = 2 Octaves**
- **Q of 1 = 1 1/3 Octaves**
- **Q of 1.4 = 1 Octave**
- **Q of 2.8 = ½ Octave**

Generally speaking, a bandwidth of half of an octave or less is used for removing problematic frequencies while a width of a couple of octaves is suitable for shaping a large area of frequencies. For most work, though, and especially for those new to working with EQ mixing, a non-constant Q of 1.0 is perhaps the most suitable. This is often referred to by many engineers as the 'magic Q', since it can be used to solve most problems within mixing and sounds more natural to the ears.

It is of course possible for the producer to calculate his or her own Q values using the following mathematical theorem of Q=[CF/(TF–BF)]. Here, CF relates to the centre frequency, whilst TF and BF relate to the top and bottom frequencies that are being carved out.

For example, if a central cut were placed at 1,000 Hz, and it resulted in affecting the outermost frequencies of 891 Hz and 1,123 Hz; these would be the top and bottom frequencies (Figure 9.11).

In this situation, the producer would begin with the simple math of TF – BF (1,123 Hz–891 Hz=Q of 232 Hz). With the central frequency placed at 1,000 Hz, it's a simple case of dividing 1,000 Hz with the Q frequency of 232 Hz that produces a Q of 4.31 (1,000 Hz/232 Hz=4.31). This is approximately a ¼ of an octave.

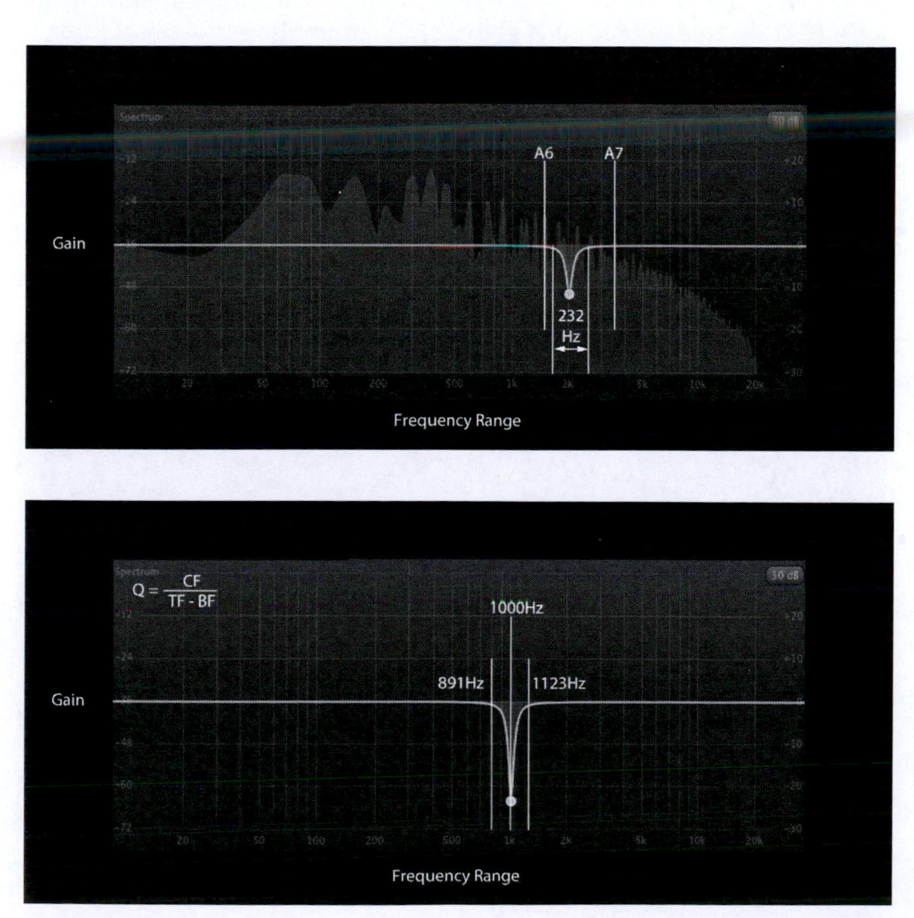

FIGURE 9.10
232 Hz cut applied between octave A6 and A7

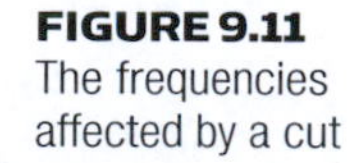

FIGURE 9.11
The frequencies affected by a cut

In addition to the standard parametric features on a plug-in EQ unit, they will often also permit the use of shelving filters. These operate in a similar method to the low-pass and high-pass filters than adorn many synthesizers. As previously discussed, a low-pass filter will attenuate the higher frequencies that are above the cut-off point, whilst a high-pass filter will attenuate all the lower frequencies that are below the cut-off point. Within EQ, a low-shelf filter can be used to either cut or boost any frequencies that occur below the cut-off point, whilst a high shelf can be employed to cut or boost any frequencies that occur above the cut-off point.

The specifications of the plug-in will determine the amount of control offered over shelving filters (Figure 9.12), but typically they will offer two parameters; one to select the frequency (sometimes termed the knee frequency) and a secondary parameter to either cut or boost the frequencies above or below the chosen frequency.

Since many speaker systems are incapable of producing frequencies below 20 Hz, it is good practice to remove these frequencies by "rolling off" all the frequencies below 20 Hz. By doing so the producer can ensure that the speaker does not loose energy (and thus volume) by attempting to replicate frequencies beyond its frequency range.

Further EQ

Both parametric and shelving filters are the most commonly used EQ systems in the production of electronic dance music, since they offer the most in-depth control and during mixing it is likely the only ever EQ system that will be employed. However, there a secondary form of EQ is available known as the Graphic Band EQ.

Graphic Band EQs, or more simply Graphic EQs, are different to parametric EQs in that all the bands are fixed, and it is only possible for the producer to

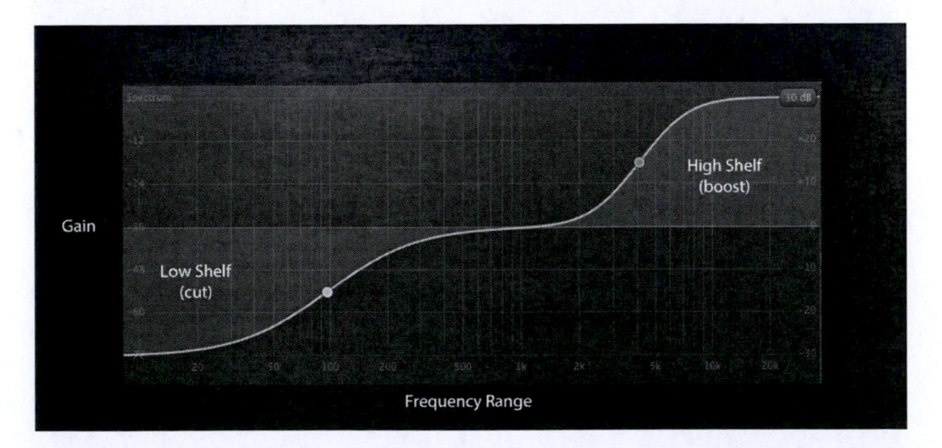

FIGURE 9.12
The action of shelving filters

either apply gain or attenuate these fixed bands. Obviously, this means the more bands that are available the smaller the frequency bands that are affected will be.

Although these may seem particularly restrictive when compared with parametric, most graphic EQs employ a constant Q in that regardless of how much gain or attenuation is applied, the Q remains constant throughout. Provided they have a large number of bands to cover the frequency range, this makes graphic EQs particularly surgical and is the reason as to why they have been the mainstay of professional mastering suites since it permits the engineer to contour the tonal content of a mix accurately.

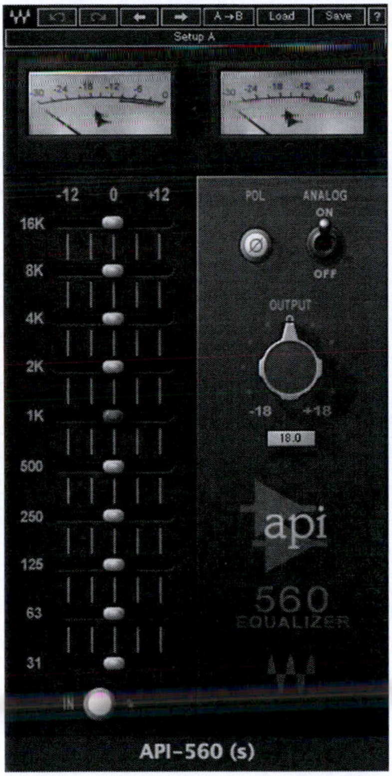

Although these forms of EQ may seem particularly limited, their uses should not be underestimated. The beauty of any good graphic band EQ is that they are immediate and do not require the producer to first home in on a specific frequency. This immediacy permits the producer to quickly experiment and concentrate on the "sound" rather than the technical theory behind its use and, therefore, plays a strong role in sound design.

FIGURE 9.13
The API 560

Notably, not all graphic EQs are constant and some of the more revered and pursued units will feature 10 or less bands but employ an non-constant Q. An impressive example of this style is the API-560; a unit that is one of the most respected and admired 10-band EQ systems. This acts in a similar manner to parametric/paragraphic EQs with the Q becoming progressively thinner, the more the specific frequencies is attenuated or boosted.

Above all, the potential audience do not see EQ settings; they only hear them. One of the biggest problems facing any producer working on 'visually enhanced' EQ is the tendency to trust their eyes more than their ears and this can lead to disastrous results.

EXPANDERS

Expanders are simply compressors working in opposite. Whereas a compressor works to reduce the dynamic range of an audio signal, an expander works to increase it. Since these are essentially opposite compressors they feature the same parameters as a typical compressor with a threshold, ratio, attack and release time.

The threshold parameter on an expander is used to configure which parts of the incoming signal that the expander should process. That is, any signals that breach the current threshold setting will be increased in volume (i.e. expanded) compared with those that remain below that are left unmolested, hence increasing the dynamic range between the two.

The level of expansion applied when a signal breaches the threshold is expressed via the ratio control and this operates in exactly the same manner as the parameter discussed in compression by allowing the producer to set the ratio of expansion compared with the level of the signal level breaching the threshold. Similarly, both the attack and release parameters operate in much the same way permitting the producer to set both time it takes for the expand to respond to signals that exceed the threshold and the time it takes for the expander to stop processing when the signal begins to fall. Finally, an output gain can be used to increase or more usually attenuate the signal at the expander's output stage.

CHAPTER 10
Effects

'iLok is still the worst thing ever invented'.

Calvin Harris

With the most common processors discussed in previous chapters, we can now turn attention to the most common effects. This is by no means exhaustive as to discuss every effect in detail would require a book in itself rather this chapter discusses the most fundamental effects that any producer should have readily available for the production of electronic dance music.

REVERB

Reverberation is a term used to describe the natural reflections a person expects to hear in different environments and provides essential auditory clues as to the space, furnishings and general structure. As described in an earlier chapter, when an object vibrates, the resulting changes in air pressure emanate out in all directions but only a small proportion of this pressure reaches the ears directly. The rest of the energy strikes nearby objects, floors, ceilings and walls before then reaching the ears.

These secondary reflected waves result in the creation of a series of discrete echoes that follow the direct sound, and it's from this 'reverberant field' that we can decipher a large amount of information about our surroundings. In fact, this effect is so fundamental to our survival that even if a person were blindfolded, they would still be able to determine a large amount of information about a room from the sonic reflections alone.

Much of this perception is a result of the sound absorption properties of different materials. Since each surface features a distinct frequency response, different materials will absorb the sounds energy at different frequencies. For example, a bare brick wall will rebound high-frequency energy more readily than a wall covered with pictures or wallpaper. Similarly, in a large hall, it would take longer for the reverberations to decay away than in a smaller room. In fact, the further away from a sound source a person stands the more reverberation they will determine to the point that if stood far enough away, the reflections would be perceived as a series of distinct echoes rather than reverberation.

Consequently, while compression could be considered the most important processor, reverb is perhaps the most important effect since plug-in or hardware synthesizers and audio samples will not generate natural reverberations unless the signals they produce are exposed to air. Thus, to create some depth and natural space within a dance mix, the producer must employ artificial reverberation.

Artificial reverb is available in two forms, convolution and synthetic. Currently, convolution is the buzzword amongst many dance musicians since these reverb units rely on samples, or 'impulses', of the reflection properties of real spaces. Since these impulses are based on sonic fingerprints of real-world environments, they produce the most realistic acoustic spaces but as realistic as they may appear, this comes at the expense of any detailed customization whilst also retaining the realism.

The second reverb, synthetic, doesn't suffer from this problem since these reverbs employ algorithmic calculations to roughly determine how certain acoustic spaces would react to a sound source. The sonic quality and realism offered by these styles of reverb units vary greatly depending on the manufacturer and range from incredibly versatile lifelike recreations to what can only be described as the sonic reflection you might expect from a soggy cardboard box. This is not to suggest that a producer will make a choice of one over another, and typically, they will use a careful mix of both convolution and algorithmic reverbs within a production.

Regardless of the unit chosen, applying reverb is a little more complex than first meets the eye since reverb behaves differently depending on the sound source, our distance to it and the behaviour of the room and its furnishings. What's more, as reverberation plays a strong part in the human survival instinct, we are instinctively accustomed to how it should sound, and therefore, a (good) reverb unit will offer a somewhat bewildering array of adjustable parameters to emulate this real-world response.

What follows is a list of the available controls on many quality reverberation units, but depending on the manufacturer and the algorithms employed, only some of these parameters may be available.

Ratio (Sometimes Labelled as Mix)

The ratio controls the ratio of direct sound to the amount of reverberation applied. If the producer increases the ratio to near maximum, there will be more reverberation than direct sound present; whilst if this is decreased, there will be more direct sound than reverb. This is the main parameter control for adjusting how close the original sound source is from the listener. The higher the ratio (or mix), the further away the sound source will appear to the listener.

Pre-Delay Time

After a sound occurs, the time separation between the direct sound and the first reflection reaching the ears is termed the pre-delay. Configured in milliseconds this parameter permits the producer to determine the time differential between the original sound and the arrival of the very first sonic reflections.

In a typical situation, this can contribute to the distance a listener is from a sound source and/or a reflective surface. For example, if the listeners were stood in the middle of an empty room, the first sonic reflections would take longer to reach their ears than if they were positioned next to a wall whereby the reflections from the wall would reach them sooner.

Although this parameter can be used to emulate this response, typically pre-delay is commonly employed to prevent the reverberation field from smothering the transient of an instrument. This can be vital in preventing the early reflections from washing over the transient of instruments since reverb smothering over the transients can result in a mushy, muddled appearance of the sound.

Early Reflections

An early reflections parameter will often only feature on high-end equipment and offers the producer a way to control the sonic properties of the first few reflections received by the listener.

Since sounds reflect off a multitude of surfaces, it creates very subtle differences between each subsequent reflection reaching the ears. Using this parameter, it is possible to determine the type and style of surface the initial sound has reflected from.

Diffusion

Diffusion is often associated with the early reflections and is a measure of how far the early reflections spread across the stereo image. The amount of stereo width associated with reflections is determined by how far away from the sound source the listener is.

If the initial source is at a good distance from the listener, then much of the stereo width of the reverb will dissipate as it travels through the air. If the sound source or reflective objects and are close to the listener, however, then reverberations will exhibit less stereo spread. This is a parameter often overlooked by many who wash a sound in stereo reverb to push it into the background only to question wonder why it doesn't sound 'right' in context with the rest of the mix.

Density

After the early reflections have occurred, the original signal will continue to reflect off different surfaces and culminate to create a reverberant field. On a reverb unit, this field is controlled by the density parameter.

With this parameter, the producer can vary the number of reflections and how fast they should repeat. Through increasing this, the reflections will become denser giving the impression of a larger reverberant field with more complex surface reflections.

Reverb Decay Time

This parameter controls the amount of time the reverb will take to decay away. In large buildings, the reflections generally take longer to decay into silence than in a smaller room. Thus, by increasing the decay time, the producer can effectively increase the size of the 'room'.

The amount of time it takes for reverb to decay to 60 dB below the level of the direct sound is termed the RT60, and therefore, some reverb units may label this parameter as the RT60.

Nevertheless, this parameter must be used with caution since large reverb tails can quickly evolve into a complex mush of frequencies. For example, if a large decay time is employed on closely spaced notes, the subsequent reflections from previous notes may be still decaying when the next note starts. If this pattern is subsequently repeated, it will be subjected to more and more reflections, smothering transients and creating a mix with little intelligibility.

This does not mean large settings should be avoided though, and a popular trick amongst dance musicians is to swamp a timbre in reverb with a large decay and follow this reverb with a noise gate. Using a high threshold on the gate, as soon as the timbre stops, the gate activates preventing the tail from occurring and washing over the following transients. The perception of heavy reverb that is missing its tail can be an ear catching effect if employed infrequently.

HF and LF Damping

The further reflections have to travel the less high-frequency content they will exhibit due to absorption through air. In addition, soft furnishings will absorb higher frequencies; thus by reducing the high-frequency content (and reducing the decay time), the producer can create the impression that the sound source is occurring in a small enclosed area with soft furnishings.

Alternatively, by increasing the decay time and removing smaller amounts of the high-frequency content, the sound source can be made to appear further away. Moreover, by increasing the lower frequency damping, it is possible to emulate a large open space. For instance, singing in a large cavernous area would create a low-end rumble with the reflections with less high-frequency energy.

The main strategy to employ reverb in a realistic manner is to envisage a room and its furnishings and then proceed to replicate this with the parameters offered by the unit. It is difficult, if not impossible, to accurately recreate the natural response of a room precisely, but a ball park reconstruction is often more than suitable. However, not all instruments should be treated to the same amount or style of reverb, and careful consideration must also be given to any number of factors from sound design to acoustic positioning.

FIGURE 10.1
The Lexicon PCM Reverb – the most popular reverb used by producers

For example, with electronic dance music, the kick drum should remain as the forefront of a mix, but the producer may want some house style strings towards the rear. Although this movement could be accomplished with a reduction in volume, it can result in the strings disappearing into the mix. In this situation, slightly heavier reverb applied to the strings could fool the listeners to perceive that the strings are further away than the kick drum due to the stronger reverberation field surrounding them.

Similarly, the reverberant field resulting from any reverb unit should not be considered the be all and end all, and further effects or processor following the unit can produce interesting effects. For example, it is not unusual to follow a reverb with an EQ unit, permitting the producer to further sculpt a reverb tail with EQ. Or reverb may be applied to a channel and followed directly after with a noise gate that has a keyed input from a secondary channel with rhythmic elements. This would produce a rhythmical reverb. It's the experimentation and discovery of new techniques that remain central to any great electronic music producer.

DIGITAL DELAY

Alongside reverb, digital delay is one of the most important and influential effects in the production of electronic dance music, and it will find its way onto almost every single channel in an audio workstation in one form or another.

The simplest delay units permit the producer to delay the incoming audio signal by a predetermined time that is commonly referred to in milliseconds or note values. The amount of delays that are generated are termed the feedback, thereby increasing the feedback setting, it is possible to produce a multitude of repeats from a single sound.

The effect operates by dividing an incoming signal into two channels and storing one of these channels for a short period of time before sending it to the output or back through to the input. This storage and repeat cycle creates a number of discrete echoes with each subsequent repeat lower in amplitude

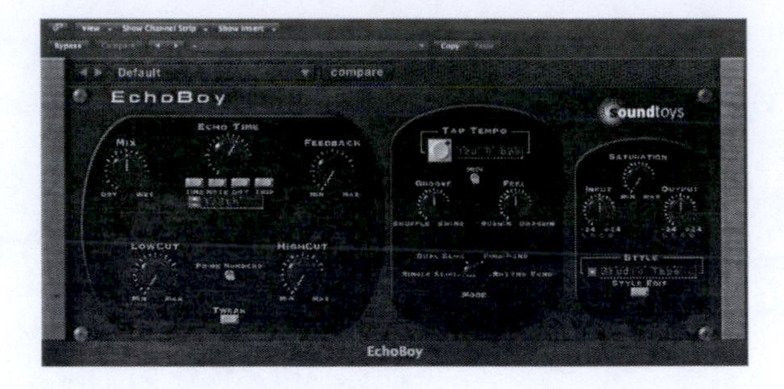

FIGURE 10.2
The ultimate in digital delays – Sound Toys Echoboy

than the previous. These echoes are often synchronized to the project tempo matching both the resolution and grid of the current project and thus creating rhythmical effects.

Whilst all delay units operate under this premise, many units offer far more parameters than adjustment of delay times and permit the producer to delay left and right channels individually or pan them to the left and right of the stereo image. They may also permit pitch shifting of subsequent delays, employ filters to adjust the harmonic content of the delays, apply distortion, permit groove adjustments or employ an LFO to modulate any number of effects that can act upon the delays.

The uses for all of these are literally limitless within the production of electronic music. It's certainly not unusual to employ delay to open and closed hi-hats to produce different rhythms or apply it to leads and basses to increase groove. It's also employed to generate its own rhythmical effects, place sounds in large spaces, enhance the stereo image or placement of instruments and thicken timbres.

The latter technique is often termed granular delay and consists of setting delays to occur at less than 30 ms since this will result in the Haas effect (the psychoacoustic theory that states any delay that occurs within 40 ms or less after the direct signal will not be perceived as a distinct event) producing a timbre that appears thicker and wider. It is often employed on leads or basses to produce powerful driving grooves.

CHORUS

Chorus effects attempt to emulate the sound of two or more of the same instruments playing the same parts simultaneously. Since no two instrumentalists will play exactly in time with one another, the result is a series of phase cancellations that thicken the sound. This is analogous to two synthesizer waveforms slightly detuned and played simultaneously together; there will be a series of phase cancellations as the two frequencies move in and out of phase with one another.

A chorus unit achieves this effect by dividing the incoming signal into two channels and running one of these channels through an LFO modulated delay line before mixing it back in with the original signal at the output. By employing a very short delay of 40 ms or less, the two channels blend into one at the output rather than creating a series of distinct echoes.

FIGURE 10.3
A typical chorus effect

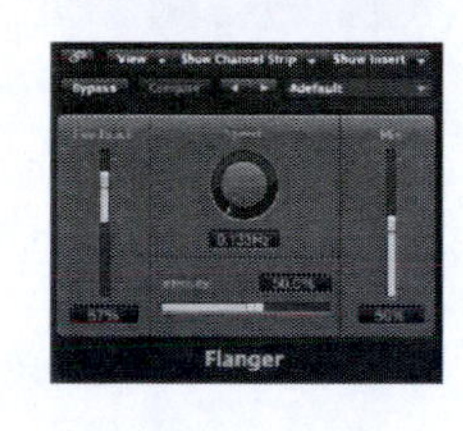

FIGURE 10.4
A flanger effect

To provide control over the chorus, the producer is typically offered control over three main parameters: intensity, rate and mix. The intensity permits adjustment of the modulation amount, as this is increased the intensity of the effect is increased. The rate is often used to define the frequency or speed of the LFO employed in the chorus unit, whilst the mix allows the producer to determine the balance between dry signal and chorus effect. Some of the more capable chorus units may also permit the user to select from a number of different LFO waveforms, although generally, a sine wave will produce the most natural sounding results.

It should be noted that because the modulation rate stays at a constant depth, rate and the LFO waveform remains unchanging, it doesn't produce truly 'authentic' results and therefore shouldn't be relied upon to reliably double vocal or instrumental performances. Nonetheless, it has become a useful effect in its own right and is often employed to make oscillators and timbres appear thicker, wider and more substantial.

PHASERS AND FLANGERS

Phasers and flangers are similar effects with some subtle differences in how they are created but work on a comparable principle to the chorus effect.

Originally, flanging was introduced in the 1960s when engineers used to record and playback studio performances on two tape machines simultaneously so that in the event when one became damaged, they could quickly switch to the second. However, it was discovered that if the engineer placed some pressure onto the 'flange' (the metal edge of the tape reel) for a short moment, it resulted in a slowing down one of the tapes and thus moving it out of sync resulting in the flanger effect. This was picked up and used on the 1967 song *Itchycoo Park* by *The Small Faces*, and after this became a hit song, everyone wanted the same effect.

Phasers are digital effects that work by dividing the incoming signal into two copies and shifting the phase of one before recombining it with the original signal again. This produces an effect that is similar in some respects to flanging but not as powerful and therefore tends to be a less popular effect.

With plug-in effects, both of these effects are produced by mixing the original incoming signal with a delayed version whilst also feeding some of the output back into the input. The only difference between the two is that flangers use a

FIGURE 10.5
A phaser effect

time delay circuit to produce the effect, while a phaser uses a phase shift circuit instead.

Nonetheless, both use an LFO to modulate either the phase shifting of the phaser or the time delay of the flanger. This creates a series of phase cancellations since the original and delayed signals are out of phase with one another. The resulting effect is that phasers produce a series of notches in the audio file that are harmonically related (since they are related to the phase of the original audio signal), whilst flangers have a constantly different frequency because they use a time delay circuit. Consequently both flangers and phasers share the same parameters.

Both units will feature a rate parameter to control the speed of the LFO effect along with a feedback control to configure the depth of the LFO modulation. Notably, some phasers will only use a sine wave as a modulation source, but most flangers permit the producer to change both the waveform and the number of delays used to process the original signal. Of the two, flanging has become a staple in the production of electronic dance music.

DISTORTION

Distortion effects are employed to replicate the effects of analogue and digital distortion. Typically, musicians will be inclined to employ analogue distortion units since these will replicate the sound of overdriven valves and valve gear since this produces a warm sound generated by second order harmonic distortion. Alternatively they will opt for the same style of distortion experienced from overdriving guitar amplifiers.

Digital distortion usually appears in the form of bit reduction, allowing the producer to give the impression that the number bits of an audio channel have been reduced significantly and thus producing a digital glitch style of distortion.

FIGURE 10.6
Sound toys decapitator distortion unit

In the early distortion units, the producer commonly only offer two parameters, one to control the amount of distortion introduced and the second to adjust the tone of the distortion (this was typically a filter). However, due to the rise and subsequent importance of distortion effects in the production of EDM, these effects have grown beyond the normal and can now offer everything from multiple bands of distortion through to emulations of distortion from overdriving classic analogue hardware.

FILTERS

The filter is one of the most powerful elements of any synthesizer and the same holds true for any modern studio producing electronic dance music. To produce the filter sweep that is present in just about every dance music record release in the past 20 years, the producer needs access to a filter plug-in that can be inserted into any (or all) audio tracks.

The action of a filter was discussed in great detail in Chapter 6 and any plug-in filter operates in the exact same manner, commonly offering switchable low-pass, high-pass, band-pass and notch filters along with other effects such as resonance and sometimes distortion.

FIGURE 10.7 Sonalksis low-pass filter

GLITCHERS

'Glitching' is a technique that has been around for many years but has only recently began to draw the attention of both producers and audiences with the introduction of genres such as *Complextro* and *Dubstep*. Fundamentally, this is a technique employed by producers whereby they slice audio file into any number of smaller events and then randomly offset the events from one another and apply different effects to each to produce a glitch style effect.

Originally this was an incredibly time-consuming technique but recently developers have released plug-ins aimed at simplifying the process. Of particular note, *Sugar-Bytes Effectrix* and *Glitch* are both multi-effect sequencers permitting the producer to divide incoming audio into individual steps of various lengths and then apply a multitude of effects to each. The effects included are dependent on the plug-in but typically they involve distortion, modulation, shuffle, reverse, crusher, gates, delays, vinyl and tape stop effects.

There are of course many more effects available to the producer than discussed in this chapter. Everything from auto-panning, ring modulations and pitch shifters are now available as effects and developers are releasing new effects plug-ins on an almost daily basis.

However, many of these effects are based around the principles already discussed in this chapter and commonly consist of a number of effects

FIGURE 10.8
Sugar-Bytes Effectrix glitch processor

chained together in various ways to produce different results. And this is a fundamental production ethic within the production of electronic dance music since the arrangement and order of effects will provide different results. For example, a delay line employed after a reverb will produce very different results to a reverb placed after a delay line. We will discuss this in more detail in later chapters.

CHAPTER 11
Mixing Structure

'Technology presumes there's just one right way to do things and there never is'.

Robert M. Pirsig

The purpose of any mixing desk whether software or hardware is to take any number of audio and/or virtual instrument channels along with any signals from external equipment and combine them together to produce a single stereo file at the output. However as simple as this basic premise is, with the introduction of audio workstations and software mixers with limitless routing and bus options, the once humble mixer has developed into a sound design tool in its own right.

Whilst the mixer is still fundamentally employed to mix together any number of tracks created within the workstation, the option to employ and organize any number of effects and processor in series of chains and the opportunity to access these chains through multiple bus configurations or subgroups has opened up a complete new area of sound mangling options that often play a central role in the creation of some genres. Consequently this chapter will discuss the basic theory and application behind a mixing desks internal bus structure along with processing and effects chains.

In most digital workstations, every time a channel is added to the working project, a mixer channel will automatically be created. Depending on the channel, this can be stereo, mono, virtual instrument, MIDI or buss. With the exception of the MIDI channel, all of these channels are designed to work with an audio signal and therefore provide a number of 'buses' with which to route that particular channels signal.

A bus is an internal signal path within the mixer that any number of audio channel signals can be routed through. For example, since main duty of mixers is to combine all of the individual channels into a single stereo file, every channel will commonly enter the mixers main stereo bus, and this particular signal

FIGURE 11.1
The stereo mix bus

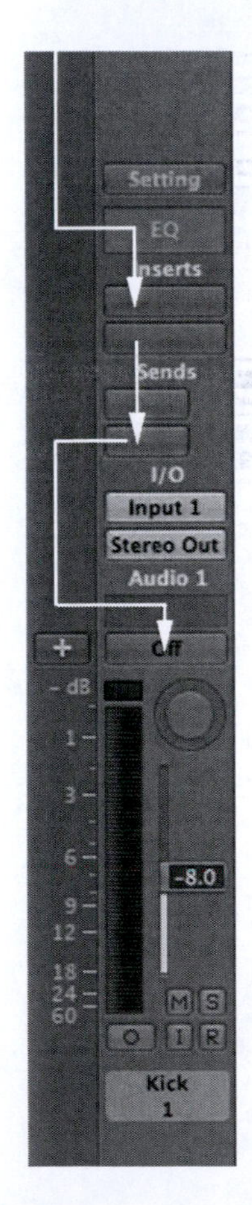

FIGURE 11.2
An audio track channel in Logic Pro

path is commonly patched through the master fader and then directly onto the audio interfaces physical outputs.

All audio workstation mixers feature a vast number of buses to permit the user to route multiple audio signal channels into different configurations such as groups, master groups and master sends. To understand this operation in more detail, we first need to examine the layout and signal path of a typical audio channel in a workstation.

Figure 11.2 shows the mix channel from Apple's Logic Pro. Here, the audio signal from a channels region enters at the top of the mixing strip and runs through the channel inserts, then into a series of auxiliary sends, onto the channels main fader and controls before finally reaching the stereo mix bus to be combined with all the other channels in the project. Although different audio workstations will display this layout differently, the path a signal takes will nevertheless remain the same.

An audio signal leaving the channels region will first enter into the inserts of the mixing desk. Both effects and processors can be inserted into a mix channels insert but traditionally inserts are reserved for processors rather than effects. This is because processors are often required to process all of the signal rather than an effect that often only requires a small percentage to be mixed with the audio signal.

By inserting a processor such as EQ into the mix channels insert, the signal will travel directly through the processor on route to the stereo mix buss. This means that the signal will travel through one insert processor and then into the next and the next and so forth in a completely consecutive manner. Therefore, if an EQ were placed into the first insert and a compressor into the second, it is the *output* signal from the EQ unit that would *enter* the compressor. Therefore, if the producer were to adjust the EQ, it would modify the signal entering the compressor and this could alter the compressors reaction. Since this could result in a different set of frequencies triggering the compressor, the producer would have to re-program the compressor every time an EQ adjustment was made.

Alternatively, if a compressor were positioned *before* the EQ, the compressor could be used to control the dynamics of the signal entering the mix channel but any EQ adjustments would not be subjected to dynamic control and therefore any boosts may result in clipping the mixer. It's because of this consecutive processing signal manner that careful consideration must be given in regards to the order of processors of effects in any chain.

In many situations, the decision on the order of which to place processors must be based around both creativity and thoughtful application. For example, placing EQ before the compressor, it would be possible to create a frequency-selective compressor. In this arrangement, the loudest frequencies control the compressor's action and therefore boosting quieter frequencies so that they instead breach the compressor's threshold it's possible to change the dynamic action of a channel.

In a typical application to maintain a more natural sound, a noise gate is commonly the first processor inserted into a signal chain. This is because they can be used to remove any extraneous noise from the audio signal before it reaches any further processing since these may only accentuate the unwanted noise.

For example, if the compressor were placed before the gate, it would reduce the dynamic ratio between noise and the signal you want to keep and the reduced dynamic range would also make it difficult to program the gate. Since a gate effectively monitors the dynamic range by removing artefacts below a certain gain, if the dynamics are reduced, first it would be more problematic to program the gate to respond accurately. These processors would then be followed with an EQ, permitting the producer to sculpt the signal before finally passing through into any effects applied on auxiliary buses.

Another possible order for processing and effects could be a gate followed by a compressor, EQ, effects and then further effects. As before, the beginning of this signal chain will produce the most natural results, but the order of the effects afterwards will determine the signals tonal modification. For instance, if the producer

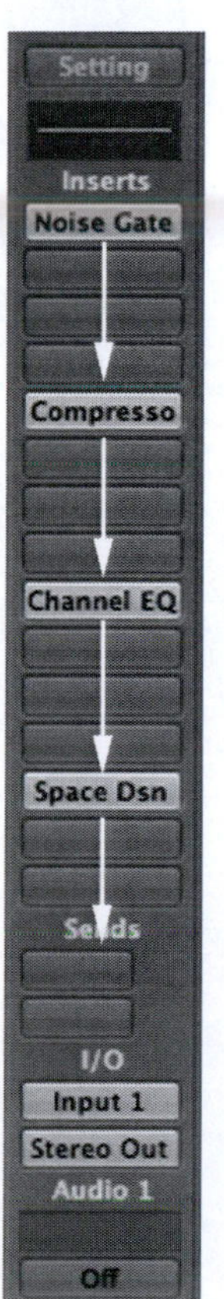

FIGURE 11.3
An effects chain (Noise Gate > Compressor > EQ > Effects)

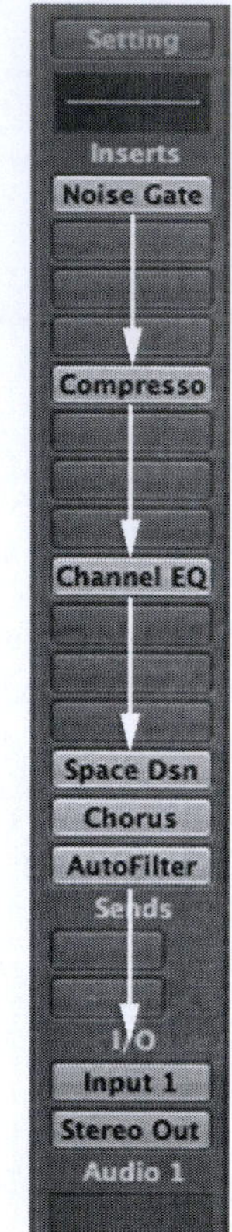

FIGURE 11.4
An effects chain (Noise Gate > Compressor > EQ > Effects > Effects)

were to place reverb before distortion, the reverb tails would be treated to distortion, but if it were placed afterwards, the effect would not be as obvious.

Similarly, if delay were placed after distortion, the subsequent delays would be of the distorted signal, while if the delay came before the distortion, the distortion would be applied to the delays producing a very different result. If flanger were then added to this set-up, the resulting sound would become more complex since this is essentially a modulated comb filter. By placing this after distortion, the flanger would comb filter the distorted signal producing a spaced out phased effect, yet if the flanger were placed before distortion, the effect would vary the intensity of the distortion.

What's more, if the flanger were placed after distortion but before a reverb, the flange effect would contain some distorted frequencies, but the reverbs tail would wash over the flanger diluting the effect but producing a reverb that modulates as if it were controlled with an LFO.

An alternative set-up to this approach would be to employ inserted effects after the gate and compressor but *before* the EQ. This approach would again maintain a natural signal but by placing the EQ after any inserted effects, the EQ could be used to sculpt the tonal qualities produced by the effects. For example, if the effect following compression were distortion, the compressor would level out the dynamics creating a more noticeable distortion effect on the note decays. In addition, since distortion will introduce more harmonics into the signal, some of which can be unpleasant, they can be removed with the EQ following the distortion.

Taking the above approach a step further, EQ could be placed after each effect on the insert bus and then finally followed with another compressor. In this arrangement, the EQ could be employed to 'clean up' any annoying frequencies that are a result of an effect before it enters the next effect in the chain. In addition by placing a compressor at the end of the chain, any peaking frequencies introduced by effects or EQ can be tamed with the compressor.

Different results can be attained by placing a gate after the effects, producing the chain of gate then compression followed by EQ, effects and finally another gate. In this configuration, by placing the gate *after* the effects, the resulting signal could be treated to a gate effect. Although there is a long list of possible uses for this, the most common technique is to apply a large reverb to a drum kick or trance/techno/minimal/Dubstep/house lead and then use the following gate to remove the reverbs tail. This has the effect of thickening out the signal without the reverb washing over the transients and blurring the image.

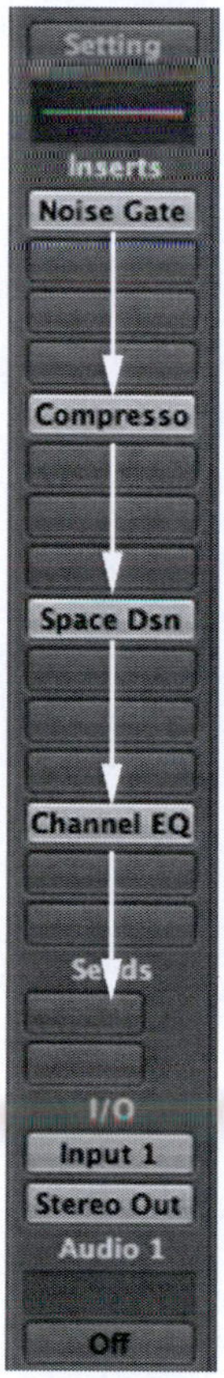

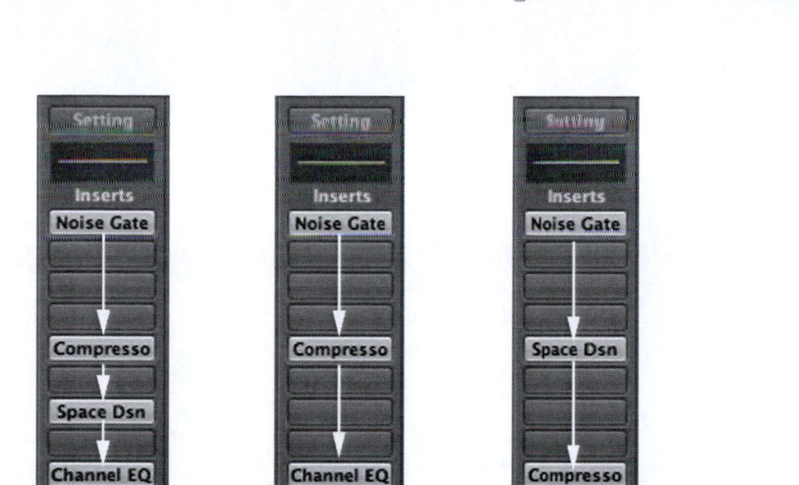

FIGURE 11.5
An effects chain (Noise Gate > Compressor > Effects > EQ)

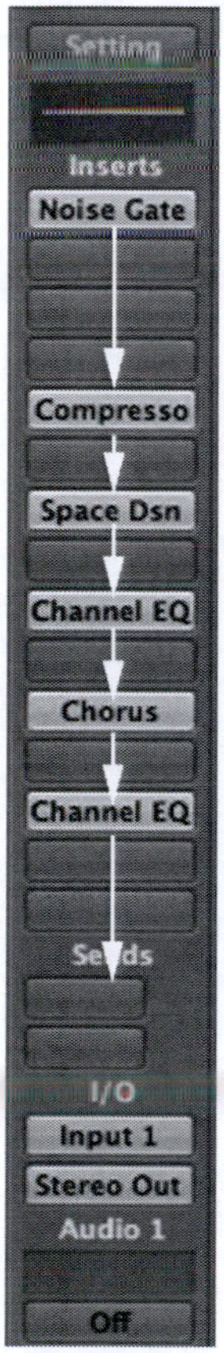

FIGURE 11.6
An effects chain (Noise Gate > Compressor > Effects > EQ > Effects > EQ)

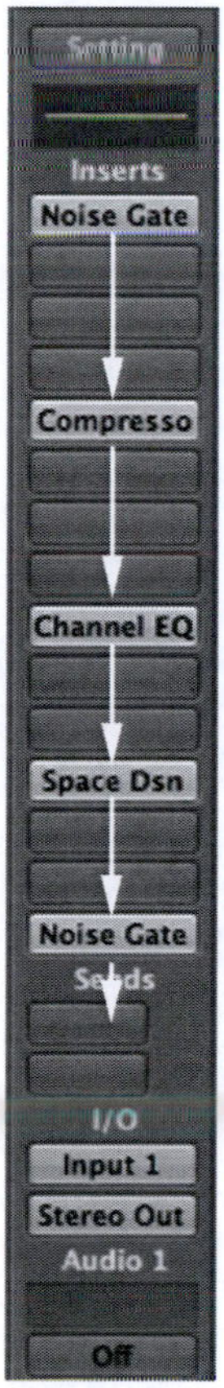

FIGURE 11.7
An effects chain (Noise Gate > Compressor > EQ > Effects > Noise Gate)

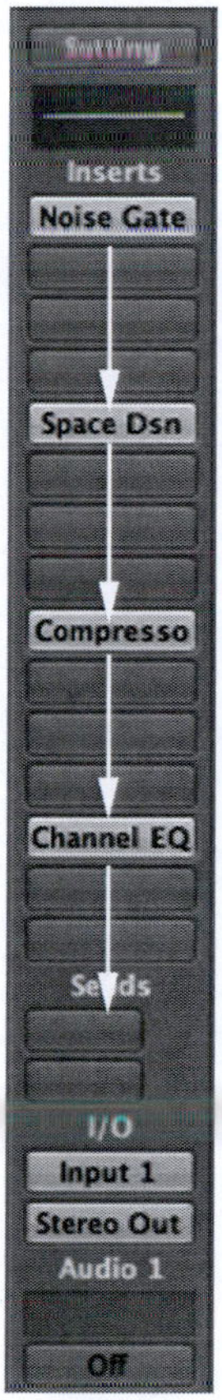

FIGURE 11.8
An effects chain (Noise Gate > Effects > Compressor > EQ)

Although it is generally accepted that the compressor should come before effects, placing it directly after can be useful. For example, if a filter effect with a high resonance has been employed, this is followed by chorus or flanger, it can result in clipping of the signal. By employing a compressor after these effects, the compressor will control the dynamics before they're shaped tonally with the EQ. Placing compression after distortion will have little effect, however, since distortion effects naturally reduce the dynamic range and thus can be used even as a very primitive compressor!

Whilst many of these examples have featured effects inserted into the audio channel, in a more traditional application, they should be accessed through an auxiliary bus rather than an insert. This is because although it is perfectly feasible to route reverb, chorus, flangers, phasers and so forth into a mix as

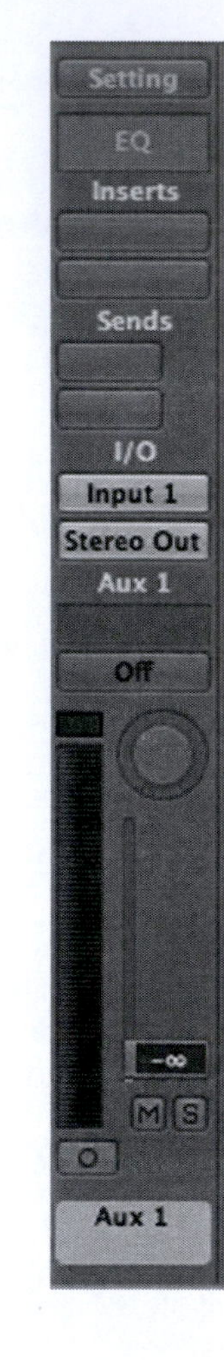

FIGURE 11.9
An aux channel in Logic Pro

inserts, chaining effects in this way results in the audio signal being processed completely by the effect.

Although all effects will feature a wet/dry parameter to control the relationship between a dry signal and the effect, they all employ a single parameter to accomplish this balance. This means that as the wet level of the effect is increased, the dry level will decrease proportionally. Similarly, increasing the dry level will proportionally decrease the effects of wet level.

While this may not initially appear problematic, if the producer opted to thicken out a lead with reverb, as the wetness factor is increased, the dry will decrease proportionally which results in a wet reverb sound. This can cause the transients to be washed over by the subsequent reverb tail from any preceding notes reducing the impact of the sound.

Consequently, a producer will often opt to employ mix sends to access effects. Here, a percentage of a channels audio signal can be routed into an auxiliary bus and onto any number of auxiliary channels in the mixer. Typically, in many software mixers, the moment a producer introduces send bus onto a channel, the workstation will automatically create the subsequent aux channel.

As shown in Figure 11.9, aux channels are fundamentally the same as a standard audio channel featuring inserts, further sends and fader parameters. The only difference is that they do not receive a signal from a channel event in the arrange window and can only receive signals that are 'sent' to them via the aux bus.

Using a send bus, with the send parameter on the original channel strip set at zero, the audio signal will ignore the send bus and move directly onto the gain fader. However, by gradually increasing the send parameter, it is possible to control how much of the channels signal is split and routed to the aux buss. The more that the send parameter increased, the more of the audio signal will be directed to the aux buss.

The benefit of this approach is twofold. First, a channels signal can be sent to an effects bus as well as continuing onto its own channels fader. This permits the producer to maintain a dry level through the standard channel and also introduce some of the desired effect through the send bus. By then adjusting the level of the signals channel *and* the aux bus channels fader, it is possible to create a more crafted balance between the two.

What's more, every channel of the mix can have its signal sent down the *same* auxiliary bus, and therefore, every channel can have the signal sent to the same auxiliary buss. A typical situation here is to employ reverb on the aux buss inserts so that every channel can access the same style and settings of reverb to

FIGURE 11.10
Signal bus path

create a more cohesive sounding mix. In this configuration, the wet/dry parameter should be configured to 100% wet since the aux send channel should apply the effect completely to any audio signal sent to it.

Typically, sends can be configured to operate in either pre- or post-channel fader. If pre-fader is chosen, the signal is split and delivered to the aux bus *before* the channels gain fader. This means that if the producer were to reduce the fader, it would not reduce the gain of the signal being bussed to the effect, and therefore, if the channels fader were to be adjusted, the aux channel would also need adjusting to balance the level between the two again.

This permits the producer to reduce the volume of the dry channels signal whilst leaving the aux bus at a constant volume but typically post-fader is a more useful approach. In this mode, reducing the channels fader will also systematically reduce the auxiliary send too, saving the need to continually adjust the aux buss send level whenever the gain fader of the channel is adjusted.

Similar to the consecutive arrangement of inserts, aux buses feature effects stacked in the same consecutive manner. Since an aux bus is fundamentally a mixer channel that can only be accessed via sends, effects placed one after the other behave in the same manner as if inserted into a standard mixer channel. This means that the output of a previous effect will flow into the input of a following effect and so forth throughout the signal chain.

Of course, there are no rules to suggest to that effects should only be placed on an aux track and processors only employed as inserts and creative adaptation have resulted in some classic genre defining effects. For example, a popular technique in dance music is parallel compression whereby a compressor is placed on an aux bus and the channels signal is sent to the compressor via the auxiliary buss.

The uncompressed source track is mixed in with the compressed signal on the auxiliary bus, and this produces a fatter sound since the compressor reduces the dynamics yet the original sound retains the peaks. Sometimes termed 'New York Compression', this form of parallel compression can be employed on any instrument but within dance music, it is commonly employed on the rhythmical elements (drum loop).

CHANNEL FADER AND SUBGROUPS

Once the signal passes the mixers send bus, it is routed into the main gain fader for the channel. This gain control is commonly always in the form of a fader, but it is important to note that these do not amplify the channels signal but instead are configured to permit more or less of the channels signal through to the stereo mix buss. This is why the faders are often configured to 0 dB near to the very top of the fader movement.

Here, alongside adjusting the gain of the channel, it is also possible to mute or solo the channel, adjust the panning of sounds across the left and right stereo spectrum and route signals via subgroups rather than through the stereo mix bus.

So far we have considered that every mix channel is routed directly into the main stereo mix bus whereby all sounds culminate at the stereo master fader at the end of the mixer and to the audio interfaces outputs. However, it is possible to route any number of channels into a series of subgroups of which the signals from the subgroups then enter into the stereo mix bus.

A subgroup permits the producer to send any number of channel outputs into a newly created channel allowing them all to be controlled universally with just one fader. A typical application for a subgroup is to route the kick, snare, claps, hi-hats and cymbal channels outputs all to the same subgroup. By adjusting each channels individual gain – mixing down the drum elements – once they are routed to a single group track, the gain of the entire drum submix can be adjusted to suit the rest of the mix.

Since a subgroup channel shares all the functions of a standard audio channel, it is also possible to apply any number of insert effects or access sends for the entire drum 'sub mix'. For example, a compressor applied on the drum sub mix could be used to universally control the dynamics all of the drum elements saving on processing power within the host computer since there may be no requirement to place a compressor on each individual drum channel. Or alternatively, a compressor may be placed on every drum instruments and on the sub mix too to help 'gel' the instruments together in the sub mix.

A subgroup channels output can be routed to the main stereo bus of the mixer or can be further routed into a secondary subgroup. In fact, with most audio workstations, there is no limit to the buses, and it is possible to route in any number of configurations including a subgroup into subgroup into subgroup into subgroup ad infinitum. Similarly, it is equally possible to send to an auxiliary bus and have that bus send to another auxiliary bus and then onto another bus and so forth.

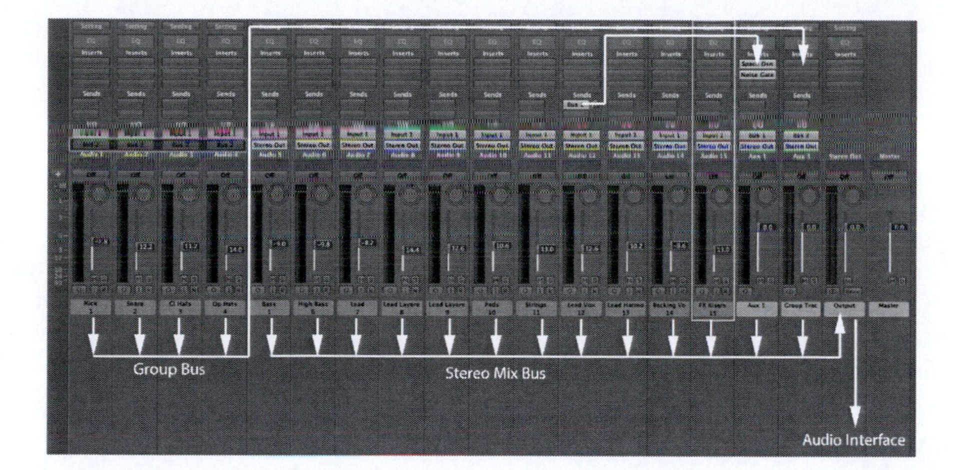

FIGURE 11.11
Sub mixes and groups

AUTOMATION

A final yet equally vital aspect of software mixing for todays dance music appears in the form of mix automation. Using this, every parameter on an effect, processor or on the mixing desk can be recorded and automated to move in real time during playback of the track. This style of automation has become a central element to the sound of most dance tracks permitting the producer to program everything from slow filter sweeps to sudden changes in effects.

Originally mix automation was carried out by a number of engineers sat by a mixing desks faders and any connected effects, adjusting the relevant parameters when they received the nod from the producer. As a result, any parameter changes had to be performed perfectly in one pass since the outputs of the desk were connected directly into a recording device such as DAT. In fact, this approach was the only option available when dance music first began to develop and we used to have to sit by the filter and time our sweeps manually to fit with the music.

This type of approach is not only defunct today but impossible to accomplish since the development of dance music has embraced the latest forms of mixer automation so much so that it isn't unusual for genres such as Progressive House and Techno to have over 20 parameters changing simultaneously and for the mixer and effects units to jump to a whole new range of settings for different areas of the same track.

Mix automation is accomplished in almost every workstation through the use of read, write, touch and latch buttons that are commonly located on the mixers channel but can sometimes be located in the sequencers inspector for that particular channel.

The creation of mix automation is commonly accomplished in one of two ways, it can either be drawn in directly onto the channel with a pencil tool or more commonly, the channel is set to 'Write' mode and playback initiated. During playback, any parameters that are moved on that particular track, regardless of what they are, will be recorded live into the sequencer.

This is the more common approach for many since after recording the automation of one part the producer can jump in an edit this automation with a pencil, and it saves having to search through a plethora of possible parameters in an attempt to find the parameter they wish to edit. Once automation has been recorded, this process can be repeated multiple times to record an unlimited number of automation parameters. After automation has been recorded and edited the mixer is set into 'read' mode. On playback, this automates all of the parameters on the current track live, using the previously recorded automation data.

Although 'write' and 'read' are the two most commonly used automation parameters, both latch and touch can be employed to further edit any recorded automation data. If 'touch' mode is activated during playback, all currently recorded automated data will be read back as in read mode.

However, if the producer adjusts any parameter that is previously automated (such as a channel fader or effects parameter), the current movements will replace any automation data until the parameter is released again whereby the automation will continue along its originally recorded values. This makes touch mode useful for small live edits to previously recorded automation data. Latch mode is very similar to touch mode but here, after releasing the parameter, the current parameter value will continue to overwrite any previously recorded automation data.

FIGURE 11.12
Mix automation

CHAPTER 12

Audio Interfaces and Connectivity

'Well, it's not really hi-fi, and not really lo-fi. It's just kind of "fi"'.

Aimee Mann

An audio interface provides the all-important link between a DAW's internal audio engine and the physical sound emitting from speaker monitors or headphones. Consequently, it proves to be one of the most essential links in the music production chain but equally also happens to be one of the most overlooked and underestimated.

With most laptops, PC and Mac computers fitted with their own internal soundcard, it's easy for a producer to believe if they're not recording instruments they can rely on connecting a set of headphones or monitor speakers direct into their computers jack socket. This, alongside the media's presentation of professional artists working on their next 'hit' through headphones in hotel rooms or aeroplanes further exasperates the situation that unless you're actually planning on recording instruments, all you need is a computer and there is no requirement for a professional audio interface. This couldn't be further from the truth. Even though many professional artists do work on headphones connected to a laptop, it's only for the initial idea generation and afterwards it's taken to a studio and played through a more reliable interface.

The fact is converting the digital numbers that occur within a DAW into analogue sound you can physically hear is a process that relies heavily on the quality of the converters and the converters used in 'factory–fitted' soundcards, whilst good for general duties, are not suited towards professional audio production. As an analogy, mixing and working within a DAW whilst relying on a factory-fitted soundcard is like working with a soft blanket covering the monitor speakers. You're not going to receive a truthful representation of what's occurring inside the DAW and all of your decisions are going to be influenced by what you can hear. To understand why this is, we can consider the process that occurs when recording real-world audio into a digital device since it's much the same whether converting digital to analogue (D to A conversion; *DAC*) or analogue to digital (A to D conversion; *ADC*).

With any digital recording system, sound has to be converted from an analogue signal to a digital format that the DAW can understand. An audio interface accomplishes this by measuring the incoming signal at a number of specific intervals and then converting these measurements into a series of numbers that can digitally represent the audio. The numbers created by the ADC are based on the two most important factors of audio: time and magnitude.

Figure 12.1 shows an analogy of this ADC process in operation. On the time axis, the waveform is sampled a specific number of times every second. The total number of 'samples' that are taken every second is termed the 'sample rate'. On this basis, the more samples that are taken every second the more accurately the audio will be represented digitally.

Of particular note, the sample rate must *always* be more than double the frequency of the audio being converted in order to avoid introducing errors. Bell Laboratories engineer, Harry Nyquist, first discovered this requirement, and it's therefore known as the Nyquist theorem. This states that to recreate any waveform accurately in digital form at least two different points of a waveform's cycle must be sampled. For example, if the converters were to sample a 400-Hz sinusoidal wave, they must take a minimum of two measurements; otherwise, the waveform could be mistaken for being at a different frequency.

As shown in Figure 12.2, if two or more points of a waveform are not sampled then the audio will not be measured accurately and the converters can be fooled into thinking the frequency of the waveform is lower than it actually is. This behaviour results in an effect termed *aliasing* and is a result of the converters being unable to accurately represent the original audio. Audibly, aliasing manifests itself as a series of spurious spikes in the audio and in severe situations can result in complete silence.

Since the highest range of the human ear is a frequency of approximately 20 kHz, it stands to reason that the sample rate should be just over double this and this is the reason that 44.1 KHz is a standard in many audio workstations.

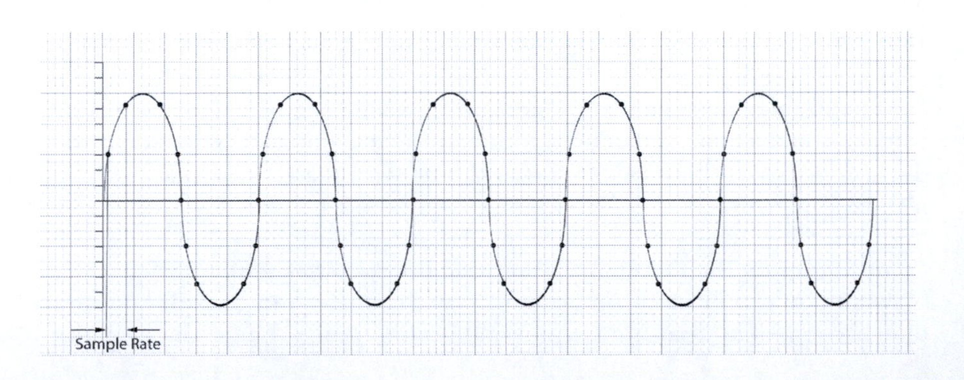

FIGURE 12.1
The sample rate

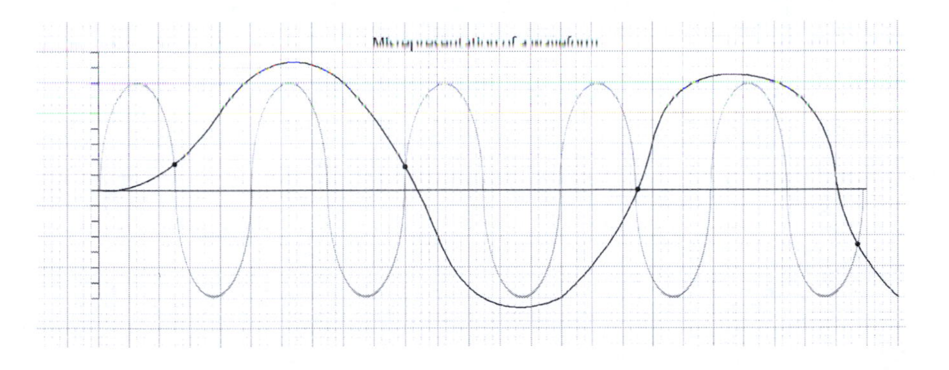

FIGURE 12.2 The misrepresentation of a waveform due to a low sample rate

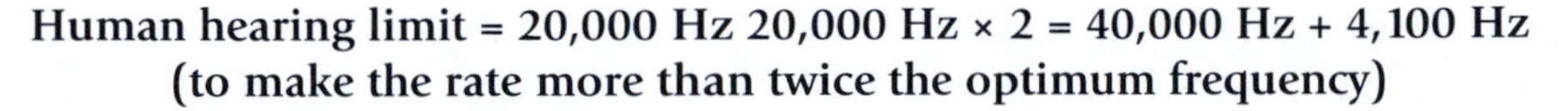

Human hearing limit = 20,000 Hz 20,000 Hz × 2 = 40,000 Hz + 4,100 Hz (to make the rate more than twice the optimum frequency)

Although this frequency response has become the de facto standard in all workstations, many producers today will operate their sequencers at a much higher sampling frequency of 88,200 Hz or 96,000 Hz. Although these sampling rates are far beyond the frequency response of human hearing, these higher sampling rates can reduce unwanted side effects such as frequency cramping.

Frequency cramping often occurs when a processor or effect is employed in an audio workstation at frequencies that are close to half the sampling rate. For example, if the current working project were set at 44.1 kHz and the engineer were to boost a wide range of frequencies at 18 kHz, the boost occurring on the higher frequency side of 18 kHz could extend well beyond 22 kHz. This is more than half the project's sampling rate and, in accordance with the Nyquist theorem, this would result in aliasing and frequency cramping, whereby the boost reduces sharply resulting in an uneven balance.

Figure 12.3 shows this effect in action and sonically it appears harsh, reducing both presence and spatial resolution of the sound. What's more, this effect can become much more noticeable when using processors and effects that emulate analogue characteristics such as analogue modelled EQ's or distortion units since these often introduce yet further harmonics either side of the frequency range being processed.

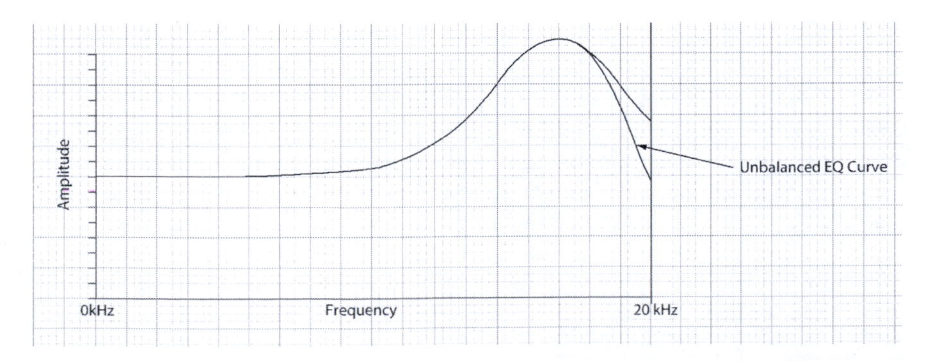

FIGURE 12.3 Frequency cramping

BIT RATES

Up until now, we have only considered one measurement of audio, that of time or frequency. However to accurately represent a waveform, a converter must also measure the magnitude or gain and this relates to the total dynamic range.

Dynamic range can be compared with the decibel measurement discussed in Chapter 4 in which it is used to express the ratio between two differing levels. When working with audio equipment on a technical level, the dynamic range refers to the ratio between the loudest possible sound and the noise floor generated by a piece of equipment.

Similar to sampling frequency whereby the sample rate determined the precision of the recorded sample, the bit rate determines the magnitude precision and dynamic range of the audio. It accomplishes this by taking periodic measurements of the waveform. As shown in Figure 12.4, bit depth can be related to a vertical axis of a grid and each vertical coordinate on the grid refers to one 'bit' measurement that equates to 6 dB of dynamic range.

The problem with this approach is if the audio signal level changes between the two coordinates of the grid, it can't be represented accurately and the ADC will simply round up to the next grid coordinate. The mathematical rounding up due to a low bit rate will produce an inaccurate representation of the audio signals magnitude and produces an effect known as quantization error. The effect of this is shown in Figure 12.5 and manifests itself as digital noise in a recording. To make matters worse, this noise can sometimes be imperceptible to human hearing, but as processing is applied within the audio workstation, it produces an abrasive sound that lacks stereo imaging and depth.

As discussed earlier, this sample and bit-rate conversion occurs both ways so even if you have no plans to record any instruments, the conversion from digital to analogue can still be heavily influenced by the capabilities and quality of the interface being used.

For example, soundcards that are supplied as standard in many computers use a monolithic chip design and rely on small amplifiers built onto these chips,

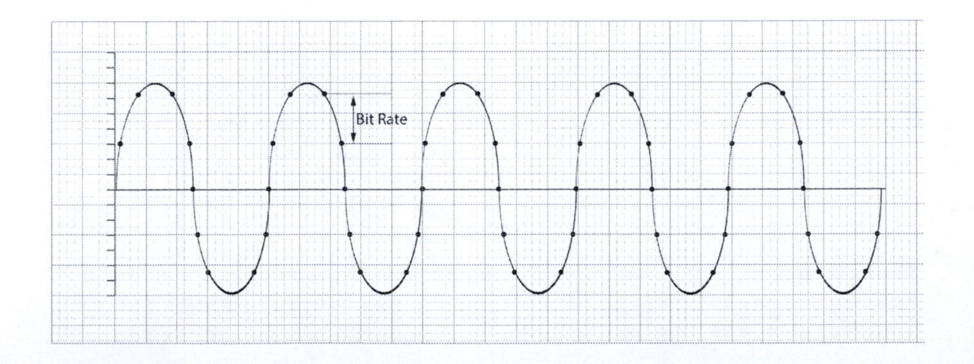

FIGURE 12.4
Bit rate measurements

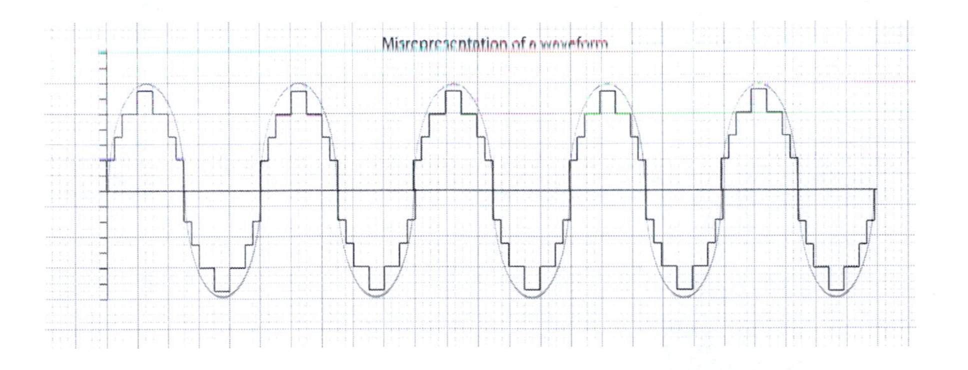

FIGURE 12.5
The effects of bit depth

termed op-amps. The voltage rails within these chipsets and the soundcard are limited to just a couple of volts, and therefore, if driven hard, it can overdrive the rails resulting in signal deficiency at the outputs. The outcome of this is that the bit rate and sample rate are not correctly converted, and there is a removal of lower frequencies and smothering or masking of the higher frequencies.

In addition, with the soundcards being part of the main motherboard, they suffer from electrostatic interference and thermal noise generated by the board, both of which can reduce the total dynamic range, introduce noise and again result in a smothering the higher frequencies of the mix. The end result is a soundcard that reproduces the audio workstations output incorrectly, presenting a mix that lacks clarity and the producer will over compensate for this lack of clarity by applying heavy-handed processing that isn't required. Consequently, if you plan to produce electronic dance music to a professional standard you should look towards purchasing a professional audio interface.

HARDWARE AND CONNECTIONS

Since most professional artists enjoy using a hybrid mix of computer *and* external hardware such as analogue synthesizers, and many choose to work on laptops, dedicated PCI soundcards are no longer as readily available and external audio interfaces have dominated the current market. Here, whilst the quality of the AD and DA convertors should obviously always be the uppermost priority in the decision when choosing an interface, few will offer just a headphone and stereo output. Indeed, most interfaces will offer a variety of connection formats beyond the standard ubiquitous jack sockets, and therefore, an understanding of all the currently available formats available is necessary.

Over the past 30 years, there have been a large number of format and connection protocols developed but of these only a few have remained and been adopted by most manufacturers. These are the analogue balanced and unbalanced connections alongside MIDI, ADAT (Lightpipe), MADI, DSUB, S/PDIF and AES/EBU.

Note that whereas USB, Firewire and Thunderbolt are also connection formats used within the audio industry, they are only employed to connect the audio

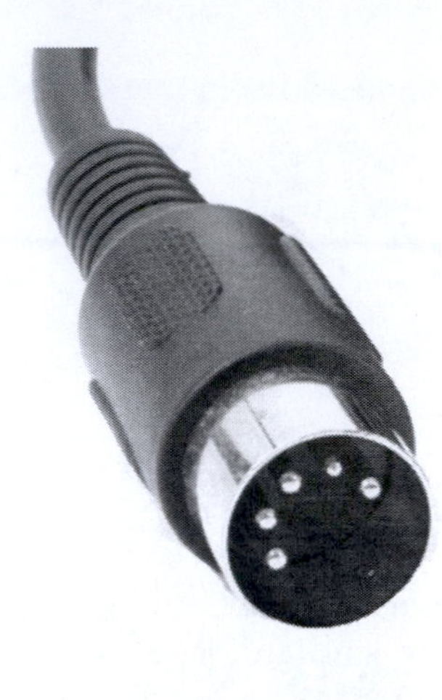

FIGURE 12.6
A standard 'DIN' MIDI connection

interface directly to the computer and the only real connection difficulties here are a result of the drivers supplied with the device. In addition, many manufacturers will no doubt evolve beyond these two formats over the coming years into the MADI format anyway.

When discussing both music and computers, and depending on your age, MIDI is perhaps the first connection protocol that comes to mind. Although MIDI was briefly discussed in a previous chapter on workstations, its origins lie with interconnecting external instruments. It's the most standardized data transfer protocol and has maintained its original specifications for over 30 years and, therefore, features on just about every audio interface available.

Originally introduced in 1983, Musical Instrument Digital Interface consists of a standardized communication protocol that can be transmitted via USB (if the receiving hardware device accepts a USB connection) or a five-pin male DIN plug as shown in Figure 12.6.

As discussed in Chapter 5, MIDI can transmit data from a workstation to any MIDI capable device and like virtual instruments, these data consist of simple commands transmitted by the DAW's piano roll editor informing the connected device when to play a note, how hard the note was struck and for how long. It can, however, also be used to transmit many more complex commands from changing parameters in the receiving device through to dumping its memory into the audio workstation to allow it be recalled at any time its required. Typically, artists will use this form of MIDI 'dump' to save instrument settings or presets when the synthesizer itself doesn't have the capability.

Typically an audio interface will feature a MIDI IN and a MIDI OUT port. The MIDI OUT of the interface would be connected to the MIDI IN of a MIDI compatible synthesizer or sampler using a suitable 5-pin male DIN cable. Similarly, the MIDI OUT of the synthesizer/sampler would connect to the MIDI IN on the computers MIDI interface using another five-pin DIN cable. This arrangement permits a simple two-way communication from workstation to synthesizer and vice versa. Figure 12.7 shows this simple MIDI set-up.

Whilst many producers will often begin with this small style of set-up, over time it will generally increase to include further synthesizers but since most audio interfaces only feature one MIDI IN port and one MIDI OUT port, to send data to multiple devices producers must either utilize THRU ports on the devices, if fitted, or expand the system to a third-party Multi-MIDI interface.

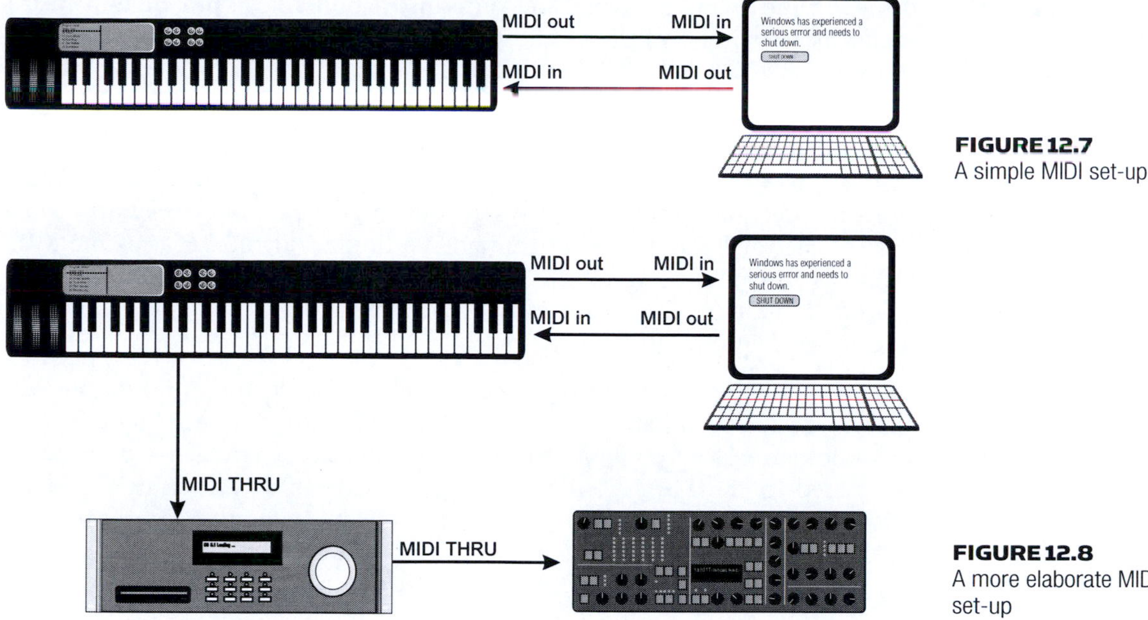

FIGURE 12.7
A simple MIDI set-up

FIGURE 12.8
A more elaborate MIDI set-up

Some devices will feature MIDI THRU ports. Using these, provided the device is properly configured, any information received at the MIDI IN will be repeated at the device's THRU port. In a typical arrangement, this THRU connection would subsequently be connected to the MIDI IN of the next device in the chain. Connecting and linking devices in this manner is termed 'daisy-chaining' and permits much more elaborate set-ups.

This arrangement is far from perfect, however, since MIDI is a serial interface and therefore messages are sent one after the other so take a finite amount of time to travel down the MIDI cable to the synthesizer. While this delay may not be discernable on the first few devices in the chain, if the data from the sequencer is destined for the tenth device, for example, it has to travel through nine devices beforehand and this can result in the information arriving late resulting in a noticeable delay. This delayed response results in *latency*.

Whilst almost all latency occurs in milliseconds, depending on the situation, as little as 20 millseconds can be discernable since the human ear is capable of detecting incredibly subtle rhythmic variations in music and even slight timing inconsistencies can severely change the entire feel of the music. Even an apparently insignificant timing difference of only 10 milliseconds can have a serious effect on the strict rhythms required in some dance music genres and can becomes particularly evident with complex breakbeat loops.

A secondary problem arises when considering the transmission and receipt of MIDI packets to specific devices in the chain. For example, simply setting the DAW to transmit instruction through its MIDI OUT would result in every device in the chain receiving the same instruction since with plug-in instruments also relying on MIDI, almost all DAWs are automatically configured to transmit MIDI over all available channels.

Due to the huge resources available inside most computers many plug-in instruments are single channel. That is, if the producer wishes to use the same instrument again in a sequence, they simply open up another instance of the plug-in. This same principle can occur in some hardware devices but in these devices the total number of instruments is limited to the 16 available MIDI channels. This multi-timbral operation permits an audio workstation to transmit MIDI to, say, a piano instrument set to channel 1 and a bass instrument set to channel 2 and so forth.

The individual channel configuration can also be used to communicate with specific devices within a MIDI chain. By setting the device to ignore channel 9, for example, any signal reaching that device on channel 9 would simply be passed to the MIDI THRU port and into the next device in the chain. Disregarding the problem of latency for the moment, this approach obviously limits the MIDI set-up to 16 instruments, and therefore, a more efficient solution is to employ a multi-MIDI output device.

Multi-MIDI interfaces are external hardware interfaces connected to the sequencer via USB and offer a number of individual MIDI IN and OUT ports. This can range from as little as two MIDI OUTS and MIDI INS through to 4 and often 8. By employing these, the DAW can utilize each separate MIDI output to feed different devices. This prevents the need to daisy chain devices together and also permits each device to accept the complete 16 channels if required. Many of these multi-buss interfaces also feature a number of MIDI inputs too saving you from having to change over cables if you want to record information from different synthesizer.

Unfortunately, only a few manufacturers now produce multi-MIDI output devices since many of the more recent devices communicate both MIDI and audio via USB, such as the Virus Ti. Nonetheless, some companies such as Steinberg, MOTU and iConnectivity are still producing multi-MIDI interfaces. The latter produce an interface called the iConnectMIDI2, and iConnectMIDI 4 that permit the Apple iPad to be connected for transmission and receipt of MIDI *and* audio.

If devices are connected through MIDI and the audio is not transmitted to the workstation via internal structure or a USB cable, then most audio will need to be connected in the old-fashioned way via cables. In fact, even if a piece of hardware can transmit its audio output via USB direct into the workstation, it will still most likely also feature analogue audio outputs or/and inputs in the form of unbalanced jacks and RCA's or balanced jacks, XLR or even DSUB.

The distinction between balanced and unbalanced cable can be determined by examining the termination connections at each end of an audio cable. Cables that are terminated with mono jack or phono connectors are the unbalanced

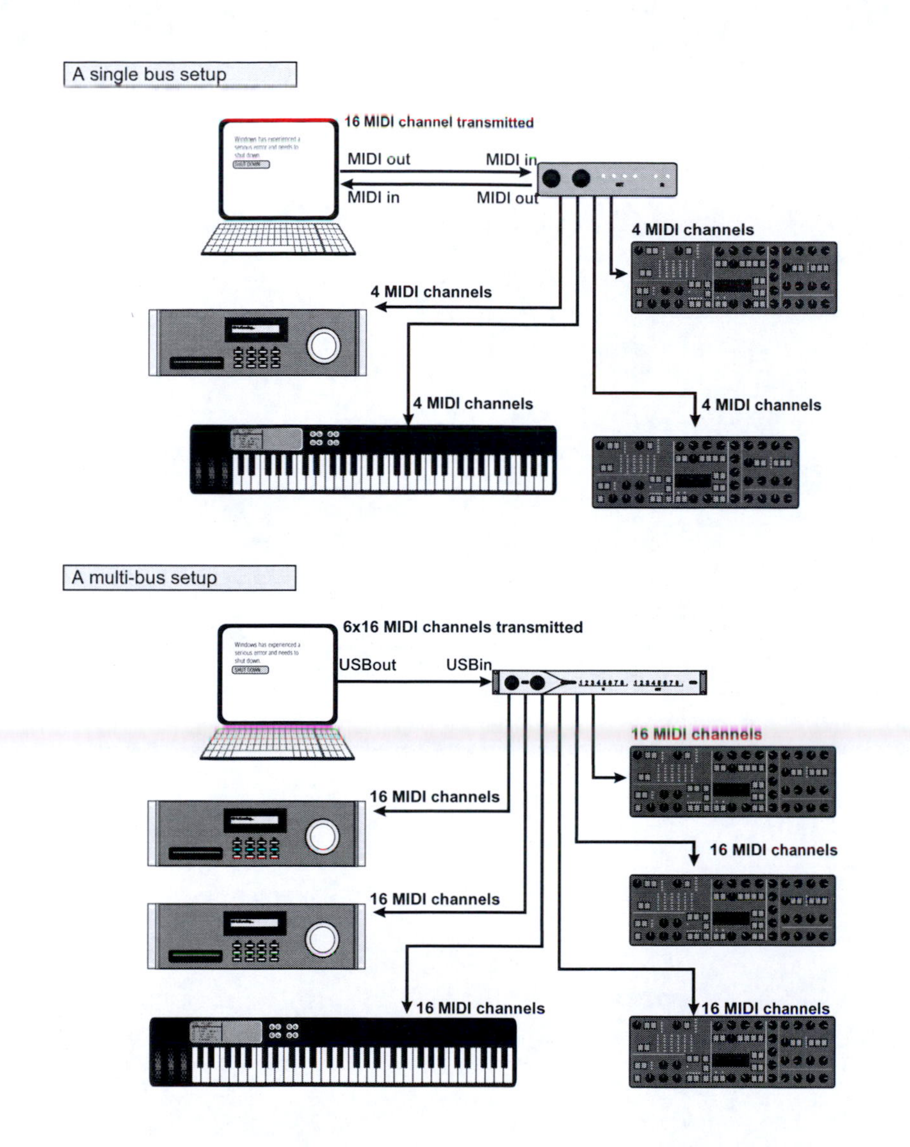

FIGURE 12.9
A multi-MIDI set-up

type, whilst those that terminate as a stereo TRS (Tip – Ring – Sleeve) jack connection or an XLR connection are often balanced if used on a mono output. Examples of TRS, XLR and mono jack are shown in Figure 12.10.

DSUB connections differ slightly from the aforementioned connections in that these offer a way to transmit eight balanced analogue channels through a single cable. Originally introduced by Tascam, the DSUB utilizes a 25-pin 'D' style male or female connection and are commonly employed in multi-channel professional equipment to permit easier interconnection between equipment.

A typical example of this is with many multi I/O professional audio interfaces and the inputs on summing mixers. Since these can employ multiple inputs or outputs, it is easier to use DSUB connections than rows of jack or XLR sockets.

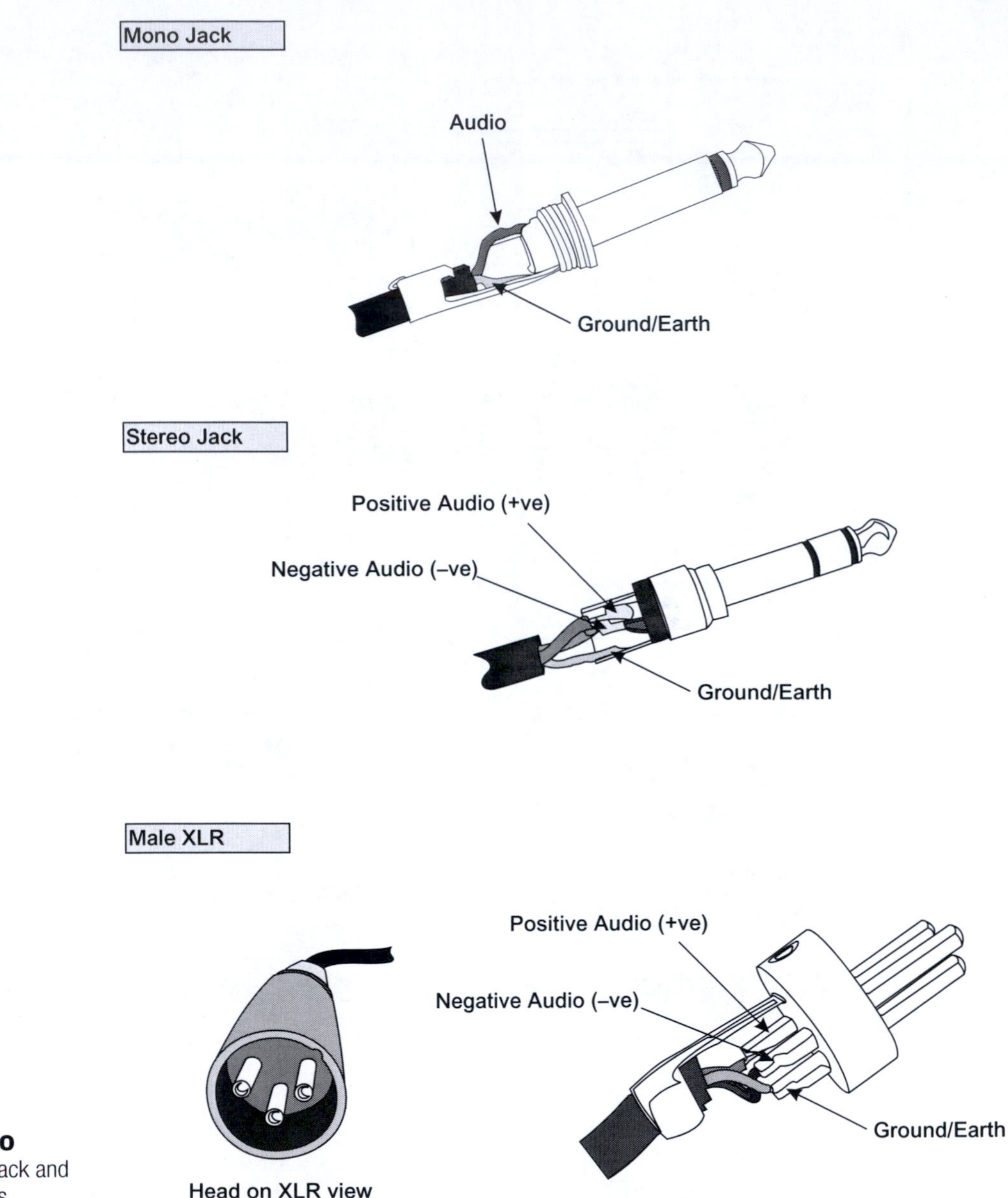

FIGURE 12.10
Mono, Stereo jack and XLR connectors

This not only helps reduce the size of the device but also makes multiple I/O connectivity much easier to manage.

A majority of professional hardware equipment, with the exception of many hardware synthesizers, will employ balanced connections. This is because in many professional studio applications, the equipment is spread around the room and, therefore, employs long cable runs. This makes the audio signal travelling through the cables susceptible to electromagnetic interference.

To prevent this from occurring, balanced cables contain three wires within a screen. In this configuration, one wire carries an earth that is grounded at both ends to prevent any possible radio interference whilst the other two are used to carry the audio signal. Using two wires to carry the audio signal, one is phase inverted at the source so that when the signal arrives at the receiving device it can then be put back into phase and summed with the original signal. By doing so, any interference is cancelled out because when both signals are put back in phase, any extraneous signal is removed. This cancellation process is similar in some respects to the theory of when two oscillators 180° out of phase with one another are mixed together.

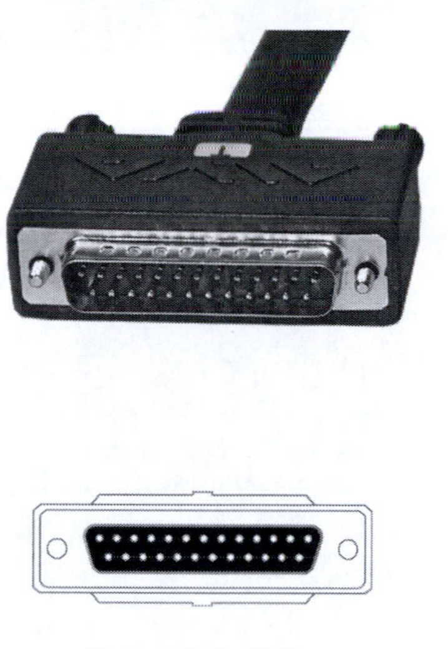

FIGURE 12.11
Tascam 25 DSUB connection

This interference cancellation, however, does not apply to unbalanced connections. Whilst unbalanced wires and connections do contain an earth terminated at both ends to prevent radio interference, only one wire carries the audio signal and, therefore, are much more prone to ground hum. Consequently, in any studio set-up that utilizes unbalanced equipment (such as many hardware synthesizers), they should be kept in close proximity to the mixing desk or audio interface in order to reduce the length of the cable run and also away from any other possible sources of electromagnetic interference. This includes power cables (these create an even larger electromagnetic field due to higher current), transformers, microphone pre-amplifiers and loudspeaker amplifiers.

Digital Interfacing

Whilst both MIDI and audio have maintained themselves as steadfast formats since the 1980's, this is not the case with digital interfacing and there are now numerous formats available, none of which are compatible with one another without the requirement for additional digital conversion interfaces. Also digital interfacing is more complex than analogue interfacing because the transmitted audio data must be encoded and decoded properly by both devices. This means that the bit rate, sample rate, sample start and end points, and the left and right channels must be coded in such a way that the receiving device can make sense of it all.

Nonetheless, digital connectivity does still provide a preferable alternative to analogue connectivity since it doesn't introduce noise, there is no electromagnetic interference, and some of the formats are capable of transmitting multiple channels simultaneously. An excellent example of this type of connectivity is Alesis ADAT *lightpipe*, a connection format that appears on almost all quality audio interfaces and proves the most useful for adding an additional eight channels of I/O cheaply.

ADAT lightpipe was developed by *Alesis* for use with their digital multi-track recorders and over time became a well-recognized industry standard. Lightpipe employs fibre-optic cables that commonly terminate at each end with *TOSLink* connectors and is a format capable of transferring up to eight channels of uncompressed audio at 48 kHz and at 24-bit resolution.

Higher sample rates of 96 kHz are possible using a process known as SMUX but in this mode, the lightpipe only permits up to four channels of transfer. Similar to the USB interface, lightpipe is also a *hot plug* device in that the sending and receiving devices do not need to be switched off before connections are made, although it is advisable to mute both devices since connecting whilst devices are switched on can result in large signal spikes.

Lightpipe has become one of the most popular formats in the industry since it makes sending and receiving multiple audio channels between different devices simple. With just one thin light cable, eight channels of audio can be transferred to or from a device digitally. Note, however, that lightpipe is a one-way communication protocol and cannot send *and* receive data in the same pipe, it is either one or the other, and therefore, many audio interfaces will feature two TOSlink connectors. One connection will be used to output eight channels, whilst the secondary connection will be used to receive 8× channels thereby permitting simultaneous I/O connectivity.

The ease and relatively cheap application of lightpipe has enabled many manufacturers to add this type of connectivity to their interfaces to permit an additional eight channels of I/O into an audio interface. This could be used to connect directly to another digital hardware device such as a mixer or alternatively, there are a number of convertors available that will convert analogue

FIGURE 12.12
TOSLink connections

inputs to ADAT and vice versa. Using these, it makes it possible to add extra analogue I/O to a device.

Manufacturers such as Presonus and M-Audio have also released Firewire interfaces featuring eight lightpipe connections (four lightpipe in and four lightpipe out) permitting an audio workstation to simultaneously transmit and receive 32 channels of digital I/O. This allows the workstation to communicate directly with digital mixers or, if employed with analogue convertors, permits the user to access a total of 32 simultaneous analogue inputs and outputs to and from individual channels in their workstation. These particular units have now been discontinued but due to their practicality still demand reasonably high prices on the second-hand market.

Employing ADAT into a studio set-up involves more than simply purchasing a few lightpipe cables, however, since for any two device to communicate with one another properly they must also be *clocked* together. Clocking offers a way to ensure that digital devices can communicate reliably with one another and is accomplished by one device being configured as the *master* and all other connected devices acting as *slaves*.

Using this arrangement, the master sends out a *wordclock* signal that is received by all the slave devices informing them of the current clock rate, left and right channels and ensuring they all remain in sync with one another. Without this clock, the devices become confused as to when a signal is expected to arrive at the input or should be sent to the output. This results in an effect termed jitter, whereby any number of audio artefacts can be introduced from spurious spikes, a lack of stereo definition and lack of detail through to an ear-piercing racket.

Although it is possible to transmit clock via lightpipe cable, it is far more preferable to deliver it via an individual cable to prevent any possible jitter from devices not clocking properly. Because of this, each device that features ADAT will commonly feature an additional BNC wordclock connection port as shown in Figure 12.13.

Some of the higher spec devices will feature two BNC connections, an *input* and a *through*, allowing the wordclock to enter the device through the input and then mirroring the wordclock in the *through* port thus permitting a number of devices to be daisy chained together. If the device doesn't feature a through port and, provided the daisy chaining of devices remains quite small, you can employ BNC splitter instead.

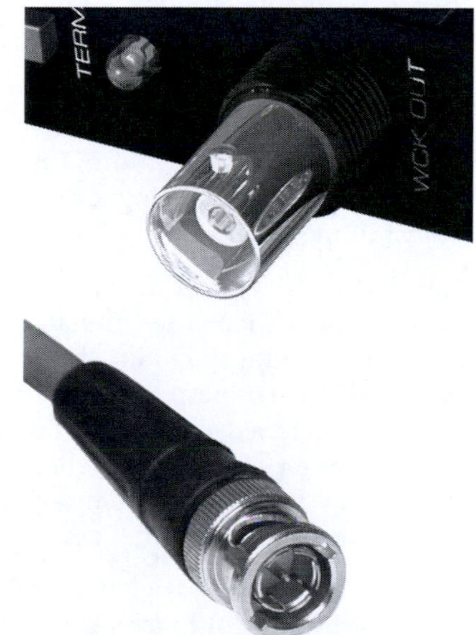

FIGURE 12.13
A BNC cable and connection

These units are designed to accept one input signal that is then divided across a small number of outputs. Although these distribution units are typically designed for surveillance camera use (the exact same BNC connections are used in video/visual) if the wordclock signal is only distributed across a small number of devices these can prove a cheap alternative to systems designed specifically for musical applications.

It should be noted, however, that wordclock signal grows significantly weaker as it travels through different devices and the device sending the master wordclock must be carefully chosen. Not all wordclocks are truly accurate and minor amounts of deviation can result in small jitter that although not immediately perceptible to the human hear can result in a weaker or unbalanced stereo image. Consequently, if the studio relies heavily on lightpipe or a number of devices are to be clocked together a much better alternative is to employ a specific wordclock generator and distribution device. In my experience, the best example of a standalone wordclock generator/distribution device is Apogee's *Big Ben*.

S/PDIF

A similar and often confused connection with lightpipe is Sony and Philips S/PDIF (Sony/Philips Digital Interface). This is often mistaken for lightpipe since, although it sometimes uses RCA connections, it often employs the exact same TOSLink connection and cable as lightpipe.

Whilst the cable is the same, however, the format is completely different and only permits the transmission of two channels of digital audio (commonly a stereo L/R pair) or compressed surround sound formats such as Dolby Digital. S/PDIF used to feature on many of the earlier samplers and synthesizers but has more recently remained confined to smaller audio interfaces and consumer hi-fi and satellite products.

AES/EBU and MADI

A more professional alternative to S/PDIF is the AES3, more commonly known as the AES/EBU connection. This requires a three-pin XLR connection, similar to the balanced analogue equivalent, but here the connection is specific to transmitting digitally.

Similar to the S/PDIF format, AES/EBU is only capable of transmitting two digital audio channels simultaneously but can utilize several different transmission mediums including balanced and unbalanced cables. AES3 provides a more reliable connection when running over long lengths and for a long time was the preferred connection in many studios for digital devices. However, being limited to only two channels of audio it again makes multichannel audio transfer difficult to employ in devices and therefore more recently, most professional multi-channel devices will employ the MADI format.

Multichannel Audio Digital Interface or AES10 was developed by Sony, Solid State Logic, Mitsubishi and AMS Neve and permits the transmission of multiple

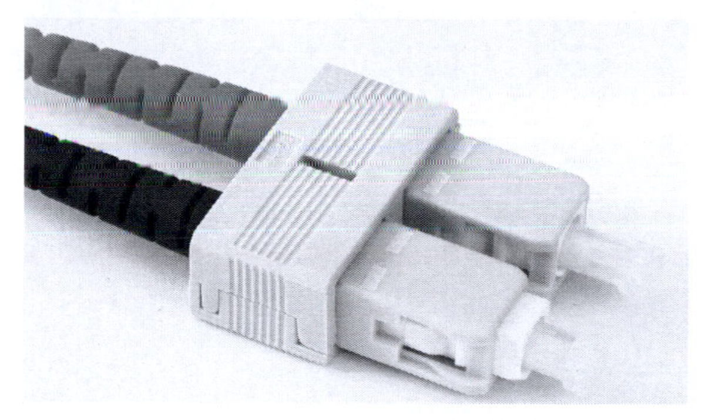

FIGURE 12.14
The MADI connection format

audio channels simultaneously at both high bit and sample rates. Capable of transmitting up to 64 channels of audio at 24-bit and 96 kHz and at lengths of over 100 m, it has become a standard in most professional studio devices and is currently employed by manufacturers such as Allen & Heath, Solid State Logic, AMS Neve, Fairlight, Yamaha, Avid (Pro Tools) and Ferrofish.

When connected to the computer, MADI permits 24 I/O and above, audio interfaces to connect direct to a PC or Mac through a single cable and offer instantaneous and simultaneous input/output steaming of up to 64 audio channels to and from a DAW. As appealing as this is, this connection standard is currently considered prohibitively expensive and therefore commonly remains within professional production circles although it is just now beginning to leak into the prosumer market.

CHAPTER 13

Gain Structure

'...And now he's flanging the VU meters...'
A&R (attempting to impress a client)

Although it is perfectly possible to produce and mix a complete track with nothing more than a laptop and some carefully chosen software, in many instances, this is rarely the case. In fact, only a proportionately small amount of professional dance music artists produce their music entirely within the confines of a computer environment and many will rely on a hybrid mix of both in the box (ITB) and external analogue synthesizers, processors and effects.

Hardware instruments such as the Moog and Access Virus continue to sell in high numbers as they remain unparalleled in software for both sound and character, whilst processors such as the Empirical Labs Distressor and Fatso along with the Smart Research C2, API 525 and Chandler TG-1 continue to be some of the most desired hardware to run a mix through in order to produce the best analogue sound. What's more, the increasing popularity of external hardware summing units such as the Dangerous 2 Buss, SSL Sigma, Neve 8816 and Neve Satellite continue to show there is a continuing trend towards using a knowledgeable mix of both software and hardware that shows no signs of abating.

The fact is that although music can be produced entirely in the box, most professionals have learnt it's a hybrid mix of both great software and hardware that provides the most creative avenues and elevates the sonic potential of the artist. However, when employing any number of external instruments or effects it is of vital importance to maintain a healthy gain structure within the signal chain.

In a typical studio situation, gain structure pertains to maintaining a suitable gain level between any number of interconnected synthesizers, processors or effects. For example, before audio workstations became the mainstay of dance music production we relied on MIDI sequencers such as the Atari STE running Cubase or the Alesis MMT-8 hardware sequencer. These were connected via MIDI to numerous hardware synthesizers such as the Juno 106, TB303 and TR909, alongside various samplers such as the Roland S10, Akai S950 and Akai S1000. Finally, somewhere in the chain there would commonly be a few select processors such as compressor or limiter along with a small number of guitar effects pedals and/or multi-effects units. The audio outputs of all these devices were often patched into one another in numerous configurations before they all eventually culminated together into a mixing desk at the centre of the studio.

With this style of hardware configuration some forethought must be given to the gain arrangement of the entire set-up. For example, if the Juno 106's audio outputs were routed directly into a distortion guitar pedal and the output of the effects unit were routed into the S950 sampler before finally running into the mixing desks input, each individual device in this serial configuration must have an appropriately set gain.

For example, the producer must ensure that the output gain of each device in the chain is not set too high that it overdrives the inputs of the receiving device otherwise unwanted distortion could occur. However, the producer must equally ensure that the output gain of each device is not set too low; otherwise, the thermal noise generated by each device would culminate together through the chain and may become noticeable.

In order to ensure a good ratio is maintained between the audio signal and the unwanted thermal noise generated by the equipment, each device in the chain must have its gain finely managed. This subsequent fine-tuning of the output and input gain on each device in the signal chain is termed 'gain structure', and this ensures that the producer receives the best signal quality throughout the entire signal chain resulting in a good signal to noise ratio in both the mixer and eventually out into the monitor speakers.

Of course, maintaining a gain structure may appear meaningless when working entirely within an audio workstation since there are no multiple hardware devices to contend with and thus no thermal noise. However, even when working entirely in the box the same style of connectivity still occurs, it just happens to be in software form. Indeed, when working entirely in the box the

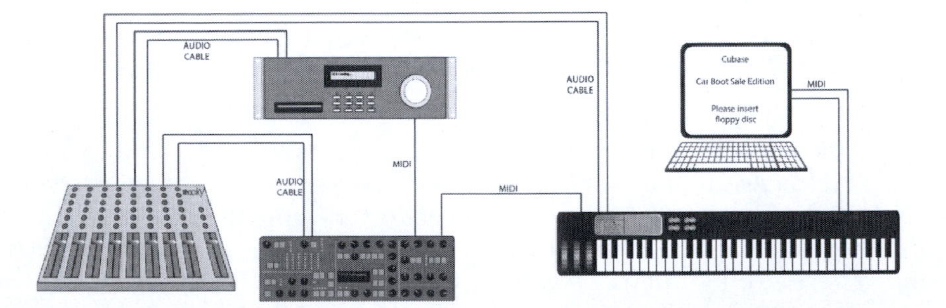

FIGURE 13.1
A typical serial configuration in a project studio (circa 1990's)

processors and effects will often emulate their hardware counterparts and the workstation itself will often imitate external hardware connectivity and thus gain structure still plays a fundamental role.

A common complaint amongst the dance music fraternity is that music composed, produced and mixed entirely in a computer environment often sounds 'flat' or lacking spatial resolution when compared to music mixed through an analogue console. But in many instances, this missing resolution is not a result of lacking any expensive analogue equipment it's the natural outcome of not understanding the theory of gain structure and how it pertains to a digital workstation. In fact, one of the most common reasons for the lack of musical soundstage is often because the producer has employed a poor gain structure in the digital domain.

LEVELS AND METRES

Maintaining a good gain structure relies on a thorough understanding of signal levels and how level metres are calibrated. Most audio devices feature a visual notification in the form of a signal-level metre that is calibrated in decibels. This informs the producer of the current outgoing or incoming signal levels (or in some cases both).

However, in order for this arrangement to be successful all of the devices in a studio set-up must share the same reference level so that the signal metres all provide the same metre reading for the signal level regardless of who manufactured the device. If not, different equipment would provide different readings.

When audio travels through a cable it is little more than a voltage. This voltage increases as the gain increases, and therefore, engineers use voltages as a reference for signal levels. Many years ago the reference level chosen was milliwatts, and from this, it could be determined that if one millwatt were travelling through the audio cable the VU metre would read 0 dBm, a reference level that meant the audio signal was approximately 75 dB above the thermal noise of the equipment.

If several devices were chained together and all used the reference level of dBm the producer could adjust and maintain a good gain structure by doing little more than watching the VU metres and ensuring that the average signal level on each device remained at around 0 dBm. This would result in a good signal-to-noise ratio throughout all of the devices and would produce the best possible sonic results.

As the years progressed and more valves were employed into equipment, they required greater voltages, and therefore, as the power levels in equipment increased milliwatts were no longer considered to be a suitable reference level. Consequently engineers – using a convoluted mathematical theorem (shown below) – converted the older reference levels and came to the conclusion that 0.775 V should be the new reference level for audio equipment. This new reference level became known as dBu.

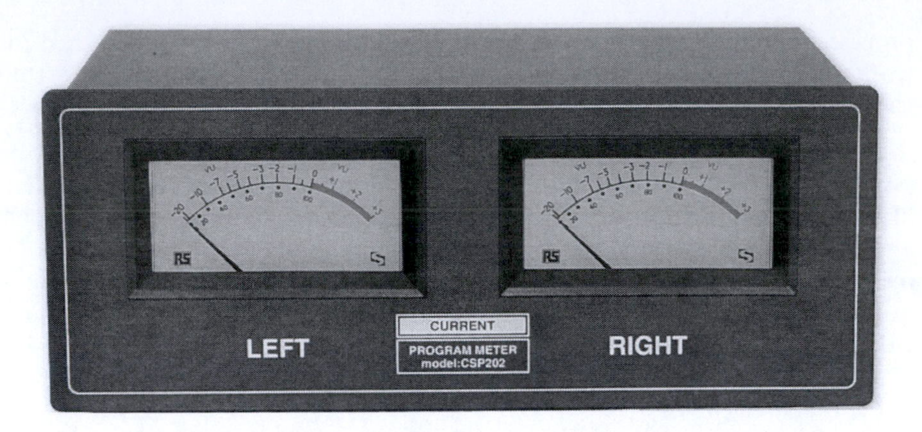

FIGURE 13.2
A VU unit

$$(P = V2/R\ (001\ W = V2/600\ W\ (V2 = 0.001\ W * 600\ W$$
$$(V = sq\ (0.001\ W * 600\ W)))$$

Since a large amount of studio equipment relied on valves (i.e. tubes), the plates in the valves were particularly noisy and, therefore, to achieve the best signal-to-noise ratio, engineers further boosted the voltage reference to +4 dBu (1.23 V). This meant that a device must receive the full 1.23 V before it would register 0 VU on its metre, a reference level otherwise termed +4 dBu. This is also perhaps the most common reference level in professional equipment today and is subsequently termed the 'Professional Standard' or 'Balanced Equipment'.

This may beg the question as to why dBu is used and not dBV and the answer comes in the form of unbalanced equipment. Since balanced equipment utilizes a higher voltage, it is more expensive to manufacture, therefore, a secondary reference level of dBV was introduced. This uses a reference level of 1 V RMS (Root Mean Square – an overall average), so the audio equipment outputs 0.316 V – equivalent to –10 dBV – to register 0 VU on the receiving device. This is sometimes termed 'Consumer Standard' or 'Unbalanced Equipment'.

There are no special requirements for connecting unbalanced *outputs* to balanced *inputs*, and in most instances, a standard balanced cable can be used but more care must be taken if connecting a balanced *output* to an unbalanced *input*. This is because a balanced output can be electronic or transformer. If the balanced output is electronic then generally, a standard balanced cable can be used again, but if its transformer the cable must be specifically balanced to unbalanced to prevent damaging the receiving equipment.

More important, however, is the obvious signal level and metre difference. If unbalanced equipment is connected directly to balanced inputs, or vice versa, there is a significant voltage difference between the two resulting in inaccurate readings on the signal metres. In fact, this difference is almost 4:1, creating an 11.8 dB difference between a consumer and professional device. In this situation, the producer can do little more than ignore the metres, listen carefully, and try to ensure the levels remain reasonable throughout the chain.

If interconnecting devices such as a balanced output to an unbalanced input the signal being received by the unbalanced device will obviously be much hotter than expected but the gain level can often be reasonably reduced on the balanced unit and many unbalanced devices do tend to employ a large threshold before distortion. However, it's far more likely that the receiving device will be balanced whilst the output device (usually a synthesizer) will be unbalanced.

In this configuration, the signal received at the balanced inputs will be 11.8 dB less than expected and thus requires the output of the unbalanced device to be increased significantly in order to capture a healthy signal level. Generally, many of the later hardware synthesizers such as the Moog Voyager feature reliable output amplifiers with a low-noise floor, and therefore, output gain can be increased significantly without introducing too much low-level noise; however, older analogue synthesizer will tend to introduce a large amount of noise as the gain is increased.

In this situation, a common workaround is to feed the output of the synthesizer into a microphone pre-amplifier since these often feature instrument inputs the pre-amp and can be used to increase the signal gain to a suitable level for the balanced receiving device. Here, however, it is strongly recommended that a quality amplifier is used and typically one that employs valves since this can add to the character of the recording. The Focusrite ISA One and ISA Two, or the SSL XLogic Alpha Channel, are a popular and relatively budget choice for amplification of unbalanced to balanced equipment.

GAIN STRUCTURE IN AUDIO WORKSTATIONS

Of course, these reference levels do not exist within the digital domain of an audio sequencer because there are no cables and therefore no voltages to travel through them. There are, however, digital mathematical versions of this very same procedure occurring that are determined by both bit depth and sample rates, and therefore the theory of gain structure also exists in the digital realm too.

Almost all audio interfaces today will operate at 24-bit since this provides a total dynamic range of 144 dB, a value greater than that of human hearing (130 dB). However, whether recording numerous channels through an audio interface or simply replying on samples and plug-in instruments, each channel that produces audio in the sequencer is routed via the mix buss and summed into the final stereo image within the workstations virtual mixer. This, alongside the producer applying numerous processing and effects to each audio channel sums together to increase the overall volume of the stereo mix and this, in effect, reduces the 'virtual' headroom of the software mixer.

Due to the need for increased headroom, many sequencers now operate at 32-bit floating point. This increases the word length in the calculations and permits the sequencer to access much larger headroom to prevent overloading when summing channels. However, even with this extended headroom, the sequencer has a finite way of representing the audio within a project. That is, audio within any digital device is represented using nothing more than a series of numbers and these are not infinite, they end abruptly at 0 dBFS. Termed 'Decibel Full Scale', this is the audio sequencers version of +4 dBu, 0 dBm and 0 dBv levels but unlike in the analogue domain, if a digital audio workstation is pushed beyond 0 dBFS it results in a particularly unpleasant distortion whilst also reducing audio quality. This leads to many of the complaints that digital doesn't have the 'headroom' of analogue, or that analogue mixes sound much better than cold digital. This couldn't be further from the truth.

As previously discussed, signal metres in professional analogue devices are commonly referenced in +4 dBu. That is, when 1.23 V are travelling through a cable to receiving device it will register 0 VU on its metre and this is considered to produce the most acceptable signal-to-noise ratio between the two pieces of equipment. Consequently, whilst applying processors, effects and mixing, an engineer will continually maintain the gain structure with an average signal level of 0 VU for the body of the music. As further audio is added to the mix, further gain staging takes place as the engineer reduces the gain of individual channels, processors, effects and the master fader to ensure that the average signal level remains around 0 VU.

Note, however, that 0 VU is *not* the uppermost limit of the devices or the mixing desk in the analogue domain. Indeed, there is often another 20 dB of headroom *above* 0 VU on an analogue desk and any further devices but this is treated by the engineer as additional headroom so that on the occasion the odd high transient does slip by compression, it doesn't create distortion in the mix because there is a good 20 dB headroom above the main mix level.

If this approach is compared to a digital audio sequencer, 0 dBu roughly translates to −18 dBFS, not 0 dBFS. I state roughly since it is not possible to convert a relative voltage to digital numeracy and therefore it is only possible to determine results from the calibration of audio converters (the SSL Alpha AX in this case). Therefore, if the producer begins balancing and mixing at 0 dBFS, it could be compared to deliberately pushing an equivalent analogue hardware mixing desk as hard as possible so that every channels VU metre is at the extreme limit of +20 dB.

Even on a £500,000 analogue-mixing desk, this extreme range mixing would produce a mix that sounds harsh and lacking in any spatial resolution. In addition, since many plug-in effects and processors in use today are modelled on their hardware counterparts, these are programmed to work there best at the equivalent 0 VU (−18 dBFS) and not a signal that's being driven at 0 dBFS.

Consequently, gain structure is not the domain of an analogue hardware-based studio and has an equal importance within the digital domain. By giving the math in the sequencers and the converters in the audio interface additional headroom, the bandwidth is not being driven hard and this produces a clearer, cleaner mix with a stronger resolution.

CHAPTER 14

Vocals and Recording

'I recorded one girl down the phone from America straight to my Walkman and we used it on the album. If we'd taken her into the studio it would have sounded too clean…'

Mark Moore

Although not every genre of dance music features a vocal track, a proportionate amount of dance tracks do, even if these are little more than heavily processed vocal snippets such as in Techno, Minimal or Tech House. Whilst these snippets could be assembled from sample CDs or perhaps even sampled from other records, it is simply no substitute for recording your own. Not only does this approach avoid any potential copyright issues, it also permits the producers to record vocals exactly to their requirements and therefore it is beneficial to learn how to record them proficiently.

Recording professional vocals is often presented by a lot of media as being potentially expensive and overtly convoluted, requiring specific vocal booths with thousands more lavished on microphones and pre-amplifiers. However, this is far from the case and it is possible to record studio quality vocals on a fairly modest budget.

THE TECHNOLOGY

The first link in the production chain for capturing great vocals lies with the producer's choice of microphone. Simply put, a microphone converts sound into an electrical signal through the use of a diaphragm, but the way in which this conversion is performed will dictate the tonal quality of the results.

Whilst there are a number of styles of microphone available that can be employed to record vocals, ultimately there are two main options, the dynamic and the capacitor or electrostatic microphone. Both of these employ a moving diaphragm but have different methods for converting the diaphragms

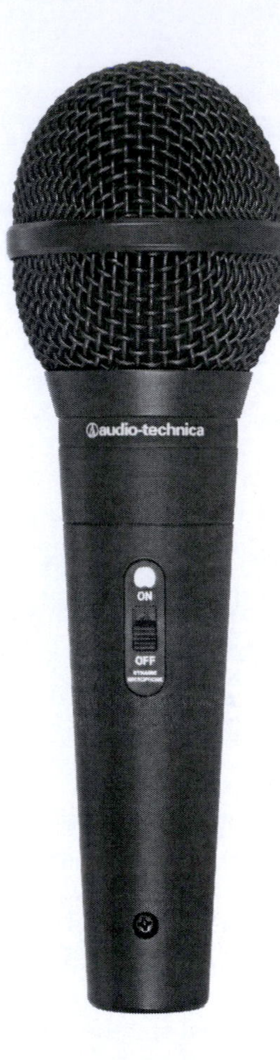

FIGURE 14.1
A dynamic microphone

movement into an electrical voltage and thus provide a different tonality from one another.

The dynamic microphone is perhaps the most instantly recognizable microphone since it's used in most live situations, from TV interviews to live performers.

Dynamic microphones employ a very thin plastic film (the diaphragm) at the head with a wire coil attached to the rear. This coil is suspended over a tubular magnet so that when sound strikes the diaphragm, it is forced into motion and the coil of wire vibrates over the tubular magnet. As taught in basic high school physics, when a wire is moved over a magnetic field it results in the generation of a small electrical current and in the case of a microphone, this small electrical signal replicates the sound waves received by the diaphragm.

Since there is a metal coil permanently attached to the rear of the diaphragm, it produces a relatively heavy assembly at the head of the microphone and therefore it's not particularly sensitive to changes in air pressure. This means that it readily captures the lower and mid-range frequencies rather than higher and thus they typically exhibit a nasal quality.

This is not to suggest that dynamic microphones are inappropriate for vocals and they are sometimes employed in studio recordings, especially if the vocalist is young since the microphones exhibit an increased bass presence and can increase the presence of the vocalist. That said, however, due to the nasal quality they are often reserved for live use since the diaphragm arrangement is particularly sturdy and will survive the odd knock or drop plus many can run on batteries and therefore do no require a pre-amplifier.

Nevertheless, the preferred choice of microphone for studio recording is the capacitor or electrostatic microphone. Sometimes referred to as a 'condenser' microphone, these employ a different diaphragm arrangement that makes them much more sensitive to changes in sound pressure.

Capacitor microphones employ a similar diaphragm to the dynamic microphone but rather than have a coil attached to the rear, it is suspended and separated from a metal back plate by a few microns. When an electrical charge is applied to the back plate, an electrical 'capacitance' field is generated between the diaphragm and back plate.

As changes in air pressure strike the diaphragm, it alters the distance between the diaphragm and back plate thus changing the capacitance that results in changes in electrical current. These changes in the current reproduce the audio signal at the end of the chain. The capacitance approach means that manufacturers can employ much lighter diaphragm assemblies than the dynamic variants and therefore these microphones are far more sensitive to changes in air pressure. Consequently, these microphones are the first choice for many professional studios but they do require a power source to function.

FIGURE 14.2
A capacitor microphone

Since an electrical charge is required to provide the capacitance between the diaphragm and back plate, all capacitor microphones require either batteries or an external +48 V power supply. In almost all situations, this power is provided to the microphone through a pre-amplifier and is delivered through an XLR cable connecting the microphone to the pre-amplifier. This is often termed as 'phantom power' since the voltage can be delivered through *any* standard three-pin XLR cable into the microphone and therefore you can't physically see where the microphone is receiving its power. What's more, if the microphone doesn't require the voltage, it simply ignores it and therefore will not damage the microphone.

Alongside delivering the voltage to power a microphone, the pre-amplifier's main purpose is to increase the signal voltage to a more appropriate level for recording. This is because the capacitance voltage produced by any capacitor microphone is so small (commonly –60 dBu or lower) that it provides an insufficient signal level for recording. By employing a pre-amplifier, the average signal level can be increased greatly.

Although many audio interfaces do have pre-amps installed, these are commonly part of an integrated circuit (IC) design and rarely offer the quality required for recording a great vocal take. Therefore it is recommended that the producer look for an independent pre-amp of the audio interface. Perhaps unsurprisingly, there is a vast range of pre-amps available that can range in price from £100 to over £6,000 and above depending on the design and parameters they have on offer. Some pre-amps appear in the form of a complete channel that offers the most common parameters you would require when recording.

These include de-essers, EQ, compressors and limiters but whilst these *can* be necessary when recording a better solution is to employ a basic standalone pre-amplifier and apply any processing *afterwards* through plug-ins whilst the signal is in the workstation. This is because in many pre-amplifier channels, the signal will travel through all of the processors whether they're being used or not and this can increase the thermal noise levels. In addition, unless the pre-amp is a very high quality unit, the included processors may not be sub-standard and therefore could do more harm than good on the signal and any processing applied *during* recording cannot be undone afterwards.

The choice of pre-amplifier is down to the producer's personal tastes but the most lauded are the Focusrite ISA range of pre-amplifiers. These are available in various configurations and price ranges but perhaps the most suitable for the home producer is the Focusrite ISA One as shown in Figure 14.3. This pre-amplifier, fitted with the digital I/O expansion card was used in all of the books examples.

It should be noted that different combinations of microphones and pre-amplifiers produce different sonic results. For example, two different microphones connected to the same pre-amplifier will sound very different to one another in terms of tonal quality and therefore producers will often match a microphone to a pre-amplifier.

Further tonal changes in a microphone can result from the size of the diaphragm fitted to the unit. A larger diaphragm, for example, exhibits a heavier mass and therefore it reacts relatively slowly to changes in air pressure when compared to that of a smaller diaphragm. Indeed, due to the smaller overall mass of these, they respond faster to changes in air pressure that results in a sharper and more defined sound.

FIGURE 14.3
The Focusrite ISA One

In theory, this would suggest that to capture a perfect performance the producer should look towards a smaller diaphragm model but in practice this is rarely the case. To understand why, we must examine the relationship between the wavelength of a sound and the size of the diaphragm.

Typically, large diaphragm microphones will feature a diaphragm diameter of approximately 25 mm. If this is exposed to a wavelength of equal size (10 kHz has a wavelength of 25 mm), there is a culmination of frequencies at this point. This results in more directional sensitivity at higher frequencies. If this is compared to the reaction of a smaller diaphragm microphone that typically feature a 12 mm diameter, the build up occurs much higher in the frequency range (20 kHz has a wavelength of 12.5 mm) and this is beyond the threshold of human hearing.

Consequently, whilst a smaller diaphragm will capture a more accurate recording, the higher directional sensitivity of a larger diaphragm tends to impart a warmer signal that is more suited and flattering to the human voice.

Note, however, that the distance the vocalist is from the microphone can also heavily influence the frequency response of the microphone. Regardless of the size of diaphragm, as the performer moves closer to the microphone, the bass frequencies will become more pronounced. This effect is known as the proximity effect and can be used to make smaller diaphragm microphones sound like the more expensive larger diaphragm models. However, care has to be taken in regards to the proximity since if the vocalist is too close to the microphone and they perform loudly, the bursts of air could damage the sensitive diaphragm.

Consideration must also be given to the polar pattern of the microphone since different patterns will affect the sensitivity to sound based on its directional axis. Often termed the 'polar pattern', this determines how sensitive the microphone is to signals arriving from different directions. The most common types of patterns are cardioid, omni-directional, figure of eight, and hyper-cardioid.

Cardioid

The most commonly used microphone pattern for vocals is the cardioid microphone. This is most sensitive to sounds arriving from the front with its sensitivity gradually reducing as it moves towards the back of the microphone where it is at its least sensitive.

Whilst these microphones are best suited for vocals, they are most prone to the proximity effect (the vocals experience a strong increase in bass) and therefore the vocalist must try to maintain an equal distance from the microphone during the performance. The term cardioid is used because when its pick-up pattern is drawn on a graph, it takes the form of a heart shape.

Hyper-cardioid

Hyper-cardioid microphones are similar to the cardioid models but have a much narrower pattern in that they do not pick up as much from the sides.

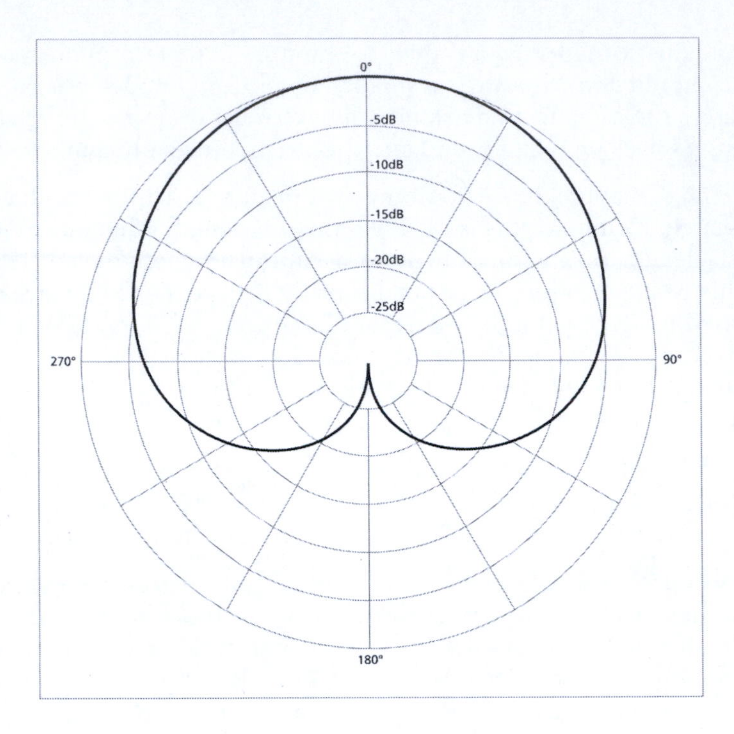

FIGURE 14.4
Cardioid pattern

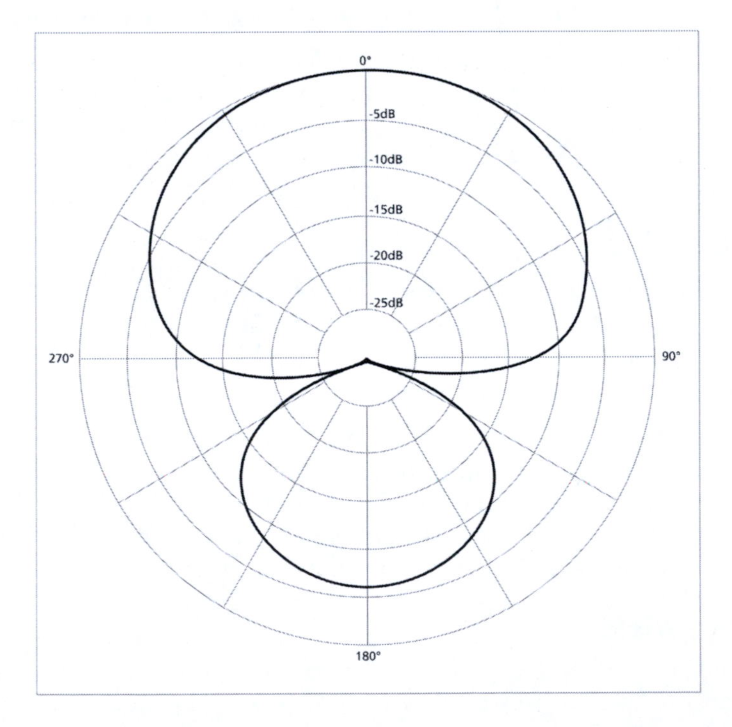

FIGURE 14.5
Hyper-cardioid pattern

However, whilst the side sensitivity is reduced, they do also accept sound from the rear of the microphone.

Since these styles of microphones are more focused, they tend to be employed when recording instruments or drum kits since they are less susceptible to overspill from the sides. The signal received at the rear of the microphone often adds to the ambience when recording instruments or drum kits.

Figure of Eight

The figure of eight microphone accepts sound from both the front and rear but rejects sound from the sides. Typically, these employ a ribbon rather than a diaphragm assembly and permit recording from both sides simultaneously. Some engineers prefer these to cardioid microphones since the ribbon assembly is lighter and thus they are far more sensitive. These microphones also do not suffer from a proximity effect. They are, however, expensive and very sensitive and can easily be broken if mistreated in any way.

Omni-directional

The omni-directional microphone has an equal sensitivity through its entire axis and therefore can pick up sound from any direction. Sports interviewers often employ this style of microphone since it doesn't require any prior knowledge of where to speak and will capture all sound. It is sometimes used to record small string sections provided the room has a good ambience.

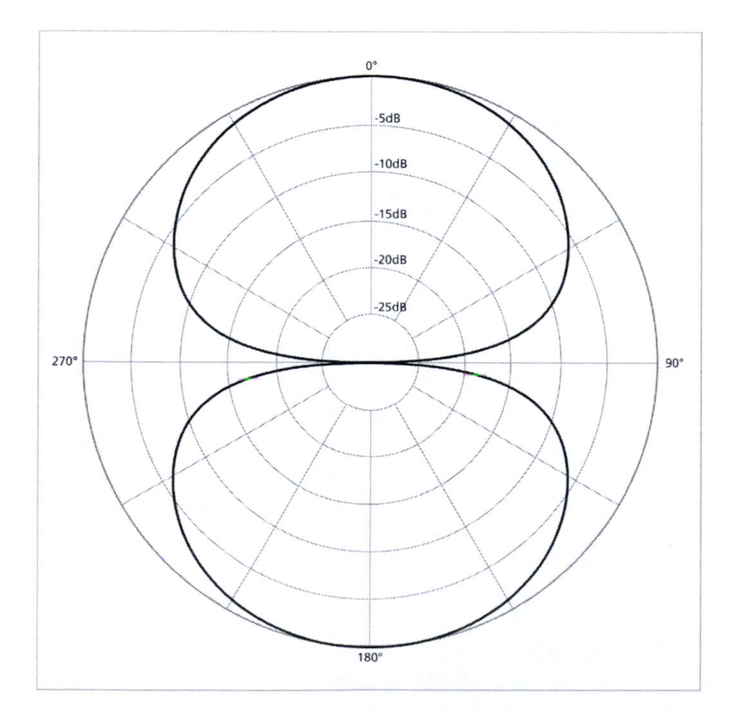

FIGURE 14.6
The figure of eight

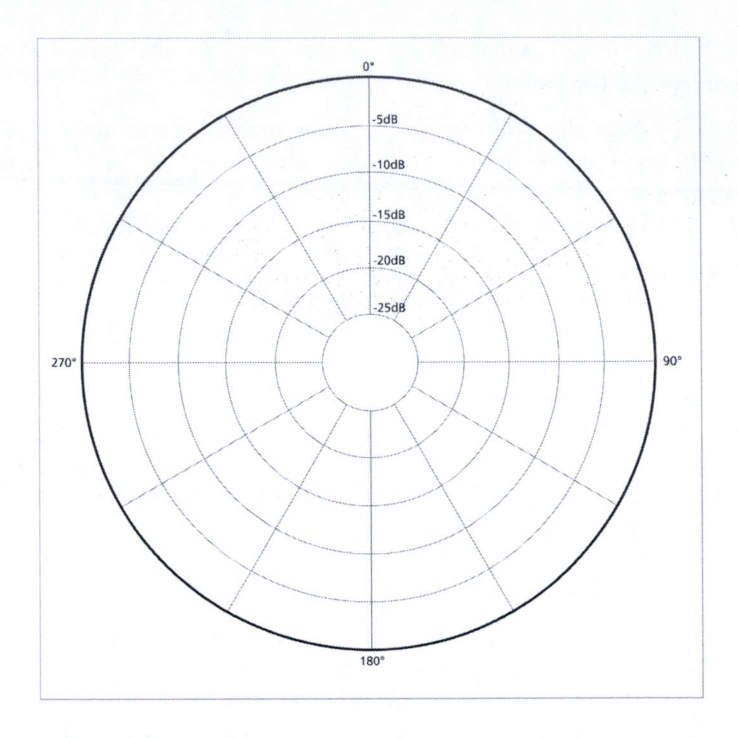

FIGURE 14.7
Omni-directional pattern

Typically, for most genres of dance, the preferred choice of microphone is a capacitor with a cardioid pattern and a large diaphragm. This is generally accepted as providing the best overall warm response that's the characteristic of most genres, but any large diaphragm will suffice provided that the frequency response is not too flat.

Note that some large diaphragm models will deliberately employ peaks in the midrange to make vocals appear clearer but this can sometimes accentuate sibilance with some female vocalists. This is the result of over emphasized 'SH' sounds that can create unwanted hissing in the vocal track.

RECORDING THEORY

Beyond the actual recording equipment, any great recording begins with the room. Ideally, the producer should aim to record vocals as dry as possible with as little of the room's natural ambience as possible. This is because if a large amount of a rooms natural ambience is captured during recording, it cannot be removed and any further effects applied at a later stage in production will not only be applied to the vocals, but also to any naturally recorded ambience.

There are various methods to capturing dry vocals but the most widely publicized include hanging drapes or curtains behind the vocalist through to placing the vocalist in a small cupboard or portable sound booth. Whilst both these approaches can work, neither offers the perfect solution for a great vocal take. For instance, it can be difficult to hang heavy drapes behind your vocalist, unless you make plans to record near a window with the curtains drawn. This can be fraught with problems, though, since drapes cannot successfully absorb all of the acoustic energy from the room and vocalist and therefore the sound waves strike a flat, thin glass object and create sonic reflections stronger if the vocalist had nothing but a bare wall behind them.

Similarly, whilst a portable vocal booth may appear to provide the perfect solution, they can be particularly expensive, require a lot of room, are only portable if you happen to own a large van to transport them and more importantly, a vocalist is a human being and packing them into a small enclosed area can result in psychological stress, so whilst you may capture a great recording there is no guarantee it will be an equally great performance. Instead, a simpler solution for capturing a dry vocal is to employ a reflection filter and, if possible, some carefully positioned gobo's too.

A reflection filter is a semi circular portable acoustic absorber that is mounted on a stand and sits *behind* the microphone. Although placing an absorber behind a microphone may appear incongruous, it is based around the theory that when a vocalist performs into a microphone, the acoustic energy is captured by the absorber and therefore unable to escape into the room and thus reduces any acoustic reflections. More over, since many capacitor microphones will pick up sound from their sides, by placing the microphone within the semi circular design, less of the room reflections are captured.

FIGURE 14.8
The SE reflexion (reflection) filter

Whilst a reflection filter will certainly help towards controlling the acoustics of a recording, they will work best when combined with acoustic gobo's. These are 'transportable' acoustic isolation panels that can be either purchased or constructed by the producer. Three panels are commonly placed behind a vocalist in a widened U shape to prevent reflections from the side and behind of the vocalist being captured by the microphone. Figure 14.9 shows a typical gobo arrangement.

Although many professional gobos are commonly 6 ft. to 7 ft. tall, reaching from the floor upwards behind the vocalist's head, it is possible to construct much smaller panels that are more easily transported. A simple wooden frame measuring 3 ft. tall by 2 ft. wide packed with Rockwool RW3 insulation and covered by a breathable material can act as an excellent acoustic absorber for recording vocals. These can be mounted onto speaker or microphone stands behind the vocalists head to prevent any rear or side reflections from being captured by the microphone.

With acoustic panelling in place, the microphone should be mounted on a fixed stand and in a reliable shock mount. These are mechanical elasticated units that the microphone sits in and are designed to isolate the microphone from any vibrations working their way up the stand from the vocalist accidentally knocking the microphone stand or tapping their foot. Almost all capacitor microphones are supplied from the manufacturer with a shock mount (aka, a Cats Cradle) but the quality varies greatly. Some can perform particularly well whilst others can actually intensify the problems of vibration from the stand and cable, which could cause resonance and thus alter the tonality of the recorded material.

The microphone itself should be positioned so that the diaphragm is approximately level with the performers' nose, whilst they stood with their

To demonstrate the difference a reflection filter can make to a recording, I recorded a session vocalist in a typical environment faced by a home producer, in this instance a large kitchen with a wooden floor. I recorded the vocals through an SE Z5600a II microphone into a Focusrite ISA One pre-amplifier. The signal from the pre-amplifier was delivered to my MacBook through ADAT. Since I knew the track Amy was to perform, I avoided application of compression or limiting and chose to ride the gain to avoid these processors from influencing the results.

COMPANION WEBSITE Chapter 16 is the recording without any filters
COMPANION WEBSITE Chapter 16 is the recording with a reflection filter
COMPANION WEBSITE Chapter 16 is the recording with no reflection filter and just the gobo
COMPANION WEBSITE Chapter 16 is the recording with both a reflection filter and a gobo.

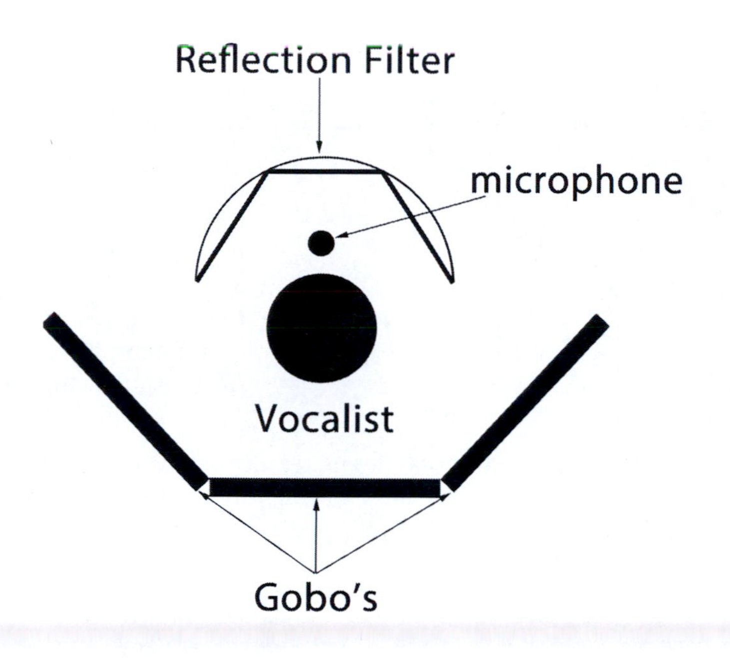

FIGURE 14.9
A gobo arrangement for recording

FIGURE 14.10
A shock mount

FIGURE 14.11
The Rycote pop shield (regarded by many as one of the best pop shields)

back straight. This will ensure a good posture while also helping to avoid heavy plosive popping or breath noises. Plosive popping is perhaps the most common problem when recording vocals and is a result of the high-velocity bursts of air that are created when performing words containing the letters P, B or T. These create short sharp bursts of air that force the microphones diaphragm to its extremes of movement that result in a popping sound.

Plosives are difficult, if not impossible, to remove afterwards so they must be controlled during the recording session through the use of a pop shield. These consist of a circular frame approximately 6 to 8 in. in diameter that are covered with a fine nylon mesh and are available as free standing units or with a short gooseneck for attaching to the microphone stand. By placing these approximately 2 in. away from the microphone, any sudden blasts of air are diffused by the pop shield preventing them from reaching the diaphragm.

Though pop shields are available commercially, some producers choose to make their own from metal coat hangers or embroiders' hoops and their girlfriends/wives nylon stockings. However, I don't recommend this approach. Good pop shields are not expensive and will perform much better than any home-made devices. Plus, a professional pop shield will exhibit some professionalism unlike turning up with a bit of a coat hanger with a stocking wrapped around it.

The microphone should be positioned the correct way up so that the cable exits the bottom of the microphone and runs down the stand. Although there are many images of studio microphones mounted upside down, this was originally to prevent the warmth of the tube inside the microphone from heating up the diaphragm or to prevent the vocalist from kicking the stand.

This positioning is not necessary today unless you happen to be employing a particularly expensive and classic Neumann, or similar, microphone or are recording a performer who is unable to control their body while performing. Indeed, hanging a microphone upside down today is more to do with looking 'gangsta' and cool than offering any sonic advantages and only increases the risk of the microphone falling out of its mount and striking the floor or your vocalist. If you must look cool or are 'concerned' about the heat from the tube warming the capsule, it is preferable to mount the microphone sideways rather than upside down.

With the positioning set, the ideal distance between the performer's mouth and microphone is approximately 4 to 6 in. At this distance, the voice tends to sound natural, plus any small movements on the vocalist behalf will not create any serious changes in the recording level. If you a recording an amateur vocalist or one who is used to live performances, they will often edge closer to the microphone during the performance and this only serves to increase the proximity effect resulting in unwanted low-frequency energy.

In this case, some engineers and publications suggest that positioning the microphone above the performers head and angling it downwards can prevent this from occurring, but personal experience has taught that if vocalists raise their chins to perform, the throat tenses and it affects the tonality of their voice. A much better alternative is to increase the distance between pop shield and microphone. If the performer is then informed to brush their lips against the pop shield during the performance, it ensures a good constant distance throughout.

The next consideration is how the mix should be delivered to the vocalist for them to perform too. In almost all situations, this mix is delivered through headphones and these should offer good isolation in order to prevent the mix from the headphones being picked up by the microphone. This means that open headphones should be avoided due to their leakage but equally closed headphones should also be avoided too. This is because the closed design clamps around the ears resulting in a rise in bass frequencies that can make it difficult for a performer to intonate. Thus, the semi-closed designs offer the best solution and although these do leak some of the mix onto the microphone, it can easily be disguised when mixed with the track.

Some vocalists may require their own voice to be fed back into the headphones along with the mix. This is termed as a 'foldback' mix and is often employed so that the vocalist can hear himself or herself in order to intonate correctly. In this instance, the vocals should be treated to a small amount of reverb since this can help the performer intonate whilst also making them feel a little more confident about their own voice (reverb can make anything sound a little better than it is). This only needs to be lightly applied and, typically, a small room with a long pre-delay and short tail of 50 ms is enough.

I personally haven't come across many performers who require a foldback and most have preferred to cup one headphone over their ear and leave the other off so they can intonate. Of course, this does mean that if the monitoring volume is quite high, it can result in the mix spilling out of the unused earpiece and working its way onto the recording. However, as long as the mix isn't ridiculously loud, then a little overspill doesn't present too much of a problem as it will be masked when dropped into the mix. Indeed, even professional artists such as Moby, Britney Spears, Kylie Minogue and Madonna have overspill present on many of their individual vocal tracks and it's something you will have to deal with if you have the opportunity to remix a professional artist.

Some engineers also suggest compressing the vocal track for the performer's foldback mix but I would recommend against this. I find that an uncompressed foldback mix tends to persuade the vocalist to control their own dynamics that invariably sounds much more natural than having a compressor perform it.

RECORDING VOCALS

Ideally, unless the producer has access to an impressive array of hardware compressors, I would strongly suggest avoiding the application of any processing during recording and instead ride the gain. As old fashioned as this approach may be, the producer should have a very good idea of the dynamics of the performance and therefore has foresight on both louder and quieter passages.

Any processing applied during the recording stage cannot be removed, so if compression is applied too heavily, or incorrectly, it cannot be removed from the performance. Moreover, if the compressor is poor, the thermal noise and character of the compression will be permanently stamped onto the recording. However, if the producer rides the gain during recording, plug-in compression can be used afterwards in the workstation to even out dynamics. Here, there will often be a wide variety of compressors, all that exhibit a different character and each that can be tried and tested to find one that suits the vocalist the most.

If a compressor must be used, then ideally it should be applied very lightly. A good starting point for recording compression is to set a threshold of –12 dB with a 3:1 ratio and a fast attack and moderately fast release. Once the vocalist has started to practice, reduce the threshold so that the reduction metres are only lit on the strongest part of the performance.

Generally, a performer will require some warm up time and this can range from 30 minutes to over an hour. Most professional vocalists will have their own routine to warm up and many prefer to perform this in private so it may be necessary to arrange some private space. If they are inexperienced, they may insist that they do not require a warm up and therefore the producer should exhibit his charm and ask them to perform a number of popular songs that they enjoy while he/she 'adjusts the recording levels…'

During this time, the producer should listen for any signs that the performer has poor intonation or vocal straining and if this occurs, ask thee performer to take a five-minute break. If they need a drink, fruit juices are better than water, tea or coffee as these can dry the throat but you should avoid any citrus drinks. Alternatively, Sprite or 7-Up (the soft drinks) seems to help many of my vocalists but avoid letting them drink too much as the sugar can settle at the back of the throat.

Before recording begins, it is essential that the vocalist has practiced and knows the lines off hand without having to read them off a sheet of paper. Reading from a sheet can severely reduce the power and feeling of a performance plus, if they're-holding the lyric sheet, the voice can reflect off the paper and create phase issues in the recording. If the performer absolutely insists they have a lyric sheet to hand, position them to the rear of the microphone and slightly off axis to prevent any possible phasing issues.

Once the vocalist has warmed up, their voice should be at peak performance and they should be able to achieve depth and fullness of sound. It's at this point the producer should attempt to capture the complete performance from beginning to end. The performer should be persuaded to avoid stopping for any mistakes and just to keep going since more takes will be recorded later. There is no music today that is completed in just one pass and even the most professional artists will make six to nine passes but the first performance is commonly the most energetic and powerful so the producer should aim to capture as much of it as possible with any further takes being little more than edits.

During the recording, the producer should not wander off mentally but sit and listen intently to the performance being recorded and try to identify any problems. Typical problems to identify here are very poor intonation, the inability to hold high notes, uneasiness and tenseness, phasing effects or too much bass presence. If the microphone has been placed correctly, there shouldn't be any phasing or bass problems but if there are, then the microphone placement should be reconsidered. If there is too much bass presence, wait until the performer has finished and then ask them move further away from the microphone.

Severe intonation problems, the inability to hold high notes, uneasiness or tenseness are more difficult to resolve as they reside in the psychological state of the performer. Assuming that the vocalist can actually perform and is in tune, the most common reason for poor intonation derives from the performers monitoring environment.

As previously touched upon, many vocalists will monitor the mix through headphones and this can result in an increase in bass frequencies. This is especially the case if the headphones are tight fitting to the performers head since the increase in bass frequencies will make it difficult to perceive pitch resulting in a loss of intonation.

The vocalist involuntarily moving further away from the microphone and performing louder is a common sign of this problem since it reduces the bass presence of their voice in the foldback mix. If this occurs, a wide cut of a few dB at 100 to 300 Hz (where the bass usually resides) or increasing the volume of a higher pitched instrument may help.

If the problem is an inability to hold high notes or a tense or uneasy voice, then this is the result of a nervous performer not breathing deeply enough and this can only be resolved through tact, plenty of encouragement and excellent communication skills. A vocalist feels more exposed than anyone else in a studio so they require plenty of positive support and plenty of compliments after they've completed a take. They should always receive positive reinforcement and 'wrong' or 'don't' should not be a part of the producer's vocabulary.

This means that eye contact throughout is absolutely essential but the producer should refrain from staring at the vocalist with a blank expression and instead maintain an encouraging smile.

Once the producer has successfully recorded the first tonally correct, emotional charged take, there will undoubtedly be a few mistakes throughout. Very few vocalists today can perform a perfect take in one go and it certainly isn't unusual for a producer to record as many as eight takes of different parts to later 'comp' (engineer speak for compiling not compressing) various takes together to produce the final, perfect, vocal track. At this point, it's unwise to ask them to perform the whole song again since it may exhaust them, so instead it can be beneficial to have them perform the verses (and perhaps chorus sections) singularly with breaks in between.

EDITING VOCALS

With all of the vocal parts recorded, the producer must now be willing to spend as much time as required to ensure that the vocal track is the best it can possibly be. Typically, this process begins by compiling all the best vocal takes together to produce one complete take throughout. This involves the producer listening to each different take, choosing the best vocal parts from each, and then pasting them together. This can involve anything from comping individual lines through to individual words and even syllables to ensure the take sounds the best it possibly can. It's certainly not unusual to spend a number of days doing little more than comping a vocal line.

During the comping process, the producer may come across further errors in the recording such as the odd plosive that crept through or small amounts of sibilance. This latter effect can sometimes be removed by placing a thin EQ cut somewhere between 5 and 7 kHz or through employing professional de-esser such as the SPL unit. These are frequency-dependent compressors that operate at frequencies between 2 and 16 kHz and reduce any sibilance in recordings. If the problem is the odd plosive, these can sometimes be cured with small amounts of corrective, surgical EQ but if this approach fails, there may be no choice but to recall the performer and record it all again.

Once the vocals have been compiled into verse and chorus, the next step is to cure any intonation problems using pitch correction software. This form of software often unfairly receives a bad reputation for correcting vocalists who can't hold a note but this couldn't be further from the fact. Unless the producer is specifically looking for 'T-Pain' effect, pitch correction software can only correct a vocalist that is very close to the key of the music, it cannot completely recover a poor performance.

However, no matter how great the performer, a lack of pitch correction is one of the main reasons why a vocal does not appear to 'sit' *in* with the music but instead gives the impression it sat *on* the music. Even vocals that are only one or two cents away from the key of the composition will prevent them from seating correctly and this often results in heavy EQ and application of effects in an effort to cure the problem.

After intonation, it's common practice to compress the vocals to restrict their dynamic range for them to sit at a constant level throughout the mix. Indeed, the most widespread processing used to obtain the upfront sound is compression. Notably, most producers will employ more than one compressor and the chain will commonly consists of two or three, depending on the vocals. By employing more than one compressor, the signal can be gradually restricted in dynamics and this often results in less-evident compression.

The first compressor in the chain is commonly a solid-state system that's employed to smooth out the overall level, using a ratio of 4:1, a fast attack and release and threshold set so that the gain reduction metre reads approximately 4 dB. The signal then enters a second compressor and this is usually a valve system to impart some character onto the vocals. This is applied more heavily with a ratio of 7:1 with a fast attack and release, along with a threshold adjusted so that the gain reduction metre reads 8 dB.

Of course, these are general settings and should be adjusted dependent on a number of other factors including the vocals character, the microphone and pre-amp used to capture them. Note that applying compression may reduce the higher frequency as it squashes down on the transients of the attack but if the high end 'air' seems lacking, it can be restored with small amounts of sonic enhancement or EQ.

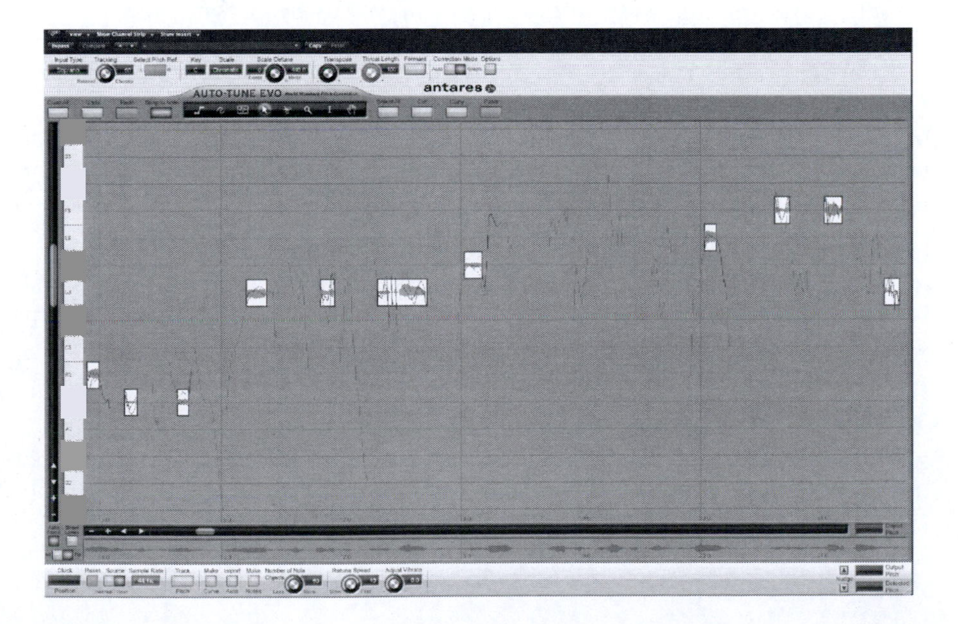

FIGURE 14.12
Correcting intonation with pitch software

Following compression, the next processor inserted is typically EQ. This is employed to remove any low-end rumble and to fine-tune the mid range so that it sits well within a mix. Since all vocals are different, it is impossible to offer guidance on EQ but generally it results in a low shelf to remove the lower frequencies with small boosts applied between 2 and 5 kHz.

After EQ, the signal is commonly sent to a reverb unit to help thicken and place the vocals into the mix. The vocals should be sent through a bus to reverb and not inserted since this permits the producer to keep the majority of the vocal track dry. Typically, the reverb employed is a small room with a short tail and long pre-delay but like EQ, this is heavily dependent on the mix and the producer's plans for the vocals.

The basic technique behind applying reverb here is to set the effect so that it's only noticeable when it's removed by keeping the decay time short for a fast tempo and increasing this for slower tempos. A pre-delay of 40 ms with a 70 ms decay will provide a good starting point and by then increasing the pre-delay, it is possible to separate the effect from the dry vocal to enable the voice to stand out more. It is also common practice to employ a different reverb than used in the rest of the mix in order to give the vocals a slightly different tonal character to help draw the listener's attention.

If the vocals do not appear up front enough after compression, EQ and reverb has been applied a popular technique in some genres of dance is to bus the vocal track to a very subtle chorus or flange effect. Employing a slow chorus or flange rate, the producer can send the vocals to the effect by small amounts. This maintains the stability of the vocals while also applying enough of an effect to spread it across the stereo image to make them appear more upfront.

An alternative method is to double track the vocals by copying them onto a second track before applying a pitch shift effect to both. This should be applied as an insert effect otherwise they will exhibit phased effect but by adjusting one channel to pitch up by 2 to 5 cents and the other down by 2 to 5 cents it result in a wider, upfront sound. What's more, if the second vocal track is moved forward in time by a few ticks so that it occurs slightly later than the original, this too can help to thicken the overall sound.

If all of these techniques fail to produce the upfront vocal sound, then it is more likely a result of poor recording techniques or equipment. If the vocalists were positioned too far away from the microphone during recording, this would result in a thin sound that no EQ could repair or recover.

VOCAL EFFECTS

It would be amiss to end a chapter on vocals without first discussing some of the more popular effects that are employed within electronic dance music. In fact, it's not uncommon to now treat a vocal as any other instrument in the mix and heavily EQ or process them through effects chains to create interesting variations.

Naturally, there are no definitive approaches to processing vocals and usually it's the producer who pushes the envelope and experiments that reaps the greatest rewards. Nonetheless, there are some vocal effects that are still in wide use throughout the genres today and knowing how to achieve these may aid towards further experimentation, so here I will discuss a few of the popular techniques.

Telephone Vocals

A popular effect is the telephone vocal, so named because it sounds as though the vocal line is being played over a long-distance phone line or through a small speaker transistor radio.

The effect can be accomplished in two ways. The first employs a band-pass filter inserted onto the vocal track and tuned into the vocal so that frequencies below 1 kHz and above 5 kHz are removed from the signal. If the producer doesn't have access to a band-pass filter or its functions are limited, the same effects can be accomplished by chaining a low-pass and high-pass filter together on the inserts. With the low-pass inserted first, this can be employed to remove all frequencies above 5 kHz. The output of this filter then feeds into a high-pass filter that is adjusted to remove all the frequencies below 1 kHz. Generally, two-pole (12 dB) filter slopes will invariably produce better results than four-pole but this is open to experimentation.

An alternative approach to employing filters is to insert an EQ unit. Through the use of shelving filters, it is possible to remove frequencies on either side of the vocals and then add further presence by boosting with a thin Q at 2 to 4 kHz. Once this effect is accomplished, it is not unusual to automate the filters or the EQ, slowly modulating back and forth by small amounts to add rhythmical modulation and interest to the vocal.

Intonated Vocals

The moving pitched vocal effects made famous through Cher's Believe and T-Pain have become commonplace in music, so much so, that they should be applied cautiously rather than zealously, otherwise the results may appear particularly cliché.

The effect is typically accomplished by inserting a pitch auto-tuning plug-in such as Antares Auto-Tune on the vocal track. A scale is chosen for the vocals (most commonly C major is chosen, even if the vocals are not in C major) and then the 'retune speed' is set as fast as possible. If there is a mode to select how 'relaxed' the tuning plug-in is, this is set to relax and the tuning is set to automatic.

When playback occurs and the tuning program detects the voice is not in tune with the current scale, it immediately corrects the intonation dragging the pitch of the vocals upwards and downwards. The effects can be made more noticeable by changing the scale even further away from the original key of the vocals.

Since this has become a very formulaic effect, it is often best employed very occasionally and only lightly on the occasional words so that it doesn't draw too much attention to itself.

The Vocoder

The voice encoder is perhaps the most powerful vocal effect to own since it offers a huge array of potential beyond its normal robotic voice. Vocoders are relatively simple in design and permit the producer to use a modulator – commonly the human voice – to modulate a secondary signal known as the carrier, which is usually a sustained synthesizer timbre.

A vocoder operates on the theory that the human voice can be subdivided into a number of distinct frequency bands. For example, plosive sounds such as 'P' or 'B' consist mostly of low frequencies, 'S' or 'T' sounds consist mostly of high frequencies and vowels consist mostly of mid-range frequencies. When the modulator (vocal) signal enters the inputs of a vocoder, a spectrum analyzer measures the signal properties and employs a number of filters to divide the signal into any number of different frequency bands. Once divided, each frequency band is delivered to an envelope follower that produces a series of control voltages based on the frequency content and volume of the vocal part.

This exact same process is also employed on the carrier signal and these are tuned to the same frequency bands as the modulators input. However, rather than generate a series of control voltages, these are connected to a series of voltage-controlled amplifiers. This means that the modulating signal and its subsequent frequencies act upon the carrier's voltage-controlled amplifiers and therefore attenuate or amplify the carrier signal, in effect, superimposing the modulator onto the carrier (the instruments timbre).

What's more, because the vocoder only analyses the spectral content of the modulator and not its pitch, there is no requirement to 'sing' into a vocoder. Instead, the modulating signal will change pitch with the carrier signal. This means the performer could speak into a microphone connected into the vocoder and whatever notes are played on the carrier signal (the synthesizer), the vocals will move to that pitch. This makes vocoding a particularly powerful tool to employ, especially if you're unable to sing and it's an effect used by countless dance artists including Daft Punk.

It should be noted that the quality of the vocoder will have a huge influence over the end results. The more filters that are contained in the vocoders bank, the more accurately it will be able to analyze and divide the modulating signal, and if this happens to be a voice, it will be much more comprehensible. Typically, a vocoder should have an absolute minimum of six frequency bands but generally 32 or more if preferable. However, the number of bands available isn't the only factor to take into account when employing it with vocals.

The intelligibility of natural speech is centred between 2.5 and 5 kHz. Higher or lower than this and we find it difficult to determine what's being said. This means that when using a vocoder, the carrier signal must be rich in harmonics around these frequencies since if its any higher or lower then some frequencies of speech may be missed altogether.

To prevent this, it's prudent to employ a couple of shelving filters to remove all frequencies below 2 kHz and above 5 kHz before feeding them into the vocoder. Similarly, for best results, the carrier signals sustain portion should remain fairly constant to help maintain some intelligibility.

For instance, if the sustain portion is subject to an LFO modulating the pitch or filter, the frequency content will be subject to a cyclic change that may push it in and out of the boundaries of speech resulting in some words being comprehensible whilst others become unintelligible.

More importantly, vocal tracks should be compressed before they enter the vocoder. If not, the vocals may change in amplitude and this will create a number of different control voltages within the vocoder. This can result in the VCA levels that are imposed onto the carrier signal following the changes in amplitude producing an uneven effect that can sound distorted. In addition, breath noises, rumble from the microphone stand and any extraneous background noises, can also trigger the carrier and therefore alongside compression, a noise gate should be employed to remove any superfluous noise.

Obviously due to the effect, a vocoder must be employed as an insert effect. With the voice acting as a modulator, the producer then needs to program a suitable carrier wave. It is the tone of this carrier wave that will produce the overall effect and therefore this is the key element to producing a great vocoding effect. For robotic voice effects similar to Daft Punk's 'Get Lucky', two sawtooth waves detuned from each other by varying amounts with a short attack, decay and release but a very long sustain provide a close approximation. If the vocals appear a little too bright, sharp, thin or 'edgy', replacing one of the oscillators with a square or sine wave, detuned by an octave will add more bottom end weight.

Unsurprisingly, much of the experimentation with a vocoder derives from modulating the carrier wave in one way or another and the simplest place to start is by adjusting the pitch in time with the vocals. This can be accomplished in any workstation by programming a series of MIDI notes to play out to the carrier synth, in effect creating a vocal melody.

Similarly, an arpeggio sequence used as a carrier wave can create a strange gated, pitch-shifting effect while an LFO modulating the pitch can create an unusual cyclic pitch-shifted vocal effect. Filter cut-off and resonance can also impart an interesting effect on vocals and in many sequencers this can be automated so that it slowly opens during the verses, creating a build-up to a chorus section. What's more, the carrier does not necessarily have to be created with saw waves. A sine wave played around C3 or C4 can be used to recreate a more tonally natural vocal melody that will have some peculiarity surrounding it.

Vocoders do not have to be used on vocals and great result can be achieved by imposing one instrument onto another. For instance, employing a drum loop as a modulator and a pad as the carrier, the pad will create a gating effect between the kicks of the loop. Alternatively, using the pad as a modulator and the drums as a carrier wave, the drum loops will turn into a loop created by a pad.

CHAPTER 15

Kicks and Percussion

'The great thing about drum machines is that, unlike a real drummer, you only have to punch the rhythm in once'.

Anonymous

The most fundamental element in the creation of electronic dance music lies with the production and sonic design of the bass/kick drum. This along with the surrounding percussive instrumentation forms the ubiquitous four to the floor drum loop of almost all-electronic dance music. Typically one or four bars in length, the EDM drum loop not only provides the grounding element to the music, but also the underlying timing and rhythm that the potential listeners will dance too.

However, whilst both drum loops and kick drums can appear particularly simple to the uninitiated, in order to maintain interest throughout the entirety of the record, they employ a vast number of carefully crafted production and compositional techniques. For this chapter, I'll examine the common approaches for programming and designing the kick drum and ancillary percussion and then in the next chapter, examine how to combine these timbres to produce professional sounding drum loops.

For many years, EDM relied on a kick drum direct from the Roland TR808 and Roland TR909 drum machines. Whilst the TR808 earned its place in hip-hop and house music, it was the TR909 that became the more pervasive of the two, most likely due to it being the only machine of the two that featured MIDI connectivity.

Released in 1983, the Transistor Rhythm 909 was a partial analogue, partial sample-based drum machine with a 16-step sequencer. Originally introduced as a rhythm accompaniment section for solo gigging musicians, the sounds produced were often considered too artificial sounding and the TR909 was dropped by many in favour of other rhythm composers available at the time.

FIGURE 15.1
The infamous Roland TR909 drum machine

The lack of interest resulted in a sharp decline in second-hand prices and consequently landed them in the hands of many of the early dance music pioneers. Gaining popularity through the use by these pioneers, the TR909 soon formed the fundamental sound of EDM and became one of the must-own instruments for any EDM producer.

Much of the TR909's signature sound was a result of the analogue synthesis engine behind the kick drum. By generating a kick drum through analogue synthesis, it permitted the user to control multiple parameters such as the tuning (pitch), decay and attack of the kick drum.

Just these three basic parameters permit an incredibly diverse range of kick drums and resulted in the TR909 being the most sampled drum instrument of all times. The love affair with this drum machine, and its predecessor the TR808, is still very much alive and both are very highly prized by many dance producers despite neither actually playing such a fundamental role in dance music today.

Indeed, in more recent years, many EDM producers place a much heavier focus on the kick drum aiming to create a kick low enough to move bowels whilst also punching the listener hard in the chest. Consequently, the TR909 is no longer considered to contain enough energy to produce this effect even with judicious application of effects, EQ and compression and is therefore only sometimes used to form *part* of the overall kick timbre.

BASS/KICK DRUM CREATION

Within current production ethics, there are two approaches to the creation of an EDM kick drum. It can either rely on a number of samples that are layered together or with synthesis programming and layering. Both of these approaches are then heavily processed to achieve the desired deep and heavy results.

The choice of which to use is largely down to the producer and their skill sets, since both methods can produce the requisite kick timbres. Typically, I've found that many of the younger producers tend to work with audio editing whilst those who are older and with more experience are more liable to synthesize the timbre.

Layering Kicks

The principle behind layering a number of kicks may seem apparently simple but it involves some very careful consideration, experienced listening, audio editing and processing.

Typically, the process begins by choosing a three kick drum samples based on their different sonic characteristics. Typically, the first kick chosen is the character kick. This is the kick that displays the overall tonal quality and shape the producer wants, whilst the other two kicks are often employed to augment this initial kicks tone, attack, decay or depth. For example, the second kick may exhibit a deep heavy sub-bass body that could be used to enhance the body of the original, whilst the third could be employed to enhance the transient or decay of the original kick.

A common mistake by the novice is to layer sample after sample atop one another in an effort to produce a powerful kick drum but this often ends in disaster. Employing too many samples will only result in a confused mash of timbres that lack any sonic definition. Instead, almost all kicks are often only ever created from a combination of three well-chosen samples and the talent is being able to spot which samples would be best suited for the creation of the final kick drum.

Once the samples are chosen, they are each placed on their own separate audio channel in the workstation and the producer proceeds to zoom right into the waveforms to ensure that the phase and starting points of each kick line up accurately. This can be an incredibly time-consuming process relying on both auditory and visual senses.

Examining the waveforms from each kick is important to ensure that the phasing is roughly aligned. Where the phase of the kicks is aligned, it results in an increase of gain at that point that can increase the presence, attack or body of the kick. If, however, the phase of two separate kicks is misaligned, it can result in phase inversion and cancellation of certain frequencies in the overall timbre.

Both these occurrences can be beneficial to the overall timbre but at this stage, the producer often aims towards increasing the gain rather than creating any phase inversion. So much so that if there are occurrences of phase inversion, the audio may be cut at that particular point and the sequencers phase inversion tool used to invert the phase to ensure they all line up to increase the overall gain.

Once all of the kicks are successfully aligned, balancing, effects and processing are applied whilst listening to the kick. Ideally, the producer will balance

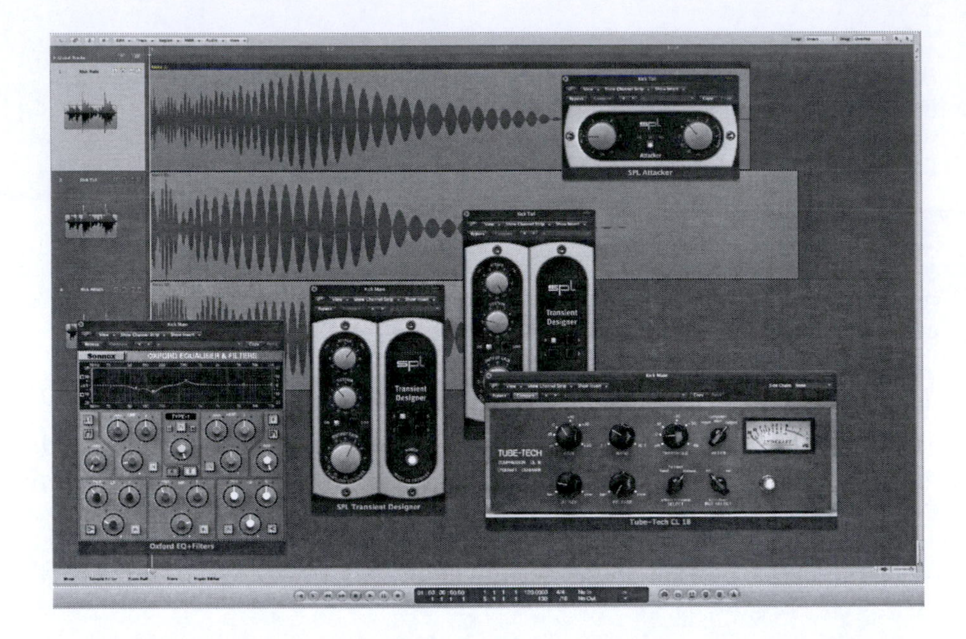

FIGURE 15.2
Layering kicks in the sequencer

the gain levels of each individual kick first in order to reach close to the sound they may have envisioned at the beginning of the project. This is followed by application of effects such as EQ, filters, transient designers, compression, distortion, tape distortion and pitch processing.

The choice of effects and processing is an entirely creative endeavour but the goal is to produce a single coherent kick timbre and this can only result from practice, experimentation and knowledge of effects and processing. For example, if the body of one of the kick layers is heavily distorting or incorrectly influencing the end results, the producer may employ a low-pass or high-pass filter to reduce the frequency content of that particular kick channel. Alternatively, EQ may be used to sculpt out chunks of an offending frequency to help all three kicks gel together and produce a coherent whole. EQ is rarely used to boost at this point and ideally should be avoided since it may create signal overloads and distort the sound.

Individual kicks may be treated to a transient designer to modify the initial attack stage or may be bussed to a single group where the transient designer is used to modify the complete timbre. A compressor may be employed across the bus channel, set to a high ratio and low threshold with a fast attack to help all the individual channels gel together or a compressor may be placed over each individual kick on separate channels and used to model each kick to produce a more coherent overall kick timbre.

For instance, if the kick suffers from a heavy attack transient with little body, a compressor set to capture the transient and release before the body would result in a change of dynamic ratio between the kick's transient and its body, in effect, increasing the gain of the body.

Similarly, the producer may apply some pitch processing to change the pitch of one, or even all of the kicks to better match their idea, this is in an effort to help the different layers gel or perhaps increase the body by lowering the pitch of a kick channel. Alternatively, judicious application of distortion or analogue style distortion may add further character to the kicks.

There are, of course, many more processing and effects routes possible but these are born from hours of experimentation. A good understanding of the working principles of a number of effects and processors such as compression is obviously beneficial but there is *no* universally accepted principle for kick layering since it depends entirely on the samples being used and the type of sound the producer wishes to accomplish. It is the individuality of the artist and their ideas that create the best results so experimentation is the key. Typically, a professional producer will spend 20 to 30 hours programming and processing a kick drum and this should be doubled for the novice producer.

Whatever methods are employed, once the sound is complete, all the channels are typically bounced into a single audio file and this audio is reimported into the current working project as a single kick drum sample. However, whilst this sample layering approach is certainly a feasible technique for many producers, it can often introduce some potential difficulties.

The process of layering sampled kicks.

Since EDM kicks are now being carefully engineered to induce bowel movement and/or give the listener a particularly large punch in the chest, a kick will often feature a discernable pitch at some point during its cycle. Identifying the point at which this notable pitch occurs is difficult to quantify since there are many variables depending on the creation of the kick, but if this pitch does not match the key of the music and in particular the bass, it can produce a low-end groove that appears dissonant and unsettling.

In order to prevent this dissonance, it has become common practice to tune the pitch of the kick drum to the music to help substantiate the groove of the record. This is often accomplished by *bouncing* the kick into a single audio file, re-importing it into the working project and adjusting its pitch to suit the music.

If the kick has been created from layering samples, it can be difficult to discern its original pitch and even the latest pitch recognition software is frequently unable to determine the pitch since it changes during the cycle of the kick. Because of this, the only way is to tune by ear.

In order to tune a kick by ear, you should ideally have an acoustically treated room with a good reliable monitoring system equipped with a subwoofer. Standard monitors will often struggle replicating the lowest frequencies of a kick drum accurately, so a subwoofer can be essential if tuning a kick to the bass by ear. This is not beyond the capabilities of many producers, but you must obviously have a good musical ear otherwise you could elevate the problem rather than solve it.

Synthesizing Kicks

A second approach to designing an EDM kick is to again layer a series of kicks but rather than use samples it involves layering synthesized kicks. This offers much more creative freedom than layering samples since the producer can not only specifically pitch the kick to their requirements, but also synthesize the parts required for layering. Also there is no requirement to sit through a sample CD full of kicks to find some good starting points.

There are a number of software and hardware units that are specifically structured towards programming or designing percussive instruments. In the software domain, products such as FXPansion's Geist, Vengeance's Metdrum (more sample-based than synthesis), Audio Damage's Tattoo, Linplug's RM1v or the Waldorf Attack are all perfectly suited towards synthesizing kick timbres although from using many of these instruments I don't personally feel they match the sonic presence of analogue hardware.

Hardware instruments such as the Jomox Mbase01 and Mbase11 can produce kicks with incredible depth and body that often don't require layering and can quite easily break the cones on a monitoring system if not used with care. Both these units are based entirely around analogue synthesis and are only capable of modelling kick drums but their capabilities for producing instant chest-thumping kicks have resulted in them becoming the mainstay of many EDM studios.

Of course, it is not necessary to use either a hardware or software drum synthesizer to create a kick drum and they can also be accomplished in most capable synthesizers but in almost all cases, drum-specific synths offer more synthesis options targeted towards the creation of these timbres.

Whatever synthesizer the producer chooses to employ, the basic principle behind synthesizing a kick remains the same. In essence, it consists of sweeping the pitch of a sine wave downwards and then employing further techniques to add presence, body or perhaps a stronger attack stage. In the original Roland TR909, the kick was created using an analogue sine wave oscillator with an attack/decay envelope controlling the speed of pitch descent. To add more of a transient to the kick, a pulse and a noise waveform were combined together and subsequently filtered internally to produce the initial attack stage. This process created the basic 909-kick timbre.

This method can, of course, be replicated with most synthesizers but when doing so, the first consideration should be on the frequency of the sine wave since this will determine the ultimate body or depth of the kick. An EDM kick is commonly positioned between 40 and 80 Hz (E1 to E2 on a typical keyboard) depending on the genre, style of kick and the key required. For the latter, tuning the kick to the fifth note of your working scale often produces a more consonant and pleasing harmonic result than tuning it to the root note.

An attack/decay envelope is used to positively modulate the pitch of the sine wave so that the pitch descends rather than rises. All amplifier parameters are set to zero and then the decay is slowly increased to produce the start of the timbre. Here, increasing the decay increases the time it takes for the pitch to fall resulting in a kick with a heavier boom whilst decreasing will create a kick with a snappier feel.

On many synthesizers, the decay envelope will run in a linear fashion typical of the original TR909 but ideally for today's production standards, this linear decay needs to be adjusted to behave in an exponential manner. Indeed, many producers of drum synthesizers today are well aware of the importance of modifying the decay potential and will permit you to alter this to modify the pitch characteristics of a kick.

By employing an exponential movement to the decay and controlling the speed of the pitch sweep, it's possible to create a much wider variety of kicks. For example, if the envelope were to bow outwards, as the pitch decays, it will sweep more slowly at the beginning and begin to pick up speed as it closes towards the end of the kick.

Alternatively, if the exponential decay slope is inwards, then the pitch will initially sweep faster at the beginning and slow down as it reaches the end of the decay portion. These two decay settings will provide very different results with the first producing a fatter, rounder kick whilst the second will produce a timbre with a 'slapping' characteristic. By increasing the period of decay, these effects can be increased further to enhance the characteristics of the kick.

If the synthesizer is not capable of adjusting the envelope's linear behaviour, it is possible to create an exponential decay through other methods. For example, some synthesizers permit the user to modulate the decay parameter of the envelope with itself. By doing so, the modulation destination can be used to modulate both the oscillators pitch and the decay parameter by negative or positive amounts and this results in creating exponential behaviours in the decay. This effect is known as recursive modulation but is a limited feature that only appears on very few synthesizers.

Alternatively, it is possible to create exponential behaviour using little more than a compressor. By inserting a compressor with a low threshold, high ratio and a slow attack onto the kicks channel, the compressors action will increase the body of the kick resulting in an effect similar to 'bowing' the pitch. Similarly, using a fast attack on the compressor with a quick release, high ratio, and the threshold adjusted so that only the attack breaches the compressors threshold, it is possible to recreate the inwards style rate of decay. However, this approach should be considered a last option and better results will often be attained through synthesis alone.

Even with an exponential decay, the initial synthesized timbre is rarely enough and its common practice to synthesize further kicks for layering duties. These additional kicks are often used to add further body but can also be used to add more high energy to the transient of the kick to help it pull through a busier mix. This higher, increased transient can be accomplished through synthesizing a secondary kick with either a sine or square wave but tuning them at double the frequency of the original kick. Using an immediate attack with a very fast decay setting, it produces a short transient click that can then be layered above the original kick.

An alternative to this approach is to simply layer a hi-hat over the kick. As strange as this may appear, layering a closed hi-hat over the kick and then employing a noise gate to remove the tail of the hat can produce a high transient to the kick allowing it to pull out of the mix over the bass.

Further sonic modifications are also available through the synthesizer's oscillators and filters. For instance, replacing the sine wave with a square produces a timbre with a boxier body, whilst replacing it with a triangle will produce a sound similar to that of the infamous Simmons SDS-5 drum machine. Similarly, adjustments made to the filter section of the drum synthesizer can create further variations. If a square wave was used in the original creation of the kick, a high-resonance setting will produce a kick with a more analogue character. Increasing the filter's cut-off can result in a more natural sounding kick, whilst increasing the pulse width can create a more open, hollow timbre.

More body or energy can also be added to synthesized kicks by layering further synthesized kicks, samples, or more commonly a sine wave under the original kick. A typical synthesizer used for creating and layering a sub-bass for many producers is Logic Pro Audio's ES2 plug-in. These can then be treated in a similar manner to sampled kicks with application of various processing and effects to help the different layers gel together.

FURTHER PROCESSING

Whether synthesis or sample layering is employed or a mix of both is employed to produce the kick, it will often benefit from yet further processing within the music. First, the kick is very often treated to very small amounts of reverb.

Although many publications will recommend avoiding reverb on low-frequency instruments such as the kick and bass, it is nearly always applied to the kick drum in EDM to add both character and body. This, however, must be applied with great care since over zealous application will result in a kick that lacks definition and clarity.

Usually the reverb is inserted onto the kicks channel rather than the kick sent to a reverb bus, but both approaches are equally viable provided the parameters are carefully set. The settings are dependent on individual tastes but in many typical applications, a small 'room' type reverb is employed with a lengthy pre-delay and very short tail.

The pre-delay should be set long enough so that the initial transient is not molested by the first reflections and therefore remains distinct and is able to pull through the mix. To accomplish this, the pre-delay time will depend entirely on the kick but commonly settings between 15 and 30 ms will produce the best results.

The reverb tail is usually applied with a very low time setting, typically less than 25 ms but a good way to accomplish a preferable setting is to set a long reverb time of 1 second and then, whilst monitoring on headphones, gradually and slowly reduce the reverb time until its no longer noticeable. At this point, increase the reverb time again by just a few milliseconds.

Alternatively, for a different style of timbre, some artists choose to employ a much longer, 500 ms or more, tail but then employ a noise gate directly after the reverb. Using this, the reverb is initially bypassed as an effect and noise the gate is adjusted so that the moment the kick ends, the noise gate activates. This approach ensures that when the reverb is re-engaged onto the kick, there is no reverb tail present and the reverb itself only affects the body of the kick drum.

After reverb, more judicious compression is commonly applied. This is generally applied as a finishing polish to help gel reverb and kick and create more equilibrium in gain level across the length of the kick.

The process of reverb and compression on a kick.

SNARE AND CLAPS

A secondary instrument in the drum loop that shares an almost equal relationship to the kick is the snare or clap. The choice of whether to use one or the other is often dictated by the producer's own creativity but genre considerations can also play a role in this decision. Indeed, listening to the current market leaders in any genre should never be underestimated since it will provide many aural clues onto what is considered to be part and parcel of any particular genre.

Originally, the snare drum in the earliest forms of dance music was derived from the ubiquitous TR909, the E-mu Drumulator or the Roland SDS-5. All of these vintage machines employed a very similar synthesis method to create a snare drum by employing a triangle wave mixed with pitch modulated pink or white noise. Similar to the kick drum, however, this approach is no longer applied and both snares and claps are created in pretty much the exact same way as the previously discussed layering of kicks.

With snares, three are commonly chosen, layered, aligned, phase inverted (if required) and followed with judicious use of compression, EQ and transient designers to help the three individual samples gel together in a cohesive whole.

Unlike kicks, however, layering various snare samples can be much more difficult since they exhibit a more individual and complex character. Consequently, layering only two samples and then synthesizing a final sample can often help in this respect and cut down a few days of sample editing into a few hours of programming. Although not always necessary, by using synthesis to create the third clap, the timbre can be specifically designed in time, frequency and timbre to help the other two-snare/clap samples merge in a more fluid manner.

Snares can be programmed in most synthesizers but generally it is preferable to use drum-specific synthesizers since these feature parameters that are obviously more useful for the creation of drum timbres.

The creation of a snare requires a minimum of two oscillators, for the first a triangle wave is used whilst either pink or white noise is employed for the second. The decision between pink and white noise is down to the producers creative choices but generally, pink noise produces a more accurate and realistic sounding snare that can prove more suited to help gel together layers. This is because pink noise contains more low-frequency noise and therefore produces a wider range of frequencies.

To produce the initial snare timbre, much of the low- to mid-frequency content of the noise should be removed through a high-pass, band-pass or notch filter depending on the type of sound required to help gel the samples together. Notching out the middle frequencies will often create a 'clean' style of snare timbre commonly employed in breaks and minimal whilst a band-pass can be employed to add a sonic crispness to the timbre making it more suitable for tech-house and techno-style projects. Alternatively, employing a high-pass filter with a medium resonance setting can be used to create a more house- or trance-style timbre.

As with kick drum synthesis, snares should start immediately on key press and remain fairly short so the amplifiers envelope generator (EG) should employ a zero attack, sustain and release while the decay can be used to control the length of the snare. Ideally, a different amplifier EG for both the noise and the triangle wave should be used so that the triangle wave can be kept quite short with a fast decay while the noise can be made to ring a little further by increasing its decay parameter. The more this noise decay is increased the more atmospheric snare will appear. It is often worth making the noise ring out for a little longer than the layered samples snares to help gel them together and end with a single tonal ring out of noise.

If two amplifier EGs are not available then small amounts of reverb can be used instead to merge and lengthen the timbre. Room style settings are a favourite of many producers although euphoric and uplifting trance aficionados tend to favour hall settings for that larger more atmospheric off beat strike. In either case, the pre-delay is adjusted so that initial transient creeps through unmolested and the decay time is adjusted to between 250 ms and 1000 ms. Noise gates with a fast attack almost always follow this and by employing a short hold time with a fast release the threshold can be set to the producers taste. Low thresholds times can create a more ambient feel to the snare whilst higher thresholds will cut the gate off earlier producing a more snappy timbre. If the producer plans to cut the gate early, it is recommended that it be employed on a bus with all three samples fed into the bus. This ensures that all three samples are cut at the same time.

If the snares are still not gelling, further tonal modification may be possible in the form of pitch envelopes. By employing positive or negative pitch modulation, the pitch of the synthesized snare can be forced to sweep through a wider range of frequencies that can often help gel a number of samples together. Alternatively, an LFO set to a sine or triangle wave could be employed to modulate the pitch, although the frequency of the LFO must be set by ear in order to ensure the pitch sweep occurs at an acceptable rate for gelling all three snares. After this, small amounts of compression applied so that only the attack of the synthesized hit is squashed (i.e. fast attack) will bring up the decay in volume so that it can be employed to gel the samples together.

Claps

Claps follow the same layering principles as snares in that three samples are carefully chosen and then layered together. This mixed with judicious sample alignment, phase inversion and thoughtful application of compression, EQ, and effects all aid in the three samples gelling together. However, as with the snare drum, a clap timbre can exhibit a complex sonic character and layering can be difficult, therefore an oft-used technique is to layer two clap samples and synthesize a third.

Claps are perhaps the most difficult 'percussive' element to synthesize because they consist of a large number of 'snaps' all played in a rapid, sometimes pitch shifting, sequence. In many of the older analogue drum machines claps were created from white noise passed through a high-pass filter and modulated with a sawtooth envelope before being treated with reverb.

This can be emulated on most synthesizers by employing a filter and amplifier envelope onto a white noise oscillator. Both envelope's should employ a fast attack with no sustain or release and the decay used to set the length of the clap. The old analogue style sawtooth envelope can be emulated through employing a sawtooth LFO to modulate the filter frequency and pitch of the timbre. Increasing or decreasing the LFO's frequency can change the sonic character of the clap significantly.

FURTHER PERCUSSION

Typically, beyond kicks, snare and claps many of the further percussive instrumentation in an electronic dance music mix are gleaned from the multitude of sample CDs that are now available on the market. These are almost always available in WAV and AIFF format so they're simply imported into the workstations audio channel and processed depending on the producer's own creativity. This can range from applying compression, transient designers, and EQ, through to the more common practices of employing reverb and following the reverb with a noise gate to remove the tail. However, the ability to synthesize percussive elements should not be underestimated.

Many genres such as minimal, techno, tech-house and progressive house depend heavily on any number of strange, evolving percussive sounds within the rhythms and therefore a basic understanding of how to program and synthesize these elements is fundamental to the production of electronic dance music. Consequently what follows is a run through of how to synthesize some of the more common percussive elements within a mix. By following these, it should provide you with some ideas on how to program any number of percussive style elements.

Hi-Hats

The ability to program a hat should be considered as fundamental as the ability to program a kick, snare or clap since there will often be occasion where a hat is layered atop a kick, snare or clap to introduce a tighter transient stage to permit the timbre to pull through the mix.

In the original analogue drum machines, hats were created through little more than filtered white noise. This means it can be accomplished quickly in most synthesizers by employing a white noise oscillator with a high-pass filter. The filter should be adjusted so that the lower frequencies are rolled off leaving only the higher frequency content. Both amp and filter envelopes should be adjusted to a fast attack with no sustain or release leaving the decay parameter to control the length of the hat. If the decay is set suitably long enough, the hat will change from a closed hat to an open hat.

Alongside employing analogue synthesis to create a high hat its possible to employ ring modulation or frequency modulation to accomplish much the same thing. Creating a hat through ring modulation involves feeding a high- and low-frequency triangle wave into a ring modulator and then adjusting the pitch of each oscillator until it achieves the desired results. Again, the amplifier envelope should have a fast attack with no sustain or release permitting the decay parameter to determine the length of the hat.

Similarly, creating a hat through FM synthesis consist of modulating a sine wave with a high-frequency triangle. This will result in a high-frequency noise waveform that can be sculpted via the amplifier EG with the usual hi-hat settings. An advantage to employing FM to produce hats is that a pitch modulation can be used to vary the timbre and therefore once the basic timbre is constructed it is beneficial to adjust the oscillators pitch to achieve a number of different sonic styles.

Whatever method is employed, hats often benefit from experimentation with EQ, transient designers and reverb followed by a noise gate to sculpt the final timbre. Dynamic restriction via a compressor, however, should be avoided. Even if the attack parameter on the compressor is adjusted to bypass the attack stage, since the cycle is so short, many compressors will still capture the transient and thus remove the high-frequency detail. Consequently, compression is usually avoided on any high-frequency timbres such as hats and shakers.

Shakers

Shakers, like hi-hats are more often than not gleaned from the multitude of sample CDs on the market but they can be synthesized in a very similar fashion to analogue style high hats. Indeed, to program a shaker, the producer creates a hi-hat by employing a white noise oscillator modulated by the filter and amplifier envelopes. These employ a short attack no release or sustain so that the decay can be used to determine the length of the timbre. Here, the decay is lengthened considerably to create a timbre that is longer than an open hi-hat that is then LFO modulated with a high-pass filter. Typically, a sine wave LFO with a high frequency and medium depth produces the more typical shaker effects but changing the LFO waveform and frequency can produce a multitude of different shaker effects. As always, this can then be further sculpted with EQ, transient designers and reverb followed with a noise gate.

Cymbals

Cymbals are created in a fashion similar to hi-hats and shakers as they can again be synthesized from a basic noise waveform, but a better alternative is to employ ring modulation or frequency modulation since this produces a much stronger and realistic resonance in the tail of the cymbal.

If ring modulation is available, two high-frequency square waves should be fed into the ring modulator and the output modulated with an amplifier envelope employing a fast attack and a medium decay (sustain and release should be set to 0). The producer then begins to detune the two square waves from one another to run through a vast range of cymbal-type timbres. Typically, detuning the oscillators two octaves from one another produces the most 'realistic' results but experimentation is the key and many techno tracks have made use of three octaves or more.

Alternatively, if the synthesizer permits frequency modulation of oscillators other than sine waves (such as Native Instruments FM8), a high-frequency square wave modulated by a low-frequency triangle oscillator can produce a number of cymbal effects. This can be further experimented upon by adjusting the frequencies of both oscillators and increasing/decreasing the FM amount. What's more, modulating the pitch of one or both oscillators with a slow-frequency sine wave LFO can create a crash cymbal that slowly changes during its cycle.

Tambourines

Tambourines, like claps, consist of a number of rapid successive hat strikes that are often modulated in pitch with a sawtooth envelope and have a band-pass filter applied. This behaviour can be emulated through programming a basic shaker with a longer decay but employing a band-pass filter rather than a high-pass filter. A sawtooth LFO is then used to modulate both the band-pass filter and the pitch of the oscillator. For better results, the LFO's frequency should be modulated with another sine wave LFO so that the original LFO's frequency changes over time to produce more realistic results. Finally, the producer can adjust the band-pass to produce a variety of different timbres. Here, wide band-pass settings will recreate a tambourine with a large tympanic membrane whilst thinner settings will recreate a tambourine with a smaller membrane.

Cowbells

If you need more Cowbell, they can be constructed in a number of different ways depending on the style of sound required. For cowbells with a wider, more well-bodied sound it is best to employ two square waves whereas if a brighter style of timbre is preferred, a square mixed with a triangle will produce lighter results.

For a cowbell with more body, both square oscillators should be detuned by approximately five tones from one another, so if one occurs at around C5 (554 Hz) the other should occur at G5 (830 Hz). The amplifier envelope should employ a fast attack with no release or sustain and a short decay, and a band-pass filter is then employed to shape the overall tonal content of the timbre.

Alternatively, to produce a cowbell that exhibits a brighter colour that will be closer to the higher frequency elements of a mix, sit near the top of a mix it's preferable to employ a square wave mixed with a triangle wave. The frequency of the square can initially be set to around G5 and the triangle should be detuned so that it sits anywhere from half an octave to an octave away from the square. Both these oscillators should be fed into a ring modulator with the results run through a high-pass filter to remove the lower frequencies. Once these basic timbres are created, the amplifiers decay can be lengthened or shortened to finalize the timbre.

Congas

Congas are best constructed via frequency modulation synthesis by employing a sine wave and noise waveform. The sine oscillators amplifier EG should initially employ a fast attack and decay with no release or sustain. This will produce a basic clicking style timbre that is then employed to modulate the noise waveform via frequency modulation. The noise waveforms amplifier EG should be set to a fast attack with zero release or sustain and the decay set to taste. By increasing the FM amount and detuning the frequency of the sine wave its possible to create a number of different styles of congas hits and strikes.

There is usually no requirement to employ a filter on the resulting sound but if it exhibits too much lower or upper energy a high- or low-pass filter can be employed to remove any offending frequencies. By employing a high-pass filter and reducing it slowly it is possible to create any number of muted congas while adjusting the noise amplifiers decay slope to a non-linear function can produce the slapped style congas that often appear in progressive house rhythms.

Tom Drums

Tom drums are synthesized by employing the same methods as kick drums but instead utilize a higher pitch with a longer decay on the amplifiers EG. It's also advisable to mix in some white noise with the sine oscillator to produce some ambience to the initial timbre and this can employ the same amplifier envelope as the original tom drum (zero attack, release and sustain with a medium decay). An alternative method to creating a tom drum is to synthesize a typical snare with a triangle and noise waveform but then modulate just the pitch of the noise with a slow triangle wave. This results in the noise sweeping down in pitch whilst leaving the triangle unmolested.

ANCILLARY PERCUSSION

Although here I have concentrated on the main elements in percussive dance loops, genres such as tech house, minimal, techno and progressive rely on a large number of percussive style sounds that fall outside of this category. However, despite the differences in timbre, they all follow the same basic programming principles laid out here.

For example, the main oscillator will usually always consists of a sine, square or triangle wave with its pitch modulated by a positive pitch envelope. This creates the initial tone of the timbre while a second oscillator is often employed to create the subsequent resonances of the percussions skin or the initial transient. For recreating the effects of resonance, white or pink noise prove to be the most common options while if the purpose is to create a transient, then a triangle or square wave is often used.

Both amplifier and filter envelope of the first oscillator is commonly set to a zero attack, zero release and medium decay. This is so that the sound starts immediately on key press (the drummer's strike) while the decay controls how ambient the surrounding room is. If the decay is set quite long, the sound will take longer to decay away producing an effect similar to reverb on most instruments.

If the second oscillator is being used to create resonances to the first oscillator, then the amp and filter settings are generally always the same as the initial oscillator whereas if it's being used to create the transient, the same attack, release and sustain settings are employed but the decay is generally shorter.

For more creative applications, it's worth experimenting with the slope of the amplifier and filter's decay and attack parameters since non-linear envelopes can produce strikingly different results on the same timbre. Also, by experimenting with frequency modulation and ring modulation it's possible to create a host of new drum timbres. For instance, if the second oscillator is producing a noise waveform, this can be used to modulate the main oscillator to reduce the overall tone of the sound. What's more, both effects and processing chains shouldn't be underestimated. Many of the sounds in today's productions will be subjected to any number of processing and effects chains to add interest to the sound and its not unusual for the effects to be heavily modulated so that each strike is slightly different from the last in order to sustain repeated listening. This is one of the techniques we will discuss in more detail in Chapter 16.

CHAPTER 16

Composing Drum Loops

'In search of the perfect beat, I found my own rhythm...'
John Michael

Perhaps the most fundamental element to the production of electronic dance music lies with the drum loop. This is the ultimate foundation that the rest of the music will be built upon, and therefore, it is essential that it is produced well otherwise the track will not hold up well against others.

Whilst on casual listening a typical EDM drum loop may appear to be fairly simplistic, it often belies a far more complex production ethic. Indeed, doing little more than dropping drum timbres into a workstations grid will often result in a loop that sounds flat, insipid and uninspiring. To replicate the professional feel from released records requires a lot more attention, which is paid to the loop in terms of timbre, effects, positioning and timbre modulation.

Depending on the genre of music being produced, drum loops are, of course, all composed differently in terms of timbre choice and quantize positioning, and the second part of this book is dedicated to discussing these production techniques but there are nevertheless a number of fundamental techniques that are employed within *all* professional drum loops regardless of genre.

THE BASIC LOOP

The most common drum loop in EDM is the *four to the floor*. This is produced in a 4/4 time signature and features a kick/bass drum positioned on each beat of the bar alongside a snare or clap positioned on beats two and four. This is further augmented with a 16th closed hi-hat pattern and an open hi-hat positioned on every 1/8th.

The synthesis programming of these percussive elements has been discussed in the previous chapter, but it is especially important that the producer pays close attention to the timbres and ensures they all sit well with one another. Almost every percussive instrument will exhibit a pitch with a specific frequency range.

FIGURE 16.1
Logic Pro audio displaying the EDM drum arrangement (these are all contained in the same event for the purpose of this picture)

Although you may not be able to determine the exact pitch on listening, it does nevertheless mean that each instrument must be carefully chosen so they seat well together. This is almost impossible to accomplish by simply picking out a single sample from a hard disk, and an often approach is to employ *sample lanes*.

Sample lanes consist of loading a single sampler instrument with a large collection of one specific instrument and mapping each different sample to a different note in the sampler. For example, one sampler may consist of 40 different closed hi-hat samples mapped to the individual notes on the samplers keyboard, whilst another sampler may contain nothing but snares (or claps) and another nothing but open hats.

The producer lays down the four to the floor kick first since this is the primary instrument and most commonly a programmed or layered timbre, and then inserts the snare/clap sampler on a secondary instrument channel. After programming the MIDI for the snare pattern in the piano roll, playback is initiated so the producer can listen to the kick and snare combination. By selecting all the snare MIDI in the piano roll, it can then be moved up and down the piano roll editors keyboard, and by doing so, the connected sampler will play a different snare sample.

If this movement is mapped to a keyboard shortcut, the producer can close his eyes and listen intently to the relationship occurring between the kick and the different snare samples until a sample is found that appears to seat well with the kick. This same process then occurs with the closed hi-hats and the open hi-hats.

The advantage of this approach is that the producer can carefully reference both the pitch and frequency relationship between the instruments already employed in the loop and any further instruments that are being added to the loop. This careful selection ensures that the instruments all seat well together and is a major contributing factor to producing a professional sounding drum loop. The result of choosing samples using this lane approach can be heard on the COMPANION WEBSITE.

The basic EDM drum loop.

Whilst this approach will certainly produce a more professional, cohesive sounding loop than if the producer randomly choose the odd sample and dropped it into the loop, it will not permit sustained listening periods. In order to accomplish this, the producer must employ a number of techniques.

The first technique is in direct contrast to the advice from many publications and involves applying reverb to the kick drum. Whilst numerous publications will warn the producer from applying reverb to lower frequency elements, for EDM production, it is almost common practice to apply small amount of controlled reverb to a kick.

This must be applied cautiously since too much will wash the kick out resulting in it loosing its central focus in the mix, but if applied lightly, it will generate the typical kick drum character heard in many EDM records. Typically, a very small room setting with a pre-delay of 20 ms and a tail of less than 30 ms provides a good starting point, but this must be carefully monitored. As a rough guideline, the tail of the reverb should have completely decayed away before the next occurring 1/8th hi-hat to prevent the loss of focus.

An alternative method to this is to employ a longer reverb tail of 80 or 90 ms but then follow this immediately with a noise gate. In many genres, the gate will be set to close the moment the kick timbre ends. This not only allows the reverb to add the textural character to the kick but also produces a transient dynamic edge to the kick drum. The noise gate must be set carefully, however, since if the attack and release are too fast, some gates will track the individual cycles of a layered kick and this can result in distortion. Notably, some genres such as Techno, Minimal and Tech house will deliberately influence this distortion to add to the kicks character.

Snares and claps can also benefit from a similar application of reverb although often the reverbs tail is permitted to run rather than cut off at the end of the percussive strike. The reverb settings to employ here are entirely dependent on the producer's creative goals, but typically it should remain fairly short so as not to wash over the loop completely. Long drum sounds will mask the silence between each of the drum strikes, and this can lessen the impact of the rhythm. The sudden strike from silence to transient on a snare/clap can have a dramatic impact on a rhythmical loop, and therefore, caution should be exercised to ensure the loop doesn't become too washed out with tails as this will result in sloppy sounding drums that lack dynamic action.

A popular technique in some genres is to apply small amounts of reverse reverb on a snare or clap so that it draws up into the strike. This is accomplished by first duplicating a sampled clap or snare event and then reversing it in the workstations editor. Once reversed, reverb is inserted into the track and applied with a short pre-delay and a tail of approximately 100 to 300 ms. This effect is printed onto the audio (by bouncing the file to audio and re-importing it into the arrange page), and the audio file is reversed again so it now plays the correct way around. The result of this exercise is a reverse reverb effect that draws up into the transient of the snare of the lap.

The reverb draw into the note can be modified if required through the use of a transient designer, EQ and the workstations audio fade in/out parameters. Ideally, the draw in should be kept fairly short so as not to smother the loop or change the appearance of the snare/claps timing. Indeed, the simple action of moving the timing (or employing effects that give the appearance of timing shifts) of the clap or snare by just a few milliseconds can make a significant difference to the feel of the record and the groove in general. For example, if the snare/clap is made to occur slightly behind the beat, it will result in a laid back feel. Whilst if it occurs just ahead of the beat, it will often produce a more intense surging feel.

Reverb and Gate applied on the Kick with reverse reverb on the clap.

Both open and closed hi-hats may have small amounts of reverb applied, but like other instrumentation in the loop, this must be applied with caution and the producer must take into account the composition, tempo and timbre of the hi-hats. For example, if the closed hi-hats occur on every 1/16th and reverb is applied, the tail of the reverb can wash over each consecutive hat transient resulting in an increasing reverberation field that can seriously affect the transient dynamics of the loop. Consequently, if reverb is to be applied to such a close-knit pattern, the producer will typically employ a noise gate or transient designer in serial after the reverb so that it can remove the reverbs tail.

This problem rarely occurs with the open hat due to their larger quantize spacing but ideally the reverb tail should expire before the next open-hat transient occurs. A good technique here to ensure clarity is to employ reverb on a send buss and follow it with an EQ. By doing so, the producer can sculpt the reverb, removing any low-frequency elements in the reverbs tail.

A more common alternative to the application of reverb on hats is to treat them to small amounts of tempo-synced delay. The settings for the delay are dependent on the producer's creativity, the tempo of the track and whether the open and closed hats are rhythmical or positioned specifically at regular occurrences on the quantize grid.

Since delay can often wash over the hats with faster paced genres of music, a common technique is to side chain compress the delayed hi-hat signal to produce a cleaner sound. Here, the delay unit is placed on a bus and followed with a compressor set to a fast attack and release with a high ratio and low threshold. The compressor is then fed with the original hi-hat signal into its side chain input, and the hi-hats are *sent* to this delay bus where a short to medium delay setting is employed.

The result of this configuration is that every time an original closed hat occurs, it triggers the compressor through its side chain input. As the compressor is triggered, it restricts the dynamics and lowers the volume of the delayed hi-hat. However, during the silence between the original hats, no compression is applied, and therefore, the delay occurs at full volume.

This technique prevents the delay from washing over the hi-hats and creating a mush of frequencies. What's more, if the producer then inserts a noise gate onto the original hat channel, the attack and release settings of the gate can be used to shorten the hats and in turn affect how the delay behaves producing a large variety of effects. The noise gate is often automated throughout the track to produce slow modulating effects as the rhythm track progresses.

The effect of a noise gate on a hi-hat channel that is employing a side chained delay.

SUBLIMINAL MODULATION

With the basic rhythm programmed and effects applied, it is common practice to employ a form of subliminal modulation to many of the timbres. Here, the snare or clap is first to be treated, and this commonly involves microtonal pitch or filter modulation so that the timbres occurring on the second and fourth beats of the bar exhibit an almost insignificant tonal difference between one another. This effect can be applied in a workstations audio editor or via a modulating effect.

To apply a microtonal change in an audio workstation, the producer will duplicate the snare/clap event in the audio editor (to not affect all snares or claps in the project!) and then pitch this timbre up or down by a few cents. A typical application is around 7 to 8 but it should *not* exceed 12 cents. The purpose here is not to make the pitch movements between the two evident, and the just noticeable difference occurs in humans anywhere between 7 and 12 cents. Therefore, it is often best to remain below 12 cents to achieve this style of effect.

An alternative to employing pitch shifting between the two is to employ a modulating filter effect. *SoundToys FilterFreak* is the preferred effect by many dance musicians here since it offers a multitude of modulation options for the filter but it can be accomplished through any modulating filter or even automation. Here, the modulation is cyclically applied so that first snare/clap of the bar passes through a low-pass filter unmolested but the second strike occurring on beat 4 is subjected to a very light low-pass filtering that removes a very small portion of the higher harmonics of the timbre.

The purpose of this the effect is that it introduces textural variance to the timbres. This is a key element to maintaining interest and focus because the mind naturally expects to hear some kind of variance, no matter how small, and if it doesn't, it will simply switch off to it. By employing these small fluctuations in texture and pitch, the subconscious mind pick up on the harmonic variance, and therefore, it can maintain sustained listening periods.

Of course, this modulation effect is not limited to filters or pitch, and almost any harmonic altering effect can be employed including distortion and bit rate reduction. However, whatever effect is applied, it must be employed in such a way that the harmonic differences between the strikes on the second and fourth beats are almost imperceptible to the human ear.

Subliminal filter modulation applied to the snare.

Alongside the snare/clap, the open and closed hi-hats are often treated to a very similar form of cyclic filter modulation. Typically the open hats will be treated to a low-pass filter that is modulated via an offset sine wave (it is possible to offset a sine wave modulation within *FilterFreak*) or via a sample and hold waveform. When this effect is applied *very lightly* over the length of two bars, very slight harmonic changes occur over a period of two bars, and since the modulation source is offset or random, it prevents it from appearing as a cyclic repeated motion.

The result is that whilst the snares/clap are receiving a cyclic modulation every bar, the open hats are modulated every two bars, and this creates a syncopated modulation between the two different timbres. This prevents the listeners psyche from specifically identifying a fixed cyclic pattern in the rhythm.

FIGURE 16.2
Cyclic modulation applied via SoundToys FilterFreak

Applying cyclic filter modulation to the closed hi-hats over three or six bars can further augment this effect. Three or six bars are chosen since this works against the standard structural downbeat of dance music and results in a cross syncopation of modulation. Structural downbeats will be discussed in more detail in Chapter 19 but suffice to say that by employing this odd numbered modulation technique, it is difficult for the subconscious to disseminate and rhythmically position and therefore produces a loop that can maintain prolonged listening periods.

The effects of cyclic modulation applied to the snare, open and closed hats.

Once the basic rhythm is complete, further interest can be added to the loop by employing some displaced, syncopated hits. This approach is common in nearly all examples of dance music and helps to give the rhythms an additional edge. Syncopation was touched upon when discussing the music theory of EDM rhythms in chapter but to revisit the theory again here, it is a result of placing stress off the beat. Although typically this would involve changing the accent of notes or percussion that are already present, here the producer places a number of percussive hits on unequal positions of the bar.

Figure 16.3 shows the typical 16th subdivision of the bar that is employed with many drum loops. In this configuration, all of the drum instrumentation is placed on an equal subdivision of this bar. The kick appears equally subdivided on position 1, 5, 9 and 13 (every four step divisions), snares and claps occur on equally subdivided 5 and 13 (every eight step divisions) and the

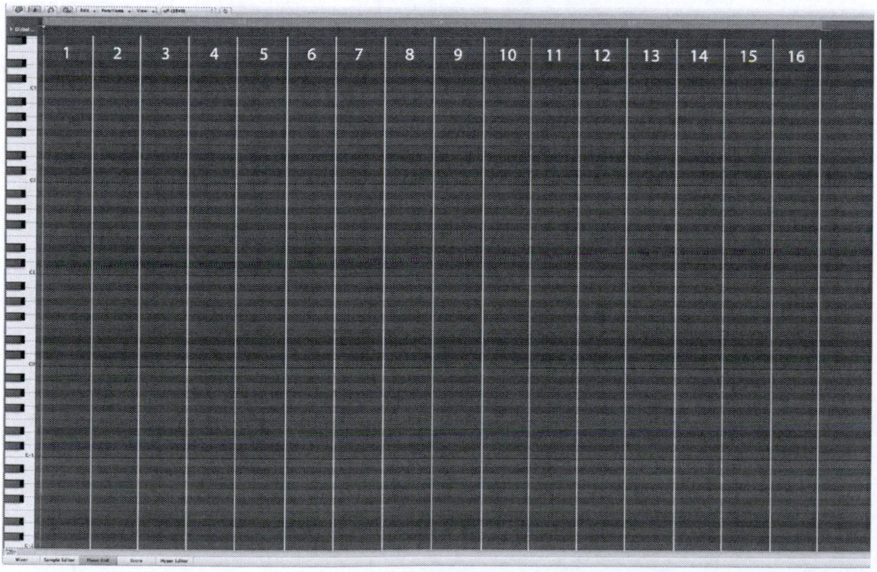

1 2 3 4 5 6 7 8 9 10 11 12 13 14 15 16

FIGURE 16.3
Equal subdivision of the bar

closed hi-hat occurs on each of these 1/16th positions. If, however, a percussive hit is placed on an unequal subdivision of the bar such as position 12 and 15, it results in a rhythmical syncopation. This produces the effect that two conflicting rhythms are playing.

The effects of unequal divisions.

In the audio example 16.3, I positioned two short percussive strikes on position 12 and 15 in the bar and then subjected them to small amounts of delay followed by a small amount of subliminal modulation through a modulating low-pass filter.

This was set over a bar and modulated with an offset sine wave (in *FilterFreak*). To add more complexity to the two delayed hits, the delay was placed on a send channel and EQ was applied after the delay. This results in the delayed signals sounding different again to the original percussive hits. Just this simple exercise of displaced positioning and delay treated to EQ immediately results in a more complex rhythm that is difficult to disseminate.

Further to syncopated hits, triplet rhythms can be introduced at specific intervals during the drum rhythms. These are rarely applied throughout the entirety of a tracks rhythmical section, since the effect can overcomplicate a steady rhythm but employed occasionally, it can introduce more interesting results and add excitement to rhythms. This form of Hemiola was discussed in detail in Chapter 3 but to quickly return to its principle here it can be used to produce the effect of cross meter modulation.

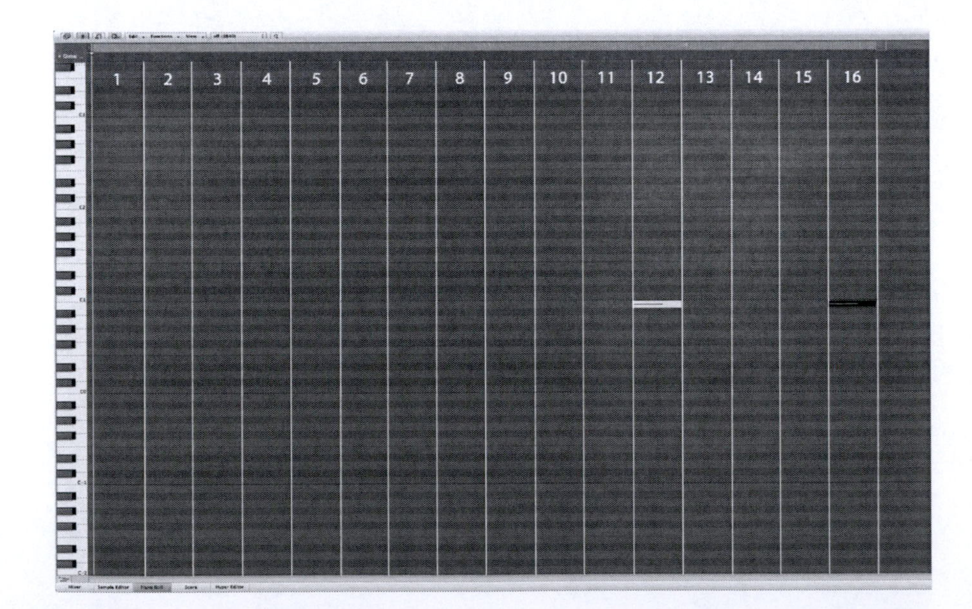

FIGURE 16.4 Hits positioned on unequal positions of the bar

For this effect, the workstation should be set to operate in a triplet grid creating three equal subdivisions of the bar. By then placing short percussive hits such as toms, congas, hi-hats or, in the case of techno and tech-house, synthesized percussive timbres or bass hits on these triplets, the pulses occur against the evenly distributed kick and snare rhythm resulting in a complex rhythmical interpolation.

Introducing Hemiola.

In the COMPANION WEBSITE example, I programmed some short percussive style timbres to sit on the triplet rhythm but changed the pitch of each microtonally so that the second note is –5 cents lower than the first and the third moves +3 cents above the second strike (making the third a total of –2 cents lower than the first note). Note that since this triplet effect clashed with the previously introduced syncopated strikes, I removed them.

This triplet can, however, be further manipulated, and a common technique for EDM is to maintain this triplet pattern over a number of bars, and then after a fixed number of bars, it is reduced to a single note for one bar. This *asymmetrical phrasing* creates a denial of expectation in the listener. This is because the listener becomes accustomed to the triplet rhythm, so when it is removed, it surprises their consciousness and increases interest in the rhythm section.

Similarly, employing techniques such as polymeter can add further interest to drum rhythms throughout a track. This again was discussed in detail in

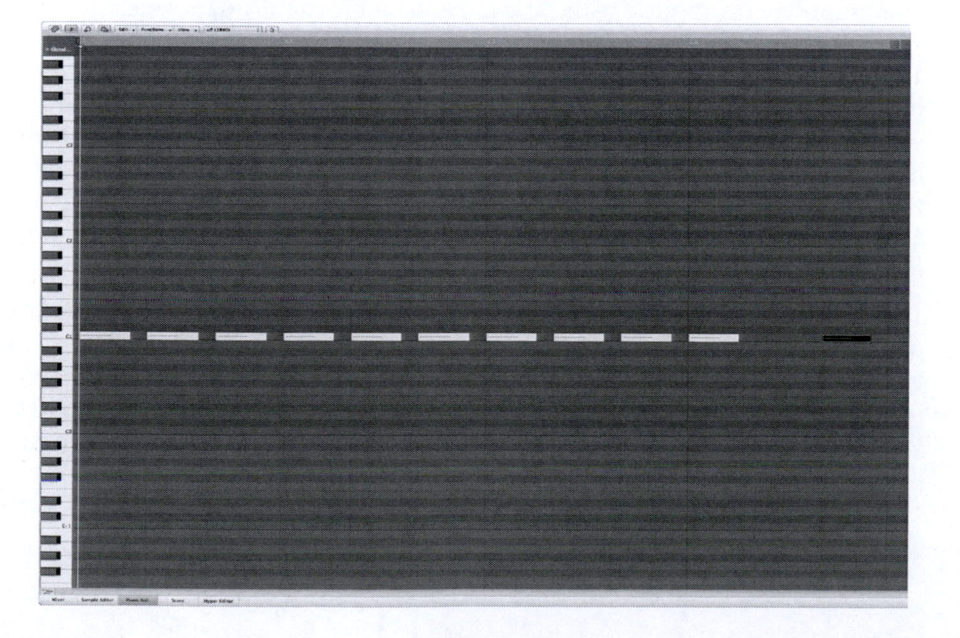

FIGURE 16.5
Hemiola

Chapter 3 on the fundamental of EDM rhythm and simply involves moving from the standard 4/4 to another such as 5/4 for just one percussive instrument. By doing so, it produces patterns of different lengths that move in and out of sync with one another over a specific number of bars. In order for this to work, however, the 5/4 time signature must employ a specific rhythmical pattern, since if it consists of equally spaced transients, it would not be possible to determine the polymeter.

SWING

Previously, the drum loop was treated to textural variance via an almost insignificant filter and pitch modulation. As mentioned, this variance helps to keep the subconscious mind interested in the rhythm, but alongside the textural development, it can also benefit from time-sensitive development.

In other words, if every single percussive hit strikes dead on the quantize grid every time, the mind is more likely to switch off from it. Therefore, alongside textural variance, the producer should also introduce some time-sensitive variance via swing.

Swing is perhaps one of the most overlooked factors in creating a great dance drum loop since it is often contributed to the laid back hip-hop style beats. Although many of these tracks did employ swing, most commonly from the classic AKAI MPC range, its application reaches further afield and into almost every genre of electronic dance music.

Swing is typically applied automatically via a workstations swing quantize option. Here, the producer can choose a percentage value up to 100% and the amount of swing applied with be relative to the current quantize setting. For example, 50% swing is considered snap quantize in that no swing is being applied. Thus, if the swing quantize value is set at 54% and the current quantize is set to 1/16th notes, the selected notes or channel will all be moved forwards or backwards in time by 4% of a 1/16th. Typically, unless the producer is deliberately aiming for a laid-back or syncopated feel to the music, in most genres, quantize should be small enough so as not to be immediately noticeable and therefore settings of between 51% and 59% will provide the required results.

Typically, the kick drum is left dead on the beat, and it is only the surrounding percussion that is treated to swing (claps, hats and further percussion). This allows the kick to determine the exact timing and provides a context for the instruments that are affected by swing. In the example on the COMPANION WEBSITE, I applied a 9% swing over a 1/16th to the claps, and open and closed hi-hat rhythms.

The effects of swing.

PARALLEL COMPRESSION

A final technique that is often applied to the rhythm section for almost all genres is parallel compression. This is sometimes termed *New York Compression*, but this is not because it was first employed or discovered there, rather it is so named because it became a popular effect with many New York mix engineers.

Fundamentally, parallel compression is a form of *upward* compression. Typically a compressors action is *downward* in that it will restrict the dynamics of any signals that breach the threshold whilst leaving those below the threshold unmolested. With upward compression, the opposite occurs so that the quieter signals are increased in gain whilst the normally restricted remain the same. This type of effect is accomplished by first sending the rhythm component tracks to a single group track so that they can all be controlled via a single mix fader. A compressor is then inserted onto a bus, and the group track is sent to the bus by a small amount for compression.

By employing this technique, both the uncompressed group signal and the compressed signal can be merged together at the desk. By doing so the transient peaks of the original group track remains, but the low-level detail of the compressed track is increased (due to the compressor reducing the dynamic range) and the result is a thicker more defined rhythm without increasing the transients.

The settings to use on the parallel compressor are heavily dependent on the loop in question but a good starting point is a ratio of 3:1 with a fast attack and an auto release (or a release of around 200 ms). With these basic settings, the threshold should be lowered until it registers approximately 5 to 8 dB of gain reduction. Once this is set, either the bus fader or the compressors make-up gain can be increased until it achieves the desired effect.

The effects of parallel compression on a drum loop.

Compression can also be used to produce a crunchy distortion that is particularly suited towards Techno and Tech House. Here, two compressors are employed in a serial configuration. The drum loop runs into the first compressor set to a high ratio and low threshold mixed with a fast attack and release. If the output gain of this compressor is pushed high enough, it will result in distortion of the midrange that adds to the character of these genres. By then running this distorted signal into a second compressor, the distortion can be controlled to prevent it from clipping the outputs of the workstation.

Of course, here, we have only touched upon the four to the flour rhythms, but the exact same techniques can be, and often are, applied to *any* style of rhythmical percussive loop. Regardless of the rhythmical positioning of any of the instrumentation, these same techniques if employed carefully will breathe life and action into the loop, creating a rhythm with a more polished professional sound whilst also helping to maintain prolonged listening periods.

CHAPTER 17

Sound Design I – The Theory

'It's ridiculous to think that you're music will sound better if you go out and buy the latest keyboard. People tend to crave more equipment when they don't know what they want from their music...'

A Guy Called Gerald

Sound design is one of the most fundamental elements in the creation of electronic dance music. The sounds programmed and employed will often determine the genre as well as the quality of the music. Indeed, poor sound design is *the* main contributing factor that separates an amateur production from a professional one. If time, experience, knowledge and effort are not put into the initial design of all the timbres the final record will inevitably suffer.

Generally, sound design can be divided into three distinct categories: audio design, synthesis design and processing/effects chain order. Although all three methods can be mixed and matched to produce the final results, the most complex, yet most important of these is basic synthesis.

Almost every synthesizer whether virtual or hardware now come supplied with any number of presets ranging from 30 to over 600. These are typically programmed to show the extent of the synthesizers capabilities, but many novice producers tend to rely entirely on them for their productions. The problem with this approach, though, is that the producer becomes little more than a preset surfer, spending countless hours flicking through one preset after another attempting to find a sound they like.

If they are unable to find a preset they enjoy, they open yet another synthesizer and the chain of moving from one preset to another continues. This is not music production, its painting by numbers and the end results will be as unexciting as the surf for the right sound, an effort that results in a novice type of production.

Indeed, the separation between amateur and professional is a passion for the music and refusing to settle for anything less. The producer has an idea of the style of timbre they require and they will forgo the preset surfing and instead get down to programming their idea from the very start of the production. Without this form of drive and focus, the producer will end up surfing and continually settling for second best resulting in an amateur sounding production.

One of the fundamental applications of synthesis is to know the type, or style of sound the mix requires, that is, what you want even *before* you open the synthesizer. You must have a good idea of the direction and style of sound you want to accomplish and from that make a decision on the synthesizer you're going to employ. Just as you wouldn't jump into a car to undertake a journey with no idea of where you want to go, the same applies with synthesis.

Despite a manufacturer's claims that their synth is capable of producing any sound, different synthesizers do exhibit very different tonal characters, and therefore, some are more suited to programming some styles of sounds more than others. Just as you wouldn't expect different models of speakers/monitors to sound exactly alike the same is true of synthesis.

Although oscillators, filters and modulation are all based on the same basic principles, they do all sound different between synthesizers. This is why some synthesizers are said to exhibit a certain character and is the reason why many older analogue keyboards demand such a high price tag on the second-hand market. Producers are willing to pay for the particular sound characteristics of a synthesizer.

Consequently, before the producer even begins programming their own timbres, they will often set aside a good amount of time to experiment and learn the characteristics of their synthesizers. Simple exercises such as listening to each oscillator in isolation followed by mixing sine waves with square waves, a square with a saw, a sine and a saw, a triangle and a saw, a sine and a triangle and actually *listening* to the results are absolutely fundamental techniques for programming. After all, if the producer has no idea how the basic oscillators sound then it's going to be quite a challenge to program a timbre with them.

Provided the producer is familiar with the synthesizer being programmed, there are two ways synthesis that can be approached. The producer can use careful analysis to reconstruct a sound they have envisaged for the music or they can find a patch close to their initial idea and then tweak the available parameters to sculpt it more favourably to their own mix. Both of these options are equally viable options for sound design, but before the producer takes either approach they must consider a number of variables.

Perhaps first and foremost, unless the producer is 'scavenging sound' – that is they are looking to build a sound to inspire them creatively – it is more usual that the producer will have a melody or motif that requires a timbre. This approach is important since a common mistake is to amend parameters whilst doing little more than striking a random note to audition the changes. While this will provide aural clues towards the character of the timbre, it does not provide it within the context of music.

For instance, the envelope generators, LFOs, filters and effects of a synthesizer will all heavily influence a motif, not just the timbre. A motif constructed of 1/16th notes will appear very different if the amplifiers release were altered. If it were shortened, the motif would become shorter and more defined, perhaps permitting longer delay effects or reverb settings, whereas if the same parameter were lengthened, the notes would flow together and delay or reverb would only create havoc within the mix.

Alternatively, lengthening the amplifiers attack parameter would severely affect the timing of the actual MIDI notes, shifting them in time in relation to the rest of the music. Similarly, low-frequency oscillators restart their cycle on patch auditioning, but may not do so through a motif as the note length changes and if filter key follow is employed, the filters action will change according to the current pitch.

Similarly, whether modifying a preset or programming a timbre from a blank canvas it is important to take note of available frequency ranges and carefully consider the application of stereo and effects. Indeed, when it comes to introducing new timbres into a mix, the producer must take the frequency range of all the other instruments currently in the mix into account.

Each timbre in an arrangement will occupy a specific part of the frequency range. If you were to program each of these individually without any regards to the other instruments that will play alongside it, you can end up programming or modifying a host of harmonically complex timbres that whilst sound great in isolation will create a cluttered complex mix when they are played alongside one another. Consequently, whenever approaching synthesis, it is important to prioritize the instruments, giving those that are genre defining the most detail and simplifying any further parts that sit around them.

For example, in a genre such as uplifting trance, the main focus of the track is on the lead instrument and its melody. Therefore, when approaching this genre, the lead should be treated as the priority instrument. This means that the lead should be programmed to be as 'wide' and 'large' as necessary but the instruments that surround this such as chords or pads should be programmed to sit behind the lead rather than be equally large and compete with it for frequency room within the mix. Without taking this frequency considerate approach and instead believing it can be 'fixed in the mix' later with EQ, the producer will end up with a poor, cloudy and cluttered mix. As discussed in later chapters, mix EQ should be considered as a fine chisel for fine-tuning and detailing a mix, not a jackhammer to completely reorganize and repair it.

The ability to hear and determine the available frequency range when programming is a result of ear training and practical experience. Both these can come with plenty of practice, but for beginners, it is prudent to employ a spectral

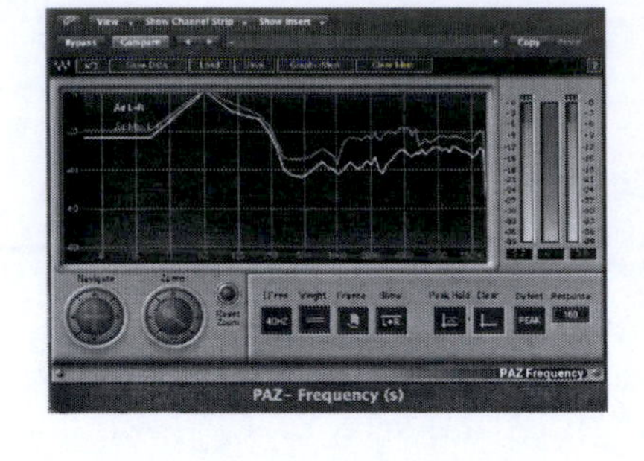

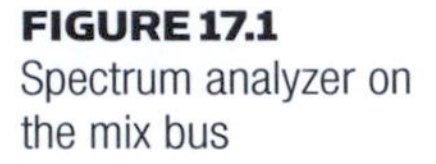
FIGURE 17.1
Spectrum analyzer on the mix bus

analyzer on each channel and also on the main mix bus. Many audio workstations provide spectral analyzers as a free plug-in and by employing them, it is possible to view how much of the frequency range each instrument is consuming and on the mix bus, the frequency areas that still remain free.

It should be noted that sounds consist of numerous overtones that do not contribute the body of the timbre, so this must be taken into account when viewing the frequencies in a spectral analyzer. For instance, a bass timbre and melody could easily consume frequencies from 20 Hz to 12 kHz, but in many instances, the body and character of the bass may only contribute a 1/8th of this. Thus, provided the bass is not the driving element of the music, the producer could employ an EQ to shelve off everything below 80 Hz and everything above 1 kHz.

Although this may result in the bass loosing some of its character when played in isolation, when played in the context of the mix with all other instrumentation, it will not be noticeable. Moreover, this 'progressive mixing' approach not only creates frequency room for further instrument programming and design but also makes the final mix-down much easier.

Similar consideration should be given to the application of any effects *during* programming. Almost all synthesizer patches are designed to sound fantastic in isolation but when these are all combined into the final mix, the effects tails, delays, chorus and so on, will all combine together to produce a cluttered and indistinct result. However, much of the impact from electronic dance music is created through contrast, and therefore, washing each timbre in numerous effects should be avoided.

For example, the defining timbres of the music will often benefit greatly from effects and a lead instrument will have a much larger impact if many of the other instruments are 'dry' in comparison to it. If all the instruments are soaked in effects, this impact is significantly lessened and the contrast is lost. Additionally, it is best to avoid applying effects direct from the synthesizer and instead apply third-party effects *after* the fact. Indeed, any effects that are supplied within an instruments interface should be considered as included only to make the synthesizers preset patches appear attractive. In many synthesizers, these effects are generally below par in regards to sonic quality, particularly in terms of reverb and delay and employing these can often make a mix appear clouded or lacking in definition.

Instead, it is preferable to employ third-party effects onto the synthesizer via the workstation. This way the producer can ensure that each effect is of good quality, and it also permits the effects to be applied in a specific creative order rather than in serial as with many synthesizers or if in parallel, through a fixed configuration.

What's more, when programming timbres, it is sensible to create timbres in mono rather than stereo. Most plug-in and hardware instruments will exaggerate the stereo spread to make individual timbre appear more impressive. This effect is often created through the use of effects or layering two different timbres together that are spread to the left and right speakers. While this makes timbres sound great in isolation, they will accumulate across the stereo field and reduce the available sound stage when mixing. Therefore, the producer should look towards programming sounds in mono, unless it is priority instrument that would benefit from being in stereo. This, however, does not apply if programming a kick drum. While this does form an essential part of the music it should be mono, as it should sit dead centre of the mix, so that both monitor speakers can share the energy.

As discussed previously, there are two viable ways to approach sound design, the producer can either program a timbre from a blank canvas or find a preset that shares some similarities to the sound they require and then proceed to modify it. Since instruments within a dance mix share many similarities, for example, a bass sound is just that – a bass sound – they are generally approached and programmed in roughly the same manner. Therefore, if you are new to synthesis and wish to construct a bass, dial up a bass preset, strip all the modulation and effects away from the preset patch, so it is left with little more than the mixed oscillators and then build upon this foundation.

With both modulation and effects stripped, the most commonly assaulted parameter is the filter and resonance. However, whilst these do strongly affect the character of the timbre, it is generally preferable to modify the amplifier envelope to first suit the melody. This is because the amplifiers EG will have a significant influence on the MIDI motif and the tonal shaping of the sound over its period.

The initial transient of a note is by far the most important aspect of any timbre. The first few moments of the attack stage provide the listener with a significant amount of information and can make a substantial difference to the timbres characteristics. We perceive timbres with an immediate attack to be physically louder than those with a slower attack. Equally, we perceive any short transient style sounds to be quieter than sustained sounds with the same amplitude and frequency content. Therefore, if a timbre appears 'slack' when played back via a MIDI motif, simply reducing the attack and/or release stage can make it appear much more defined.

Notably, the transient behaviour of the different melodies will strongly affect the energy and flow of the music in question. Short-transient notes will make music appear faster whilst longer notes will slow the pace of the music down. This is not to suggest that if you want to produce a fast track all instruments should exhibit a short transient behaviour, however.

Electronic dance music is based through contrasting behaviour of the timbres and the producer needs to not only think in terms of tone, but also time. By employing a fast attack and release on some instruments with a fast attack and slow release or a slow attack and fast release on others, it will create a mix that exhibits contrasting rhythms. For example, a fast-1/16th pattern bass line can be made to appear even faster not by shortening the envelope times but by simply introducing a slow-evolving pad into the background. Typically, dance music that exhibits a fast passed groove will employ a transient style bass whilst slower paced genres such as chill-out will employ longer, drawn out bass notes to pull the groove of the record back.

Both amplifiers decay and sustain can be employed to add further time-based characteristics to the timbre. Decay is possibly best viewed as part of the creation of the transient, since this controls the time it takes for the timbre to fall in amplitude before it reaches the sustain amplifier level. It should be noted here that sustain can effectively be considered a gain parameter, which determines the amplitude of a note for as long as the note is held. This means that if sustain is set to maximum there can be no decay, since there is nothing to decay too. Similarly, if the sustain is set to minimum, the release parameter would be rendered redundant, since there is no gain to release from. In this situation, the decay acts as the release parameter.

Typically, with shorter transient style sounds and motifs, both amplifier decay and sustain will have little influence over the timbre and both parameters are more suited towards the creation of long, developing timbres such as pads and strings. Here, the decay and sustain are often employed with a long attack stage to create a timbre that rises and falls gently in volume, whilst being modulated or augmented with the filter envelopes.

Once the overall shape of the amplifier has been modified, the filter envelopes and filter cut-off/resonance can be employed to modify the tonal content of the timbre. Typically, here it is beneficial to determine the filters envelope and then adjust the filters cut-off and resonance since the envelope will determine the cut-offs action.

A filters envelope operates similar to the amplifier envelope but rather than modify volume, it modifies the filters action over time. Here, if the attack is set

FIGURE 17.2
The basic modification parameters in FM8

to immediate then the filter will act upon the sound on key-press, whereas if it's set slightly longer the filter will sweep open into the note. More important, however, is the filters decay setting, since this determines how quickly the filter falls to its sustain rate.

In the context of a timbre, this is employed to control the 'pluck' of a sound and is one of the most important character-defining parameters on a synthesizer. Indeed, the modulation of this decay parameter whether via automation, or a secondary envelope is responsible for a proportionate number of progressive house lead timbres, but it is equally responsible for the squelch and plucked basses and the countless dance pluck style timbres throughout all the genres. Similarly, with longer progressive pad sounds, the filters attack and decay are commonly adjusted, so they are longer than the amplifiers attack and decay rates creating some of the more complex and typical filter swept pads. This is because as the amplifier envelope begins the sound, the filter slowly sweeps in creating harmonic movement throughout the sustain portion of the amplifiers envelope.

A filter envelopes sustain and release behave in the same manner as the amplifiers sustain and release, but here the envelopes sustain will determine the filters cut-off setting. For example, if the sustain parameter is set to maximum, the filter will open to its current cut-off and remain there. This means that the filters decay will have no effect on the sound since there is nothing to decay to. Alternatively, if the sustain is set to minimum then the decay would act as the release parameter since there is no sustain to release from.

Since the filters envelope will react on the current filter settings, during the modification of the envelope, it is also necessary to simultaneously modify both the cut-off and resonance. This is because the filter envelopes action will be determined by the current cut-off. In other words, the attack parameter of a filter's envelope will gradually increase from completely closed to the filters current setting but it is unable to exceed it. Therefore, if a low-pass filter is employed, and this was set halfway to 10 kHz, the attack parameter would gradually open the filter from completely closed (0 Hz) and sweep up to the maximum of 10 kHz by the time determined by its attack parameter.

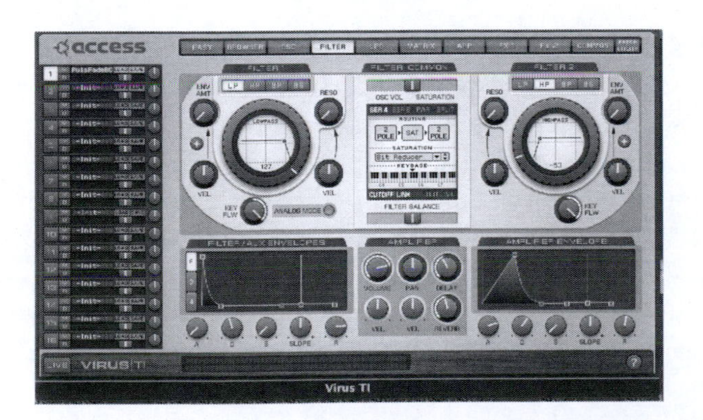

FIGURE 17.3
The filter and amplifier section from the Virus Ti

Most timbres will tend to utilize a low-pass filter with a 12 dB transition, as this tends to produce the most instantly musical results but a 24 dB transition used on other timbres can be employed to introduce contrast. A good example of this is by double tracking a MIDI file and using a 12 dB and 24 dB on the same timbre on the two different channels. By doing so, the two transitions interact creating a more complex tone that warps and shifts in harmonic content.

Additionally, while the most common filter is a low-pass, experimenting with other filter types such as band- and high-pass can aid in the creation of thinner timbres that can be layered. For example, using a low-pass on a timbre but then double tracking this to a secondary instruments and employing a band-pass or high-pass can result in any number of bright overtones. If these overtones are mixes with the original timbre, it will produce a sound that is much richer in harmonic content that can be further filtered with a plug-in.

Further modifications are available with filter key tracking. One of the more overlooked features on a synthesizer, it is one of the more powerful functions. Employing this it is possible to change the way in which a filter behaves, or opens, when playing higher or lower notes on the keyboard. This style of action was originally introduced into synthesizers to supposedly enable producer to recreate realistic instruments, but its purpose is well suited for electronic music sound design, since it can be used to add subtle difference to the different pitches in a melody.

For example, if the filter key tracking is applied to a timbre by a positive amount, the filter will open more as the pitch increases. This can be incredibly useful in the creation of moving chords (especially fifths and sevenths), but also plays a significant role in the creation of dub-step style timbres. Since these rely heavily on the manipulation of pitch bend through the course of a note, as the pitch is increased, the filter also opens creating the powerful 'growl.' Similarly, employing filter key tracking on a low-pass filtered bass with a high resonance can produce a bass that exhibits more energy whilst employing negative tracking can produce bright bass notes and darker high notes.

MODULATION

If we must pinpoint one defining element to any professional timbre then it is the continual time-based development that is accomplished through the modulation matrix. Although the filter envelope can be employed to create harmonic movement throughout the period of the timbre it is the synthesizers modulation possibilities that define its true potential.

All synthesizers will feature a modulation matrix but the more complex this matrix is, the more complex the sounds that can be created. The matrix can be viewed as a patching bay permitting the programmer to specify sources and destinations.

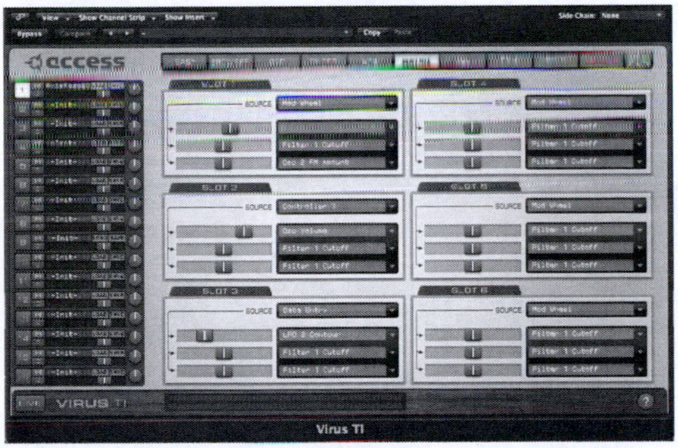

FIGURE 17.4
The modulation matrix from the Virus Ti

In many synthesizers, the most common source is an LFO whilst the most common destinations for the LFO to modulate are oscillator frequency, filter cut-off and resonance. For example, by employing a sawtooth LFO to modulate a filters cut-off, the harmonic content will rise sharply but fall slowly and the speed at which this takes place would be governed by the LFOs rate. As the tempo of dance music is of paramount importance, it's often sensible to sync this ratio modulation to the tempo, so that the rate remains in synchronization with the tempo.

This application, however, should be viewed as the most minimalistic approach and in order to program professional timbres, it is not unusual to modulate multiple destinations from multiple sources. Indeed, modulation quite simply offers the most crucial and significant contribution to the creation of a professional timbre.

It would be impossible to list all of the available modulation options in a book and experimentation and an examination of your favourite patches will provide the real key to successful programming. This is quite simply something that can only be born from practice and no publication can ever replace the producers willingness to spend as much time as necessary experimenting with, and examining any number of presets to gain an understanding of how the modulation affects the overall character of a timbre. But this understanding is absolutely fundamental to sound design and like playing any instrument, without practice, you'll never succeed at it.

While customizing synthesizer presets to fit into a mix can be a rewarding experience it nevertheless a basic approach and is only suitable provided that the producer can actually find a similar timbre to the one he/she is considering. Even then, this approach still requires the producer to embark on a mind-numbing journey through countless preset, and this search can quickly extinguish any creative impulses.

However, if a considerable amount of time is spent modifying and examining presets, modulations and effects, then the core of programming a timbre from a blank canvas consists of little more than an understanding and experimentation of the oscillator section of the synthesizer.

The oscillators and the various augmentations available in the mix section were discussed in detail in Chapter 6, and once this basic knowledge is affirmed, the rest is down to the producer listening and experimenting with the available oscillators and augmentation available in the mix section. Indeed, I cannot stress enough that in order to be competent with any synthesizer it is fundamental that the producer listens to each oscillator and the results of experimentation from mixing and detuning oscillators, along with the application of oscillator sync, frequency modulation and ring modulation.

CHAPTER 18

Sound Design II – Practice

'I see dance being used as communication between body and soul, to express what it too deep to find for words'.

Ruth St. Denis

In Chapter 17, we discussed the theory behind sound design, but as discussed, understanding the theory is only the first step to successful programming and sound design. It is essential that the producer takes this knowledge and applies the theory to all their synthesizers because it is only by actually *hearing* the results that he/she will be able to knowledgeably select and use the right synthesizer for the timbre.

To help the producer to familiarize themselves with synthesizers, this chapter consists of a number of programming examples. These are not devoted to how to program a precise timbre from any one specific track, as it would most likely outdate the book before I've completed this chapter. Indeed, on that note, it is important to understand that there are no 'hit' sounds. I've lost count of the novice producers continually requesting how to emulate a certain timbre from their peers records, but they fail to realize that these artists became their peers by refusing to emulate anyone else and charting their own course. Using the same timbres as Sasha, Skrillex or Tiesto will not make you a great producer or result in you selling just as many records.

Consequently, this chapter concentrates on constructing the basic sounds that are synonymous with dance music and as such will create a number of related 'presets' that the producer can then manipulate further and experiment with.

PADS

Pads rarely form the central focus of dance music, but understanding how the basic form is created will help the producer understand how both LFOs and envelopes work together to produce interesting evolving timbres. Naturally, there are no pre-determined pads to employ within dance music production, and therefore, there are no definitive methods to create them but there are a number of guidelines that can be followed.

Typically, pads are employed to provide one of three things:

1. To enhance the atmosphere in the music, this is especially the case with chill out and ambient music.
2. To fill any 'holes' in the mix between the rhythm and groove of the music and lead or vocals.
3. To be used as a lead itself through gating.

Depending on which of these functions the pad is to provide will determine how it should be programmed. Although many of the sounds in dance employ an immediate attack stage on the amp and filters envelope generators, it is rare to employ this approach on a pad unless the sound happens to be providing the lead of the track.

We determine timbres that start abruptly to be perceivably louder than those that do not but we also tend to perceive sounds with a slow attack stage to be 'less important' even though this may not be the producers intention. As a consequence when pad sounds are employed as 'backing' instruments, they should not begin abruptly but slowly fade in, whilst if they are being employed as a lead, it would be preferable to employ no attack at all. This abrupt start to the timbre will help it cut through the mix and also provide the impression that it is an important part of the music.

A very popular technique to demonstrate this is the gated pad. This consists of an evolving, shifting pad that is played as one long progressive note throughout the track. A noise gate is then employed to rhythmically gate the pad to create a stuttering effect. This is commonly accomplished by placing the noise gate on the track and side chaining a secondary track into the noise gate. A rhythmical pattern is then produced for the secondary track that will subsequently rhythmically pulse the pad.

This is not the only method available, however, and a compressor can be employed to perform the same function by setting the compressors output gain low enough so that it cannot be heard when activated, or more commonly, there are now a number of LFO-based plug-ins, such as the Xfer LFO tool, that can be placed onto tracks and the pulsing behaviour can be drawn in by the user.

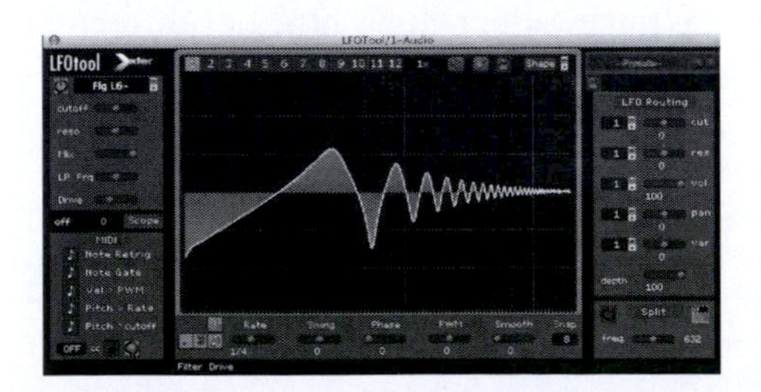

FIGURE 18.1
The XFer LFO tool

For this pulsing gated pad to be truly effective, the pad must evolve texturally throughout, since if this technique were employed on a sustained timbre with no textural movement, the producer may as well simply retrigger the timbre whenever required. Indeed, it's this textural movement and development that is key to produce a great sounding pad. If the timbre remains static throughout without any timbral variations, then the ear quickly tires and the listener switches off.

Typically, analogue style synthesizers are recommended for the creation of pads since the gentle, random fluctuations of the oscillators pitch can provide an additional sense of movement when augmented with LFOs and/or envelopes produces an attractive timbre. There are various ways to employ this movement ranging from LFOs to envelope generators, but the principle remains the same; to gradually increase or decrease the harmonic content during the progression of the pad. To better explain this, we'll look at the methodology behind how the envelopes are used to create the beginnings of a good pad.

By employing a fast attack and a long decay long on an amplifier envelope, we can determine that the sound will take a finite amount of time to reach the sustain portion. Provided that this sustain portion is set just below the amps decay stage, it will decay slowly to sustain and then continually 'loop' the sustain portion until the note is released whereby it will progress onto the release stage.

This creates the basic premise of any pad or string timbre; it continually plays until the note is released. Assuming that the pad has a rich harmonic structure from the oscillator section, textural development can be introduced by gradually increasing a low-pass filter cut-off while the pad is playing. By doing so, there will be a gradual increase in harmonics.

If a positive filter envelope is employed to control the filters action and the envelope amount was set to fully modulate the filter, by using a long attack, short decay, low sustain and fast release, the filters action would be introduced slowly before going through a fast decay stage and moving onto the sustain. This would create an effect whereby the filter would slowly open through the course of the amplifiers attack, decay and sustain stage before the filter entered a short decay stage during the 'middle' of the amps sustain stage. Alternatively, by employing the same filter envelope but applying it negatively, the envelope is inverted creating an effect of sweeping downwards rather than upwards.

The important practice here, however, is that the amp and filter envelopes are programmed to operate over different time periods to create a pad that evolves in harmonic content. Naturally, this arrangement can only be one way since if the function of these envelopes were reversed (in that the filter begins immediately but the amplifiers attack was set long), the filter would have little to no effect since there would be nothing to filter until the pad is introduced.

Obviously, to texturally evolve a pad, it must initially be programmed to exhibit a high harmonic content that is accomplished by carefully choosing the oscillators and detuning them heavily. The purpose at the oscillator mix section is to create a sound that is rich in harmonics for the filters to sweep.

In many synthesizers, this would equate to employing saws, triangles, noise and pulse wave oscillators. The choice of oscillator depends on the style of pad required but as a general starting point for practice, two saws, a saw and triangle or saw and pulse, would provide a harmonically rich timbre. These oscillators should be slowly detuned from one another in cents and semitones whilst carefully listening to the results. The producer should stop detuning when the timbre appears 'thick' enough or just before the oscillators separate into two distinct pitches.

If a pad appears lightweight then further weight can be added with a third oscillator set to a sine or triangle wave and detuned so that it sits an octave below the other oscillators. Alternatively, introducing unison may help to thicken up the sound if the synthesizer has this feature.

Alongside the previously discussed envelopes, further interest can be applied to a pad through LFO modulation. Typically, a saw, triangle or sine wave is employed to slowly and gently modulate the pitch of one or all of the oscillators whilst a further LFO could be employed to gently modulate the pulse width of the oscillator employed. Followed with any number of effects, this can all work to produce a timbre with motion and prevent it from appearing too static. This style of approach provides the foundation of many pad timbres, but this is only a starting point and it's up to the designer to experiment and modify the envelopes, modulation routings and LFO waveforms used for the LFOs to create a pad sound to suit the track.

To help lay down some foundations here, what follows is a guide to how some of the pads employed in dance music are created. This is by no means the definitive list but rather some basic starting points, and it's always experimenting to produce different variations.

Rise and Fall Pad

This pad is constructed by employing two saw oscillators detuned from one another. The amount of detuning depends on the synthesizer in question, but the producer should look to detune in cents until the pad becomes thick and wide. An envelope generator is then applied to this with a fast attack, short decay, medium sustain and long release. The filter should be a 12 dB low pass with the cut-off and resonance set quite low to produce a static saw buzz timbre.

From this, set the filters envelope to a long attack and decay but use a short release and no sustain and set the filter envelope to maximum positive modulation. Finally, use the filters key-follow so that it tracks the pitch of the notes being played. This results in the filter sweeping up through the pad before slowly settling down.

If the pad is to continue playing during the sustain portion for a long period of time, then it's also worth modulating the pitch of one of the oscillators with a triangle LFO and modulating the filters cut-off or resonance with a square wave LFO. Both these should be set to a medium depth and a slow rate.

Resonant Pads

Resonant pads can be created through mixing triangle and square, or triangle and pulse oscillators and detuning them from one another. The amount of detuning depends on the synthesizer but typically, between 5 and 50 cents will produce the best results. Similar to the previous rise and fall pad, the amplifier attack should be immediate with a short decay, medium sustain and long release, but the filters envelope should employ a long attack, long release, high sustain and a short decay.

A 12 dB low-pass filter should be used with the cut-off low but a resonance of around ¾ so that the timbre exhibits a high resonant quality at the cut-off point. Finally, modulate the pitch of the triangle oscillator with a sine wave LFO set to a slow rate and a medium depth and employ a positive filter key-follow.

The LFO modulation creates a pad that exhibits the natural analogue 'character' while the filter tracks the pitch and sweeps in through the attack and decay of the pad and then sustains itself through the amps sustain. Again, if the pads sustain is to continue for any length of time it is worthwhile employing a sine, pulse or triangle wave LFO to modulate the filters cut-off and/or oscillators pitch whilst if a pulse wave has been employed, modulating the pulse width can provide further motion.

Swirling Pads

The swirling pad is typical of some of Daft Punk's earlier work and consists of two detuned saw oscillators. As always, the detuning amount depends on the synthesizer and the style and texture that the producer requires but typically settings of 5 to 20 cents will produce the basic results. A square LFO is applied to one of the saws to gently modulate the pitch whilst a third oscillator set to a triangle wave is pitched 10 to 20 semitones above the two saws to add a 'glistening' tone to the sound.

The amp envelope employs a medium attack, sustain and release with a short decay while the filter envelope employs a fast attack and release with a long decay and medium sustain. A 12 dB low-pass filter is used to modify the tonal content that is set mid to low with the resonance approximately half way.

Finally, an effects chain consisting of Reverb > Noise Gate (to remove the reverb tail) > Chorus > Flanger is applied to the timbre to produce a swirling effect. It's important that the flanger is inserted at the end of chain so that it freely modulates the chorus effect.

Thin Pads

All of the pads discussed thus far can be particularly heavy but sometimes they may be required to simply sit in the background, and for this, it is beneficial to employ a lighter style of pad.

Many thinner style pads simply consist of one pulse oscillator that employs a medium attack, sustain and release with a fast decay on the amp envelope. A 24 dB low-pass or high-pass filter is used, depending on how bright the producer wishes the sound to be but there is commonly little need for a filter envelope since gently modulating the pulse width with a sine, saw or noise waveform produces the movement required.

Since the pad is designed to sit in the background, heavy textural movement is generally best avoided since strong movement will draw attention whilst also absorb more of the mix frequencies, but if it is required, a very slow triangle LFO set to modulate the filter cut-off or resonance set to a medium depth will provide some movement.

Examples of these timbres being constructed can be heard on the COMPANION WEBSITE.

BASS

For many dance genres, a synthesizer's bass sound is actually relatively simple since its main purpose is to simply pin the groove and offer a foundation for the lead. Consequently, they are not particularly complex to program, and it is possible to design some classic bass timbres by employing just one or two oscillators. Indeed, the secret, if there was one, to producing a great bass does not derive from the oscillators but from the filters and careful implementation of modulation to introduce textural development.

When approaching sound design for a bass, it is sensible to have the bass melody programmed and playing in the sequencer alongside the kick drum and any other precedence instruments. By doing so, it is easier to discern if the bass is interfering with the kick or other priority instruments while you manipulate the parameters. For example, if during the programming the bass disappears into the track or looses definition, then the producer can modify both the amp and filter envelopes to provide a more prominent attack.

For many bass timbres, the amplifiers attack should be set to its shortest time so that the note starts immediately on key-press and the decay should be adjusted so that it acts as a release setting (sustain and release are rarely used in bass timbres). This is also common with the filters envelope. Setting the filters attack stage too long will result in the filter slowly fading in over the length of the note that can demolish the transient of the bass.

Bass timbres often have a fairly complex attack, so if it needs a more prominent attack, it can be practical to layer a secondary instrumental 'pluck' style timbre over the transient. This pluck can be created in the same manner as creating a pluck for a drums kick but more commonly percussion sounds such as cowbells and wood blocks pitched down are used to add to the transient. If this latter approach is employed, it is advisable to reduce the length of the drum sample to less than 30 ms since we find it difficult to perceive individual sounds if they are less than 30 ms in length.

After the transient, the body of the bass should exhibit some textural or pitch momentum. No matter how energetic the bass melody, if the tone remains monotonic or feature no movement, the ears can quickly turn off. In fact, this lack of movement is a contributing factor as to why some dance rhythms don't appear to exhibit any real groove.

Movement in bass can be accomplished in a number of different ways; Dubstep employs CC pitch commands to create the warping moving basses whilst genres such as deep house, techno and tech house will employ rhythmical or micro-tonal modulation through the different notes of the melody. Many producers will often assign the modulation wheel to control parameters such as filter cut-off, resonance, LFO rate or pitch. By doing so, after programming a timbre, the wheel can be moved in real time to experiment with sonic movement and can be also subsequently recorded as CC data and later edited in the sequencer.

Deep Heavy Bass

The heavy bass is typical of Drum 'n' Bass tracks and is the simplest bass timbre to produce as it is essentially a kick drum timbre with the amplifiers decay and release parameter lengthened.

To construct this timbre, a single sine wave oscillator has its pitch positively modulated by an attack decay envelope. A transient is then applied by adding a second oscillator employing a square wave that is pitched down to suit the bass. This employs an immediate amplifier attack and decay with no release or sustain. Once this initial timbre is laid down, you can increase the sine waves amplifier EG decay and sustain until you have the sound you require.

Sub-Bass

The sub-bass is popular in most styles of electronic dance music and is often created through nothing more than a basic sine wave. This, however, is one area where the capabilities of the synthesizer are paramount since not all synthesizers will produce a great sub-bass. The preferred synthesizer by many artists for the creation of a sub-bass is Logic's ES2 instrument (often rated more highly than the requisite Access Virus Synthesizers for sub-basses). If the producer isn't using Logic Pro, then similar results can be attained through the creation of a self-oscillating filter. Here, however, the filter cut-off must be modulated via controller messages to change the pitch of the sub to match the music.

In either case, the amplifiers attack stage should be immediate, but the decay setting will depend entirely on the sound required and the current bass motif (a good starting point is to use a fairly short decay, with no sustain or release). If a click is required at the beginning of the note, then a square wave with a fast amplifier attack and decay and a low-pass filter with a medium cut-off and resonance can be layered over the transient of the sine wave.

This will produce the basic 'preset' tone typical of a deep sub-bass but is open to further modification. For example, by modulating the sine waves pitch by two cents with an envelope generator set to a slow attack, medium decay and no sustain or release, the note will bend slightly each time it is played. If an additional amplifier or modulation envelope generator isn't available, then a sine wave LFO with a slow rate that is set to restart at key press will produce much the same results.

In addition, changing the attack, decay and release of the amp or/and filter envelope generator from a linear behaviour to convex or concave will introduce new variations. For example, by employing a concave decay, it is possible to create a 'plucking' style timbre while using convex will produce one that appears more rounded. Similarly, small amounts of controlled distortion or very light flanging can also add movement.

Moog Bass

The Mini-Moog was one of the proprietary instruments in creating basses for electronic dance music and has been used in its various guises throughout almost every genre of music. Generally, if the producer wants the original moog sound, there is simply no substitute for purchasing a hardware moog. Although there are software emulations available, many artists agree that they don't match the sonic stage or presence of their hardware counterparts, and the hardware systems such as the Sub Phatty and Slim Phatty are unrivalled for producing bass timbres in dance music.

Nevertheless, the one timbre that many associate with a Moog can be constructed on many analogue modelled synthesizers and although it may not exhibit the presence of hardware, it is still a useful timbre to program. The timbre can be constructed through employing either a sine or triangle wave mixed with a square wave. The oscillators are detuned from one another by +5 to +30 cents until the right tonal character for the producer is achieved.

The amplifiers envelope is adjusted to a fast attack and medium decay with no sustain or release. The filter envelope parameters depend on the producer, but a good starting point is to employ a 12 dB low-pass filter set to medium with a low resonance. The filters envelope should then be set to a fast attack with a medium decay. By then lengthening the attack or shortening the decay of the filters envelope, it is possible to create a timbre that 'plucks' or 'growls'. Filter key-follow can be employed if the bass melody is particularly energetic in pitch, but if not, some character can be applied by modulating the pitch of one, or both, of the oscillators with a sine wave LFO set to a slow rate.

Acid House Bass

The acid house bass was a popular choice during the late eighties and early nineties scene, and although not as popular today, it still appears in a number of records from different genres.

This style of bass was first created using the Roland TB303 Bass synthesizer that, like the accompanying TR909 and TR808, is now out of production. It is, however, very highly prized and demands an incredibly high price on the second-hand market, even more so if it has the devilfish modification that adds MIDI In and Out ports.

The acid house (aka the 'donk') bass such as *Klubbheads 'Kickin' Hard'* can be created on most analogue modelled synthesizers through employing either a square or saw oscillator. The choice between the two depends on whether the producer wants a woody style sound or a more raspy result. The amplifiers attack stage is set to zero whilst the decay is set to suit the style of sound required, but a medium decay with no sustain or release provides a good starting point. The filter should be a 12 dB low-pass with the resonance set very high but just below self-oscillation and the cut-off should be set low to begin with.

The filter envelope should be set to suit the musical style, but a good starting point is a zero attack, sustain and release with a decay that's slightly longer than the amplifiers decay. This will produce the beginnings of the typical house bass timbre. Filter key-follow is often employed in these sounds to create movement, but it's also worth assigning filter cut-off to velocity so that the harder a note is struck, the more the filter will open.

Resonant Bass

The resonant bass is similar to the tone produced by the Acid Bass but has a much more resonant character that produces the TB303 squeal. This particular timbre cannot be accomplished on every synthesizer, however, since the timbre relies on the quality of the filters in the synthesizer.

The sound is constructed with a saw oscillator employing a fast amplifier attack with no sustain but a medium release and short decay. This creates a timbre that starts on key-press and quickly moves from its short decay period into the release, in effect, producing a bass with a 'pluck'. This pluck is augmented with a filter envelope with no attack or sustain, a short release and a short decay. Both these latter parameters should be set to be slightly shorter than the amplifiers parameters. The filter should be a 24 dB low-pass with the cut-off and resonance set about halfway. These settings on the filter envelope should introduce resonance to the 'pluck' at the beginning of the note so may need adjusting slightly to accomplish this effect.

This approach creates the basic timbre but it's worth employing positive filter key-follow so that the filters action follows the pitch helping to maintain some interest in the timbre. What's more, modulating the saw oscillator with a positive or negative envelope applied over a two-semitone range can help to further enhance the sound.

Sweeping Bass

The basic sweeping bass was typical of UK garage but has recently been morphed by producer to form the Dubstep style bass. It consists of a tight yet deep bass that sweeps in either pitch and/or frequencies depending on the style required by the producer.

This bass can be created with two oscillators, one set to a sine wave to add depth to the timbre while the other is set to a saw oscillator to introduce harmonics that can be swept with a filter. These are commonly detuned from one another, but the amount varies depending on the type of timbre required. Hence, it's worth experimenting by first setting them apart by 12 cents and increasing this gradually until the sound becomes as thick as required.

The amplifiers envelope generator is commonly set to zero attack and sustain with both release and decay adjusted to approximately midway. The decay setting here provides the 'pluck' while the release can be modified to suit the motif being played from the sequencer.

The filter should be a low-pass 12 dB as you want to remove the higher harmonics from the signal (opposed to removing the lower frequencies first), and this along with the resonance are adjusted so that they both sit approximately halfway between fully exposed and fully closed. Ideally, the filter should be controlled with a filter envelope using the same settings as the amp EG but to increase the 'pluck' of the sound you can adjust the attack and decay so that they're slightly longer than the amplifiers settings. Finally, positive filter key-follow should be employed so that the filter will track the pitch of the notes being played which helps to add more movement.

These settings will produce the basic timbre but it will benefit from pitch shifting and/or filter movements. The pitch shifting is accomplished by either sending pitch shift commands to the synthesizer (or recording the modulation live into the sequencer and editing later) or through modulating both oscillators with a pitch envelope set to a fast attack and medium decay. The pitch bend range on the synthesizer should be limited to two semitones to prevent the pitch from moving too high.

Further movement can be applied to the timbre via LFO modulation. Here, a saw generally provides the best results so that the filter opens, rather than decays, as the note plays. The depth of the LFO can be set to maximum so that it is applied fully to the waveform and the rate should be configured so that it sweeps the note quickly. If the melody consists of notes being played in quick succession, it's prudent to set the LFO to retrigger on key press, otherwise it will only sweep properly on the first note and any successive notes will be treated differently depending on where the LFO is in its current cycle.

Tech House Bass

In tech house, the bass is often particularly deep and powerful and is typically created with four oscillators, so this style of timbre will not be possible on many synthesizers.

For this bass, all four oscillators employ the same waveform. This is typically a saw waveform but squares and triangles can also be used to change the character of the sound. One oscillator remains at its original pitch, whilst the other three are detuned as far as possible without turning into individual timbres. The amplifier envelope employs a zero attack, release and sustain with a medium to ¾ decay setting.

The bass commonly exhibits a 'whump' at the decay stage, and this can be introduced by modulating the pitch of all the oscillators with an attack/decay envelope. This uses a fast attack so that pitch begins at the start of the note but the decay setting should be set just short of the decay used on the amplifiers envelope. The pitch modulation is commonly applied positive so that the pitch sweeps downwards but it can sweep upwards through negative modulation if it suits the music. Further, the synthesizer offers the option to adjust the slope of the envelopes, a convex decay is commonly employed, but experimentation is the real key with this type of bass, and in some cases, a concave envelope may produce more acceptable results. Filter key-follow is rarely employed, as the bass tends to remain within microtonal pitch adjustments but if your motif moves by a complete pitch, it's prudent to use a positive key-follow to introduce some movement into the riff.

Trance 'Block' Bass

This bass is typical of those used in many trance tracks, and while it doesn't exhibit a particularly powerful bottom end, it does provide enough of a bass element without being too rich in harmonics so that it interferes with the characteristic Trance lead.

This style of bass employs two oscillators, both set to square waves and detuned from each by 5 to 40 cents depending on the synthesizer. A 12 dB low-pass filter is used with the cut-off set so that it's almost closed and the resonance is pushed high so that it sits just below self-oscillation. The amplifier attack, sustain and release are all set to zero, and the decay is set approximately midway. The filter envelope emulates these amp settings with a zero attack, sustain and release but the decay should be set so that it's slightly shorter than the amps decay to produce a resonant pluck.

Finally, the filter key-follow is applied so that the filter follows the pitch across the bass motif. If the bass is too resonant, it can be condensed through reducing the filters decay to make the 'pluck' tighter or alternatively you can lower the resonance and increase the filters cut-off.

'Pop' Bass

The pop bass is commonly used in many popular music tracks, hence the name, but it is equally employed in some dance music genres where the producer requires some bottom end presence without taking up too much of the available frequency range of the mix.

These can be created in most synthesizers by mixing a saw and a triangle oscillator together. The triangle is transposed up by an octave from the saw oscillator and the amp envelope is used as an on/off trigger. In other words, the attack, decay and release are all set at zero and sustain is set just below its maximum level. This results in a timbre that immediately jumps into the sustain portion that produces a constant bass tone for as long as the key is depressed.

If a 'pluck' is required, a 24 dB low-pass filter with a resonance just below self-oscillation and a low cut-off controlled by a filter envelope with no attack release or sustain but a long decay will provide the results. By increasing the depth of the filter envelope along with increasing the filters cut-off and resonance, it is possible to control the resonance of the bass. Depending on the motif that this bass plays, employing filter key tracking so that the filter follows the pitch of the bass can add some extra textural interest to the timbre.

Examples of these timbres being constructed can be heard on the COMPANION WEBSITE.

BASS LAYERING AND EFFECTS CHAINS

Although we have examined the main properties that contribute towards the creation of bass timbres and discussed how to construct the most common style of bass in dance music, doing little more than just basic sound design will not always produce the best results. Indeed, a bass may end up being too heavy and be lacking in upper harmonics or it may be too light, lacking the required weight. In this event, the producer will often employ effects chains and different sonic elements from different synthesizers to construct a patch. This process is similar to the construction of a kick drum discussed in Chapter 15 and consists of layering different timbres together.

Typically, the different layers are sourced from different synthesizers so that each consecutive layer has a different sonic signature. This is because synthesizers will sound tonally different from one another even if they use the same parameter settings. For example, if an analogue modelled synthesizer is employed to create the initial patch, using the same style of synthesizer from another manufacturer to create, another bass will create a timbre with an entirely different character. If these are layered on top of one another and the respective volumes from each are adjusted, it is possible to create more substantial bass timbre.

Ideally, to prevent the subsequent mixed timbres from becoming too overpowering in the mix, it's quite usual to also employ different filter types on each synthesizer. For example, if the first synthesizer is producing the low-frequency energy but it lacks any top end, the second synthesizer should use a high-pass filter. This allows you to remove the low-frequency elements from the second synthesizer so that it's less likely to interfere with the harmonics from the first. This form of layering is often essential in producing a bass with the right amount of character and is one of the reasons why many professional artists and studios will have a number of synthesizers at their disposal.

Alongside layering, effects and processing chains also play a central role in the creation of bass timbres. Typically, these consist of distortion followed by compression then EQ and then finally flangers or chorus style effects. The distortion is employed to create additional upper harmonics to provide the bass with presence whilst the following compression is used to control the dynamics and prevent any peaks or unwanted distortion.

The EQ follows compression since this can be used to sculpt the distortion into the required texture whilst finally the flanger is employed to produce more textural variation. In this configuration, the bass is commonly sent to the effects bus rather than inserted since by doing so the producer can carefully mix the original bass weight with the effected results.

Care should be exercised with the application of stereo effects though, particularly those that are designed to widen the stereo image. A bass should sit in the centre of the mix, and if stereo widening effects are applied, the mix can loose its central focus resulting in a mix lacking in bass presence. Although some mixes do have a stereo bass employed, they often layer this over a mono sub-bass to ensure that there is sufficient energy at the centre of the mix.

PROGRAMMING LEADS

Lead instruments are much like the previously described bass instruments in that they are equally difficult to encapsulate. Every track will employ a different style of lead timbre, and there are no definitive methods to creating a great lead. However, it is of vital importance that the producer takes plenty of time producing one that sounds right for the music. The entire track will rest on the quality of the lead and therefore time taken is time well spent.

While it's impossible to suggest ways of designing a lead to suit any one particular track, there are nonetheless some generalizations that can be applied. First, most lead instruments will employ a fast attack on the amp and filter envelope so that it starts immediately on key-press with the filter introducing the harmonics to help it to pull through a mix. The decay, sustain and release parameters, though, will depend entirely on the type of lead and sound you want to accomplish.

For example, if the sound has a 'pluck' associated with it, then the sustain parameter, if used, will have to be set quite low on both amp and filter EGs so that the decay parameter can fall to produce the pluck. In addition, the release parameter of the amp can be used to determine whether the notes of the lead motif flow together or are more staccato (keep in mind that staccato notes may appear quieter). If the release is particularly long to create notes that flow together, it is common practice to employ portamento on the synthesizer so that timbres rise or fall into the successive notes.

Since lead is the most prominent part of the track, it usually sits in the mid to high range and therefore is designed to occupy this area. A common approach here is to construct a harmonically rich sound through saw, square, triangle and noise waveforms along with employing the unison feature if the synthesizer has it available. This is a form of stacking a number of the synthesizer's voices together to produce thicker and wider tones but also reduces the polyphony available to the synthesizer so you have to exercise caution as to the polyphony available to the synthesizer.

Once a harmonically rich voice is created, it can then be thinned, if required, with the filters or EQ and modulated with the envelopes and LFOs. These latter modulation options play a vital role in producing a lead timbre since they require plenty of textural momentum to maintain interest. Typically, these methods alone will not always provide a lead sound that is rich or deep enough, and therefore, the producer should consider further techniques such as layering, doubling, splitting, hocketing (explained in a moment) or residual synthesis.

We have already examined the basics of layering when discussing basses but with leads these techniques can be developed further as there is little need to keep the lead under any strict frequency control, after all its purpose is to sit above every other element in the mix!

It is certainly not unusual to layer different synthesized timbres atop one another but with each employing different amp and/or filter envelopes. For example, one timbre may employ a fast attack with a slow-release or -decay parameter whilst the second layer employs a slow attack and a fast decay or release. When the two timbres are layered together, the harmonic interaction between the two produces complex timbres that can then be mixed, compressed and treated to EQ and effects chains.

Doubling is similar to layering, but rather than employ the timbres from a number of synthesizers, it consists of duplicating the track (MIDI and instrument) but then transposing the copied MIDI track up by a third, fifth, seventh, ninth or an octave. A variation on this theme is to make a copy of the original lead and then only transpose some notes of the copy rather than all of them so that some notes are accented.

Hocketing consists of sending the successive notes from the same musical phrase to different synthesizers to give the impression of a complex lead. Commonly, the producer will determine which synthesizer receives which note through velocity. For instance, they may set one to not accept velocity values below a certain value such as 64 while the second synthesizer will be configured to only accept velocity values above 64. This way, the producers can easily hocket between two synthesizers using little more than velocity commands in the sequencer.

Splitting and residual synthesis are the most difficult to implement but often produce the best results. Splitting is similar in some respects to layering but rather than produce the same timbre on two different synthesizers, a timbre is broken into its individual components that are then sent to different synthesizers.

For example, you may have a sound that's constructed from a sine, sawtooth and triangle wave, but rather than have one synthesizer accomplish this, the sine may come from one synthesizer, the triangle from another and the saw from yet another. These are all modulated in different ways using the respective synthesis engines, and the audio workstations mixer is employed as an oscillators mixer. The three synthesizers are sent to a bus for volume control where filters, envelope modifiers, LFO tools, transient designers and compressors are all employed to help modify the parts to construct a timbre. Residual synthesis, on the other hand, involves creating a sound in one synthesizer and then using a band-pass or notch filter to remove some of the central harmonics from the timbre. These are then replaced using a different synthesizer or synthesis engine and recombined at the mixer.

Finally, effects and processing chains also play a reduce space part in creating a lead timbre. The most typical effects are reverb, delay, phasers, flangers, distortion and chorus along with noise gates, compression, EQ and envelope modifiers. There are far too many possibilities to list but prior knowledge of each effect and experimentation is the key.

PLUCKED LEADS

Plucked leads are employed in most genres of music from house through trance but commonly appear in progressive house genres. They can be constructed in numerous ways depending on the type of sound the producer wishes to accomplish and the genre of music it is aimed at. Thus, what follows are two of the most commonly used basic patches for plucked leads in dance which the producer should manipulate further to suit their music.

Plucked Lead 1

The first plucked lead consists of three oscillators, two of which are set to saws to introduce plenty of harmonics whilst the third set to either a sine or triangle wave to add some bottom end weight.

The choice between whether to use a sine or triangle is a production choice, but a sine will add more bottom end and is useful if the bass used in the track is rather thin. Alternatively, if the bass is quite heavy, a triangle may be better as this will add less of a bottom end and introduce more harmonics. The saw waves are detuned from each other by 5 to 40 cents (the further these are detuned, the richer the sound), and the third oscillator is transposed down by an octave to introduce some weight into the timbre.

The amplifiers attack and sustain are set to zero but both decay and release should be adjusted to a medium depth depending on the amount of pluck required and the motif playing the timbre. A 12 dB low-pass filter is used to remove the higher harmonics of the sound, and this should be set initially at midway while the resonance should be set quite low.

This is controlled with a filter envelope set to a zero attack and sustain, but the decay and release are set just short of the amplifiers decay and release settings. Finally, some reverb is best applied to the timbre to help widen it a little further, but if you take this approach, it is prudent to follow with a noise gate to remove any tail.

Plucked Lead 2

The second plucked lead consists of two oscillators, a saw and a triangle. The saw wave is pitched down to produce a low end, while the triangle wave is pitched up to produce a slight glistening effect. The amount that these are detuned by depends on the timbre required experimentation through detuning is required until the frequencies sit into the mix.

As a general starting point, detune them from one another as far as possible to the point it is about to produce two different timbres and then step back a little. Due to the extremities of this detuning, sync both oscillators together to prevent the two from beating too much and then apply a positive pitch envelope to the saw wave with a fast attack and medium decay so that it pitches up as it plays.

The sound should begin on key-press so the amplitude envelope should have no attack, sustain or release but the decay should be set to halfway. This envelope is copied to the filter envelope, but here, the decay is set to zero and the sustain parameter is increased to midway. Finally, using a 24 dB low-pass filter, the cutoff is set to around three-quarters open and the resonance to about a quarter.

Euphoric Trance Lead

The uplifting trance lead is probably the most elusive lead to design, but in many cases, this is simply because it cannot be recreated on any synthesizer, and in many cases, they are very often created on the 'classic' trance instruments (the Access Virus, Yamaha CS6R or the Novation Supernova).

The real technique behind creating the uplifting trance lead is through clever use of effects and noise. As previously discussed, noise produces a vast number of harmonics, and this is essential to creating the hands in the air vibe.

A basic trance lead timbre can be constructed with four oscillators. Two employ pulse waveforms whilst the third is often a saw wave and the fourth is noise to add some high-end harmonics. The two pulses are detuned from one another by a varying amount that is dictated by the synth and the producer's tastes. Typically, they are detuned to the point where they are just about to separate into two distinct timbres. The saw wave is then detuned from these by an octave, whilst the noise is detuned an octave higher. Generally, pink noise provides the best frequencies for a trance lead, but white noise can suffice if pink isn't available.

The amplifiers attack should be to zero, but to create a small pluck, it often employs a medium decay with a small sustain and a fairly short release. This release can be lengthened or shortened further to suit the programmed melody. The filter envelope can employ the same settings as the amplifier, although it is worth experimenting with the decay to produce a sharper pluck. The filter key-follow should be applied so that it tracks the pitch of the notes being played and treats the filter, respectively. A 12 dB low-pass filter with a low to mid resonance along with a mid to high cut-off will produce the basic timbre.

Further interest can be applied to the timbre through pulse width modulation, but if at all possible, two LFO's should be employed so that each width can be modulated at a slightly different rate. Here, the first LFO modulates the pulse-width of the first oscillator with a sine wave set to a slow to medium rate and full depth, while the second modulates the pulse-width of the second oscillator and employs a triangle wave with a slighter faster rate than the first.

Finally, the timbre is commonly always washed in both reverb and delay to provide the required sound. The reverb should be applied heavily via room or hall setting and applied as a send effect. This should employ 50 ms of pre-delay so that the transient pulls through unmolested and the tail set quite short to prevent it from washing over the successive notes. Ideally this should be followed with a noise gate to remove the subsequent reverb tails, as this will produce a heavier timbre that cuts through the mix. Delay is generally also applied but again this is commonly used as a send effect so that only a part of the timbre is sent to the delay unit. The settings to use will depend on the type of sound but the delays should be set to less than 30 ms to produce the granular delay effect to make the timbre appear big in the mix.

TB303 Leads

Although we have already discussed the use of a TB303 during bass programming, they remain a versatile machine and are equally at home creating lead sounds by simply pitching the bass frequencies up by a few octaves. The most notable, if old school classic, example of this was on Josh Wink's 'Higher State of Consciousness'. This same effect can be recreated on most analogue modelled synthesizers since it only requires the use of a single saw oscillator.

The amplifiers attack, sustain and release are set to zero, and the decay is then adjusted to suit the style of sound the producer wishes to achieve. A 12 dB low-pass filter is employed with a low cut-off setting but the resonance set just below self-oscillation. The filter envelope is a copy of the amplitude envelopes setting although the decay should be set just short of the amplifiers decay stage. Filter key-follow is often employed to create additional movement but further modulation is often applied through velocity modulation of cut-off so that the harder the note is struck, the more the filter opens.

Finally the timbre is run through a distortion unit, and the results are filtered with a plug-in filter. Depending on the melody, it may also benefit from increasing the amplifiers decay so that the notes overlap one another and employing portamento on the synthesizer so the notes slur into one another when played.

Distorted Leads

There are hundreds of distorted leads but one that often appears is the distorted/phased lead that is often employed in some of the harder house tracks. The basic patch can be created with a single saw oscillator but typically, two saw oscillators detuned from one another produce a wider, more upfront style of timbre.

The amp envelope is set to a zero attack with a full sustain, a short release setting and the decay to around a quarter. As is common, it is this decay that will have the most influence over the sound, so it is worth experimenting by increasing and decreasing to produce the style of sound required. The filter is generally a 24 dB low-pass with the cut-off quite low and the resonance pushed high to produce overtones that can be distorted afterwards with effects. The filters envelope is typically set with zero attack, sustain or release, but the decay is set slightly longer than the amplifiers decay parameter.

Distortion followed by phasers or flangers is commonly applied, but the distortion should come first in the chain so that the subsequent phaser is also applied to the distortion. It is important here not to apply too much distortion, otherwise it may overpower the rest of the mix. Ideally, the producer should aim for a subtle but noticeable effect.

Theremins

A theremin consists of a vertical metal pole approximately 12 to 24 in. in height that responds to movements of your hands and creates a warbling low-pitched whistle dependent on the position of hand. The sound was used heavily in the 50s and 60s sci-fi movies to produce an unsettling atmosphere but has made many appearances in electronic dance music.

This same effect can be reproduced on any synthesizer by employing a saw wave oscillator with a 12 dB low-pass filter. Here, the producer gradually reduces the cut-off until it produces a soft constant tone. Some synthesizers may require an increase in resonance to reach the correct tone, but many should produce the soft tone with no resonance at all.

This is perhaps one of the few examples where the tone does not start immediately on note press and instead employs a longer attack setting of about half way. Decay is not used, and instead the sustain parameter is set to maximum with a ¾ release too.

Theremins are renowned for their random variations in pitch, and this can be emulated by holding note at around C3 and using the pitch wheel to introduce small amounts of pitch variation. If the synthesizer permits portamento, then it's also advisable to use this to re-create the slow movement from note to note.

Hoovers

The hoover sound originally appeared on the Roland Juno synthesizers and has featured consistently throughout all genres of dance music including techno, tech house, acid house and house. In fact, these still remain one of the most popular sounds employed in dance music today. Originally in the Juno, this was somewhat aptly named 'What the…' but due to their dissonant tonal qualities, they were renamed 'hoover' sounds by dance artists since they share similar tonal qualities to a vacuum cleaner.

The sound is best constructed in an analogue modelled synthesizer since the tonal qualities of the oscillators play a large role in creating the right sound. Two saw oscillators are detuned as far from one another as possible but not so far that they become two individual timbres.

The amplifiers attack is commonly set to zero with a short decay and release and the sustain parameter set to just below maximum so there is a small pluck evident from the decay stage. The filter envelope employs a medium attack, decay and release but no sustain is applied. This augments a 12 dB low-pass filter with the cut-off set quite low and a medium resonance.

To add the dissonance feel that 'hoovers' exhibit, the pitch of both oscillators is modulated with a pitch envelope set to a fast attack and a short decay. This, however, should be applied negative rather than positive so that the pitch bends upwards into the note rather than downwards. Finally, depending on the synthesizer recreating the timbre, it may need some widening, and this can be accomplished by washing the sound in chorus or preferably stacking as many voices as possible using a unison mode.

'House' Pianos

The typical house style piano that is making a comeback in some of the more recent releases is often drawn directly from the Yamaha DX range of synthesizers and is usually left unmodified and simply treated to a series of effects chains including reverb, noise gates, compression, EQ and delay. Consequently, if the producer is after this particular timbre, the best option is to either purchase an original Yamaha DX7 synthesizer or alternatively invest in Native Instruments FM-8 VST instrument. This is a software emulation of the FM synthesizers produced by Yamaha that can also import the sounds from the original range of DX synthesizers.

Acquiring the FM piano is difficult on most analogue synthesizers and often impossible of most digital synthesizers because of the quirks of frequency modulation. Nonetheless it can be accomplished by employing two sine wave oscillators with one of the oscillators detuned so that it's at a multiple of the second oscillator. These are then frequency modulated to produce the general tone.

The amp envelope employs a fast attack, short decay and release and a medium sustain. To produce the initial transient for the note, a third sine wave pitched high up on the keyboard and modulated by a one shot LFO (i.e. the LFO acts as an envelope – fast attack, short decay, no sustain or release) will produce the desired timbre.

As a side note, if you want to produce the infamous bell like or metallic tones made famous by FM synthesizers, use two sine oscillators with one detuned so that its frequency is at a non-related integer of the second and then use frequency modulation to produce the sound.

Organs

Organs are commonly used in the production of house with the most frequent choice being the Hammond B-4 drawbar organ. The general timbre, however, can be emulated in any subtractive synthesizer through using a pulse and a saw oscillator.

The amplifier envelope employs a zero attack with a medium release and maximum sustain (note that there is no decay since the sustain parameter is at maximum). A 12 dB low-pass filter is employed to shape the timbre with the cut-off set to zero and the resonance increased to about halfway. A filter envelope is not employed since the sound should remain un-modulated, but if you require a 'click' at the beginning of the note, you can turn the filter envelope to maximum, but the attack, release and sustain parameters should remain at zero with a very, very short decay stage. Finally, the filter key-follow should be set so that the filter tracks the current pitch, which will produce the typical organ timbre that can then be modified further.

Examples of these timbres being constructed can be heard on the COMPANION WEBSITE.

SOUND EFFECTS

Sound effects play a fundamental role in the production and in particular the arrangement of electronic dance music. Indeed, their importance should not be underestimated since although they spend much of their time in background, if they are removed from a mix, it can appear motionless and insipid.

The key to the creation of sound effects within synthesizer lies with the LFO as this is employed to modulate the parameters to produce the different sound effect. For instance, simply modulating a sine wave's pitch and filter cut-off with an LFO set to a sample and hold or noise waveform will produce strange burbling noises while a sine wave LFO modulating the pitch can produce siren type effects. There are no limits to this application, and therefore, no real advice on how to create them as it comes from experimentation with different LFO waveforms modulating different parameters and oscillators.

To generalize the application, however, the waveform employed by the LFO will contribute a great deal to the sound effect the producer receives. Triangle waves are best suited for creating bubbly, almost liquid sounds, while saws are suitable for zipping type noises. Square waveforms are useful for short percussive snaps such as gunshots, and random waves are particularly useful for creating burbling, shifting textures. All of these used at different modulation depths and rates, modulating different oscillators and parameters will create wildly differing effects.

Many sound effects can be acquired from two or three waveforms and some thoughtful (and creative thinking). The real solution to creating effects is to experiment with all the options at your disposal and not be afraid of making a mess. Even the most unsuitable noises can be tonally shaped with EQ and filters and further processing tools.

What follows is a small list of some of the most popular effects and how they're created. Unfortunately, though, there are no specific terms to describe sound effects so what follows can only be a rough description of the sound they produce.

Siren FX

The siren effect is possibly the easiest sound effect to recreate since it only involves a single sine wave oscillator with its pitch modulated via an LFO. The amplifier envelope employs a fast attack and release with a short sustain and a medium decay. A triangle wave LFO is then employed to modulate the pitch of the sine wave oscillator at full depth. Here, the faster the LFO rate becomes, the faster the rise and fall of the siren becomes.

Whooping FX

To create the 'whooping effect', only one sine wave oscillator is required with the amplifier envelope set to a fast attack and release with no sustain and a long decay setting. The pitch of the sine wave is then modulated with a secondary envelope employing a fast attack and long decay. If a secondary envelope isn't available, the same effect can be achieved with a saw wave set to a slow rate so that the LFO only produces a single movement throughout the length of the note. The producer then programs a series of staccato notes into a MIDI sequencer that trigger an arpeggiator set to one octave or less so that it constantly repeats in a fast succession to create a 'whoop, whoop, whoop' effect.

Explosion FX

To create explosive type effects requires two oscillators. One oscillator should be set to a saw wave with the second set to a triangle wave. The triangle is detuned from the saw by +7 to +30 cents, and then a 12 dB low-pass filter is applied with a high cut-off and resonance.

The amplifiers envelope is set to a medium attack and release but with a long decay and no sustain. These settings are then copied to the filter envelope with the decay set a little shorter than the amplifiers decay. A saw LFO set to a negative amount is used to modulate both the pitch of the oscillators and the filters cut-off. Finally, the saw is used to frequency modulate the triangle. When the note is played at C1 or lower, it will produce a typical synthesizer explosion effect.

Zipping FX

This effect is popular in many form of electronic dance music as a background atmospheric timbre. Two oscillators are commonly employed here, one set to a saw and the second employing a triangle. These are detuned from one another from +5 to +20 depending on tastes and the synthesizer being employed. The amplifier envelope employs a fast attack and release with a medium to long decay and a medium sustain, and a 12 dB low-pass filter is used with a low cut-off and high resonance. Finally, a saw LFO slowly modulates the filters cut-off at full depth to slowly modulate the filters cut-off.

Rising Speed/Filter FX

Another popular effect is the rising speed filter, whereby the rise speed increases as the filter opens further. You can use a saw, pulse or triangle wave for this effect, but a saw tends to produce the most interesting results.

Both the amplifier envelope and filter envelopes employ to a fast decay and release but with a long attack and high sustain. Use a triangle or sine LFO set to a positive mild depth and very slow rate (about 1 Hz) to modulate the filters cut-off. Finally, use the filters envelope to also modulate the speed of the LFO so that as the filter opens the LFO also speeds up. If the synthesizer doesn't allow you to use multiple destinations, you can increase the speed of the LFO via the workstations automation parameters.

Falling Speed/Filter FX

This is basically the opposite of the previously described effect so that rather than the filter rising and simultaneously speeding up, it falls while simultaneously slowing down. As before, both the amp and filter envelopes are set to a fast decay and release with a long attack, but this time no sustain is used.

A triangle or sine wave LFO is set to a positive mild depth and fast rate to modulate the filters cut-off. Finally, the filters envelope also modulates the speed of the LFO so that as the filter closes the LFO also slows down.

Arrangement Sweeps

Arrangement sweeps are generally best created with saw oscillators as their high harmonic content gives the filter plenty to work with but triangle and pulse can also be used depending on the result the producer requires.

Two oscillators are required for this effect, both set to the same oscillator waveform and detuned from one another as far as possible without becoming individual timbres.

The amplifier envelope should be configured to a fast attack, decay and release with the sustain set just below maximum. A slow saw or triangle wave LFO is then used to modulate the pitch of one of the oscillators. A band-pass will produce the best results if set to a medium cut-off, but a very high resonance and a saw, sine or triangle LFO is then used to modulate the filters cut-off to produce the sweeping effect.

Ghostly Noises

Ghostly noises can be designed in a synthesizer with two triangle waveform oscillators. Using a 12 dB low-pass filter, set the cut-off quite low but employ a high resonance and set the filter to track the keyboards pitch (filter key-follow). The amplifier envelope should be set to a fast attack with a long decay, high sustain and medium release and the filters envelope should be a fast attack, long decay but with no sustain or release. Finally, an LFO sine wave is used to slowly modulate the pitch of the oscillators. Play chords in the bass register to produce the timbres.

Computer Burbles

The classic 70s computer burbling noises can be created in a number of ways, but by far the most popular method is to employ two oscillators both set to triangle waves.

Detune one of these from the other by +3 to +12 cents and then set the filter envelope to a medium attack with a long decay with a medium release and no sustain. Do the same for the amps EG but use a high sustain and set the filters envelope to positively but mildly modulate a low-pass filter with a low cut-off and a high resonance. Finally, ensure that filter key tracking is switched on and modulate the pitch of one, or both, of the oscillators with a noise or sample and hold waveform. This should initially be set quite fast will a full depth, but it's worth experimenting with the depth and speed to produce different results.

Swoosh FX

To create the typical swoosh effect (often used behind a snare roll to help in creating a build-up), the synthesizer requires a third modulation envelope that can be used to modulate the filter as a destination.

To produce this timbre, first switch all the oscillators off and then increase the resonance so that the filter breaks into self-oscillation. Use a filter envelope with no decay, a medium release and attack and a high sustain and set it to positively affect the filter by a small amount and then use a second envelope with these same setting to affect the filter again, but this time set it to negatively affect the filter by the same amount as before.

Set the amps EG to no attack or sustain but a small release time and a long decay and use a saw or triangle wave LFO set to a medium speed and depth to positively modulate the filters cut-off.

This will create the basic 'swoosh' effect but if possible employ a second LFO set to a different waveform from the previous one to negatively modulate the filter by the same amount.

Examples of these timbres being constructed can be heard on the COMPANION WEBSITE.

PROGRAMMING NEW TIMBRES

Throughout this chapter I've discussed the creation of some popular sounds within electronic dance music, but there will come a time when the producer wishes to extend further and create a sound that they have in their head. To accomplish this, however, they must be experienced in synthesis and in particular with identifying the various sounds produced by oscillators and the effects they have when mixed together.

Armed with this knowledge, constructing sounds isn't particularly complicated and just requires some careful consideration and experimentation. First, when attempting to invent timbres, it's important to be able to conceptualize the instruments character and momentum. This means the producer must try to imagine the complete sound (this is much more difficult than you may think) before mentally stripping away the character and turning it into a series of questions such as:

- Does the timbre start immediately?
- How does it evolve in volume over time?
- Is the instrument synthetic, plucked, struck, blown or bowed?
- What happens to its pitch when it's sounded?
- Are there any pitches besides the fundamental that stand out enough to be important to the timbre?
- Does it continue to ring after the notes have been sounded?
- How bright is the sound?
- How much bass presence does the sound have?
- Does it sound hollow, rounded, gritty or bright and sparkly?
- What happens to this brightness over time?
- Is there any modulation present?
- What does the modulation do to the sound?

The following graph and programming tips have been included to help with the conceptualization of new timbres.

Table 18.1 Table of Sound Design Considerations

Generalization	Technical Term	Synthesizers Parameter
Type of sound	Harmonic content	Oscillators waveforms
Brightness	Amplitude of harmonics	Filter cut-off and resonance
Timbre changes over time	Dynamic filtering	Filters envelope and/or LFO modulation
Volume changes over time	Dynamic amplitude	Amplifier envelope
Pitch	Frequency	Oscillators pitch
Sound has a cyclic variation	LFO modulation	LFO waveform, depth and rate
Tremolo (cyclic variation in volume)	Amplitude modulation	LFO augments the amplifier
Vibrato (cyclic variation in pitch)	Pitch modulation	LFO augments the pitch
Sound is percussive	Transient	Fast attack and decay on the amplifier
Sound starts immediately or fades in	Attack time	Lengthen or shorten the attack and decay stages
Sound stops immediately or fades out	Release time	Lengthen or shorten the release on the amplifier
Sound gradually grows 'richer' in harmonics	Filter automation	Programmed CC messages or a slow-rate LFO augmenting the filter cut-off

GENERAL PROGRAMMING TIPS

- Avoid relying on one note while programming, it will not give you the full impression of the patch. Always play the motif to the synthesizer before programming.
- Ears become accustomed to sounds very quickly, and an unchanging sound can quickly become tedious and tiresome. Consequently, it's prudent to introduce sonic variation into long timbres through the use of envelopes or LFOs augmenting the pitch or filters.

- Generally, the simpler the motif, the more movement the sound should exhibit. So for basses and motifs that are quite simple assign the velocity to the filter cut-off, so the harder the key is hit, the brighter the sound becomes.
- Don't underestimate the uses of keyboard tracking. When activated this can breathe new life into motifs that move up or down the range as the filters cut-off will change respectively.
- Although all synthesizer share the same parameters, they do not all sound the same. Simply copying the patch from one synthesizer to another can produce totally different results.
- Whilst the noise oscillator can seem useless in a synthesizer when compared to the saw, sine, triangle and square waves, it happens to be one of the most important oscillators when producing timbres for dance and is used for everything from trance leads to hi-hats.
- To learn more about the character of the synthesizers you use, dial up a patch you don't like, strip it down to the oscillators and then rebuild it using different modulation options.
- If you have no idea of what sound you require, set every synthesizer parameter to maximum and then begin lowering each parameter to sculpt the sound into something you like.
- Many bass sounds may become lost in the mix due to the lack of a sharp transient. In this instance, synthesize a click or use a woodblock timbre or similar to enhance the transient.
- The best sounds are created from just one or occasionally two oscillators modulated with no more than three envelopes: the filter, the amplifier and the pitch. Try not to over complicate matters by using all the oscillators and modulation options at your disposal.
- Try layering two sounds together, but use different amplifier and filter settings on both. For example, using a slow attack but quick decay and release on one timbre and a fast attack and slow decay and release on another will 'hocket' the sounds together to produce interesting textures.
- Never underestimate the uses of an LFO. A triangle wave set to modulate the filter cut-off on a sound can breathe new life into a dreary timbre.
- Remember that we determine a huge amount of information about a sound from the initial transient. Thus, if you replace the attack of a timbre with the attack portion of another, it can create interesting timbres. For instance, try replacing the attack stage of a pad or string with a guitars pluck.
- For more interesting transients, layer two oscillators together, and at the beginning of the note, use the pitch envelope to pitch the first oscillator up and the second one down (i.e. using positive and negative pitch modulation).
- For sounds that have a long-release stage, set the filters envelope attack longer than the attack and decay stage but set the release slightly shorter than the amps release stage. After this, send the envelope fully to the filter so that as the sound dies away, the filter begins to open.

Of course, sounds in many EDM tracks are not solely produced in a synthesizer and sound design through sampling other artist's records is equally commonplace. Indeed, it's not uncommon for a dance artist to sample another artists track to produce the basis of the song and for other artists to then sample that track. Skrillex, for example, has sampled 31 tracks so far, and he has subsequently been sampled 12 times. Daft Punk has sampled 52 different tracks whilst they have been sampled 149 times. This style of musical 'recycling' is extensive and occurs throughout every genre of dance music from trance to techno to minimal to house and tech house.

The sampling can undertake numerous forms depending on the genre. It's not unusual for tech house and techno to employ samples from other records, but here, it is typically such small parts that are sampled and consequently mangled that they become almost unrecognizable.

For instance, the artist may sample a beats length of music, trim it down further, time-stretch or time compress it, apply EQ, reverb and a variety of other effects chains to produce a single stab hit. This is placed into the track and then micro-tuned to produce slight timbral and pitch variations throughout the music. On the other end of the scale, French house will likely sample 8 to 16 bars of a complete record and repeat these over and over whilst applying numerous automation effects to create an arrangement. Both of these are valid approaches, and each requires a different skill set to be developed by the producer. However, it does also mean that you cannot always 'emulate' a sound from your peers on nothing more than a synthesizer simply because they may not have used a synthesizer to create the original timbre.

Regardless of how the producer chooses to approach sound design, it is a time-consuming process. There is no 'quick fix' to creating great timbres, and it is only a result of practical experience and experimentation. Indeed, many professional artists have admitted that some of their characteristic timbres were a result of experimenting and striking upon serendipity. Thus, if you want to program great timbres, patience and experimentation is the real secret. You must be willing to set time aside from writing music and learn your synthesizers.

CHAPTER 19

Formal Structure

'It sounds a lot more like it does now than it did ten minutes ago'.

Sony A&R Representative

Beyond the sound design, processing, effects and composition of melodies, a final hurdle the producer must overcome is that of formal structure. This is the terminology used to describe the overall plan, structure or general arrangement of the music, as it plays from beginning to end. Whilst this may appear insignificant compared with the challenges of initially getting a great idea down it is a lot more difficult that it may first appear.

The problem is that electronic dance music is unlike most other styles of music since it is often produced from little more than a single distinct musical pattern. In many production examples of EDM, the producer initially concentrates on producing the main hook of the record. This is the 16- or 32-bar pattern that he/she hopes will light up the dance floor and drives the club wild. However, this hook only results in 20 or so seconds of music and since most dance tracks will run anywhere from 4 minutes to 8 minutes, it requires a lot more than simply repeating the same pattern over and over. Indeed, to maintain a listener's interest when constructing an entire track from little more than a 20-second loop requires careful planning and a range of finely honed techniques.

Before any arrangement is approached, it is important to first ensure that the initial idea – the hook pattern for which the entire track will be based upon – contains enough musical information to construct a track around. Without this, any arrangement will fall flat since the producer is fundamentally trying to construct a house on dodgy foundations. Understanding what constitutes 'enough' amount of musical information is difficult to quantify since it depends on the genre of music and the producer's creativity, but it can roughly be summed up into a hook that features some melodic or rhythmical diversity, or both.

Melodic or rhythmical diversity is accomplished through binary phrasing. This is a theory that was discussed in the first chapter of this book and is fundamental to *all* genres of music, since it provides a distinct separation between two parts of a melodic or rhythmical idea. As discussed earlier, in electronic dance music, a binary phrase often consists of two rhythmical or melodic ideas of equal duration but with pitch or rhythmical differences in the second phrase from the first. For instance, if we examine Eric Prydz's Pjanoo (composed in G minor), it is the bass phrase and not the lead that provides the phrasing, consisting of:

G – F – D# – D# – F – C

And then:

C – D – D# – D# – F – G

Eric Prydz Pjanoo MIDI.

Note how both parts of the melodic bass employ the exact same rhythmical motions, and it is only the pitch that changes. Also, although there is a significant amount of pitch movement in this particular example (or at least there is for dance music), it is offset with the main lead piano that consists entirely of a single phrase by maintaining the same rhythmic and pitch movements throughout all four bars of its motif.

The use of a single-phrase piano offset with a binary phrase bass helps to maintain the requisite repetitive feel for dance music but also introduces

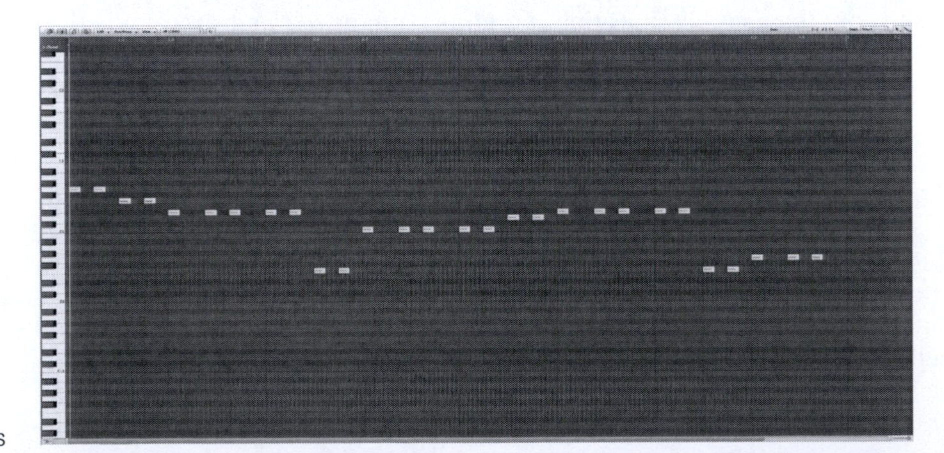

FIGURE 19.1
Eric Prydz Pjanoo Bass

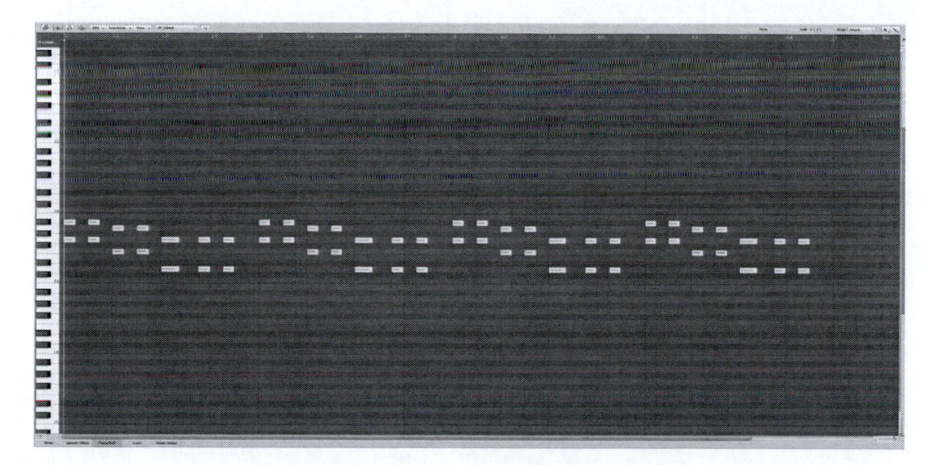

FIGURE 19.2
Eric Prydz Pjanoo Lead

enough difference to maintain interest through resolve. Different artists and genres will adopt different methods for this style of phrasing, but all music will employ some form of binary phrasing somewhere in the music in order to offer resolve.

Binary phrasing is not limited to melodies within dance music, and it will also occur within almost all the other elements of the music too and in particular within rhythmical elements. This technique was employed in Chapter 16 when constructing a drum loop. A modulating filter was used to manipulate the frequencies on the hi-hats, whilst pitch shifting and/or frequency modulation were employed on the snare.

Both of these techniques constitute a form of binary phrasing and result in the loop permitting sustained listening. Indeed, genres such as tech house, techno and minimal often employ a more complex form of this by employing slow modulating effects and micro-scale tuning on the drum rhythms and single instrument hits. By modulating timbres and the pitch of, say, a bass drone micro-tonally throughout it can aurally appear to maintain the same pitch yet subconsciously attract the mind resulting in sustained listening of what, on the surface, *appears* to be little more than a repeating, sustained loop.

CALL AND RESPONSE

A similar technique to binary phrasing is the call and response. This technique originates from African music and involves one instrument making the call and a secondary instrument responding. Commonly, the secondary instrument is a different timbre than the one making the call but this is not always the case.

For example, in tech-house and techno, it is common practice for the snares to exhibit the call and response behaviour. Here, typically a percussive element will play a specific pattern in the first bar that is answered by the same percussive element responding to this call with a slightly different pattern in the second bar.

Example of percussive call and response.

In Dubstep, two different instruments are often employed to produce this call and response. Typically, these will consist of two different styles of bass timbres but some may use a bass to provide the call that receives its response in the form of a higher-itched lead sound. In COMPANION WEBSITE example CPT19_4, a Dubstep example taken from a genre chapter later in this book, the call is provided by the first bass and then responded by a secondary, different bass timbre.

Example of Dubstep track call and response.

There is no requirement for a call and response or binary phrase to occur one directly after the other, and it is not uncommon for the phrase or call to repeat a number of times before it receives its response or is resolved. This patience technique was employed in the house example of this book. In this example, the phrase is repeated three times before it is finally resolved.

House Riff example.

Both these forms of binary phrasing and the call and response are fundamental starting points for the production of an arrangement. If these do not exist in the initial hook pattern of the music, then the music cannot resolve and will fail as a result. *All* music is based around this question and answer structure, whether it is binary phrasing, call and response or both. Without it, the music has no resolve and, therefore, will sound incomplete. Indeed, resolve is the ultimate reward in all forms of motion art and is the key ingredient to making any art satisfying to watch, read or listen too.

It is human nature to find enjoyment in the build up of tension followed by the release from that tension. This cycle of tension and release forms the basis of all great films, books, computer games and music. For example, movie directors and authors will deliberately elevate the tension levels of the viewer or reader, so that when the resolve occurs there is great relief and hence enjoyment.

In motion art, the cycle is accomplished through narrative or dramatic structure. The Greek philosopher Aristotle is believed to be the first to have recognized the importance of structure by declaring all stories must have a beginning, middle and an end. Gustav Freytag, a German Novelist, developed upon this theme by stating that any great story will commonly consist of five stages.

As shown in Figure 19.3, a typical narrative structure will consist of the following five parts.

- Exposition: This stage introduces the main protagonist and additional characters and sets the scene for the story.
- Incident: A problem will be introduced or a conflict entered that will serve to drive the rest of the story to its conclusion.
- Rising action: Events will grow ever more intense and the problems will increase exponentially. The protagonist will be faced with numerous difficult situations that go towards increasing the tension of the story.
- Climax: Here, there will be a major turning point. Any number of situations will change and the odds will often stack significantly against the protagonist. This will create the highest suspense in the story.
- Resolution: The protagonist will triumph and normal service is resumed. Here, the reader or viewer receives the largest reward, and the story is often typically resolved to a happy ending.

If you consider any great story, you'll be able to identify with this story arc. Whether it's a classical play, a best-selling novel or a box office smash, they are all based around this underlying structure. Without a beginning, middle and end, there can be no story.

Whilst the majority of the tension and eventual resolve is created through this story arc, all stories will introduce additional smaller periods of tension and resolve throughout to maintain the audiences' attention. In an action film, for example, this is typically accomplished through some kind of chase scene part way through. Here, the suspense is built through the thrill of the chase, as the protagonist chases the antagonist. At this early development, the antagonist will commonly escape the situation but this situation offers both resolve and further tension. There is resolve from the chase as it ends with the protagonist

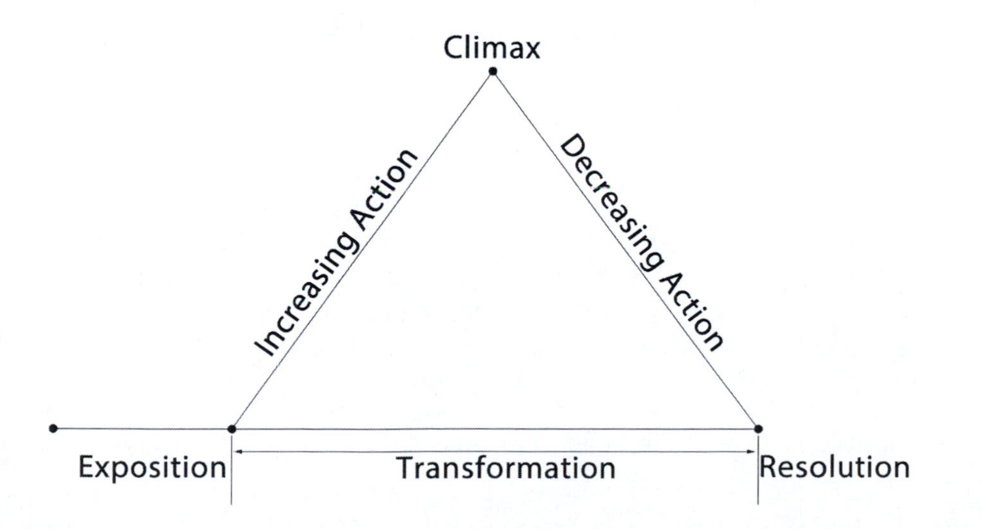

FIGURE 19.3
The dramatic structure

surviving but background suspense is increased to the main story arc as the antagonist escapes.

This same aspect of development applies within musical arrangements but rather than use dramatic or narrative structure, it employs a sonata form.

Exposition (Sonata Form)

Similar to the exposition in a story, within many forms of music this sets the scene. This is commonly accomplished by introducing the main musical theme to the listener. Within pop music, the exposition typically introduces the complete chorus or part of it. In electronic dance music that is destined for radio and iPod play this commonly involves the lead melody but lacking in full presence or missing the complete drum section or bass. Examples of this can be heard with Avicci's Levels and Superlove, Eric Prydz Valerie and Pjanoo and countless other radio friendly versions of EDM records.

Of course, not every dance music track destined for the radio will follow this form of exposition, Greyhound by Swedish House Mafia is a good example of a different style of exposition, but this should be considered an exception to the norm.

Development (Sonata Form)

During the development stage, the music will commonly move away from its defining hook or melody and develop on the song. In pop music this often consists of the verse that tells the story of the music whilst in classical music the composer would often take this opportunity to improvise on the melody, through using different instruments to changing key or pitch whilst maintaining the same melodic rhythm.

Within EDM, however, a much higher focus is placed on timbral development, textural motion and production ethics so here the lead may be thinned with EQ or filters, or it may only play a part of the lead, it may be exchanged for a different timbre or dropped altogether to focus on the groove of the music.

Recapitulation

For the recapitulation, the main melody returns once again but with all instrumentation in place. In pop music, this is commonly the final chorus to drive the message home, whilst in dance music, it is the complete reveal of the 16- or 32-bar focus of the track that was initially conceived by the producer.

Of course, on casual listening, a 6- or 10-minute club-based music may not appear to follow this sonata form but more critical listening to its structure will reveal it is essentially the same. Although the music may begin with percussive elements, the exposition could be considered the introduction of groove, whilst development is the build upon that groove and recapitulation is the final reveal of the hook.

Indeed, regardless of whether the music is a 3-minute radio mix, a 5-minute iPod mix or a 10-minute club opus the purpose is the same as dramatic or narrative structure – to create a series of events that will result in build and releasing of tension. Within all music, this is created through teasing the listener with the prospect of playing the 'best' part of the record, the complete hook.

In pop music, for example, the hook is commonly revealed at the very early stages and if the listener likes the hook, they feel a sense of anticipation waiting for it to return. Similarly, with EDM, the producers work to tease the listeners with the main hook and employ techniques such as snare rolls of filter sweeps in order to build the anticipation in the listener before offering this resolve and exposing the complete hook with all instrumentation.

However, whereas pop music employs a verse and chorus progression, dance music is based around the continuous development on a theme and this is where many novice producers struggle. Developing on a theme within a sequencer is difficult because the producer is working with nothing more than a series of 'events' or 'regions'. Since these are displayed as graphical blocks on an arrange page, it is easy to fall into the trap of treating these as *fixed* musical ideas that can be repeated across the arrange page as and when required. Whilst this technique can work for pop music with a verse and chorus structure, it is not applicable for the development of a theme.

Indeed, one of the most common pitfalls of many novice producers is to oversimplify an arrangement by repeating the same pattern over and over but with reduced instrumentation. For example, they will duplicate the same event pattern over the length of the arrangement and then systematically unmute channels as the arrangement progresses to its apex. This approach fails to produce an interesting arrangement because it is unable to create any anticipation within the music. The only anticipation created in the listener is that a new instrument will be introduced in the next so many bars.

In order to develop a great arrangement the producer has to break free of this event-based approach and rather than think vertically in terms of channels, think horizontally in terms of time. To accomplish this, the producer must carefully manipulate both the energy and contour of the music.

ENERGY AND CONTOUR

Energy in music can be associated to both pitch and frequency content. If the pitch of a melody rises or the frequency content of the track or timbre is increased, then listeners will determine this as an increase in the overall energy of the music. Examples of this can be found in all forms of music.

For instance, many pop records will move to a higher pitch for the last chorus of the record, the vocalist will perform higher and often their vocals or other instruments will be layered to increase the frequency content of the final part of the record to drive the message home and provide the ultimate release.

Similarly, many dance music records will employ a low-pass or high-pass filter, removing a significant amount of the frequency range and then slowly opening the filter to build up in energy and hence anticipation in the listener.

The EDM rising filter used to control energy and anticipation.

The control of a songs energy by employing techniques such as this form one of the key ingredients to a successful arrangement, since it can create anticipation in listeners. The control or manipulation of different energy levels through an arrangement is termed contour and any successful structure will have a carefully manipulated contour. This can be evident by loading any EDM track into a sequencer and examining its waveform since careful examination will reveal the energy contour of the music.

As shown in Figure 19.4, despite the track being based on the development of a theme, a relatively very small amount of time is spent playing the complete hook. Indeed, the only area for this example record where the entire hook is exposed is close to the end that comparatively, is an insignificant 32 bars of music. The rest of the music is spent developing towards this finale, and this is where many novice producers fail. After spending days, weeks or months, creating the hook of the record they try to hit the listener with their idea at every opportunity when it should be played as little as possible and the majority of the track should be spent teasing the listener with the *prospect* of playing it.

Teasing a listener with the prospect of playing, the complete hook and maintaining this for the length of an arrangement employs a variety of techniques from filters to syncopation, denial of expectation and canonic behaviour. How these are employed is down to the producer's own creative choices and will be discussed in more detail shortly but a more important question for the moment is *where* these should occur.

If you have listened to electronic dance music for any length of time, you'll probably instinctively know when to expect something to occur within the track but typically any instruments or textural changes will occur at specific metrical cycles in the music. These positions are termed structural downbeats and refer to the first downbeat of the bar, where a new instrument or textural development occurs.

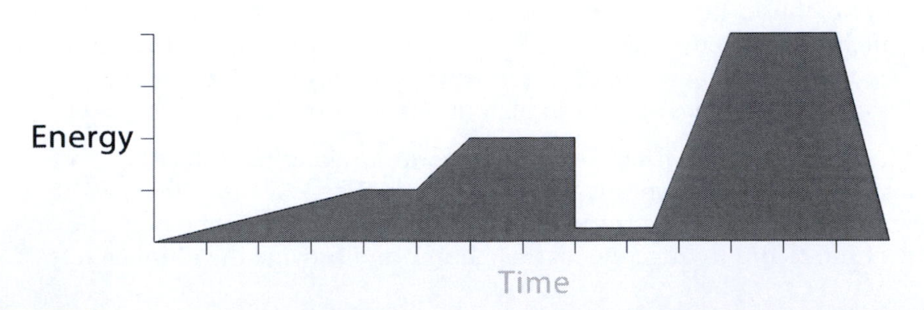

FIGURE 19.4
The energy contour of an EDM track

In almost every instance of electronic dance music, a structural downbeat will occur at multiples of two, and therefore, they will arrive on the 8th, 16th, 32nd or 64th beat of the bar. At one of these positions, it is common practice to introduce change, whether this is in the form of a new instrument, removing instruments or employing a major textural change to the arrangement. Of these, with radio friendly music, the changes will often occur at every 8 or 16 bars, whilst in club based music they will typically occur at every 16th, 32nd or 64th bar.

DEVELOPMENT

Once you have determined the point at which to introduce a new instrument, it can be accomplished in any number of ways. One of the most common techniques for the introduction of a new instrument is to precede it with a reversed cymbal crash.

This creates a short sharp build to the introduction of a new instrument and can be a particularly powerful effect if used cautiously. However, if the same process is employed to introduce each and every instrument into the mix, it becomes second nature to the listener and anticipation is lost. Similarly, consistently using this approach produces a mechanical feel to the music and reduces any emotional impact. Instead, techniques for the introduction of different instruments is to employ a mix of reverse cymbal crashes, reverse reverb and slow filtering of an instrument.

The reverse reverb effect consists of taking a single note of audio from the instrument and reversing it in the workstations sample editor (the producer must first duplicate the event to prevent *all* audio in the project from being reversed!). This new reversed event is placed onto another audio channel and reverb is inserted onto the channel with a short pre-delay and very long tail, commonly in excess of 3 seconds. The audio is then bounced to an audio file, re-imported into a new channel in the workstation and reversed again, so that the audio is the correct way around. This results in the previously stamped reverb now being reversed and sweeping upwards *into* the note.

Similarly, filters can be employed to slowly introduce an instrument over any number of bars. This is often accomplished via a high-pass filter but band-pass and low-pass are also employed depending on the mix and the producer's own creative decision. In the examples on the COMPANION WEBSITE, I began by employing a crash to introduce both the hi-hats and the bass. I then followed this by using the crash just to introduce the hi-hats followed by high-pass filtering the bass and finally, I employed reverse reverb on the bass and a high-pass filter to sweep in the hats. These are all techniques that can be employed to create a more interesting introduction to instruments.

Introducing instruments with a crash, reverse reverb and filters.

Whilst introducing instruments is the most proficient way to build energy within an arrangement, if it is employed each time, the producer feels the mix requires more energy, the music will quickly run out of instruments. Therefore, a number of other techniques must be employed to maintain interest and the first of these can be accomplished through syncopation.

Syncopation was discussed in detail in earlier chapters of this book but it is often employed during the course of an arrangement since it produces a strikingly powerful effect. A typical approach here is to modify the velocity via either MIDI or if the channel is audio, via the sample editor, to different notes every four or eight bars. This results in a continual shifting emphasis within the arrangement and helps to maintain sustained listening.

In the example on the COMPANION WEBSITE, I employed velocity syncopation to the bass line but it could equally be applied to any other melodic or non-melodic instrument in the mix. In tech-house, for example, a common technique is to employ a simple pulsing bass but syncopate the velocity of this bass differently for each bar.

The effects of syncopation on a bass line.

Alongside syncopation, another technique for melodic elements is to reduce the melody of the event but maintain its rhythm. This effect is evident in some house records but is most typical in the different forms of trance. If the bass or lead is heavily melodic, it is reduced to a single pitch. This maintains the original rhythm of the melody but at a single note that is then played through a number of bars before some of the original melody is allowed through by introducing the original pitch to some notes.

A reverse cymbal crash can be used to 'introduce' a new pitch element to the melody. Alternatively, if the track happens to be rhythmically based, the only part of the rhythm will be revealed with any empty 'spaces' filled with careful application of delay.

The effect is a sense of anticipation in the listener since they want to hear the complete melodic or rhythmical element. This must applied cautiously, however, since denying the expectation of the listener for too long can turn them off the music whilst introducing the complete melody too early can have the same effect. This is where experience, practice and critical listening of your peers should not be underestimated. In the following example, I maintain a single pitch in the bass but then introduce some pitch after a short reverse cymbal strike…

Pitch development on a bass.

An alternative to introducing pitch movement early in an arrangement is to employ a 'canonic' drone to determine the pitch movement before it actually occurs. This is very popular technique in the Swedish House scene but is slowly being developed into different styles of dance music.

The term 'canonic' is derived from musical canon, used to describe the effects of following one melodic line shortly after with an exact imitation of the same melody. An example of this is where one vocalist sings 'row, row, row your boat' whereby a second vocalist enters repeating this first line whilst the original vocalist begins to perform the second line of 'gently down the stream'. This overlapping imitation of melodies from numerous vocalists results in the musical 'canon' effect.

Within dance music, the canonic effect involves a single note drone, commonly a bass drone, playing what would essentially be the underlying motions of the bass or lead. For example if the first note of the bass were at C on the first bar, then E on the second bar, G on the third bar and then B on the fourth bar, regardless of the pitch of the rest of the notes in each bar, the canonic would play C for one bar, E in the next, followed by G and then B. Figure 19.5 shows this relationship between the canonic and a bass melody.

This canonic behaviour is similar in many respects to harmony but on a more simplified scale with no requirement to cadence. What's more, the canonic will typically occur whilst the bass and/or melody is remaining at a single pitch. This produces anticipation in the listener since they are made aware of the upcoming pitch movements of the music but are not yet presented with them in the bass or lead.

What's more, this effect can be further augmented if the canonic is syncopated in its action. For instance, in the previous example of C, E, G and B over four bars, the bass could remain at C throughout but repeated over eight bars.

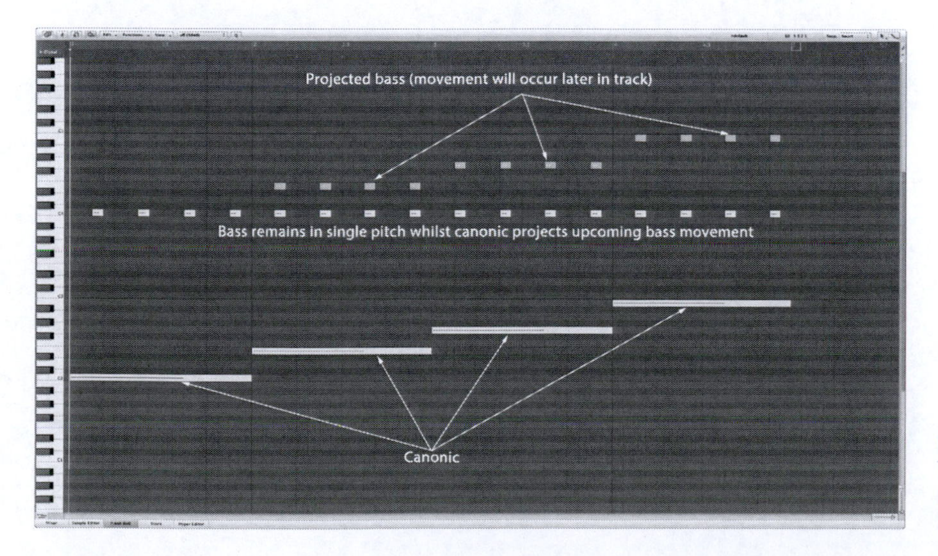

FIGURE 19.5
The canonic underlying the bass melody

Here, the canonic could play a sustained C for the first four bars, before moving to E for two bars and then finally G and B for two bars. This provides a lengthened aural clue as to the upcoming note movements of the bass but with the canonic expressing this over eight bars rather than four, it creates a syncopated relationship to the upcoming bass motion. Figure 19.6 shows the effect of a syncopated canonic and the effect can be heard on the COMPANION WEBSITE.

A syncopated canonic example.

A further technique in arranging is to employ processor or effect development on a timbre. Although all timbres within any dance music should exhibit some form of textural development over their length, the same also occurs over the course of an arrangement.

Slow-modulated changes in filter, distortion or almost any other effect can be used to slowly reveal the timbre of the music. For example, a lead may play the complete melody but be treated to a high-pass filter that slowly opens to allow more low frequencies through over a period of 32 or 64 bars.

Alternatively, a heavy feedback on a delay unit or automated reverb may wash over the sound, and these parameters are reduced gradually via automation until the instrument becomes distinct. This effect is very popular in progressive house, the timbre is exposed to vast amounts of reverb that is increased or reduced to introduce motion and interest into the arrangement. In the following example, I employed a heavy reverb tail and high-pass filter to slowly bring attention to the timbre.

Automated reverb and delay on a lead.

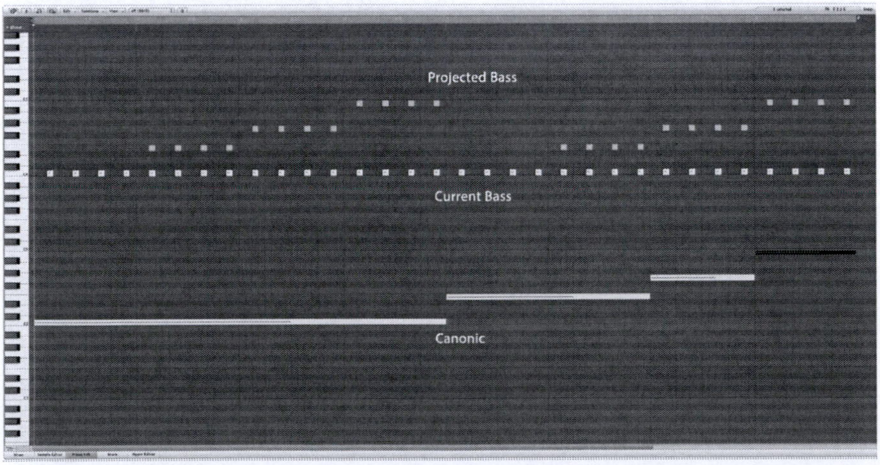

FIGURE 19.6
A Syncopated Canonic

All of these techniques work to create some anticipation or excitement in the listeners but perhaps the most powerful technique to achieve this is to deny an expectation. This can lead to an increased tension and a more powerful release when the music is finally resolved. In order to accomplish this denial, however, the listener must be able to form an expectation of what is about to, or should occur.

There are a variety of ways to achieve a denial but one of the most common is to introduce the complete binary phrase to the listener but with limited backing instruments and employing a different voice or heavily filtered lead. By doing so, the listener is presented with the complete phrase but not in the complete context of the music and therefore the tension is increased, they want to hear it with all instrumentation.

This tension can be increased further by slowly introducing the full timbre but as this occurs, reducing to only its first part by withholding the resolve. Through careful manipulation of effects and filter such as increasing a reverbs tail or increasing the delay time on the lead or bass instrument, it's possible to construct a small build in the music. If this is followed by slowly increasing a filters cut-off and then playing the complete binary phrase again, it will create a form of expectation in the listener, they will expect the music to return in full form.

However, rather than return with the complete music, the music often returns with additional instrumentation whilst still not revealing the complete hook of the record. I've employed this denial technique in the example on the COMPANION WEBSITE.

Creating a denial and increasing tension.

Of course, this is just one instance of how to accomplish denial and there are many more. A small snare roll or a snare skip or a riser can be employed to make the listener believe its building up to something more whereby the track drops instruments rather than introduces more. Simply listening to your favourite artists' music and identifying the techniques they employ to control your emotion will reveal the most common application for the genre of your choice.

Whatever method is employed, however, it must be applied cautiously because denying a listener their expectations too many times or for too long a period it can break tension of the record and it'll loose its cohesion and stability.

THE MAIN BREAK

All these techniques combined should permit the producer to construct an arrangement that continually develops on a theme and leads towards the breakdown before the finale. The overall contour of the music up to this point is down to the producer's tastes and the genre of music but typically the energy will be constructed to build to its highest level thus far before the introduction of the main drop occurs.

The drop is the precursor to the finale and is often accomplished by dropping most of the instrumentation. The lead up into this drop can be controlled in any number of ways from snare rolls, snare breaks, synthetic risers, filters, increasing effects parameters or all of the above.

Snare rolls and breaks can be easily accomplished in any workstations Piano Roll editor. A roll can begin with snare hits positioned on every beat for the first bar or two, then become more frequent occurring at every beat and 1/8th, then at every 16th and then for the final bar at every 32nd. During this time, a low-pass filter may be automated to open or the velocity may be used to increase the volume of the snare as it progresses. Figure 19.7 shows the development of a snare roll in a MIDI editor.

A basic snare roll.

MIDI OF THE SNARE ROLL

Further interest can be added to a snare roll through the application of effects such as delay, phasers, flangers and reverb. These are all typically automated, so that the parameters such as delay time and reverb tail time gradually increase as the snare roll reaches its apex. By doing so, when the roll ends, the reverb and delay can echo across the mix as the instruments are removed for the final drop.

A snare roll with automated delay, reverb and phaser.

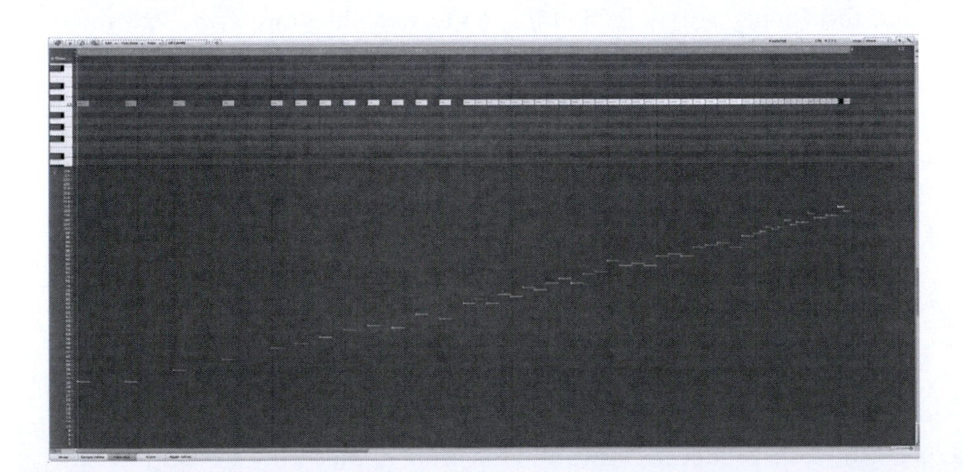

FIGURE 19.7
A snare roll

This is of course only one approach and sometimes a snare roll may not be employed at all. A more relaxed technique is to simply introduce a snare skip. This involves placing a few snare hits just before the fourth beat of the bar to produce a skip in the rhythm.

Alternatively, a popular practice in progressive house is to automate a filter on the lead, so that it opens allowing more frequencies through, whilst reverb and delay times are gradually increased. This can produce introduce a powerful building effect, so that when the music reaches the main drop, the reverb and delay can echo across the mix or stop abruptly behind a falling riser.

When the music reaches the drop, the most commonly applied effect is the low-pass filter since this offer the most powerful way to control the energy and contour of the build. How the drop is handled depends entirely on the genre of music in question and close listening to your favourite styles should reveal the commonly accepted techniques. Typically, though, the drop will consist of only a few instruments that are filtered or all the instruments filtered. This filter is typically automated to open slowly to create the ultimate tension and build and may be backed with another snare roll or a riser until it reaches its apex where the hook is finally introduced and the resolve is presented to the listener.

A typical build up to a finale (with a single bar denial).

Of course, everything discussed here is open to modification, experimentation and creative input. Indeed, it's a close analysis of your favourite music that will reveal the most common arrangement techniques that are employed in that particular genre. Having said that, if you now analytically listen for the above techniques in any EDM track, there is no doubt that most if not all will be employed in the arrangement. They will often be mixed with a variety of other applications drawn from the producer's own creativity, experimentation and experience, but they will typically include most of the aforementioned techniques.

Having now discussed, the theory and application of formal structure, the following genre chapters will concentrate on the basic principles behind composing the initial 16- or 32-bar hook. With this down, and the techniques discussed in this chapter mixed amongst careful analysis of your peer's music, it is possible to construct complete arrangements from this limited starting point.

CHAPTER 20

Trance

'The "soul" of the machines has always been a part of our music. Trance belongs to repetition and everybody is looking for trance in life... in sex, in the emotional, in pleasure... So, the machines produce an absolutely perfect trance...'

Ralf Hütter (Kraftwerk)

Trance can be an ambiguous genre since it manifests itself in a large number of forms and few will agree on exactly what makes the music 'trance'. However, it can be generalized as the only form of dance music that's constructed around exotic melodies that are vigorous, laid-back or pretty much anywhere in between. Indeed, it's this 'feel' of the music that often determines whether it's 'Progressive', 'Goa', 'Psychedelic', 'Acid' or 'euphoric'. Even then, some will place euphoric trance in with progressive and others believe that acid trance is just another name for goa. To make matters worse, both DJs and listeners often interchange and mash up the genre terms. For example, euphoric trance may be subdivided into 'commercial' or 'underground' of which many believe to be different whilst others consider them to be one and the same.

Much of how trance music is written can be determined by examining its history, and one thing that is certain is that trance, in whatever form, has its roots embedded in Germany. During the 1990s, a joint project between DJ Dag and Jam El Mar resulted in Dance2Trance. The first track entitled 'We came in Peace' is considered to be the first ever 'club' music in the trance genre. Although by today's standards it was particularly raw, consisting solely of repetitive patterns (as Techno is today), it nevertheless laid the basic foundations for the genre with the sole purpose of placing clubbers into a trance-like state.

The ideas behind this were nothing new; tribal shamans had been doing the same thing for many years, using natural hallucinogenic herbs and rhythms pounded out on log drums to induce the tribe's people into trance-like states. The only difference with Dance2Trance was that the pharmaceuticals were man-made, and the pounding rhythms were produced by machines rather than skin stretched across a log. Indeed, to many trance aficionados (if there is such a thing), the purpose of placing clubbers into a trance state is believed to have formed the basis of goa and psychedelic trance.

Although both these genres are still produced and played in clubs to this day, the popularity of 3,4-methylenedioxy-*N*-methylamphetamine (MDMA or 'E') amongst clubbers inevitably resulted in new forms of trance being developed. Since this pharmaceutical stimulates serotonin levels in the brain, it's difficult, if not impossible, to place clubbers into states of trance with tribal rhythms and instead the melodies became more and more exotic slowly taking precedence over every other element in the mix. The supposed purpose was to no longer induce a trance-like state but emulate or stimulate the highs and lows of MDMA. The basic principle is still the same today in that the chemicals, arguably, still played a role in inducing a state, albeit a euphoric one, and the name euphoric trance was given to these tracks.

This form of music, employing long breakdowns and huge melodic reprises, became the popularized style that still dominates many commercial clubs and the music charts today. As a result, when the term trance is used to describe this type of music, most view it as the typical anthemic 'hands in the air' music that's filled with large breakdowns and huge, exotic melodic leads. As this *still* remains one of EDM's genres with the largest following (although many will argue this is due to its 'pop' music feel), it will be the focus of the chapter.

MUSICAL ANALYSIS

Possibly the best way to begin composing any genre of dance music is to find the most popular tracks around at the moment and break them down to their basic elements. Once this is accomplished, we can examine the similarities between each track and determine exactly what it is that differentiates it from other musical styles. In fact, all music that can be placed into a genre-specific category will share similarities in terms of the arrangement, groove and sonic elements.

In the case of euphoric trance, it exhibits an up-tempo, uplifting feel that's accessible to most mainstream clubbers. As a result, it can be best illustrated as consisting of exotic melodic synth and/or vocal hook line laid over a comparatively unsophisticated drum pattern and bass line. The drums usually feature long snare rolls to signify the build to the reprise and breakdowns, alongside small motifs and/or chord progressions that work around the main melody. Uplifting trance can be written in any scale, but a large proportion of trance tracks do tend to be written in A minor, B minor, G# major, A# major and E major.

Most will also employ the 4/4 time signature, and whilst the tempo can range from 130 to 150 BPM, a large proportion remain at 137 or 138 BPM. This, of course, isn't a demanding rule, but it is important not to go too fast as it needs to remain anthemic, and the majority of the public can't dance to an incredibly fast beat.

THE RHYTHM SECTION

With most uplifting trance, the rhythm section is invariably kept simple and will almost always rely on a four to the floor pattern. With this, a kick sits on all four beats of the bar with snares or claps resting on the second and fourth beat. This pattern is further augmented with a 1/8th off beat open hi-hat and a closed hat strike occurring on every 1/16th of the bar. In many examples, any closed hi-hats that occur in the same position as the open hats are removed to prevent any frequency clashes between the closed and open hi-hat.

Trance kicks are commonly kept 'tight' rather than 'boomy' and often exhibit a bright transient stage to help them cut through the bass and lead instruments in the mix. The kicks can be created through sample layering, taken directly from a sample CD or programmed in a synthesizer. If programmed, a good starting point for the main kick (for which further samples can be layered onto) is to employ a 90 Hz sine wave and modulate it with a positive pitch envelope with no attack and a medium release. If you have access to the pitch envelopes shape, then generally a concave release is more typical than convex. Finally, experiment with the decay setting to produce the timbre required for the mix. It is important that the kick is 'tuned' to the key of the music, and for the best results, it should be tuned to the fifth note of your chosen scale. In the example track, which is composed in A minor, the kick is tuned to E.

The kick is commonly treated to small amounts of reverb. It is preferable to employ a room style reverb with a long pre-delay that will skip the transient to ensure it can cut through the mix. A typical setting is a pre-day of 90 ms with a tail of approximately 40 ms, but it does depend entirely on the kick being used. If a more powerful kick is required, a hall setting tail can be employed provided it's followed with a noise gate to remove any tail.

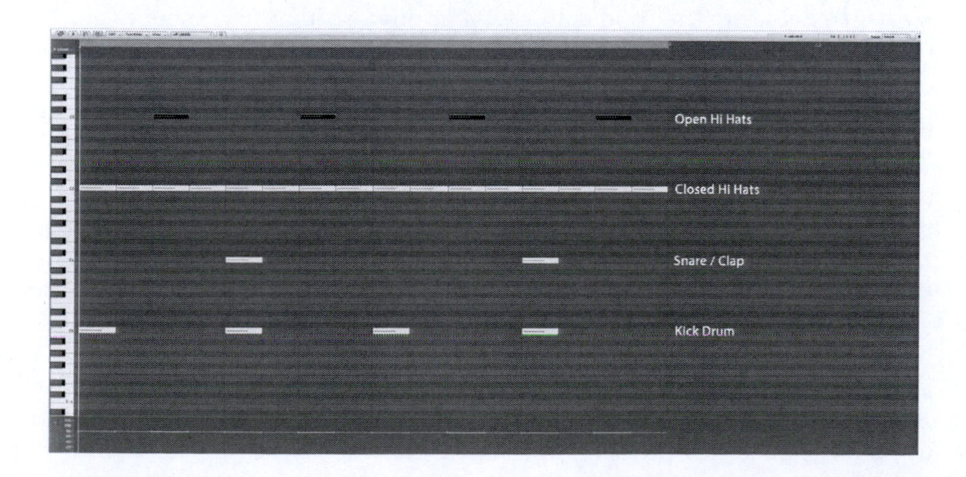

FIGURE 20.1
Typical trance drum loop

Following reverb, compression is often applied since this can play a large role in achieving the correct timbral style. Here, the attack of the compressor should be configured so that it skips the initial transient but captures the body and tail. The ratio is set at approximately 3:1 to 5:1, and whilst listening to the kick, the threshold is reduced until the required sound is achieved. Bear in mind the reduction of the compressors threshold will reduce the dynamic range of the transient that may prevent it from pulling through a mix. If this occurs, a simple hi-hat timbre can be layered over the kicks transient to help it pull through. Alternatively, some artists will apply small amounts of controlled distortion to the kick, following this distortion with EQ so that it can be further sculpted.

Trance will employ either snares or claps to augment the second and fourth beats, and like the kicks, these can be produced through sample layering, or more typically they are sourced from a sample CD or synthesized. This is because unlike many other genres, the snares and claps are kept very light, and it's not unusual to roll off all frequencies below 800 Hz to 1.5 kHz. In fact, these instruments are often made so 'light' in frequencies; hence, they can be often confused for high-hats.

Typically, the snares or claps are treated to a room style reverb although often the reverbs tail is longer than a kick, allowing it to tail off gently. The reverb settings to employ here are entirely dependent on creativity, but typically, it should remain fairly short so as not to wash over the loop completely. Some artists will employ reverse reverb to some of the snare/clap hits, but this is typically limited to either one hit per bar (commonly the second beat) or will only occur at structural downbeats. The snares or claps will, however, be treated to modulation on every bar.

With uplifting trance, filter modulation appears more popular than microtonal adjustments on the snares/claps, but both will produce the requisite feel of movement between the two beats. In the example track, I used *SoundToys FilterFreak* on the snare with an offset sine wave modulation to control a low-pass filters cut-off. The modulation was applied so that first snare of the bar passes through a low-pass filter unmolested, but the second strike occurring on beat 4 is subjected to a very light low-pass filtering that removes a very small portion of the higher harmonics of the timbre. This introduced enough of a textural difference between the two beats to maintain interest.

Finally, the hi-hats can be programmed or taken direct from a sample CD. In many trance tracks, the hats are commonly sourced from a sampled, original or software emulation of a TR909 (the 909 has a very distinctive timbral character); however, some artists do program their own by ring modulating a high-pitched triangle wave with a lower pitched triangle. This produces a high-frequency noise that can be modified with an amplifier envelope set to a zero attack, sustain and release with a short to medium decay. If there isn't enough noise present, it can be augmented with a white noise waveform using the same envelope. Once this basic timbre is constructed, shortening the decay creates a closed hi-hat, while lengthening it will produce an open hat. Similarly, it's also worth experimenting by changing the decay slope to convex or concave to produce fatter or thinner sounding hats. As both these timbres depend on the high frequency content to sit at the top of the mix, compression should be avoided but adjusting the amplifier decay of synthesized hi-hats can help to change the perceived speed of the track.

If the hi-hats are sampled, then this can be accomplished by importing them into a 'drum sampler' such as Native Instruments Battery or alternatively you can use a transient designer such as SPL's Transient Designer. Using these to keep the decay quite short, they will decay rapidly, which will make the rhythm appear faster. Conversely, by increasing the decay, you can make the rhythm appear slower. Most trance music tends to keep the closed hit-hats quite short and then employ a delay on them. The settings for the delay are dependent on the producer's creativity and the tempo, but it should be kept fairly short to prevent too much delay wash that will cloud the signal. Typically a 1/16th or 1/8th setting with a very short delay time will be enough.

The hi-hats in this genre are often sourced from the multitude of sample CDs available on the market and are treated to a multitude of effects to the producer's own personal taste. This can include noise gates to shorten the hat, transient designers, EQ, distortion and/or reverb. However, they can equally be synthesized through ring modulation.

This is accomplished by modulating a high-pitched triangle wave with a lower pitched triangle to produce a high-frequency noise. As a starting point, set the amp envelope to a zero attack, sustain and release with a short to medium decay. If there isn't enough of a noise present, it can be augmented with a white noise waveform using the same envelope. Once this basic timbre is constructed, shortening the decay creates a closed hi-hat while lengthening it will produce an open hat. Both open and closed hi-hats will commonly be treated to small amounts of delay, but this must be applied with caution so as not to upset the syncopated feel of the hats.

Like the snare/clap, the open and closed hi-hats are treated to a similar form of cyclic filter modulation. Typically the open hats will be treated to a low-pass filter that is modulated via an offset sine wave or via a sample and hold waveform. When this effect is applied *very lightly* over the length of two bars, very slight harmonic changes occur over a period of two bars, and since the

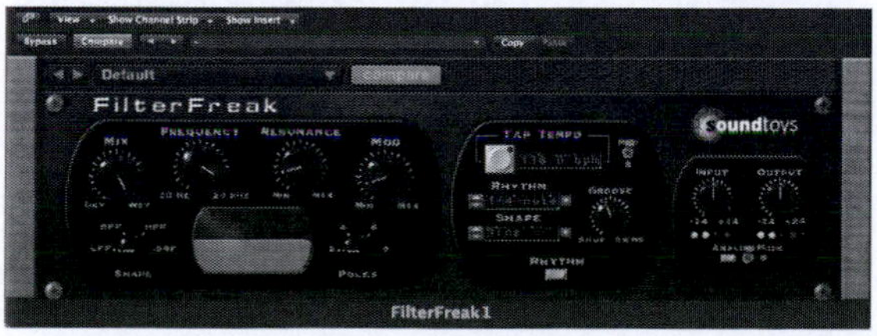

FIGURE 20.2 Modulating the drum rhythm

modulation source is offset or random, it prevents it from appearing as a cyclic repeated motion.

Of course, this should all only be considered a general guideline and I entirely open to interpretation. For instance, occasionally Hemiola closed hats may sit at the end of every few bars to add some extra variation or alternatively the snare/clap may be placed a 1/16th before the kick occurs on the last beat of the fourth bar. Another common technique is to employ a noise gate on the closed hi-hat channel and then trigger the gate with a rhythmical programmed pattern from another channel via the side-chain function. This pattern is changed every so often on a structural downbeat to inject more interest into the drum rhythms.

Finally, swing quantize is almost always applied to the snares and open hi-hats. This is applied very, very, lightly so as not to interrupt the timing too heavily and typically applications of between 51% and 53% on a 1/16th grid will be enough to maintain some rhythmical interest in the loop. After this, the loop will be often treated to some parallel compression to help the loop gel and introduce a little more energy.

BASS GROOVE

The bass rhythm in trance varies greatly and is open to artistic interpretation, but it will generally always consist of two bass lines that not only syncopate with one another but the transients also remain offbeat. This approach ensures that the bass transient does not occur at the same time as the kick or lead instruments transients. By doing so, more compositional 'room' is created in the mix since the bass transients, lead transients and kick transients are not all occurring at the exact same time. Indeed, the basic theory employed in a large amount of trance music is that the lead and kick transients can occur together, and the lead and bass transients can occur together but the bass and kick, or the bass, kick and lead should not occur simultaneously.

The two basses in this genre typically consist of a sub-bass and a high bass. The sub-bass will most commonly consist of an octave shifting bass constructed of short 1/16th notes.

Figure 20.3 shows a typical trance sub-bass arrangement in Logic Pro's Piano Roll editor. Note how each transient of the bass not only remains off the beat so as not to clash with the kicks transient but also remains at a single pitch with only a single octave jump occurring in each bar (note pitch A octave jumps to A). Moreover, additional rhythmical motion is applied to this simple rhythm by moving the position of the octave jump a sixteenth *back* in time on alternative bars.

For the high bass, the lowest sub-bass pattern is often duplicated onto another channel and transposed up by an octave. To maintain the rhythmical motion but also introduce some difference between the two basses, occasional notes are removed from the high bass pattern and then lengthened to join these notes together (i.e. eliminate any spaces between the pattern). This approach is shown in Figure 20.4.

Of course, this approach is only one example and is open to artistic interpretation, but whilst there are differences in rhythm and pitch between records, this is the *general* approach taken by many trance artists when composing the bass line for the record. Indeed, it is important that the bass remains close to a single pitch for much of a bar and only changes pitch between the different bars. With this style of trance, the lead instrument is the focus of the track, and if the bass melody continually changed pitch during the course of the bar, it would dissuade the listener's focus from the lead.

For the bass timbres, analogue synthesizers (or DSP equivalents) are invariably chosen over digital since the bass is exposed from the kick, and the 'distortion' introduced by analogue produces a warmer tone to compliment the kick. Many analogue style synthesizers are capable of producing both the bass timbres,

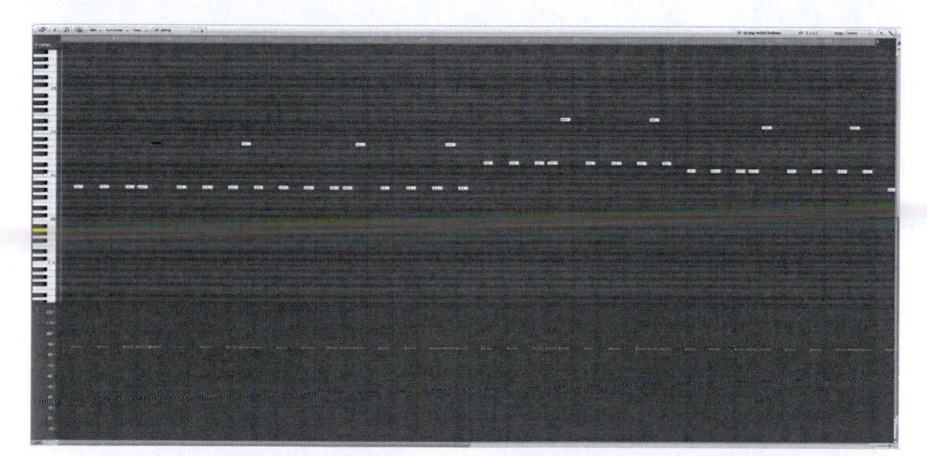

FIGURE 20.3
The sub-bass

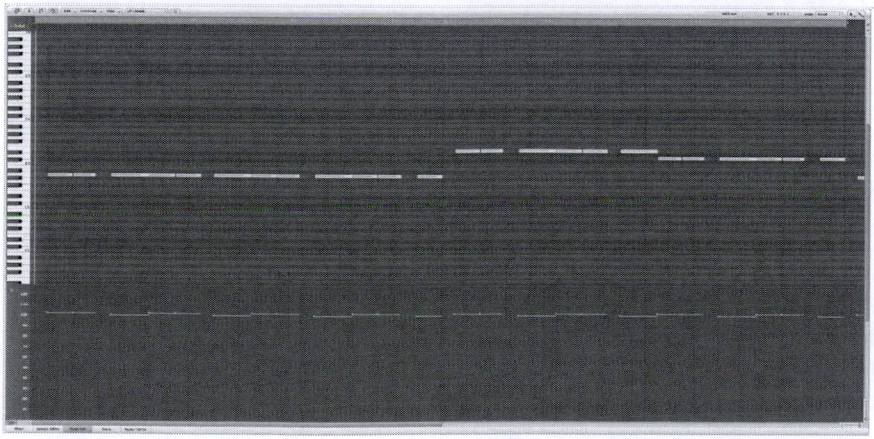

FIGURE 20.4
The high bass

but many trance musicians often rely on the Novation Bass Station (virtual instrument), the Novation SuperNova (hardware), Novation V-Station (virtual instrument), Sylenth (virtual instrument) or the Access Virus Ti (hardware).

Using any of these, possibly the easiest solution for this timbre is to dial up a preset bass in a synth and then contort the results to suit the tune. If, however, you wish to program your own, a good starting point for the sub-bass is to employ a pulse and sine (or triangle) oscillator and detune them from one another until you reach the basic tone you like. In general, a heavy transient pluck is not necessary on the sub-bass since it isn't competing with the kick and its purpose is to simply energize the groove of the music. Typically, it will employ a medium attack on the amp envelope with a slightly slower attack on the filters envelope. Both amp and filter sustain, and release parameters are rarely used, and instead the decay will act as the release parameter. This way, with the sub-bass playing back, the decay parameter can be used a fine tune control to balance the compositional relationship of the sub-bass with the kick.

The high bass can use the same oscillator waveforms as the sub-bass but the attack envelope should be set at its fastest so that is precedes the sub-bass and exhibits more presence. More pluck can be added to the character by shortening the attack and decay parameter of the filter envelope and set to modulate at a positive depth. In fact, it's essential to accomplish a good rhythmical and tonal interaction between the drums and two bass lines before moving on by adapting and changing the amplitude and filters on both synthesizers.

Once both basses are sitting together, both are commonly treated to delay. The delay settings to employ will depend heavily on the timbre but they are often treated to different delay times and the sub-bass is often run through a mid/side processing unit so that the delay only affects the few higher frequencies leaving the lower frequencies in mono. To prevent the delay from each from washing over one another, the delay unit is placed on a bus and followed with a compressor set to a fast attack and release with a high ratio and low threshold. The compressor is then fed with the original bass signal into its side chain input, and the basses are *sent* to this delay bus.

This way, every time an original bass hit occurs, it triggers the compressor through its side-chain input. As the compressor is triggered, it restricts the dynamics and lowers the volume of the delayed bass. However, during the silence between the original bass hits, no compression is applied, and therefore, the delay occurs at full volume.

Distortion and compression are also often used on basses. Here, small amounts of controlled distortion can help to pull the bass out of the mix or give it a much stronger presence whilst side-chain pumping compression is applied to both basses. To accomplish this, the producer will often duplicate the original kick track but set this one to no output in the mixer so that it cannot be heard. This track can then be used as a side-chain input channel on both basses.

With a compressor applied on both basses, the newly created kick channel is side-chained into the compressor and the compressor is adjusted so that each bass pumps rhythmically with the kick. By creating a secondary kick track to accomplish this side-chain duty rather than using the original kick channel, if the original audible kick is removed from the track during a drop, the 'pulsing' in any side-chained instruments will still be present.

The amount of compression applied will depend on the timbres used but as a general guideline start by setting the ratio to 9:1, with an attack of 5 ms and a medium release of 200 ms. Slowly decrease the threshold until every kick registers on the gain reduction meter by at least 3 dB. To avoid the volume anomaly (i.e. louder invariably sounds better!), adjust the make-up gain so that the loop is at the same volume as when the compressor is bypassed and then start experimenting with the release settings. By shortening the release, the kicks will begin to pump the basses and will become progressively heavier, the more that the release is shortened.

Unfortunately, the only guidelines for how short this should be set are to use your ears and judgment but exercise caution. The principle is to help the drums and bass gel together into a cohesive whole and produce a rhythm that punches along energetically.

UPLIFTING TRANCE MELODIES

The lead melody is the fundamental element to trance music. Although time and care must be taken producing every element of the music, with uplifting trance this is the selling point of the music. It's also the most difficult part as a good lead is derived not only from the melody but also the timbre, and both of these have to be 'accurate' so close scrutiny of the current scene and market leaders is absolutely essential in acquiring the right feel for the dance floor.

Unfortunately, in terms of MIDI programming, trance leads follow very few 'rules' so how to program one is entirely up to your own creative instincts and a careful analysis of the current market leaders, but there are some basic guidelines that apply.

The first and possibly easiest method is to producing a trance lead is to begin by fashioning a chord structure. This is because the lead is often constructed using a 'chorded' structure so that the notes alternate between two or three notes. This creates the results of jumping from the 'key' of the song to a higher series of notes before returning to the main key again. What's more, the pitch motion of the bass and subsequent lead melody throughout the bars is one of the fundamental elements of trance music and provides much of the driving force.

Unlike popular music, a chord progression in trance does not have to be particularly complex, and in many instances, the movement is kept very simple. Indeed, in some of the most popular trance tracks of the past 10 years, many have employed little more than a three-chord progression! These progressions are often as simple as I-IV-V or I-V-IV but some have extended into I-V-vi-IV, vi-IV-I-V or i-VI-III-VII (A-minor).

However, many chords are employed in the progression, typically, uplifting trance works on a structural downbeat design of four over eight bars. That is, the bass, chords and lead melodies will repeat every fourth bar but be repeated to construct eighth bar downbeats. This helps to maintain the requisite repetition and motion that forms the basis of the music.

If only three chords are employed in the composition of the music, then commonly the harmonic rhythm will consists of I for two bars, IV for third bar and then finally V for the fourth bar. With this harmonic movement, the bass and lead melody will follow the root note of the chords. This style of motion is shown in Figure 20.5.

Once this basic chord is constructed, it can be duplicated to another track and used as the basis to creating a lead. The most common way to accomplish this is to simply record a 3 or 4 note rhythm over the period of one bar and then repeat this rhythm throughout all eight bars, pitching it with the underlying chord.

To keep the relative speed and energy of trance, the chorded lead will remain short, employing 16th or 18th notes so the chord is cut into smaller notes, and then the odd notes throughout are deleted to create a 'step up in pitch'. The results of this approach are shown in Figure 20.6

Naturally, this is only one way to accomplish the goal, and there are many others. For example, by inserting a noise gate onto the chords channel and then programming a rhythmical hi-hat pattern, it can be used to control the noise gate and thus create a rhythmical lead. This gated effect, once suitable, can be applied physically to the notes (i.e. cut them up to produce the effect) so that not only does each note retrigger the synth but also it permits you to offset the

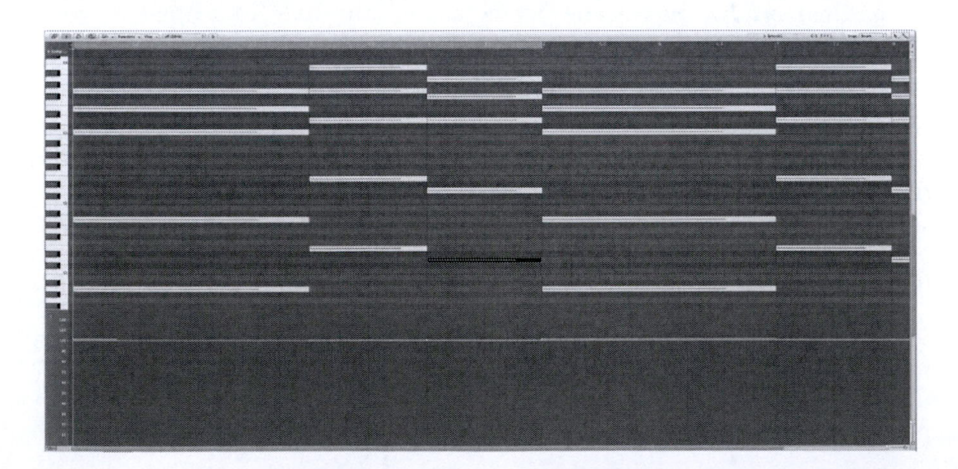

FIGURE 20.5
The pitch structure of three chords in a four over eight structural composition

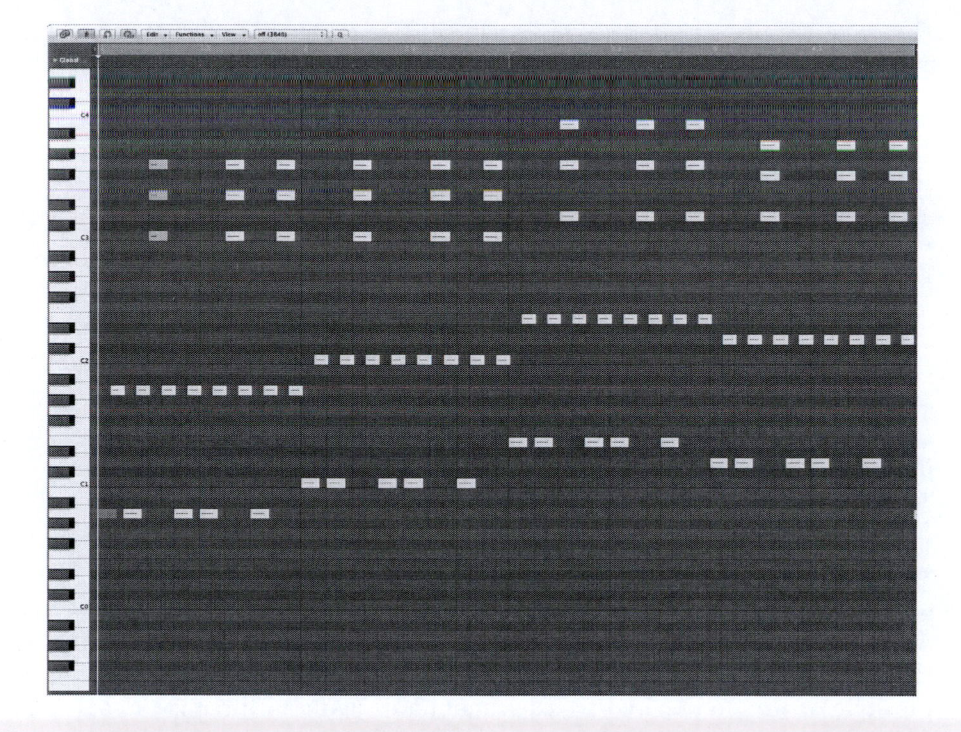

FIGURE 20.6
The rhythmical step approach

top notes of a chord against the lower ones to produce the alternating notes pattern that's often used for the leads.

Alternatively, another frequently used method to create a trance melody is with a synthesizer's arpeggiator, most commonly the Access Virus' arpeggiator. Here, a chord is held down, and the producer tweaks the available arpeggiator patterns until he or she reaches inspiration that can be developed upon further. Ultimately, although, a great trance melody is the holy grail of any trance artist, and many are simply a result of careful analysis, experimentation and serendipity on the producer's behalf.

MELODIC TIMBRES

Just as the melody for the lead is important, so is the timbre since the music can stand or fall on its quality. As discussed in the chapter on sound design, this lead is the most prominent part of a track and therefore sits in the midrange/upper midrange. Consequently, it should be rich in harmonics that occupy this area and should also exhibit some fluctuations in the character to remain interesting. This can sometimes be accomplished through modulating the pitch with an LFO on a digital synth but the phase initialization from analogue synths, more often than not, will produce much better, richer results. Consequently, the preferred synths by many artists are the Access Virus and the Novation SuperNova (now the MiniNova).

The secret (if there is one) to any epic uplifting trance lies with saws, unison and effects. Begin with three saw oscillators, or if the synth has them available, stacked saw oscillators (aka multisaw oscillators) and detune them from one another by as much as possible but not too far that the oscillators separate from one another. Once detuned, unison should be introduced until the sound becomes very thick. The amplifier envelope should be set to a fast attack with a fast decay, medium release and medium sustain value whilst a low-pass filter is modulated with a second envelope with the same settings as the amplifier. The low-pass filter should be completely open, allowing all frequencies through.

Once a lead has been synthesized, the amplifier decay and release envelope can be used to lengthen and contort the timbre as required. This gives much more control over the groove, allowing you to immediately adjust the 'feel' of the music by simply adjusting the amplifiers release envelope rather than having to 'kick' MIDI notes around.

This approach will produce the *basic* timbre and should be followed with EQ. The EQ settings will depend on the synthesizer but typically a 6 dB boost at 650 Hz with a high shelf boost of 10 dB at 5 kHz will provide the bright characteristic sound. This EQ is commonly followed with a distortion unit and chorus and then a secondary EQ unit. The settings for these are dependent on the style of timbre required so experimentation with all three is the key. The distortion is usually applied gently to add further harmonic characteristics to the higher frequencies whilst chorus can be used to thicken the sound further and the EQ can be used to remove any problematic frequencies.

A large hall setting can be applied to the lead with a pre-delay set so that the initial reflections bypass the transient. Generally speaking, the larger the reverberant field, the more "epic" the lead will become, but the reverb should be followed with a noise gate so that tail of the reverb is cut the moment the timbre ends to prevent reverb from washing over the mix. Finally, the lead is processed with a short delay. Both 1/16th and 1/8th settings are common but experimentation will yield the best results.

Although not always necessary, parallel compression can be employed on the lead to bring up its presence in the mix but if the lead is lacking it low or top end energy, additional chorded notes added to the lead can quickly resolve the problem. Indeed, it is important not to underestimate the power of simply adding more chorded notes to a melody that appears to be lacking in energy.

Notably, if vocals are employed in the track, then there may be little need for any effects, as the lead should sit under them. However, if you want to keep a wide harmonically rich sound and vocals, it's wise to employ a compressor on the lead timbre and feed the vocals into a side chain, so that the lead drops when the vocals are present.

Of course, this is all open to artistic license, and once the basics of the lead are down, it's prudent to experiment by replacing oscillators, envelope setting's and effects to create variations. If this technique still does not produce a lead that is rich enough, then it's worth employing a number of methods such as layering, doubling, splitting, hocketing or residual synthesis. All of these were discussed in the sound design chapter.

MOTIFS

With both main melody and the all-essential groove down, the final stage is to add motifs. These counter-melodies are the small ad-lib riffs best referred to as the icing used to finally decorate the musical cake and play an essential part of any dance music, adding much needed variation to what otherwise would be a repetitive track.

These are often derived from the main melodic riff as not only do they often play a role in the beginning part of the track before the main melodic lead is introduced but they are also re-introduced after the reprise.

There are various techniques employed to create motifs, but one of the quickest ways is to make a copy of the MIDI lead and 'simplify' it by removing notes to create a much simpler pattern. Once created, this pattern can be offset from the main lead by shifting them forward or later in time by a 1/16th, or alternatively, it can occur at the same time as the lead but the attack and release of the timbre are lengthened. In fact, this latter approach will often produce better results since, as discussed, dance music works on contrast and so far all the instruments have had a definite attack stage. By lengthening the attack and release, the motif will take on a whole new rhythmic character that can quickly be changed by listening back to the arrangement so far and adjusting the amplifier envelopes parameters.

Above all, while using any of these techniques it's sensible to avoid getting too carried away and it's vital that they are kept simple. Many dance tracks simply use a single pitch of notes, playing every eight, sixteenth or quarter or are constructed to interweave with the bass rather than the lead. This is because not only do simple motifs have a much more dramatic effect on the music than complex arpeggios but you also don't want to detract too much from the main melodic element.

In addition, the timbre used for any motif should be different from the lead melody, and it's at this point that you'll need to consider the frequencies that are used in the mix thus far. As touched upon, in many trance tracks, the lead is harmonically rich, and this reduces the available frequencies for a motif, so it's quite common to utilize a low-pass filter cut-off to reduce the harmonic content while the lead is playing, yet set this filter to open wider while there is no lead playing.

Moreover, as with all dance music, this filter is tweaked 'live' to create additional movement and interest throughout the arrangement. This movement obviously has to be restrictive while the lead is playing otherwise the mix can quickly become swamped in frequencies and difficult to mix properly, but during the beginning of the song, the filter can be opened wider to allow more frequencies through to fill up any gaps in the mix.

More importantly, the rhythmical pitch of the motif will have to be carefully chosen. A motif that is mostly written using notes above C4 will naturally contain higher frequencies and few lower ones so using a low-pass filter that is almost closed will reduce the frequencies to nothing. Therefore, the pitch of this motif and the timbre must occupy the low- to mid-range to fit with the bass and the lead melody, assuming of course, that the lead melody contains higher frequencies.

Indeed, with trance, it's careful to use the filter that creates the final results, and often you will need to compromise carefully by cutting some of the frequencies of the lead to leave room for the motif and vice versa. Having said that, if the bass is quite simple, an often used technique is to program a motif that contains an equal amount of low frequencies as mid, and then use a filter to cut higher frequencies to mix the motif in with the frequencies of the bass, helping to enhance the low-end groove, before proceeding to remove this low-end interaction by cutting the lower frequencies and leaving the higher ones in as the track progresses. This creates the impression that all the sounds are interweaving with one another helping to create more movement and the typical 'energy' that appears in most club tracks.

Ultimately, although, as with all the chapters in this section, the purpose here is to simply reveal some of the techniques used by producers and show how the theory and technology discussed earlier in the book combines to produce a track. However, it should be viewed as offering a few basic starting ideas that *you* can evolve from.

There is no one definitive way to produce this genre, and the best way to learn new techniques and production ethics is to actively listen to the current market leaders and experiment with the tools at your disposal. There are no right or wrong ways to approach any piece of music, and if it sounds right to the producer, then it usually is. New genres do not evolve from following step-by-step guides or from emulating peers; producers who experiment and push boundaries create them.

Nevertheless, with just these basic elements, it is possible to create the main focus of the music, and from here, the producer can then look towards creating an arrangement. The theory behind arranging has already been discussed in an earlier chapter, but simply listening to the current market leaders mixed amongst the theory discussed in the previous chapter will very quickly reveal the current trends in both sounds and arrangement.

The companion website contains an audio example of a typical trance track using the techniques described.

CHAPTER 21
Dubstep

'In human life no-one is inventing something. It's always the remix of the remix of the remix. I am the remix of my father...'

Ricardo Villalobos

Although some Dubstep is viewed as a relatively new genre of dance music, its history goes back further than many imagine, and although like most genres of dance music, its history is amorphous, it can roughly be traced back to 1998 and two-step garage.

A development from UK garage, two-step consisted of heavily syncopated rhythms veering away from the usual four to the floor rhythms of most other genres of dance music. However, it was the B-sides of many of these records that formed the beginnings of Dubstep. These contained more experimental remixes of two-step, mixing the deep bass philosophy of two-step with the more funky syncopated rhythmic undertones of Drum 'n' Bass and Grime.

DJ Hatcha, an employee at Big Apple Records in Croydon, South London, is attributed as being one of the first DJ's to play these B sides to clubbers. During his sets at 'Forward' in the Velvet Rooms in Soho, London, he used it as a showcase for this experimental two-step music that many club goers soon termed 'forward music'. Ammunition Promotions ran the 'Forward' club and promoted the night as 'b-lines to make your chest cavity shudder'.

Forward also ran as a pirate radio station in London entitled *Rinse FM* that featured artists such as DJ Hatcha, Koed9, Zed Bias and Jay Da Flex who would take the opportunity to play out the latest 'forward' tracks. It was during this time that many believe Ammunition Promotions coined the term 'Dubstep' to describe this new form of music in an *XLR8R* magazine in 2002.

Throughout the next year, DJ Hatcha is attributed to pioneering the direction of Dubstep. Working in a record store that was frequented by artists such as

Skream (who later worked in the same store), Benga and Loefah, who passed him their dub plates to broadcast on both *Rinse FM* and club *Forward*.

As Dubstep began to gain momentum on the underground scene, in 2003, a new event named *Filthy Dub* promoted by Plastician (formerly PlasticMan) and David Carlisle was established to further promote Dubstep with records from N-Type, Benga, Walsh Chef, Skream and Digital Mystikz.

Whilst Radio One DJ John Peel had been a staunch supporter of Dubstep since 2003 playing many of their records, it was actually Digital Mystikz (aka artists Mala and Coki) along with Loefah and Sgt. Pokes that are attributed to bringing Dubstep to the mainstream market. After creating the Dubstep label DMZ, they held a bimonthly DMZ nightclub at the Mass club complex in Brixton. Radio One DJ Mary Anne Hobbs discovered this new genre at DMZ in 2006 and subsequently gathered all of the top Dubstep producers together for a Radio One show entitled Dubstep Warz. This proved to be the turning point of what, so far, was largely an underground form of music.

By 2007, the sound of Dubstep began to influence the commercial mainstream market with artists such as Britney Spears employing the deep bass wobble that was typical of the genre in some of her music (in particular 'Freakshow'). Then, in 2009, La Roux asked Skream to remix their single 'In for the Kill' alongside assigning other tracks to artists such as Nero and Zinc to remix.

In 2010, Dubstep was in full flow and had become an unstoppable force. Magnetic Man reached number 10 in the UK chart with his track 'I Need Air' whilst 'Katy on a Mission', produced by Benga, debuted at number 5 in the UK singles chart and remained in the top 10 for another 5 weeks.

It was DJ Fresh, however, that scored the first Dubstep UK number one with 'Louder' in 2011 followed shortly after by Nero's 'Promises', a second-Dubstep track that reached the number one spot.

MUSICAL ANALYSIS

Dubstep can be characterized today as featuring heavy bass lines (often as low as 20 Hz) mixed with a two-step *style* drum rhythm and dark reverberated chords. Typically, it is composed in a 4/4-time signature with a tempo of 140 BPM, but due to the two-step nature of the music, it often results in the music appearing to be half of that to the listener. The music is also largely composed in a minor key, and in many examples, A minor appears to be the preferred choice.

Perhaps the best place to start in the production of Dubstep is with the drum rhythms. Originally, the beats followed the two-step drum rhythm with a kick drum occurring on the first and third beat of the bar and the snare dancing around this, but more recent examples are revealing that artists are pushing these boundaries. Indeed, many artists now tend to place a snare only on the second or third beat of each bar, and it is the kick that moves around this fixed snare position. Since the snare remains constant on the second *or* third beat of each bar, it reinforces the slower half tempo paced feel of the music.

In many examples, the kick will never occur at the same time as the snare and rarely occurs on the beat. Instead, the kick will dance around this beat in order to introduce a skippy feel to the rhythm. Moreover, this kick pattern will often run in a Polymetre fashion against the rest of the music with the kick pattern only completing after a period of five or seven bars rather than the usual one or two.

Sonically, in many Dubstep tracks, the kicks remain 'tight' with almost a woody style character and a bright transient stage to help them cut through the sweeping subsonic bass instruments. These styles of kicks are available on the multitude of Dubstep sample CD's that have crept onto the market but many producers employ Ultrabeat (Logic Pro), Fruity Drum Synth (FruityLoops) or one of the many drum synthesizers available on the market.

If programmed, a good starting point for the Dubstep kick is to employ an 80 Hz square or sine wave modulated with a positive pitch envelope with no attack and a medium release. If you have access to the pitch envelopes shape, then a concave release is more typical than convex. Finally, experiment with the decay setting to produce the timbre required for the mix. Since this kick is quite bright, it should be tuned to the key of the music and generally, the fifth note in the scale will produce the best results.

To help the kick remain prominent above the sub-bass, a square wave pitched down with a very fast amplifier attack and decay setting may also be applied. This will produce a short sharp click that will help the kick remain prominent in the mix. The amount that this square wave is pitched down will depend on the sound you want to produce, so it's sensible to layer it over the top of the

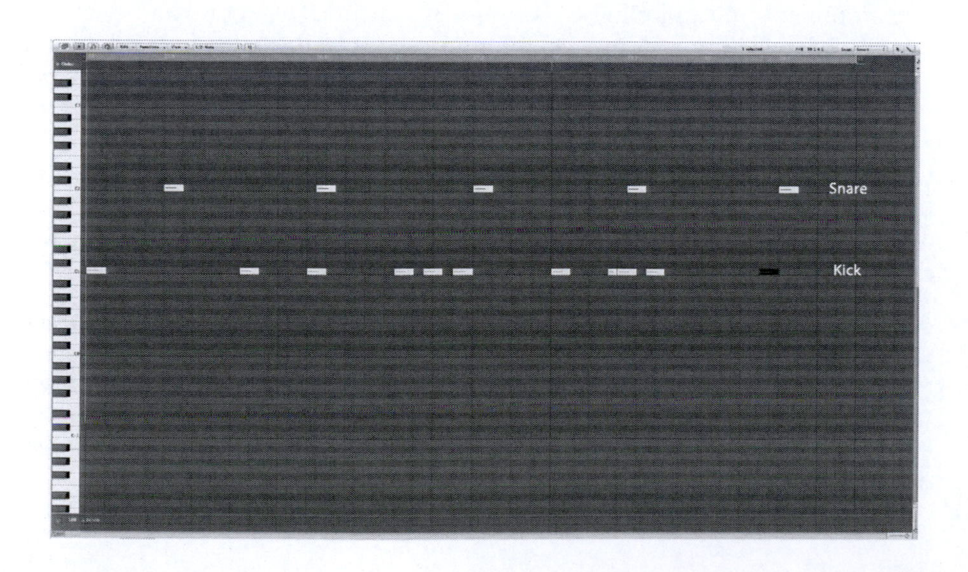

FIGURE 21.1 A typical dubstep kick and snare arrangement (note how the snare is 'fixed' occurring on the third beat of every bar and the kick pattern changes through a period of five bars before repeating)

sine wave and then pitch it up or down until the transient of the kick exhibits a pleasing tonal character.

The kick in this style remains fairly dry in appearance but will often be treated to very small amounts of reverb. Here, a small room style reverb with a long pre-delay that will skip the transient and a very short tail followed by a noise gate is a common approach. Many producers will also EQ the low-mid of the kick to bring out its presence and energy. Typically, small boosts of 1dB to 3 dB at 100 Hz alongside a low shelf to remove all of the kicks frequencies below 70 Hz. This approach increases presence and makes more frequency room for the bass and prevents conflicts.

In addition to reverb and EQ, compression is often employed since this can play a large role in achieving the correct tonal character. For this, the attack of the compressor should be configured, so that it skips the initial transient but captures both body and tail. The ratio is set at approximately 2:1 and whilst listening to the kick the threshold can be reduced until the required sound is achieved. After compression, its not unusual to layer a short transient hi-hat on top of the kicks transient to help it pull through the powerful bass lines that are introduced later.

The snares in Dubstep are often layered to create a noisy style snare that features a hard transient. The transient snare is often taken direct from a TR909 or TR808 emulation and consequently treated to processing and effects. Typically, they are treated to compression with a fast attack, so that the compressor captures the transient and a 2:1 ratio to pull down the transient of the same and produce a harder timbre. This is often followed with EQ providing 2 dB to 4 dB boosts at around 200 Hz to further pull out the transient of the snare.

The second snare consists mostly of noise and can be programmed in most synthesizers with little more than a square oscillator and some white noise. The amplifiers attack should be set to immediate and the decay can be set anywhere from medium to long depending on the sounds that are already in the mix and how slurry you want the snare to appear. This is treated with a high-pass filter to remove much of the low end, but alternatively a band-pass filter can be used if the timbre exhibits too much high-end energy. If possible, it's worth employing a different amp EG for both the noise and triangle wave. The square wave can be kept quite short and swift by using a fast decay while the noise is made to ring a little further by increasing its decay parameter. The further this rings out, the more slurred it will appear.

After programming, the snare can be boosted with EQ to bring out its noise characteristics. In many Dubstep tracks, the prominent frequencies of the snare tend to be around 3.5 kHz. Following a small boost here, they are also often treated to a room style reverb although often the reverb's tail is longer than a kick, allowing it to tail off gently. The reverb settings to employ here is dependent on creativity, but typically it should remain fairly short so as not to wash over the loop.

Some tracks in Dubstep will feature a reverse snare that crawls up to the original snare. This can be accomplished through duplicating the noisy snare and then reversing it in the workstations editor apply reverb with a short pre-delay and a tail of approximately 100 ms to 300 ms. This effect is printed onto the audio, and the audio file is reversed again, so it now plays the correct way around. This effect is often applied over a period of two bars. The first reverse snare will build up to the first snare, whilst in the second bar it may be time stretched, so it occurs much faster than in the previous bar. This produces a syncopated feel over two bars that will contribute to the half-tempo feel.

As with almost all genres of dance music, the snares are often treated to micro-tonal adjustments or filter modulation. This isn't as strict as in many other genres, however, since the snares occur over two bars and not one but it can add to the motion of the loop. For the Dubstep track, I produced for the book, and I employed pitch modulation between the two snares. Duplicating a snare event and pitching it up in Logics editor by +4 cents accomplished this. I then bounced the file and re-inserted it into the project, so that the second bars contained the pitch-shifted snares.

The closed hi-hats in this genre often exhibit a triplet feel or in some examples will remain much more spacious, only occurring every second and third bar out of a four-bar loop. Here, they may only occur six times in one bar positioned closely together off-beat and then in the next bar may appear only three times off the beat. These rhythmical patterns of the hi-hats are entirely up to the producer's own creativity, but it is important they remain offbeat to maintain the syncopated relaxed feel of the music. Constant 1/16th pattern should be avoided since this will 'give away' the much faster underlying tempo and completely remove the illusion of a slower tempo. Open hi-hats are employed

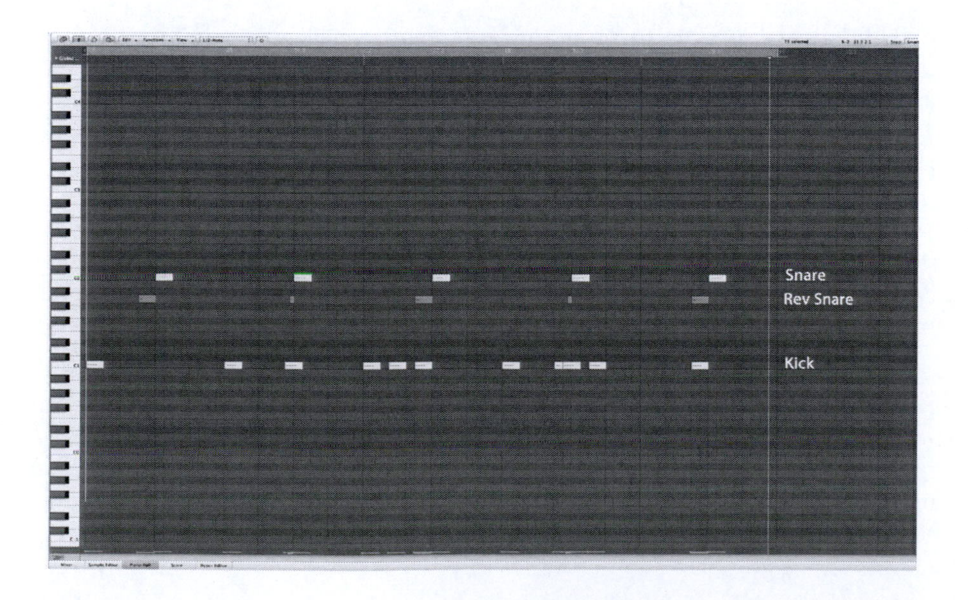

FIGURE 21.2
A reverse snare arrangement (note how the second reverse snare cuts short of its goal – this is shown in MIDI for clarification)

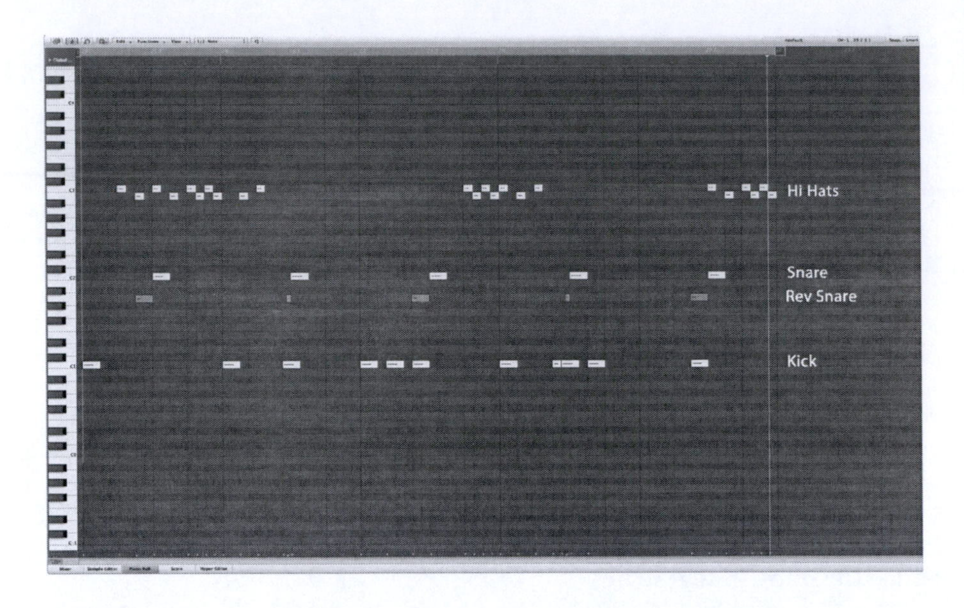

FIGURE 21.3
A typical sparse hi-hat arrangement

in this genre but they often only appear occasionally, sometimes spread out to occur once in four bars, or sometimes they will occur on each beat at the end of a structural cycle of the music.

The closed hi-hat timbres very often exhibit a dirty, almost distorted, bit reduced feel and commonly consist of alternating between two different timbres or have a strong form of pitch of filter cyclic modulation applied, so that it becomes evident. Typically, hi-hats will be taken from sample CDs and treated to numerous effects and processors such as bit-rate reduction and light distortion.

It is, however, possible to program your own hi-hats through ring modulating a high-pitched triangle wave with a lower-pitched triangle. This produces a high-frequency noise that can be modified with an amplifier envelope set to a zero attack, sustain and release with a short to medium decay. If there isn't enough noise present, it can be augmented with a white-noise waveform using the same envelope. Once this basic timbre is constructed, shortening the decay creates a closed hi-hat while lengthening it will produce an open hat. Similarly, it's also worth experimenting by changing the decay slope to convex or concave to produce fatter or thinner sounding hats. If cyclic modulation is applied, many examples appear to be treated to a low-pass filter applied over a period of four-bar loops.

Beyond these instruments, further ancillary percussive instrumentation is rare in Dubstep with many artists preferring to leave the beats sparse in order to maintain the slower paced feel of the music. Occasional instruments may be used, but these will be spaced at every 4, 8 or 16 bars and often only consists of a single percussive strike that will ring out for the duration of a bar.

THE BASS

Perhaps the key feature of Dubstep is the 'Wub'. This is commonly a low-frequency bass exhibiting an almost distorted, dirty sound that cyclically modulates or 'wobbles' in frequency content or pitch during its duration. Indeed, the 'Wub' in many tracks will remain at the same melodic pitch in the workstations piano roll, and it's the cyclic movement of the filter that creates the interest.

There are a multitude of different styles of 'Wub', but the most common can be produced with three saw waves. These are all detuned to −12 (to produce a low-end energy) after which two of three waves are detuned from one another. A good starting point for this is +20 and −17 cents. The amplifier envelope should employ a fast attack with a short decay, high sustain and a no release whilst a second envelope set to a fast attack with a medium decay, no sustain and a long release is used to modulate a low-pass filter. Here, the cut-off should be set about half way and the resonance should be set at around a quarter.

This will produce the basic bright harmonically rich timbre, but it must be modulated with an LFO filter to produce the 'Wub'. The original instrument should not be used to modulate the LFO and instead the producer will typically employ an auto-filter plug-in effect to accomplish this. Inserted onto the synthesizers channel, the filter is set to a band-pass with a low cut-off and a resonance set mid-way. This filter is tempo-synced and modulated via a triangle LFO at a ¼ triplet of the current sequencers tempo. It's this additional auto-filter that creates the 'Wub' sound.

> The companion files contain an audio example of programming a typical 'Wub' sound.

Whilst this Wub forms the foundation of many Dubstep records, it is unusual for any track to feature only one bass and most will contain a combination of three or maybe four bass lines constantly interacting with one another at different positions in the bars to create a call and response system within the music.

For example, one bass line will make the call whilst a secondary bass employing a different sonic texture and 'Wub' will reply to it. This could be followed by a call made by yet another bass line, followed by a reply in another texturally different bass line. Generally speaking, many of these single-note bass lines will be quarter, half or full notes and often straddle over the bars of the drum loop in order to maintain the half tempo feel but a brighter bass in response to a call may consist of a series of short 1/16th strikes to push the tempo of the music.

More importantly, however, almost all examples of Dubstep will feature a low-frequency sub-bass line that underpins the entire record and supplies the trouser flapping sub undertones that push the clubs speakers to their limits.

These characteristic sub-basses can easily be created in most synthesizers with a sine wave and the amplifiers attack set to zero. The decay setting is then increased whilst listening back to the programmed note until it fills the lowest frequencies of the music.

Next, modulate the pitch by a couple of cents using an envelope set to a slow attack and medium decay. This will create a bass timbre where the note bends slightly as it's played. Alternatively, you can use an LFO set to a sine wave with a slow rate and set it to start at the beginning of every note. Experimentation is the key, changing the attack, decay and release of the amp or/and filter EG from linear to convex or concave will also create new variations. For example, the decay to a convex slope setting will produce a more rounded bass timbre. Similarly, small amounts of controlled distortion or very light flanging can also add movement.

Typically, these styles of sub-bass will employ longer notes that are treated to pitch modification. This pitch movement is commonly recorded 'live' into the sequencer via the pitch wheel and then further editing in the sequencer to create bass lines that warp in pitch and can act as either a call or a response.

Like Drum 'n' Bass, the rhythmic movement and interaction with the bass and rhythms provide the basis for this genre, and therefore, it's also worth experimenting by applying effects to this bass timbre. While some effects should be avoided since they will spread the sound across the image, in this genre the

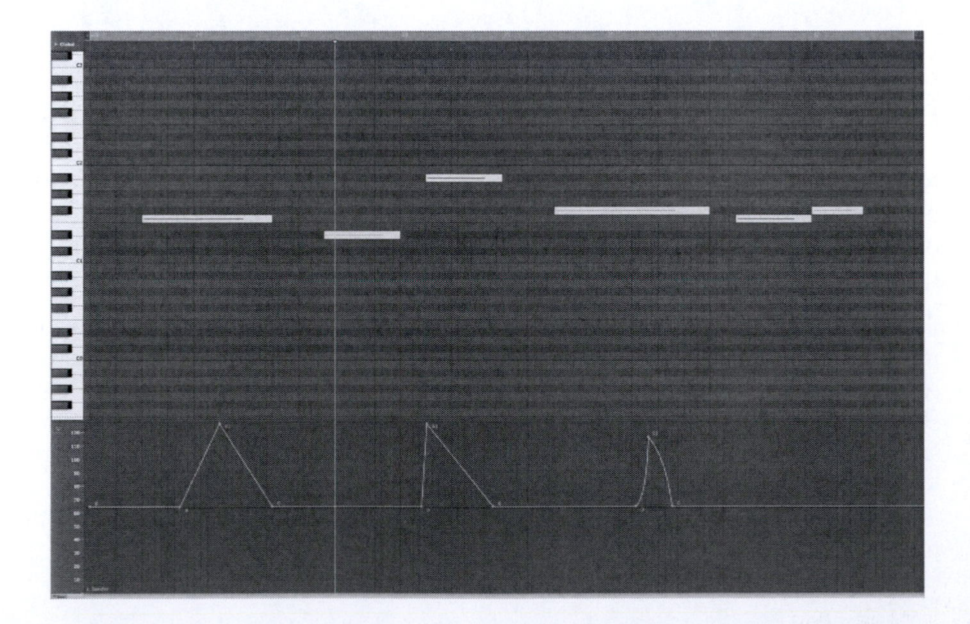

FIGURE 21.4 An example of a sub-bass rhythm in Dubstep (note the pitch bend)

bass is one of the most important parts of the music so small amounts of automated distortion can create interesting effects.

As with the drum rhythms, creative compression can also help in attaining an interesting bass timbre. As before, try accessing the compressor as a send effect with a medium threshold setting and a high ratio. The returned signal can be added to the uncompressed signal, you can then experiment with the attack and release parameters to produce an interesting bass tone. A common technique here is to send the transient snare and kick to a bus (to 0 dB) and then use this bus as a side-chain for a compressor placed onto the bass and lead instruments.

The settings to use for the compressor depend on the effect you wish to accomplish but a good starting point is a fast attack with a 50 ms release and a ratio of 2:1. Once applied, lower the threshold and increase the make-up gain until the effect is evident.

CHORDS, MELODIES AND EFFECTS

In many examples of Dubstep, there are no melodies to speak of and much of the instrumentation is in the form of very, very simple chords and progressions. Indeed, it is rare to hear a chord progression move beyond two chords, and it's only in a few examples where the chords will move beyond three.

Unlike *traditional* music theory, the chords do not resolve in cadence but are typically constructed from a random selection, most commonly the minor chords of that particular minor key. For example, the chords of III, VI and VII (in the natural minor) or III, V and VI (in harmonic minor) are often avoided.

Effects can play a vital role in creation of interesting strings and pads for the genre, although these should be used conservatively so as not to detract

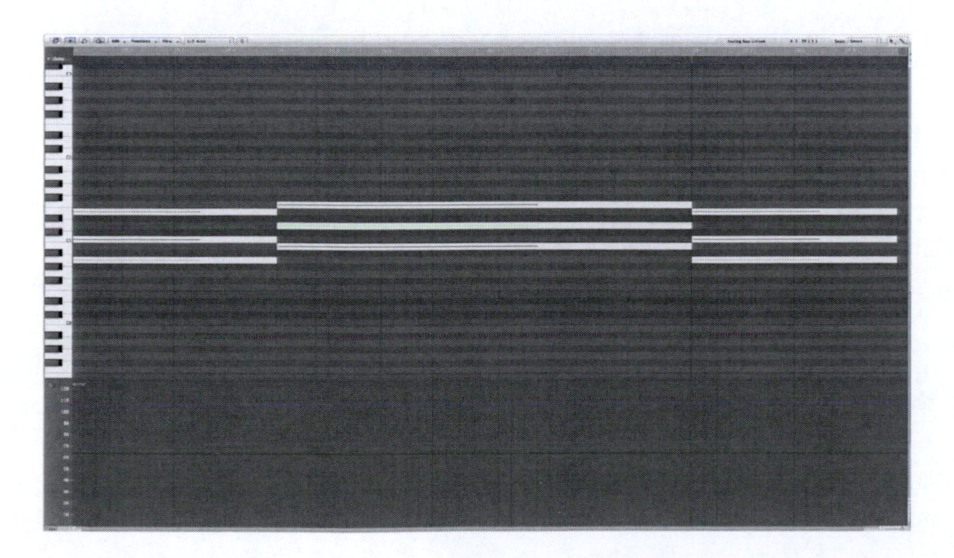

FIGURE 21.5
A typical dubstep chord (i.e. harmonically simple)

from the bass rhythm. Often, wide chorus effects, rotary speaker simulations, flangers and phasers can all help to add a sense of interest but reverb is *the* main contender. Cavernous reverb is often employed to fill out the spectrum and bring energy and space to the mix, whilst the underlying bass punches and shudders the listener's chests.

In addition to the chords, some Dubstep will feature short vocal snippets, and even though there have been some more commercial mixes that have featured a lengthy vocal takes (such as Nero's 'Promises') for many aficionados, this is considered commercial and outside the bounds of where Dubstep has originated from. The snippets are commonly sampled from TV, old vinyl records or gleaned from sample CDs and chopped/edited to suit the music.

Further sound effects can obviously be generated by whatever means necessary, from sampling and contorting sounds or samples with effects and EQ. For contorting audio, the Sherman Filterbank 2, the Camelspace range of plug-ins, Glitch, Sugarbytes Effectrix or Steinbergs GRM Tools are almost a requisite for creating strange evolving timbres.

That said, the effects and processing applied are, of course, entirely open to artistic license as the end result is to create anything that sounds good and fits within the mix. Transient designers can be especially useful in this genre, as they permit you to remove the transients of the percussive rhythms that can evolve throughout the track with some thoughtful automation. Similarly, heavy compression can be used to squash the transient of the sounds and with the aid of a spectral analyzer you can identify the frequencies that contribute to the sound whilst removing those surrounding it. Alternatively, pitch shift individual notes up and by extreme amounts or apply heavy chorus or flangers/phasers to singular hi-hats or snares or try time stretching followed by time compression to add some digital clutter and then mix this with the other loops.

Ultimately, however, as with all the chapters in this book, the purpose here is to simply reveal some of the techniques used by producers and show how the theory and technology discussed earlier in the book combines to produce a track. However, it should be viewed as offering a few basic starting ideas that *you* can evolve from.

There is no one definitive way to produce this genre, and many artists have mixed the four to the floor kicks of dance music with Dubstep's low bass, only to then slip in and out of the two-step vibe part way through the music. This is a relatively new genre in many respects and to some artists one that is still finding its feet. Indeed, it will not to be long before the genre begins to diversify like every other genre of EDM and new Dubstep forms such as Commercial Dubstep, Funky Dubstep, Hardcore Dubstep begin to flourish as artists diversify the music further.

As ever, though, the best way to learn new techniques and production ethics is to actively listen to the current market leaders and experiment with the tools at your disposal. There are no right or wrong ways to approach any piece of music and if it sounds right to the producer, then it usually is. New genres do not evolve from following step by step guides or from emulating peers; producers who experiment and push boundaries create them.

Nevertheless, with just these basic elements, it is possible to create the main focus of the music, and from here, the producer can then look towards creating an arrangement. The theory behind arranging has already been discussed in Chapter 19 but simply listening to the current market leaders mixed will very quickly reveal the current trends in both sounds and arrangement.

The companion files contain an audio example of a Dubstep track using the techniques described.

CHAPTER 22

Ambient/Chill Out

'Ambient is music that envelops the listener without drawing any attention to itself…'

Brian Eno

Ambient music has been enjoyed a long, if somewhat diverse history, and its subsequent offshoots have formed an important role in dance music since 1989. However, it re-established itself to many as part of the electronic dance music scene when beats were again dropped over atmospheric content, and it became relabelled as 'Chill Out' by record companies and the media.

The roots of ambient music are nothing if not unusual. The story goes that it first came about in the mid-1970s when Brian Eno was run over by a taxi. Whilst recovering in hospital, a friend gave him a tape machine with a few analogue cassette tapes of harp music to pass the time. These tapes were old, however, resulting in the volume of the music fluctuating and on occasion dropping so low in volume that it mixed with the rain hitting the hospital windows.

This second 'accident' formed the beginnings of ambient as Eno began to experiment by mixing real-world sounds such as whale song and wind chimes with constantly changing synthesized textures. Described by Eno himself as music that didn't draw attention to itself, it enjoyed some success but was soon relabelled as 'Muzak' and subsequently poor imitations began to appear as background music for elevators, and to many the genre was soon written off as 'music suitable only for tree huggers'.

When the rave generation emerged in the late 80s and early 90s, Alex Patterson, a rave DJ, began to experiment with Eno's previous works playing the music back to clubbers in small side rooms who were resting from the faster paced hard hitting beats of the main room. These side rooms were often known as 'chill out' rooms, a place where you could go and take a break from the more manic up-tempo beats.

As the slower paced music grew in popularity for those looking to relax, Patterson teamed up with fellow musician Jimmy Cauty to form 'The Orb' and released their album 'A Huge Ever Growing Pulsating Brain That Rules from the Centre of the Ultraworld'. This is believed to be the first ever ambient house music aimed at chill out rooms but soon after its release Patterson and Cauty went their separate ways. Jimmy Cauty teamed up with Bill Drummond to form The KLF, whilst Patterson continued to write under the moniker of The Orb and continued to DJ in the chill out rooms.

This form of 'Ambient House' began to grow into its own genre and chill out rooms grew to form a fundamental part of the rave scene. Some DJs moved from the main rooms and became Video Jockeys (VJs), mixing real-world sounds with slow drawn out drum loops whilst projecting and mixing images onto large screens to accompany the music.

By 1992, the genre reached the main stream as different artists adopted the scene each putting their own twist on the music and diversifying it into genre's such as Ambient Dub (ambient music with a bass), conventional (ambient with a 4/4 backbeat), beatless (no backbeat but following the same repetition as dance music) and soundscape (essentially pop music with a slow laid-back beat).

By 1995, the larger record companies took the sound on board and saturated the market with countless ambient compilations (although thankfully there was never any 'Now That's What I Call Ambient music volume…'). Even artists who had ignored the genre before began to hop on board in the hopes of making a quick buck out of the new fad. Eventually, as with most musical genres, ambient house became a victim of its own success. The general public became tired of the sound and, to the joy of many a clubber, it was no longer the new 'fashion' and returned to its humble beginnings where it had originated – the chill out rooms.

Fast forward to the year 2000 and a small Balearic island in the middle of the Mediterranean began to revive the public's and record companies' interest in ambient house. DJs in Ibiza's Café Del Mar began to tailor music to suit the beautiful sunsets by mixing Jazz, Classical, Hispanic and New Age together to produce laid back beats for clubbers to once again chill out too. Now repackaged and relabelled 'chill out music' it's enjoying renewed interest and has become a genre in its own right. However, while chill out certainly has its roots deeply embedded in ambient music, they have over time become two very different genres. Thus, as the purpose of this book is to cover dance music, for this chapter we'll concentrate on chill out rather than ambient.

MUSICAL ANALYSIS

As always, possibly the best way to begin writing any dance music is to find the most popular tracks around at the moment and break them down to their basic elements. Once this is accomplished, we can begin to examine the similarities between each track and determine exactly what it is that differentiates it from other musical styles. In fact, all music that can be placed into a genre specific category will share similarities in terms of the arrangement and/or sonic elements.

Generally chill out is music that incorporates elements from a number of different styles such as Electronica, New Age, Classical, Hispanic and Jazz. However, it's this very mix of different styles that makes an exact definition impossible. Indeed, it could be said that as long as the tempo remains below 120 BPM and it employs a laid back groove, it could be classed as chill out. In fact, the only way to analyze the music is to take Brian Eno's advice in that it shouldn't draw too much attention and ideally be the type of inoffensive music that most people could easily sit back and relax too. (Sometimes the best chill out can be derived from just rolling up a fat err cigarette and just playing about.)

Defining exactly what this means is difficult, but it can commonly be defined as exhibiting a slow rhythmic or trance-like nature combining both electronic and real instruments that are often backed up with drop-out beats and occasionally smooth haunting vocals. Generally, many of these real instruments will be recorded live or sourced from sample CDs, but in some instances they can also be 'borrowed' from other records.

Chill out can employ almost any time signature from the four to the floor, to a more swing orientated 3/4 but typically for many examples the signature remains at 4/4 and it's the positioning of the beats that undermine the 4/4 feel. In terms of physical tempo, it ranges from 80 BPM and move towards the upper limits of 120 BPM and although it can be written in any key, popular keys appear to be C major, D major, Eb major, A minor, E minor and G minor.

For this chapter, we will examine the production aspects that are generally considered in the creation behind a typical chill out track. Having said that, music is an entirely individual, artistic endeavour, and the purpose of this analysis is not to lay down a series of 'ground rules' on how it should be created, rather its intention is to describe some of the general principles behind how to accomplish the characteristic arrangements, sounds and processing. Indeed, in the end, it's up to the individual (i.e. you) to experiment and put a twist on the music that suits your own particular style.

THE RHYTHMS

Unlike many other genres of dance music, there are no particular distinctions as to what makes a good chill out drum loop and the only guide that can be offered is that the loop should remain relatively simple and exhibit a laid back feel.

The kick drum can occur on the beat, every beat, or it can be less common, appearing on the second and fourth, or first and third beat, or any 16th division thereof. Indeed, in many chill out tracks where the kick drum occurs on all four beats, there is commonly an additional kick drum that will occur outside of this standard beat and occur on an 1/8th or 1/16th position. This moves the music away from the rigidity of the standard four to the floor and can help to increase the relaxed feel of the music.

Like the kick, the snare drum can appear anywhere in a chill out loop but generally speaking if the kick occurs *off* the beat, the snare will occur *on* the beat in order to maintain rhythmical positioning with the listener. This same principle can work in opposite too, if the kick lands on the beat, the snare can occur off the beat, but it is important to note that this relationship should not be solely attributed to the laid back feel of the loop, rather this relaxed motion occurs more with the positioning of the hi-hats and ancillary instruments.

As shown in Figure 22.1, multiple rhythms are employed for the open hi-hats. Whilst a closed hat occurs on the somewhat standard 16th, a number of differing open hats patterns is employed that syncopate with the main beat. It's this syncopation with the beat that creates the more chilled and relaxed feel to the music. This syncopation is further augmented by another percussive instrument (in this case a heavily processed clap).

This, of course, is only a general guideline to the drum patterns used, and it's fully open to artistic license. The key is not to produce a loop that sounds rigid or programmed through experimenting by moving the kicks in relation to the snares to adjust the interplay between the two. If the tempo appears too fast, reducing the amount of snares, hi-hats or auxiliary instruments employed in the rhythm will often help slow it down and is preferable to physically slowing down the tempo since this will affect the rest of the instrumentation.

Closed hi-hats, for example, do not have to be placed on every 1/16th division, and in many examples, they will often play a rhythmical pattern of their own. In addition, auxiliary instruments such as congas bongos, toms, cabassas,

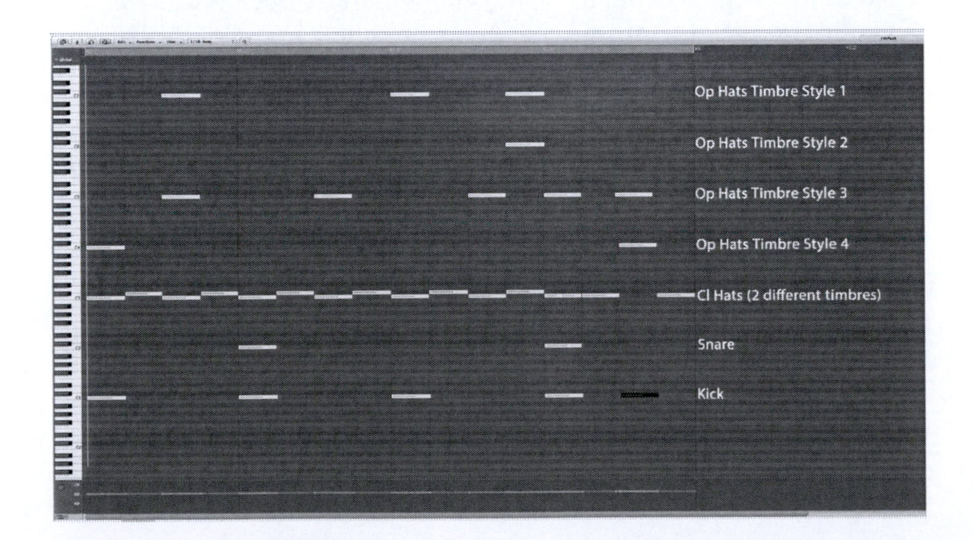

FIGURE 22.1 The basic foundations of a chill out loop (as in the example supplied with the book)

triangles, shakers and tambourines also often make an appearance and offer further syncopation from the main rhythm and the uses of both Hemiola and polymeter will often appear in chill out loops.

RHYTHMIC TEXTURES

The kicks in chill out are often programmed to remain 'loose' sounding by exhibiting more of a boom texture than a tight-controlled one. Kick layering is also not necessary with chill out since the purpose is not to punch the listener in the chest on every beat and therefore a single kick is often sourced direct from sample CDs or programmed by the producer on any capable synthesizer.

If programming, a good starting point for a chill out kick is a 40 Hz sine wave with positive pitch modulation using a fast attack and a medium to longish decay period. There is little need to employ a 'click' at the beginning of the waveform since this can make the kick appear tighter, which isn't generally required but as always experimentation is essential. If possible, experimenting with the kicks decay transition is important in the creation of creating a suitable kick drum. Here, exponential is commonly the favoured choice so that the pitch curves outwards as it falls resulting in a boomy sound.

If using a kick sample and it appears too tight or exhibits a bright attack stage, a compressor with a low threshold, medium ratio and fast attack so that clamps down on the transient may help towards the creation of a looser sounding kick. Alternatively, the SPL transient designer could be employed to remove the initial attack.

The kick is typically treated to small amounts of reverb and here a room reverb is commonly employed with a long pre-delay so that the initial reflections bypass the transient of the loop. The reverb tail can be longer than most other genres of music, ranging from 60 ms to 100 ms, but it should not be allowed to wash over other percussive instruments otherwise it may seriously affect the loop. If high reverb settings are employed to make the kick appear even 'looser', then a noise gate should follow the reverb to ensure that tail doesn't reduce the impact of any other instruments.

In direct contrast to the kicks, the snares often employ a sharp transient stage as these need to be apparent in the loop, more so if it's particularly busy loop packed with auxiliary instruments. As with kick, snares can be sourced from sample CDs and be processed with EQ to make them appear sharp and snappy or alternatively they can be programmed in most synthesizers with little more than a triangle oscillator and some pink noise.

The amplifiers attack should be set to immediate and the decay can be set anywhere from medium to long depending on the sounds that are already in the mix and how slurry you want the snare to appear. This is treated with a high-pass filter to remove much of the low end but alternatively a band-pass filter can be used if the timbre exhibits too much high-end energy.

If possible, it's worth employing a different amp EG for both the noise and triangle wave. The triangle wave can be kept quite short and swift by using a fast decay while the noise can be made to ring a little further by increasing its decay parameter. The further this rings out, the more slurred and 'ambient' it will appear. If the snares are too bright, it can be removed with some EQ or preferably, try replacing the triangle wave with a pulse and experiment with the pulse width.

The snares benefit from small amounts of reverb by sending, not inserting, them to a reverb unit. A typical snare room preset that's available on nearly every reverb unit around will usually suffice but depending on the snare positioning and tempo, the decay should be set carefully so that it does not wash over the transients of further sounds.

The hi-hats in this genre are often sourced from the multitude of sample CDs available on the market and are treated to a multitude of effects to the producer's own personal taste. This can include noise gates to shorten the hat, transient designers, EQ, distortion and/or reverb. They can, however, equally be synthesized through ring modulation.

This is accomplished by modulating a high-pitched triangle wave with a lower pitched triangle to produce a high-frequency noise. As a starting point, set the amp envelope to a zero attack, sustain and release with a short to medium decay. If there isn't enough of a noise present, it can be augmented with a white noise waveform using the same envelope. Once this basic timbre is constructed, shortening the decay creates a closed hi-hat while lengthening it will produce an open hat. Both open and closed hi-hats will commonly be treated to small amounts of delay but this must be applied with caution so as not to upset the syncopated feel of the hats.

Ancillary instrumentation is equally important in chill out for the creation of the syncopated groove, and these sounds can be sourced again from sample CD or synthesized in a variety of ways. The various synthesis techniques have been discussed in details in previous chapters but one typical ancillary instrument is the jazz brush that can be produced by introducing a small amount of pink noise over the top of the snare. This gives the impression of a brush stick being used to play a snare instrument and if treated with small amounts of reverb, distortion and EQ, it can produce a convincing effect.

The snares, hats and any ancillary percussive instruments will also be treated to a form of cyclic pitch or filter modulation. There is no defining method for any of these instruments, but the effect is lightly applied to each instrument over a different period of bars. For example, the snares may be pitched to change by a few cents fourth beat of the bar, whereas the closed hi-hats may be cyclically modulated with a filter every two bars, one open hi-hat every three bars, another open hat every four bars and ancillary instruments every six bars. This pattern is changed every so often on a structural downbeat to inject more interest into the drum rhythms. Finally, swing quantize is always applied

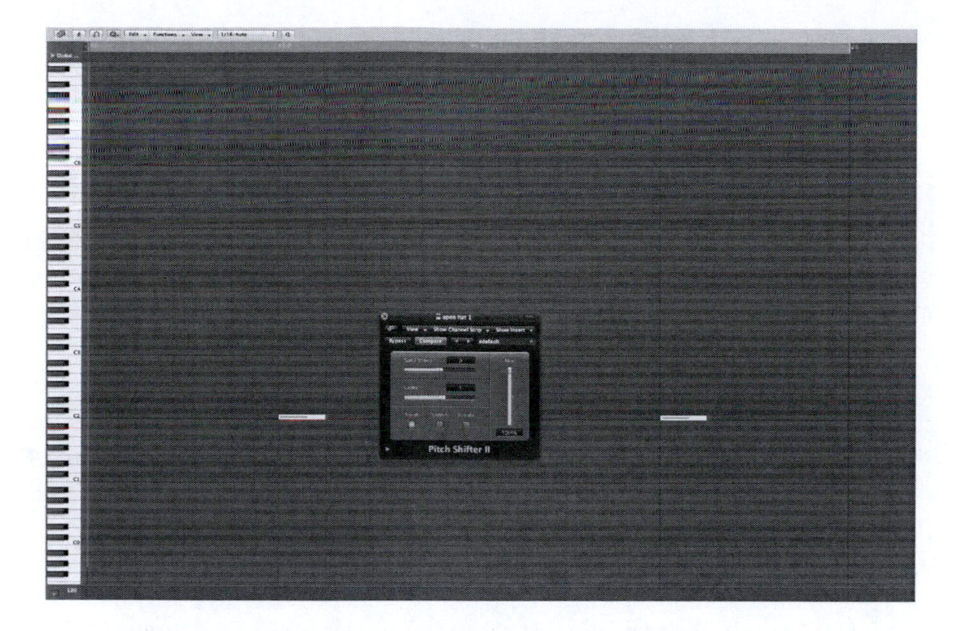

FIGURE 22.2
Pitch modulation on the snares (Pitch Shifter is automated in Logic)

to the snares and open hi-hats. This can be applied quite heavily and typical applications range from 60% to 70% on a 1/16th grid.

MELODIC LEADS

Although it isn't unusual for a chill out track to be based entirely around synthetic instruments, an equal amount will contain real instruments. These can range from acoustic guitars playing Hispanic rhythms, classical strings, pianos, wind instruments or a combination thereof.

These are often laid down *before* the bass since they're often sourced from sample CDs or other records and it's easier to form a bass around their progression than edit the audio. This is not to suggest it does not occur, however, and a good number of artists will source a lead instrument from a record and then spend a good few days changing the rhythms and pitch with the *Melodyne* audio editing suite.

Of course, samples of real instruments do not necessarily have to be used (although in the majority they do sound better when used for leads) and provided that the producer has access to either a sample-based instrument such as Nexus 2 or a Kontakt instrument, it is possible to program expressive and realistic instruments with these.

To accomplish this, it is first important to take note of the key and range you write the song in because every acoustic instrument has a limit to how high and low it can play. For instance, with an acoustic guitar, the (standard) tuning is E, A, D, G, B, E with this latter E a major third above middle C. This is important to keep in mind while programming since many sample-based instruments will stretch a sample beyond its normal range to fill the entire octave ranges. Thus, if you exceed the limitations of an acoustic instrument, the human ear will instinctively know that it's been programmed rather than played.

Creating a realistic instrument consists of more than just striking a few notes and regardless of how well the original samples may be of the instrument, without emulating the natural flow of an instrumentalist playing the melody it will never result in a realistic interpretation. For example, when programming an acoustic guitar, you should first take into account the natural movements of a guitarist.

If the style is Hispanic, then the strings are commonly plucked rather than strummed so the velocity of each note will be relatively high throughout since this should be mapped within the instrument to control the filters low-pass cut-off parameter (i.e. the harder a string is plucked the brighter the timbre becomes).

Not all of these velocity plucks will be the same, however, since the performer will play in time *with* the track and if a note occurs directly after the previous, it will take a finite amount of time to move the fingers and pluck the next string. This often results in the string not being accentuated as much as the preceding one due to 'time restraints'. Conversely, if there is a larger silence or distance between the two notes, then there is higher likelihood that the string will be plucked at a higher velocity.

In many instances, particularly if the string has been plucked hard, the resonance may still be dying away as the next note starts so this needs to be taken into account when programming. Similarly, it's also worth bearing in mind that a typical acoustic guitar will not bend more than an octave, so it's prudent to set the receiving MIDI device to an octave so you can use the pitch bend wheel to create slides. In between notes that are very close together, it may also be worth adding the occasional fret squeal for added realism.

If the guitar is strummed, then you need to take the action of the hand into account. Commonly, a guitarist will begin by strumming *downwards* rather than upwards. If the rhythm is quite fast on an upward stroke, it's rare that all the strings will be hit. Indeed, it's quite unusual for the bottom string to be struck, as this tends to make the guitar sound too 'thick' so this will need to be emulated while programming. In addition, all the strings will not be struck at exactly the same time due to the strumming action. This means that each note on message will occur a few ticks later than the previous which will depend on whether it's strummed upwards or downwards and the speed of the rhythm.

To keep time, guitarists also tend to continue moving their hands upwards and downwards when not playing so this 'rhythm' will need to be employed if there is a break in the strumming to allow it to return at the 'right' position in the arrangement. Finally, and somewhat strangely, if a guitarist starts on a downward stroke, they tend to come in a little *earlier* than the beat while if they

FIGURE 22.3
RE-FX Nexus

start on an upward stroke they tend to start *on* the beat. The reason behind this is complex, but more importantly it would involve me knowing why.

If you wish to replicate trills or tremolo, its worthwhile recording pitch bend movements onto a separate MIDI track and imposing it onto the longer notes of the original guitar track by setting them to use the same MIDI channel. These pitch bend movements are recorded onto a separate channel since once it's imposed onto the notes, it's likely that the pitch bend information will need further editing.

WIND INSTRUMENTS

Fundamentally, there are two types of wind instruments; brass and reed. However, for programming realism into both these instruments, they both follow very similar guidelines in terms of MIDI commands.

First and perhaps most important, if you wish to replicate a wind instrument, it is wise to invest in a breath controller. These can connect to the computer through a USB input of available MIDI port that measure changes in air pressure as you blow into them. This is converted into a CC2 message (breath controller) that can then be used to control the sample-based instruments. These are expensive, however, so if the use of wind instruments is only occasional, the nuances can be programmed manually (although not as realistic).

With any wind instrument, the volume and brightness of the notes are proportional to the amount of air pressure in the instrument. Nearly all sample-based instruments will employ the velocity to control the filter cut-off of the instrument so the brightness can be controlled with judicious use of MIDI velocity. The volume of the note, however, is a little more complicated to emulate since many reed instrumentalists will deliberately begin most notes by blowing softly before increasing the pressure to force the instrument to become louder.

This means that the notes begin softly before quickly rising in volume and occasionally pitch. Perhaps the best way to emulate this feel is to carefully program a series of breath controller (CC2) messages while using a mix of expression controllers (CC11) to control the volume adjustments. Alternatively, brass instruments will often start below the required pitch and slide up to it so this is best emulated by recording live pitch bend movements into a sequencer and editing them to suit later in the piano roll editor.

Many wind instruments also introduce vibrato if the note is held for a prolonged period of time due to the variations in air pressure through the instrument. While it is possible to emulate this response with expression (CC11) controllers, introducing small pitch spikes at the later stages of sustain often produces better results. These pitched 'spikes' only appear in the later stages of the notes sustain, though, and should not be introduced at the beginning of a note.

Possibly the best way to determine where these spikes should appear is to physically emulate playing a wind instrument by beginning to blow at the start of the MIDI note on and when you begin to draw short of breath, insert pitch bend messages. Alternatively, if the instrument allows you to fade in the LFO, this can be used to modulate the volume by setting it to a sine wave on a slow rate modulating the volume lightly. As long as the LFO fade in time is set quite long, it will only begin to appear towards the end of a note.

More important for realistic instruments, however, is that all musicians need to breathe occasionally. In the context of MIDI, this means that you should avoid playing a single note over a large number of bars. Similarly, if a series of notes are played consecutively remember that the musician needs enough time to take a deep breath for the next note (wind instruments are not polyphonic!). If there isn't enough time, the next note will generally be played softer due to less air velocity from a short breath but if there is too short a space, the instrument will sound emulated rather than real.

Finally, consider how the notes will end. Neither reed nor brass instruments will simply stop at the end of the note and instead they will fade down in volume while also lowering in pitch as the air velocity reduces (an effect known musically as diminuendo). This can be emulated with a series of expression (CC11) messages and some pitch bend.

SYNTHETIC LEADS

Of course, not all chill out tracks will employ real instruments (the books example is a case in point) and will instead rely solely on synthetic. If the producer chooses to employ synthesized leads, it is worth considering avoiding any sharp aggressive synthesizer patches such as distorted saw-based leads since these give the impression of a 'cutting' track whereas softer sounds will tend to sound more laid back. This is an important aspect to bear in mind, especially when mastering the track after it's complete.

Slow relaxed songs will invariably have the midrange cut to emphasize the low and high end while more aggressive songs will have the bass and high hats cut to produce more midrange and make it appear insistent.

Notably, many chill out tracks refrain from composing a melody of notes less than a 1/8th in length since shorter notes will make the track appear faster. In addition, by using longer notes, many of the tones can utilize a long attack that will invariably create a more relaxed perception of the music. In fact, long attack times can form an important aspect of the music and it's prudent to experiment with all the timbres used in the creation of this genre by lengthening the amps/filters attack and release times and taking note of the effect it has on the track.

Vocal chants are sometimes used in the place of leads as these can be used as instruments in themselves. Typically, in many popular tracks, these are sourced from other records or sample CDs but it's sometimes worthwhile recording your own. As discussed in a previous chapter, condenser microphones are the preferred choice over dynamic as these produce more accurate results but the diaphragm size will depend on the effect you wish to achieve. Many producers will use a large diaphragm for vocals but if you're after 'ghostly' chants, a small diaphragm microphone such as the Rode NT1 will often produce better results due to the more precise frequency response. Of course, this is simply the conventional approach and, if at all possible, it's worthwhile experimenting with different microphones to see which produces the best results.

Vocals will also benefit from compression during the recording stage to prevent any clipping in the recording, but this must be applied lightly. Unlike most other genres where the vocals are often squashed to death to suit the compressed nature of the music, chill out relies on a live feel with the high frequencies intact. A good starting point is to set a threshold of −9 dB with a 2:1 ratio and a fast attack and moderately fast release. Then, once the vocalist has started to practice, reduce the threshold so that the reduction metres are only lit on the strongest part of the performance.

BASS

Generally speaking the bass that tends to remain particularly minimal consist mostly of notes over an 1/8th in length to prevent the groove for appearing too fast. These are often derived from the principles encapsulated in the dub scene – by remaining minimal they allow space for the lead to breathe without having to compete with the bass for prominence.

Indeed, it's important that the bass is not too busy either rhythmically or melodically and many chill out tracks borrow heavily from the pentatonic scale used in dub and R 'n' B. That is, they use no more than a five-note scale playing a single harmony restricting the movements to a maximum five-semitone shift.

Typically, most chill out basses are synthesized rather than real since the timbre is often particularly low in frequency and doesn't want to attract as much attention as the lead. Analogue synthesizers usually provide the best results due to the pleasing phasing of the oscillators.

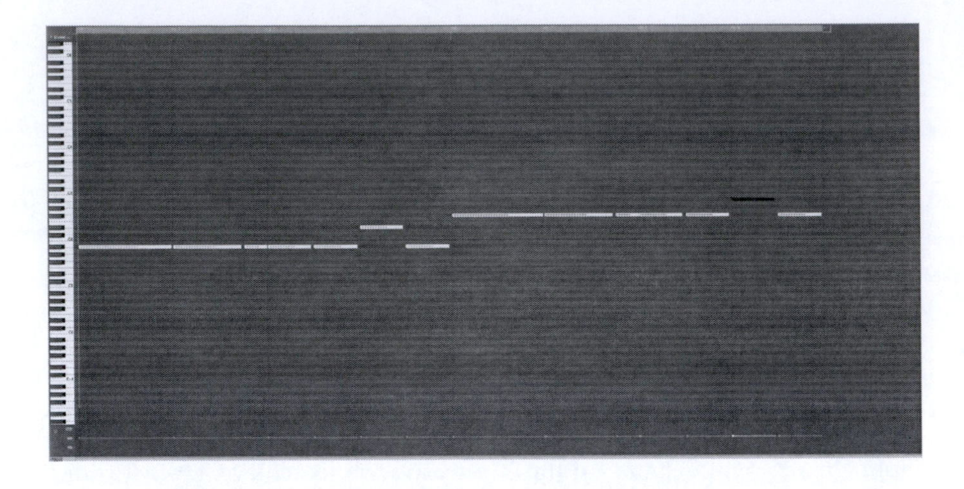

FIGURE 22.4
The bass used on the example track

A good starting point is to employ a sine wave detuned from a saw, square or triangle depending on the character you require. Unlike most of the timbres used throughout the production of chill out, the attack should be well defined to prevent the bottom end groove turning to mush but there is usually very little decay, if any, to prevent it from becoming too plucky. Indeed, in much of the music of this genre, the bass tends to *hum* rather than pluck to prevent it from drawing attention away from the chilled feeling.

The best way to accomplish this type of sound is to use a simple on/off envelope for the amp envelope. This is an envelope with no attack, decay or release but a high sustain. This forces the sound to enter directly into the sustain stage that produces a constant bass tone for as long as the key is depressed. If this results in a bass with little or no sonic definition, a small pluck can be added by employing a low-pass filter with a low cut-off and high resonance controlled through an envelope with no attack, sustain and release but a medium decay. By adjusting the depth of the envelope modulating the filters or increasing/decreasing the decay stage, more or less of a pluck can be applied. If this process is used, however, it's prudent to employ filter key tracking so that the filters action follows the pitch.

It's also prudent to experiment with the release parameters on the filter and amp envelope to help the bass sit comfortably in with the drum loop. In fact, it's essential to accomplish this rhythmic and tonal interaction between the drum loop and the bass by experimenting with both the amp and filters envelopes.

Since these waveforms are produced from synthesizers, there is little need to compress the results since they're already compressed at the source. Also, it's unwise to compress the drums against the bass to produce the archetypal 'dance music' pumping effect as this will only accentuate the drum rhythm and can make the music appear too 'tight' rather than relaxed.

The only effects that are typically applied to the bass are small amounts of reverb, delay and distortion. As always, these must be applied carefully and a preferential approach is to first use a M/S processor so that the effects are only applied above 120 Hz to ensure the bottom end of the mix maintains its solidity.

CHORDS/PADS

On the main keys to producing chill out lies with evolving pads. Since many of the tracks are thematically simple, a pad can be used to fill the gap between the groove of the record and the lead and/or vocals. What's more, slow evolving strings often add to the overall atmosphere of the music and can often be used to dictate the drive behind the track.

The basic theories of chords and structure have already been discussed in detail in an earlier chapter of this book but as a refresher, they are to act as a harmony to the bass and lead. This means that they should fit in between the rhythmic inter-play of the instruments without actually drawing too much attention (rather like the genre as a whole). For this, they should be closely related to the key of the track and not use a progression that is particularly dissonant. Generally, forming a chord progression from any notes used in the bass and experimenting with a progression can accomplish this.

For instance, if the bass is in E, then C major would harmonize since this would contain the note E (C – E – G). The real solution is to experiment with different chords and progressions until you come up with something that works. Once the progression is down, it can then be used to add some drive to the track.

Although many chill out tracks will meander along like a Sunday stroll, if you place a chord that plays consistently over a large number of bars, the feel can quickly vanish. This can sometimes be avoided by moving the chords back in time by a small amount so that they occur little later than the bar. Alternatively if they follow a faster progression, they can be moved forward in time to add some push to the music.

It should be noted here, however, that if the pads employ a long attack stage, they might not become evident until much later in the bar that can change the feel of the music. In this instance, the producer should counter the effect by moving the chords so that they occur earlier in time.

The instruments used to create chords are more often not analogue in nature due to the constant phase discrepancies of the oscillators that result in additional movement. Thus analogue or analogue emulations will invariably produce much better results than an FM synth or one based upon sample and synthesis. What's more, it's of vital importance that you take into account the current frequencies used in the mix so far.

With the drums, bass, lead and possibly vocals there will be a limited frequency range for the chords to sit and if the timbres of these chords are

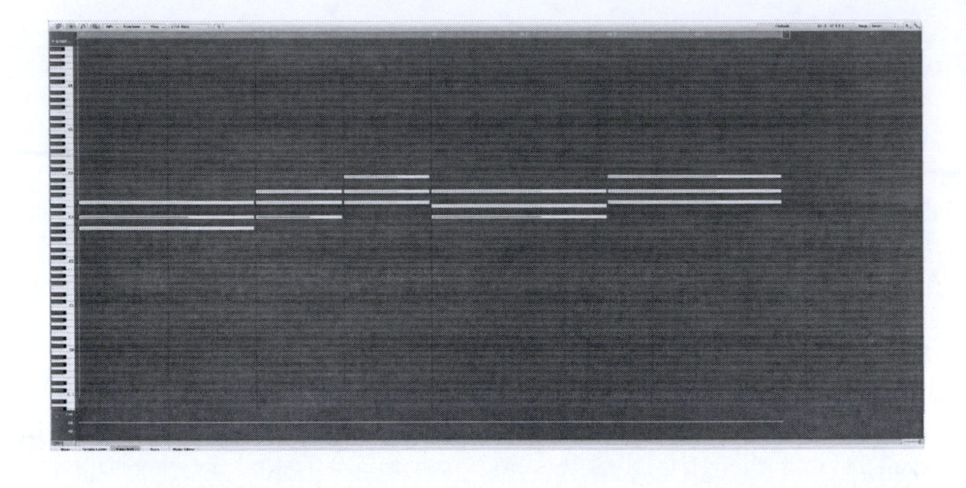

FIGURE 22.5
Basic chords

harmonically rich, it may be difficult to fit them into the mix without having to resort to aggressive EQ cuts. This can affect the sonic character of the chords and therefore it is far more prudent to *design* the chords in the synthesizer than attempt to carve a large space for them later with EQ.

If the mix is already busy in terms of frequencies, then it's prudent to build a relatively thin pad that can then be widened if required with effects such as phasers, flangers, reverb or chorus effects. This style of pad can be accomplished using pulse waves as they have less harmonic content than saws and triangles and do not have the 'weight' of a sine wave.

Generally only one pulse wave is required, with the amp envelope set to a medium attack, sustain and release but a fast decay. If this timbre is to sit in the upper mid range of the music, then it's advisable to employ a 12 dB high-pass filter to remove the bottom end of the pulse, otherwise use a low-pass to remove the top end and then experiment with the resonance until it produces a general static tone that suits the track. Following this, an LFO is to positive modulation with a slow rate on the pulse width of the oscillator and the filter to add some movement.

If the timbre appears too 'statically modulated' in that it still seems uninteresting, use a different rate and waveform for the oscillator and filter so that the two beat against each other. Alternatively, if the timbre still appears too thin even after applying effects, add a second pulse detuned from the first with the same amp envelope but use a different LFO waveform to modulate the pulse width.

If the track has a 'hole' in the mix, then you'll need to construct a wider, thicker pad to fill this out. As a starting point for these types of pads, square waves mixed with triangles or saws often produce the best results. Detune the saw (or triangle) from the square wave by the amount you deem necessary and set the amp attack to a medium sustain and release but no attack and a short decay.

Using a low-pass filter set the cut-off to medium with a high resonance and set the filters EG to a short decay and a medium attack, sustain and release.

This approach should modulate the filters positively so that the filters sweep through the attack and decay of the amp but meet at the sustain portion, although it is worth experimenting with negative modulation to see if this produces better results for the music. This can produce a basic pad timbre, but if it seems too static for the music, use an LFO set to a sine or triangle wave using a slow rate and medium depth to positively modulate the pitch of the second oscillator. For even more interest, you could also use an LFO to modulate the filters cut-off.

Effects can also play an important role in creating interesting and evolving pads. Wide chorus effects, rotary speaker simulations, flangers, phasers and reverb can all help to add a sense of movement to help fill out any holes in the mix.

A noise gate can also be used here to creatively model the pad. For instance, another channel with a hi-hat rhythm could be side chained into the gate to produce an interesting gated pad or through carefully adjusting the threshold so it lies just on the edge of the pads volume, with an immediate attack and release, the gate will chatter through starting and stopping abruptly producing a 'sampled' effect that cannot be replicated through synthesis parameters.

Ultimately, though, as with all the chapters, in this section, the purpose here is to simply reveal some of the techniques used by producers and show how the theory and technology discussed earlier in the book combines to produce a track. However, it should be viewed as offering a few basic starting ideas that *you* can evolve from.

There is no one definitive way to produce this genre, and the best way to learn new techniques and production ethics is to actively listen to the current market leaders and experiment with the tools at your disposal. There are no right or wrong ways to approach any piece of music, and if it sounds right to the producer, then it usually is. New genres do not evolve from following step-by-step guides or from emulating peers; producers who experiment and push boundaries create them.

Nevertheless, with just these basic elements, it is possible to create the main focus of the music and from here the producer can then look towards creating an arrangement. The theory behind arranging has already been discussed in an earlier chapter but simply listening to the current market leaders mixed amongst the theory discussed in the previous chapter will very quickly reveal the current trends in both sounds and arrangement.

The companion files contain an audio example of a typical chill-out track using the techniques described.

CHAPTER 23

Drum 'n' Bass

'It became the basis for drum-and-bass and jungle music – a six-second clip that spawned several entire subcultures...'

Nate Harrison

Pinpointing the foundations of where Drum 'n' Bass originated is difficult. In a simple overview, it could be traced back to being a natural development of Jungle; however, Jungle was a complex infusion of break beat, reggae, dub, hardcore and artcore.

Alternatively, we could look further back to 1969 and suggest that the very beginnings lay with a little know record by the Winston's. The B side of a record entitled *'Colour Him Father'* featured a six-second drum break – the Amen Break, taken from the title of the record, *'Amen Brother'* – and this became the sampled basis for many of the rhythms of Jungle and Drum 'n' Bass. Or it could be traced back to the evolution of the sampler, offering the capability to cut, chop and pitch rhythmic material that created the basis for Jungle and the more refined Drum 'n' Bass.

Nonetheless, we can roughly trace the complex rhythms developed in Jungle back to break beat whose foundations can be traced back to the 1970s. Kool Herc, a hip-hop DJ, began experimenting on turntables by playing only the exposed drum loops (breaks) and continually alternating between two records, spinning back one while the other played and vice versa. This created a continual loop of purely drum rhythms allowing the break-dancers to show off their skills. DJ's such as Grand Wizard Theodore and Afrika Bambaata began to copy this style adding their own twists by playing two copies of the same record but delaying one against the other resulting in more complex asynchronous rhythms.

It was early 1988 and the combined evolution of the sampler and the rave scene that really sparked the break beat revolution, however. Acid House artists began to sample the breaks in records, cutting and chopping the beats together to produce more complex breaks that were impossible for any real drummer to play naturally. As these breaks became more and more complex, a new genre evolved known as Hardcore. Shifting away from the standard 4/4 loops of typical acid house, it featured lengthy complex breaks and harsh energetic sounds that were just too '*hardcore*' for the other ravers.

Although initially scorned by the media and record companies as little more than drug-induced noise that wouldn't last more than a few months, by 1992, the entire rave scene was being absorbed into commercial media machines. Riding on this 'new wave for the kids', record companies no longer viewed it as a fad but as a cash cow and preceded to dilute the market with a continuous flow of watered down, mass consumed rave music. Indeed, the term rave and its associated media popcorn representation became so overtly commercialized that the term '*rave*' is laughed at even to this day (in the UK at least).

In 1992 and as a response to the commercialization of the music, two resident DJs – Fabio and Grooverider – pushed the hardcore sound to a new level by increasing the speed of the records from the usual 120 BPM to 145 BPM. The influences of house and techno were dropped and replaced with Ragga and Dancehall resulting in mixes with fast complex beats mixed with a deep throbbing bass line. Although Jungle didn't exist as a genre these faster rhythms mixed with deep basses inspired artists to push the boundaries further and up the tempo to a more staggering 160 to 180 BPM.

To some, the term 'Jungle' is attributed to racism but the name was derived from the 1920s. It was used on flyers to describe music produced by Duke Ellington. This featured fast, exotic drum rhythms and when Rebel MC sampled an old Dancehall track with the lyrics 'Alla the Junglists', Jungle became synonymous with music that had a fast beat and a deep throbbing bass. This was further augmented with pioneers of the genre such as Moose and Danny Jungle.

Jungle enjoyed a good 3 to 4 years of popularity before it began to show a decline. In 1996, the genre diversified into Drum 'n' bass as artists such as Goldie, Reprazent, Ed Rush and LTJ Bukem began to incorporate new sounds and cleaner production ethics into the music. Goldie and Rob Playford are often credited with initiating the move from the Jungle sound to Drum 'n' Bass with the release of the album Timeless.

Jungle still exists today and by some Jungle and Drum 'n' Bass are viewed as one and the same but to most, Jungle has been vastly overshadowed by the more controlled and careful production ethics of 'pure' Drum 'n' Bass.

MUSICAL ANALYSIS

Examining Drum 'n' Bass in a musical style is more difficult than examining its roots since it has become a hugely diversified genre. On one end of the scale, it can be heavily influenced with acoustic instruments whilst on the other it will only feature industrial or synthetic sounds. Nonetheless, there are some generalizations that all drum 'n' bass records share that can be examined in more detail.

Typically a drum 'n' bass tempo will remain between 165 and 185 BPM, although in more recent instances, the tempo has sat around 175 to 180 BPM. The music is written in any key but in particular A minor and G minor prove to be popular, and the time signature is almost always 4/4. In most examples, a two-step syncopation is involved in the drums, and here often the snare will remain on beat two and four of the bar to help keep the listener or dancer in time, whilst the kick drum will dance around this snare.

The bass often plays at one-quarter or half time of the drum tempo, again to keep the dancer 'in time'. It should be noted, however, that the more complex the drum rhythm is, the faster the music will appear to be therefore it's prudent to keep the loop relatively simple if the tempo is higher or relatively complex if the tempo is lower.

Much of the progression in this genre comes from careful development of the drum rhythms, not necessarily a change in the rhythm itself but from applying pitch effects, flangers, phasers and filters as the mix progresses. Much in the way that Techno uses the mixing desk as a creative tool so does Drum 'n' Bass with the EQ employed to enhance the rhythmic variations and harmonic content between them.

To the uninitiated, drum 'n' bass may be seen to only include drum rhythms and a bass, but very few tracks are constructed from just these elements. Indeed, although these play a significant role in the genre, it also features other instrumentation such as guitars or synthesis chords, sound effects and vocals.

PROGRAMMING

The obvious place to begin is by shaping the drum rhythm, and these are commonly sourced from old records. Although there are innumerable sample CDs on the market aimed towards drum and bass artists, only a few professionals employ them since they can be easily recognized and many prefer to rely on obscure vinyl records.

Generally, it's the break beat of the record that is sampled (i.e. the middle eight of the original record where the drummer plays his solo for a couple of bars). These samples are very rarely used 'as is' and its common to manipulate them in audio workstations or sample slicing programs such as ReCycle. The sampled loops are sliced, rearranged, and the tempo is increased to form the basis of the *background* loop.

This background loop provides the foundation of the drum and bass rhythm, resulting in a complex rhythmical 'live' pattern that would be difficult to program in a workstation. Although many drum and bass artists perform all this editing within an audio workstation, a number do still prefer to use the old samples such as the AKAI or Emu samplers to work with the loops since these introduce a different tonal feel that remains true to its original roots.

Since the sampled loops function is to provide a background, it is common to remove the kick drum from the sample. The kicks in any sampled loops can be removed via audio workstation editing to slice and delete them; however, a more common approach is to employ an EQ unit to roll off the bottom end of the loop. This removes the body of kick and much of the body of the snare just leaving the higher percussive elements to the mix. From here, a programmed or sampled kick and snare is laid on top of this sample running in a two-step syncopated fashion to create the complete drum loop.

The kick drums can be programmed or taken from a sample CD. A common approach here for some drum and bass artists is to sample a kick from a Roland TR909 or TR808 and pitch it down by any number of semitones to produce a deeper kick. It is preferable to sample rather than simply program at the pitch since pitching down in a sampler produces the different sonic character that is typical to drum and bass.

A kick drum can be created in any competent synthesizer via modulation of a sine waves amplitude. Since the bass timbre in many drum and bass tracks is particularly low in frequencies, the kick sits atop the bass rather than alongside it. Additionally, since the programmed timbre will also often be pitched down after programming, a good starting point for a sine waves frequency is around 150 Hz to 200 Hz. Using an attack/decay envelope generator, the pitch of the sine is modulated with a fast attack and short release. To help the kick remain prominent above the sub-bass, it's also worth adding a square wave that is pitched down with a very fast amplifier attack and decay setting. This will produce a short sharp click that will help the kick remain prominent in the mix.

The amount that this square wave is pitched down will depend on the sound you want to produce, so it's sensible to layer it over the top of the sine wave and then pitch it up or down until the transient of the kick sounds right to the producer. The timbre will also probably benefit from some compression so experiment by setting the compressor so that the attack misses the transient but grips the decay stage, by increasing the gain of the compressor it can make the timbre appear more powerful.

The snares are commonly sourced from sample CDs or vinyl records but if from a sample CD, they are rarely taken from a drum and bass genre specific CD. Instead an alternative genre is chosen and the snare is pitched up to create the requisite 'force pitched' characteristic timbre. Snares can, of course, also be programmed in a synthesizer through using a dual oscillator synth with a triangle wave for the first oscillator and noise for the second. The amplifier envelope generator employs a zero attack, sustain and release with the decay employed to set the 'length' of the snare. Generally, in drum 'n' bass, the snare is particularly 'snappy' and often pitched up the keyboard to give the impression of a sampled sound having its frequency increased within the sampler.

If possible, employ a different amp EG for both the noise and triangle wave. By doing so, the triangle wave can be kept quite short and swift with a fast decay while the noise can be made to ring a little further by increasing its decay parameter. This, however, should never be made too long since the snares should remain short and snappy. If the snare has too much bottom end employ a high-pass, band-pass filter or notch filter depending on the type of sound you require. Notching out the middle frequencies will create a clean snare sound that's commonly used in this genre. Further modification is possible using a pitch envelope to positively modulate both oscillators, this will result in the sound pitching upwards towards the end, giving a brighter snappier feel to the timbre.

Once these basic timbres are synthesized, the kick and snare will often be programmed in a two-step fashion. This style often features the kick on the first and third beat with further kicks and snares occurring off the beat and dancing around the first and third beat in a heavily syncopated fashion. This same approach can also be employed for the high-hats and further snares, if the producer chooses to program their own rhythms rather than rely on a sampled loop.

Programming drum and bass loops without the employment of sampled loops involves both Hemiola and polymeter on high-hats and any further percussive timbres such as congas, toms, cymbals, bongos and synthesized transient timbres. Hemiola was discussed in detail in Chapter 3 but consists of creating an unequal subdivision of beats within the bar and placing a hi-hat or percussive rhythmical pattern onto these unequal subdivisions. This effect is further augmented with the use of polymeter on further hi-hats, snares or percussive elements. A typical application of polymeter here is to

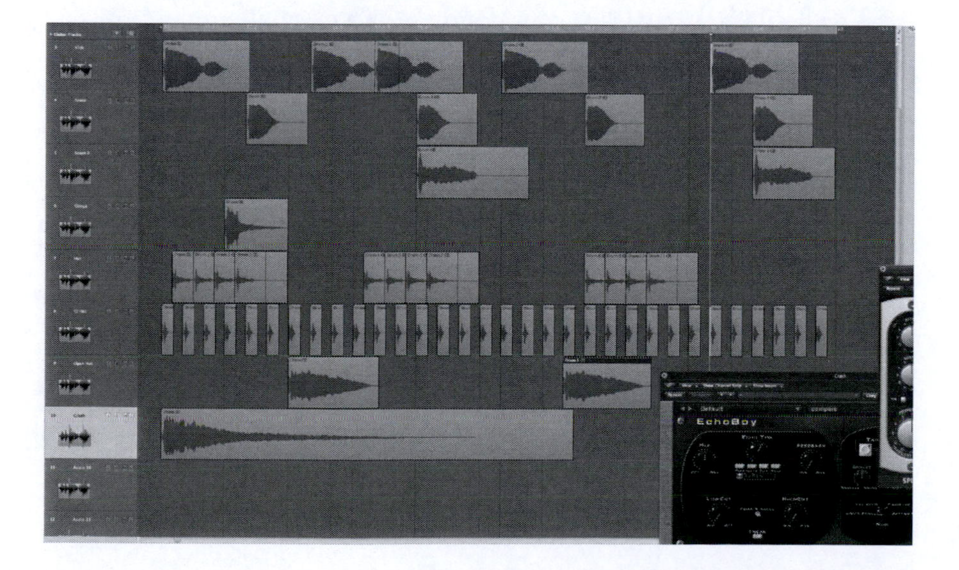

FIGURE 23.1 A drum and bass loop based on sampled loops

employ a 5/4 on a rhythmic hi-hat and 6/4 and 7/4 on further rhythmical elements such as congas or synthesized percussive hits.

The hi-hats can be programmed or taken direct from a sample CD. In many tracks, the hats are commonly sourced from vinyl or sample CDs but some artists do program their own by ring modulating a high-pitched triangle wave with a lower-pitched triangle. This produces a high-frequency noise that can be modified with an amplifier envelope set to a zero attack, sustain and release with a short to medium decay. If there isn't enough noise present, it can be augmented with a white noise waveform using the same envelope. Once this basic timbre is constructed, shortening the decay creates a closed hi-hat while lengthening it will produce an open hat. Similarly, it's also worth experimenting by changing the decay slope to convex or concave to produce fatter or thinner sounding hats. As both these timbres depend on the high-frequency content to sit at the top of the mix, compression should be avoided but adjusting the amplifier decay of synthesized hi-hats can help to change the perceived speed of the track.

Ancillary instrumentation is equally important in drum and bass for the creation of the syncopated groove, and these sounds are either sourced from sample CD or synthesized in a variety of ways. The various synthesis techniques have been discussed in details in previous chapters, and it is down to the producers own experimentation to produce different percussive timbres that can augment the loop further. Of particular note, drum and bass will also often use a call and response on the percussive elements. Here, a short conga pattern (for example) may make the call singularly or over three bars that is responded too in the second or fourth bar via a different conga pattern or a different instrument altogether.

As with most genres of EDM, each percussive instrument will be further processed with effects. The kick is commonly treated to small amounts of reverb.

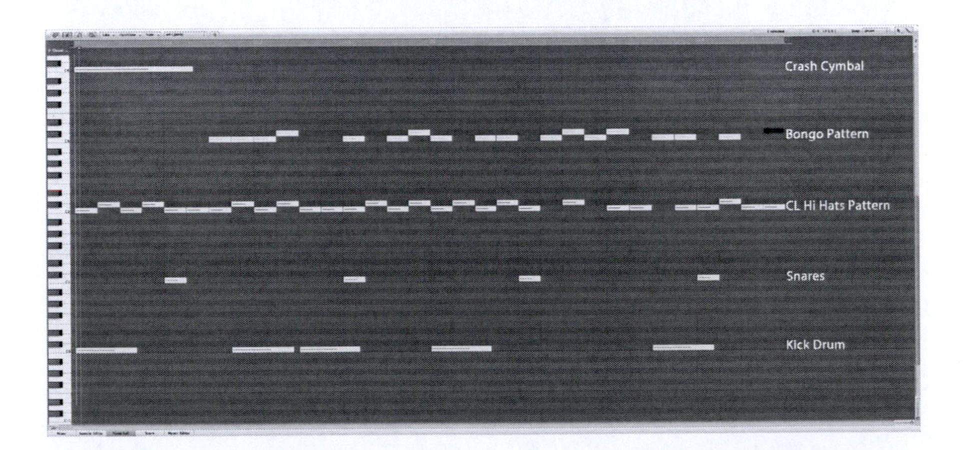

FIGURE 23.2
A programmed drum and bass loop

Here, a small-room setting is often used with a long pre-delay that will skip the transient to ensure it can cut through the mix. Following reverb, distortion is often applied. This can be applied via a distortion unit followed by EQ to sculpt the results but a number of artists simply apply a form of distortion by pushing the mixers gain fader to its limit. In fact, this is one of the few genres of dance music where digital summing is commonly preferred and the artists will often run most of the faders near maximum to produce an almost crushed style of digital sound.

Compression is often applied to the kick since this can play a large role in achieving the character of the kick. Here, the attack of the compressor should be configured so that it skips the initial transient but captures the body and tail. The ratio is set at approximately 6:1 to 9:1, and whilst listening to the kick, the threshold is reduced until the required sound is achieved.

The snares are commonly treated to both EQ and small amounts of reverb. The reverb is typically a room style with a long pre-delay to miss the transient with a slightly longer decay to add presence to the timbre. Typically, the EQ is employed to roll off any frequencies below 800 Hz in order to maintain the light snappy character of the snare.

Pitch modulation is employed on the snares but since these are syncopated and will often occur more than twice within a bar, each snare is processed to a slightly different pitch than previous whilst all remaining within 7 to 20 cents of the original. This modulation does not have to be applied in a stepwise fashion with each snare being a few cents higher or lower than the previous and more interesting results can be attained from mixing and matching the pitch adjustments. Note here, however, that unlike most other genres the snares in drum and bass have a larger cent range since a *noticeable* difference in *some* of the snare hits can help to create a more complex sounding loop. This noticeable difference should be applied cautiously, though, since if all of the hits exhibit a noticeable difference the loop will loose its character.

If the loop is programmed, delay, pitch and filter modulation are also applied to other percussive elements within the loop. The hi-hats are the first culprit to be processed with delay but this should be kept to short settings to prevent too much delay wash that will cloud the signal. Typically a 1/16th or 1/8th setting with a very short-delay time proves enough.

In regards to modulation of filter or pitch, typically the open hats are treated to a low-pass filter that is modulated via an offset sine wave or a sample and hold waveform. This is applied over a period of three bars, whilst further cyclic modulation will be applied to other instrumentation over a different period of bars. This prevents the listeners psyche from specifically identifying a fixed cyclic pattern in the different rhythms. Typically, odd numbered bars are chosen for the cyclic modulations, since this works against the standard structural downbeat of dance music and results in a cross syncopation of modulation.

Finally, swing quantize is almost always applied to all of the rhythmical and percussive elements except for the kick. This is commonly in applications of between 61% and 71% on a 1/16th grid. After this, the loop will often be treated to some parallel compression to help the loop gel and introduce a little more energy.

Here, compression should be applied lightly, with a threshold that just captures the transients, a low ratio and a quick attack and release. If applied in a more creative way, it can often breathe new life into the rhythm. With a medium-threshold setting and a high ratio, the returned signal can be added to the uncompressed signal, you can then experiment with the attack and release parameters to produce rhythm that gels with the rest of the instruments.

THE BASS

The second, vital element of Drum 'n' Bass is the bass. Generally speaking, the bass consists of notes that play at either one quarter or half the tempo of the drum rhythm. This is accomplished through using lengthy bass notes set at one quarter, half or full notes, and sometimes straddling over the bars of the drum loop. The notes of the bass usually remain with an octave and rarely move further than this to prevent the music from becoming too active and deterring the listener from the rhythmic interplay of the drums.

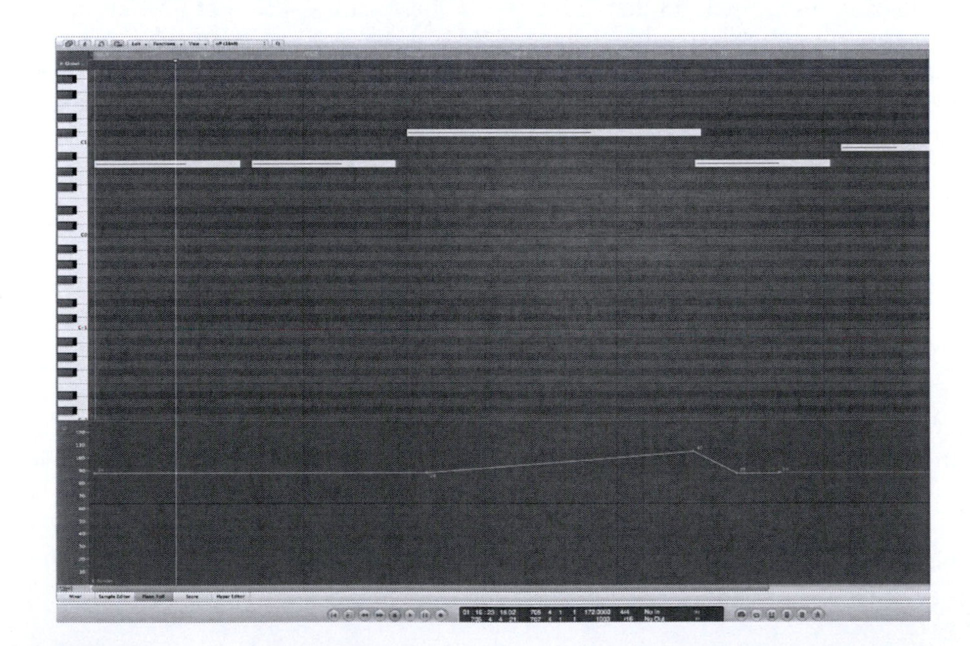

FIGURE 23.3
The Bass Melody in the example track (Note the pitch bend)

In the above example (the books example), an octave shifting bass was used. This remains within the octave but shifts from the key of the music up by an octave before returning back to the key of the music. Here, most of the notes were kept (relatively) long in order to slow the bass rhythm down whilst also crossing over the bar to create a style of polyrhythm against the single bar divisions of the drum rhythms.

Naturally, this is only one possible application, but in many tracks, the bass will overlap the bar and exhibit a slow movement to counteract the speed set by the drum rhythm. If the bass notes are kept quite lengthy, it is not uncommon to employ some movement in the bass to maintain interest and this is often accomplished through filters or pitch modulation. Since the bass in this genre is supposed to remain deep and earth shaking, a sine wave makes the perfect starting point for this timbre.

If you programmed your own kick drum, as described earlier, try copying the preset over to another bank and lengthen the decay and release parameter of the amp envelope. If you sampled the drum kick instead, use a single oscillator set to a sine wave and positively modulate its pitch with an attack/decay envelope, then experiment with the synthesizer's amplitude envelopes. This should only be viewed as a starting point and, as always, you should experiment with the various modulation options on the synthesizer to create some movement in the sound.

As an example of a typical drum 'n' bass timbre, using a sine wave, set the amplifiers attack to zero and increase the decay setting whilst listening back to the programmed motif until it produces an interesting rhythm. Next, modulate the pitch by a couple of cents using an envelope set to a slow attack and medium decay. This will create a bass timbre where the note bends slightly as it's played. Alternatively you can use an LFO set to a sine wave with a slow rate and set it to start at the beginning of every note.

Experimentation is the key, changing the attack, decay and release of the amp or/and filter EG from linear to convex or concave will also create new variations. For example, the decay to a convex slope setting will produce a more rounded bass timbre. Similarly, small amounts of controlled distortion or very light flanging can also add movement.

Alongside the synthetic basses, many drum 'n' bass tracks will employ a real bass. More often than not these are sampled from other records – commonly dancehall or Raga – but they are also occasionally taken from sample CDs or programmed via MIDI.

The key to programming and emulating a real-bass instrument is to take note of how they're played and then emulate this action with MIDI and a series of Control Change commands. In this instance, most bass guitars use the first four strings of a normal guitar E – A – D – G, which are tuned an octave lower resulting in the E being close to three octaves below middle C. Also they are monophonic, not polyphonic, so the only time notes will actually overlap is when the resonance of the previous string is still dying away as the next note is plucked.

This effect can be emulated by leaving the preceding note playing for a few ticks while the next note in the sequence has started. The strings can either be plucked or struck, and the two techniques produce different results. If the string is plucked, the sound is much brighter and has a longer resonance than if it were simply struck. To copy this, the velocity will need to be mapped to the filter cut-off of the bass module, so that higher values open the filter more. Not all notes will be struck at the same velocity, though, and if the bassist is playing a fast rhythm, the consecutive notes will commonly have less velocity since he has to move his hand and pluck the next string quickly. Naturally, this is only a guideline and you should edit each velocity value until it produces a realistic feel.

Depending on the 'bassist', they may also use a technique known as 'hammer on', whereby they play a string and then hit a different pitch on the fret. This results in the pitch changing without actually being accompanied with another pluck of the string. To emulate this, you'll need to make use of pitch bend, so this will first need setting to a maximum bend limit of two semitones, since guitars don't 'bend' any further than this. Begin by programming two notes, for instance an E0 follow by an A0 and leave the E0 playing underneath the successive A0 for around a hundred ticks. At the very beginning of the bass track, drop in a pitch bend message to ensure that it's set midway (i.e. no pitch bend) and just before where the second note occurs drop in another pitch bend message to bend the tone up to A0. If this is programmed correctly, on play back, you'll notice that as the E0 ends the pitch will bend upwards to A0 simulating the effect. Although this could be left as is, it's sensible to drop in a CC11 message (expression) directly after the pitch-bend as this will reduce the overall volume of the second note, so that it doesn't sound like it has been plucked. In addition to this, it's also worthwhile employing some fret noise and finger slides. Most good sample-based plug-in instruments will include fret noise that can be dropped in between the notes to emulate the bassist's fingers sliding along the fret board.

As the rhythmic movement and interaction with the bass and rhythms provide the basis for this genre, it's also worth experimenting by applying effects to the bass timbre. While some effects should be avoided since they tend to spread the sound across the image, in this genre, the bass is one of the most important parts of the music, so small amounts of delay can create interesting fluctuations, as can flangers, phasers and distortion. If delay or any further stereo spreading effects are applied, however, it is advisable to employ a mid/side processor beforehand to ensure that the effects only occur above 120 Hz, leaving any frequencies below this as a mono source.

As with the drum rhythms, creative compression can also help in attaining an interesting bass timbre. As before, try accessing the compressor as a send effect with a medium threshold setting and a high ratio. The returned signal can be added to the uncompressed signal, you can then experiment with the attack and release parameters to produce an interesting bass tone. Alternatively, try pumping the bass with one of the rhythms. Set the compressor to capture the transients of a percussive loop and use it to pump the bass by experimenting with the attack, ratio and release parameters.

CHORDS

Some aficionados of the genre recommend using only the minor chords from the chosen key as a harmony. This means that whilst it is generally accepted that Drum 'n' Bass is written in a minor key but the chords of III, VI and VII (in the natural minor) or III, V and VI (in harmonic minor) are often avoided. This does, of course, mean that the chords in a harmonic minor cannot cadence since the V is a major chord in the harmonic minor; however, this is entirely open to artistic interpretation and is worthy of experimentation.

Also, the chords can often work well when working in contrary with the bass line. Copying the bass line down to another workstation channel track and then converting this new track into a series of chords can accomplish this. Once created, if when the bass moves up in pitch, move the chords down in pitch and vice versa.

With a general idea of the chord structure down, you can program (or sample) a string to fit. Strings are often more popular than pads since these are often particularly heavy in harmonic structure and, therefore, take a good proportion of the mixes frequency range.

A good starting point for programming a Drum 'n' Bass string can be created by mixing a triangle and square wave together and detuning one of the oscillators from the other by 3 to 5 cents. The amplifiers envelope is set to a zero

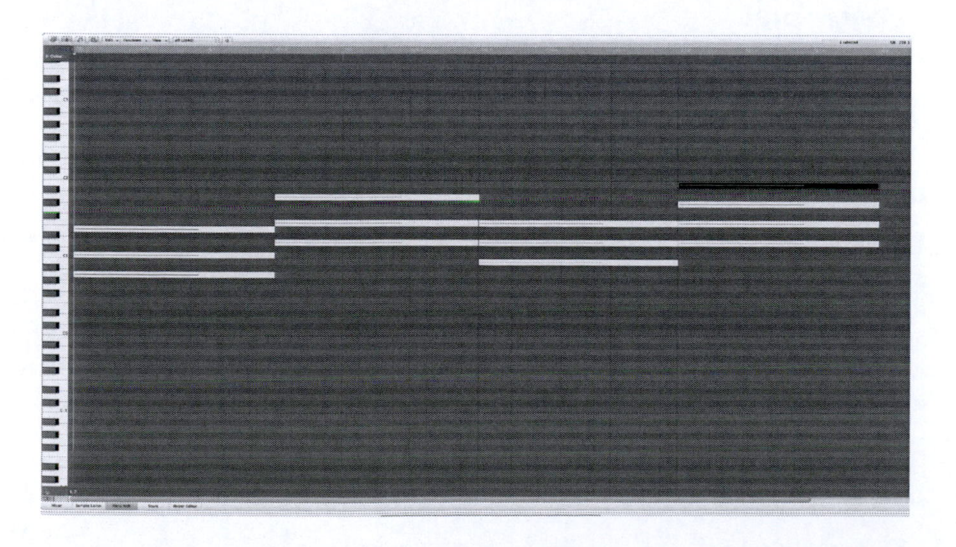

FIGURE 23.4
A typical chord progression

attack with medium sustain and release. Using the filter envelope set it to a long attack, sustain with a medium release and short decay. Finally, adjust the filter cut-off quite low and the resonance about midway and modulate the pitch of either the triangle or square wave with a sine wave set to a slow rate with a medium depth. If the string is going to continue for a length of time, it's worthwhile employing a sine, pulse or triangle wave LFO to modulate the filters cut-off to help maintain interest. As always, this should only be considered as a starting point and experimentation is the key to gaining good results.

Effects can also play an important role in creating interesting strings for the genre, although these should be used conservatively so as not to detract from the bass rhythm. Often, wide chorus effects, rotary speaker simulations, flangers, phasers and reverb can all help to add a sense of interest. Alternatively creatively pumping the string by running it through a compressor with a side-chain input channel programmed with a rhythmical element can breathe life into static sounding timbres.

VOCALS AND SOUND FX

One final aspect yet to cover is the addition of sound effects and vocals. The vocals within Drum 'n' Bass more often than not consist of little more than a short vocal snippets; however, there have been some more commercial drum 'n' bass mixes that have featured a verse/chorus progression.

It seems to be a point of contention amongst many Drum 'n' Bass producers as to whether a verse/chorus is actually part of the genre or is in fact diversifying again to produce a new genre of music. Others, however, believe that it's simply a watered down, commercialized version of the music made solely for the purpose of profit margins. Nonetheless, whether you choose to use a few snippets of vocals, some Ragga or MC'ing, or a more commercialized vocal performance is entirely up to you. It's musicians that push boundaries that reap the greatest rewards.

The sound effects can obviously be generated by whatever means necessary, from sampling and contorting sounds or samples with effects and EQ. For contorting audio, the Sherman Filterbank 2, the Camelspace range of plug-ins, Glitch, Sugarbytes Effectrix or Steinbergs GRM Tools are almost a requisite for creating strange evolving timbres.

That said, the effects and processing applied are, of course, entirely open to artistic license as the end result is to create anything that sounds good and fits within the mix. Transient designers can be especially useful in this genre as they permit you to remove the transients of the percussive rhythms that can evolve throughout the track with some thoughtful automation. Similarly, heavy compression can be used to squash the transient of the sounds, and with the aid of a spectral analyzer, you can identify the frequencies that contribute to the sound whilst removing those surrounding it. Alternatively, pitch shift individual notes up and by extreme amounts or apply heavy chorus or flangers/phasers to singular hi-hats or snares or try time stretching followed by time compression to add some digital clutter and then mix this with the other loops.

Ultimately, though, as with all the chapters in this book, the purpose here is to simply reveal some of the techniques used by producers and show how the theory and technology discussed earlier in the book combines to produce a track. However, it should be viewed as offering a few basic starting ideas that *you* can evolve from.

There is no one definitive way to produce this genre and the best way to learn new techniques and production ethics is to actively listen to the current market leaders and experiment with the tools at your disposal. There are no right or wrong ways to approach any piece of music and if it sounds right to the producer, then it usually is. New genres do not evolve from following step-by-step guides or from emulating peers; producers who experiment and push boundaries create them.

Nevertheless, with just these basic elements, it is possible to create the main focus of the music, and from here, the producer can then look towards creating an arrangement. The theory behind arranging has already been discussed in Chapter 19 but simply listening to the current market leaders mixed amongst the theory discussed in Chapter 22 will very quickly reveal the current trends in both sounds and arrangement.

The companion files contain an audio example of a drum and bass track using the techniques described.

CHAPTER 24

House

'Not everyone understands House music; it's a spiritual thing; a body thing; a soul thing.'

Eddie Amador

The development of house music has much of its success accredited to the rise and fall of 1970s disco. As a result, to appreciate the history of house music, we need to look further back than the 1980s and evolution of house, we first need to examine the rise of disco.

Pinning down an exact point in time where disco first appeared is difficult, since a majority of the elements that make disco had appeared in earlier records. Nevertheless, it can be said that it first originated in the early 70s and was a natural evolution of the funk music that was popular with black audiences at that time.

Influenced by this funk feel, some big name producers of that time such as Nile Rodgers, Quincy Jones, Tom Moulton, Giorgio Moroder and Vincent Montana began to move away from recording the 'normal' self-composed music and began to hire session musicians to produce funk inspired hits for artists whose only purpose was to supply vocals and become a marketable commodity.

Donna Summer became one of the first disco manufactured success stories with the release of 'Love to Love You Baby' in 1975 and is believed by many to be the first disco record to be accepted by the mainstream public. This 'new' form of music was still in its infancy, however, and it took the release of the motion picture 'Saturday Night Fever' in 1977 for the musical style to became a widespread phenomenon. Indeed by the late 70s over 200,000 people were attending discotheques in the UK alone and disco records contributed to over 60% of the UK charts.

As with most genres of music that become popular, many artists and record labels jumped on the wave of this new happening vibe, and it was soon deluged with countless disco versions of original songs and other pointless, and poorly produced, disco records as the genre became commercially exploited.

Eventually, disco fell victim to its own success in the late 70s and early 80s, and this was further augmented with the campaign of 'disco sucks'. In fact, in one extreme incident Steve Vahl, a rock DJ who had been against disco from the start, encouraged people to bring their disco collections to a baseball game on the 12th July 1979 for a ritual burning. After the game, a huge bonfire was lit and the fans were asked to throw all their disco vinyl onto the fire.

By 1981, disco was dead but not without first changing the entire face of club culture, changing the balance of power between smaller and major labels and preparing the way for a new wave of music. Out of these ashes rose the phoenix that is house, but it had been a large underground movement before this and contrary to the misconceptions, it had actually been in very early stages of evolution *before* disco hit the mainstream. Although to many Frankie Knuckles is seen as the 'godfather' of house, it's true foundations lie well before and can be traced back to as early as 1970.

Francis Grosso, a resident DJ at a converted church known as the *Sanctuary*, was the first ever DJ to mix two early disco records together to produce a continual groove to keep the party attendants on the dance floor. What's more, he is also believed to be the first DJ to mix one record over the top of another, a technique that was to form the very basis of dance music culture.

Drawing inspiration from this new form of mixing, DJ Nicky Siano set up a New York club known as *The Gallery* and hired Frankie Knuckles and Larry Levan to prepare the club for the night by spiking the drinks with Lysergic Acid Diethylamide (LSD/Acid/Trips). In return he taught both the basics of this new form of mixing records and soon after they moved on to become resident DJs in other clubs.

Levan began residency at *The Continental Baths* while Knuckles began at *Better Days*, to soon rejoin Levan at *The Continental Baths* six months down the line. The two worked together until 1977 when Levan left the club to start his own and was asked to DJ at a new club named the *Warehouse* in Chicago. Since Levan was now running his own club, he refused but recommended Knuckles who accepted the offer and promptly moved to Chicago.

Since this new club had no music policy, Knuckles was free to experiment and show off the techniques he'd been taught by Nicky Siano. Word quickly spread about this new form of disco and *The Warehouse* quickly became the place to be for the predominantly gay crowd. Since no 'house' records actually existed at this time, the term house did not refer to any particular music but simply referred to the *Warehouse* and the style of continual mixing it had adopted. In fact, at this time, the word house was used to speak about music, attitudes and clothing. If a track was called 'house', it meant it was from a cool club and something that you would never hear on a commercial radio station. If *you* were house, you attended all the cool clubs, wore the 'right' clothing, and listened to 'cool' music.

By late 1982, early 1983, the popularity of the *Warehouse* began to demise. This was partly due to the popularity of the club spreading to the mainstream audiences but can also be attributed to the change of musical style to cater for this new audience. The music was no longer considered 'house' as the owners not only began to play commercial music but also doubled the admission price.

Unhappy with this shift in music policy, Knuckles left and started his own club known as the *Powerhouse*. His devoted followers went with him but in retaliation, the *Warehouse* was renamed the *Music Box* and the owners hired a new DJ named Ron Hardy. Although Hardy wasn't a doctor, he dabbled in numerous pharmaceuticals and was addicted to most of them but was nonetheless a very talented DJ.

While Knuckles kept a fairly clean sound, Hardy pounded out an eclectic mix of beats and grooves mixing euro disco, funk and soul to produce an endless onslaught to keep the crowd up on the floor. Even to this day, Ron Hardy is viewed by many as the greatest ever DJ.

Simultaneously, WBMX, a local radio station also broadcast late night mixes made by the Hot Mix Five. The team consisted of Ralphi Rossario, Kenny 'Jammin' Jason, Steve 'Silk' Hurley, Mickey 'Mixin' Oliver and Farley 'Jackmaster' Funk. These DJs played a non-stop mixture of British New Romantic music ranging from Depeche Mode to Yazoo and Gary Numan along the latest music from Kraftwerk, Yello and George Clinton. In fact, so popular was the UK new romantic's scene that a third of the American charts consisted of UK music.

However, it wasn't just the music that the people tuned in for it was the mixing styles of the five DJs. Using techniques that have never been heard of before, they would play two of the same records simultaneously to produce phasing effects, perform scratches and back spins and generally produce a perfect mix from a number of different records. Due to the shows popularity, it was soon moved to a daytime slot and kids would skip school just to listen to the latest mixes. In fact, it was so popular that Chicago's only dance music store, '*Imports Etc*', began to put a notice board up in the window documenting all the records that had been played the previous day to prevent them from being overwhelmed with enquiries.

Meanwhile, Frankie Knuckles was suffering from a lack of new material. The 'disco sucks' campaign had destroyed the industry and all the labels were no longer producing disco. As a result, he had to turn to playing imports from Italy (the only country left that was still producing disco) alongside more dub-influenced music.

More importantly for the history of house, though, he turned to long-time friend Erasmo Rivieria, who was currently studying sound engineering to help him create reworks of the earlier disco records in an attempt to keep his set alive. Using reel-to-reel tape recorders, the duo would record and cut-up records, extending the intros and break beats, and layering new sounds on top of them to create more complex mixes. This was pushed further as he began to experiment by placing entirely new rhythms and bass lines underneath familiar tracks.

While this undeniably began to form the basis of house music as we know it today, no one had yet released a true house record, and in the end, it was Jesse Saunders release of 'On and On' in 1984 that land marked the first true house music record.

Although some aficionados may argue that artist Byron Walton (aka Jamie Principle) produced the first-house record with a track entitled *'Your Love'*, it was only handed to Knuckles to play as part of his set. Jesse Saunders, however, released the track commercially under his self-financed label 'Jes Say' and distributed the track through Chicago's 'Imports Etc'.

The records were pressed courtesy of Musical Products, Chicago's only pressing plant owned and run by Larry Sherman. Taking an interest in this scene, he investigated its influence over the crowds and decided to start the first ever house record label 'Trax'. Simultaneously, however, another label 'DJ International' was started by Rocky Jones and the following years involved a battle between the two to release the best house music. Many of these consisted of what are regarded as the most influential house records of all time including 'Music is the Key', 'Move Your Body', 'Time to Jack', 'Get Funky', 'Jack Your Body', 'Runaway Girl', 'Promised Land', 'Washing Machine', 'House Nation' and 'Acid Trax'.

By 1987, House was in full swing, while still borrowing heavily from 70s disco, the introduction of the Roland TB303 bass synthesizer along with the TR909, TR808 and the Juno 106 had given house a harder edge as it became disco made by 'amateur' producers. The basses and rhythms were no longer live but recreated and sequenced on machines resulting in a host of 303 driven tracks starting to appear.

One of these budding early producers was Larry Heard who after producing a track entitled 'Washing Machine' released what was to become one of the most poignant records in the history of house. Under the moniker of *Mr. Fingers*, he released *'Can U Feel It'*, the first ever house record that didn't borrow it's style from earlier disco. Instead it was influenced by soul, jazz and the techno that was simultaneously evolving from Chicago. This introduced a whole new idea to the house music scene as artists began to look elsewhere for influences.

One of these was Todd Terry, a New York Hip Hop DJ. He began to apply the sampling principles of rap into house music. Using samples of previous records, he introduced a much more powerful percussive style to the genre and released *'3 massive dance floor House anthems'* that pushed house music in a whole new direction. His subsequent house releases brought him insurmountable respect from the UK underground scene and has duly been given the title of Todd 'The God' Terry.

Over the following years, house music mutated, multiplied, and diversified into a whole number of different sub genres, each with their own names and production ethics. In fact to date, there a multitude of different subgenres of house with Progressive House, Hard House, Deep House, Dark House, Acid House, Chicago House, UK House, US House, Euro House, French House, Tech House, Vocal House, Micro House, Disco House, Swedish House, Commercial House … and I've probably missed some too.

MUSICAL ANALYSIS

The divergence of house music over the subsequent years has resulted in a genre that has become hopelessly fragmented and as such cannot be easily identified as featuring any one particular attribute. It can be funky drawing its inspiration from disco of the 70s; it can be relatively slow and deep drawing inspiration from techno, it can be vocal, it can be party-like or it can simply be pumping.

In fact, today the word house has become somewhat of a catch-all name for music that is dance (not pop!), yet doesn't fit into any other dance category. The good news with this is that you can pretty much write what you want and as long as it has a dance vibe, it could appear somewhere under the house label. The bad news, however, is that it makes it near impossible to analyze the genre in any exact musical sense, and it is only possible to make some very rough generalizations.

House music will invariably use a 4/4 time signature and is produced in almost any musical scale depending on the genre of house. For example, Disco House is invariably produced in a major key whilst the darker Swedish House will be produced in a minor key. Typically, however, many house records are produced in A minor. This is most likely because a proportionate amount of house records are produced by DJs and since many dance records are produced in A minor, it's easier to harmonically mix them. Plus, the most influential and powerful area for dance music's bass reproduction on a club speaker system is 50 to 65 Hz and the root note of A1 sits at 55 Hz. Since the music will modulate or gravitate around the root note of the scale, having the root note of the music at 55 Hz will help to maintain the bass 'groove' energy of the record at a particularly useful frequency.

In terms of physical tempo it can range from a somewhat slow 110 BPM to a more substantial 140 BPM but many of the latest tracks seem to stay around the 127 or more recently 137 'disco heaven' BPM. This latter BPM is referred to as such since this is equal to the average clubbers heart rate while dancing but whether this actually makes the music more 'exciting' in a club has yet to be proven.

Fundamentally, house is produced in one of three ways; everything is sampled and rearranged; only some elements are sampled and the rest is programmed, or the entire track is programmed via MIDI and virtual instruments. The approach taken depends entirely on what style of house is composed.

For example, the disco house produced by the likes of Daft Punk (before Random Access Memories) relies heavily on sampling significant parts from previous disco hits and dropping their own vocals over the top (Daft Punk's 'Digital Love' being a prime example).

If you write this style of music, then this is much easier to analyze since it's based around the disco vibe. It consists of a four to the floor rhythm with a heavily syncopated bass line and the characteristic electric guitar. On the other hand, deep house uses much darker timbres (deep bass lines mixed with atmospheric jazzy chords) that don't particular exhibit a happy vibe but are still danceable. Acid house relies heavily on the squawking TB303 with highly resonant timbres throughout and a pounding four to the floor beat and Swedish House leans more towards the Techno side of production with a darker feel, more complex beats and sub-basses. Because there are so many different genres of house and to cover them all would take most of this publication, what follows is simply a guide to producing the basic elements of all types of house and since you know how your choice of genre already sounds, this with the rest of the genre chapters can be adapted to suit what you hear.

HOUSE RHYTHMS

Generally, house relies heavily on the strict four to the floor rhythm with a kick drum laid on every beat of the bar. Typically, this is augmented with a 16th closed hi-hat pattern and an open hi-hat positioned on every 1/8th (the off beat) for syncopation. Snares (or claps) are also often employed on the second and fourth beat underneath the kick. This produces the basic loop for

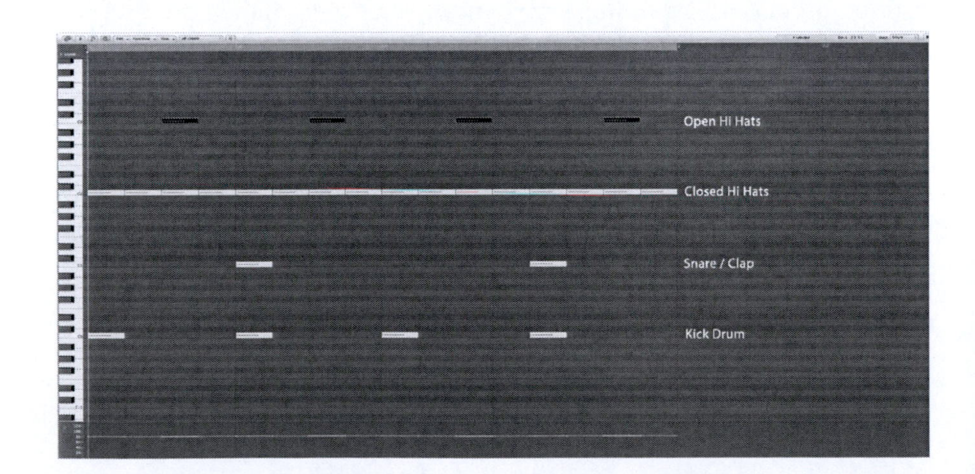

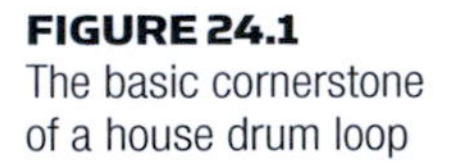
FIGURE 24.1
The basic cornerstone of a house drum loop

many genres of dance music, including house, and was discussed in detail in Chapter 16 constructing loops.

Notably, house music, more than any other dance genre relies on plenty of drive. This can be applied in a variety of ways but the most commonly used technique is to introduce a slight timing variance in the rhythm to create a push in the beat. Typically, this involves keeping the kick sat dead on all the beats but moving the snares or claps forward in time by a few ticks. This produces a small surge in the rhythm that is typical to many house genres.

For house kicks, the original synthesizer of choice was the somewhat ubiquitous Roland TR-909, but this is no longer considered to have enough low-frequency energy or power and over recent years there appears to be an unwitting competition between producers for who can produce the most powerful kick drum.

The somewhat typical powerful kick drum in almost all forms of House music is created through layering kick samples or layering a number of synthesized kicks. These techniques were discussed is great detail in Chapter 15 of this book, and therefore, I would refer the reader to that chapter rather than repeat it all again here. Having said that, even when layering samples, with House, the producer may find that the kick lacks high-end presence when placed in the mix and if this is the case, it's recommended to either lay a hi-hat sample on top of the kick or augment it with a square waveform. For this, the square wave requires a fast attack and decay stage, so that it simply produces a 'click', once this is layered over the original kick, you can experiment by pitching it up or down to produce the required timbre.

In addition, due to the often sub frequency nature of the house kick, it is highly recommended that the kick is tuned to the music and preferably the fifth note in the scale. If this tuning doesn't occur when it is played on a club system, the sub note can conflict with the bass creating a dissonant feel to the music that can easily ruin the overall effect of the underlying groove.

The house kick is almost always treated to small amounts of reverb. Here a room style reverb is the popular choice with a long pre-delay that skips the transient of the kick, so that it can cut through the mix. If the layered hi-hat or square wave approach mentioned earlier is employed, then reverb can be applied to only the kick that will leave the layered hi-hat (or square wave) unaffected. This permits the producer to use a shorter pre-delay setting on the main kick. Diffusion should be kept low and the tail should remain short so as not to wash over further instruments.

Indeed, one of the key elements of many house loops is that the individual sounds within the loop remain short and transient as this provides a more dynamic feel to the music and can often differentiate between different genres of house. Indeed, genres such as Funky House, French House and Disco House often employ rhythms that flow together whereas genres such as Minimal and Tech House will have each instrument cut short to produce a snappy, tight controlled feel to the music. To accomplish this, a noise gate can be inserted after the reverb and on each instrument channel to control the decay of each instrument.

House kicks also rely heavily on compression to produce the archetypal hardened sound. Although the settings to use will depend on the sound you wish to achieve, a good starting point is to use a low ratio combined with a high threshold and a fast attack with a medium release. The attack will crush the initial transient of the square producing a heavier 'thump' while experimenting with the release can further change the character of the kick. If this is set quite long, the compressor will not have recovered by the time the next kick begins and, dependent on the compressor, can often add a pleasing distortion to the sound.

Similar to the kicks, the snare and/or claps were often sourced from the TR909 but could also be derived from the E-Mu Drumulator due to its warmer, rounded sound. However, in order to attain the typical sound of many house records that are prevalent today, the snare (or clap) should be constructed from layered samples or layered synthesis. Again, the techniques for layering on snares and claps were discussed in detail in Chapter 15 and I would refer the reader there.

Not all house records will layer the snare, however, and some genres such as Tech House and Minimal will employ a thinner snare style timbre. This characteristic sound can be produced through using either a triangle- or square-wave mixed with some pink noise or saturation (distortion of the oscillator) both modulated with an amp envelope using a fast attack and decay. Typically, pink noise is preferred over white but if this proves to be too heavy for the loop then white noise may provide the better option. Much of this noise is removed with a high-pass filter to produce the archetypal Tech snare, but in some cases, a band-pass filter may be more preferable since this allows you to modify the low and high content of the snare more accurately.

If at all possible, it's worthwhile using a different envelope on the main oscillator and the noise since this will allow you to keep the body of the snare (i.e. the main oscillator) quite short and sharp but allow you to lengthen the decay of the noise waveform. By doing so, the snare can be modified, so that it rings out after each hit and permitting the producer to individually adjust the decay's shape of the noise waveform. Generally speaking, the best results are from employing a concave decay but as always it's best experimenting. Many Minimal and Tech snares also benefit from judicious amounts of pitch bend applied to the timbre but rather than apply this positively, negative will invariably produce results that are more typical.

If you use this latter technique, it's sometimes prudent to remove the transient of the snare in a wave editor and replace it with one that uses positive pitch modulation. When the two are recombined, it results in a timbre that has a good solid strike but decays upwards in pitch at the end of the hit. If this isn't possible then a viable alternative is to sweep, the pitch from low to high with a saw or sine LFO (provided that the saw starts low and moves high) set to a fast rate. After this, small amounts of compression so that only the decay is squashed (i.e. slow attack) will help to bring it up in volume so that it doesn't disappear into the rest of the mix.

Typically, the snares or claps are treated to a room style reverb but the length of the reverbs tail will depend on the genre of music. The reverbs tail is generally longer than a kick in the most genres but the pre-delay changes on genre. In genres such as Tech house, it's not unusual to use a very short pre-delay, so that the transient is molested. In some genres of house, reverse reverb is often used on the snare, but this is typically limited to the second beat or will only occur at structural downbeats.

Compression is sometimes used on the snares/claps, but this depends heavily on the genre of house music, since it's employed to do little more than modify the tonal character of the snare. If applied heavily with a fast attack, the compressor would clamp down on its attack stage diminishing the transient, in effect reducing the dynamic range between the transient and body. This results in a more powerful, prominent snare or clap that exhibits a 'thwack' style sound.

Additionally, depending on the genre, microtonal adjustments or filter modulation is requisite on the snares or claps to produce the motion between the two beats in the bar. In the example Swedish House style track, I made for this book, I used *Soundtoys Filter-Freak* on the snare to control a low-pass filters cut-off. The modulation was applied, so that first snare of the bar passes through a low-pass filter unmolested but the second strike occurring on beat 4 is subjected to a very light low-pass filtering that removes a very small portion of the higher harmonics of the timbre. This introduced enough of a textural difference between the two beats to maintain interest in this part of the loop. However, microtonal adjustments can accomplish much the same effect and genres such as Tech house, Glitch and Micro will lean more towards microtonal adjustments than filter.

In many examples of house music, the hi-hats are taken from sample CDs, of which there are plenty to choose from on the market. These will often be treated to numerous processing effects including distortion, reverb, transient designers and compression to produce the timbre the producer feels is appropriate for the style. All of these processors and effects have been discussed in detail and experimentation with all these effects will yield the best sonic results for the genre.

A very common effect for hi-hats in House, however, is delay. The settings for the delay will be dependent on the producer's creativity and the tempo but it should be kept fairly short to prevent too much delay wash that may cloud the rhythmic interplay. Typically a 1/16th or 1/8th setting with a very short delay time is suitable, but some producers will employ a noise gate after the delay unit or employ side chain compression. Employing a noise gate directly after the delay allows the producer to use larger settings, but also prevent them from 'stacking' up and collecting into a loud noise by reducing the threshold of the gate.

An alternative method to this is to employ the delay on a bus and place a compressor directly after the delay unit. The ratio of the compressor is set to 2:1 with a low-output gain and a low threshold. The original hi-hats are then fed into the side chain of the compressor. This way, each time a hi-hat occurs, the compressor activates ducking out any delayed hi-hats, and when no hat occurs, the compressor isn't activated permitting the delayed hats to continue. This method can provide a more open mix with more space for other instrumentation and prevents the delay from overpowering the mix.

Both open and closed hi-hats will also be treated to a cyclic microtonal pitch or filter modulation. Typically, the open hats are treated to a low-pass filter that is modulated via a sine or sample and hold waveform over a period of two or four bars whilst closed hats are treated to microtonal pitch movement over a period of three bars. Three or six bars are chosen since this works against the standard structural downbeat of dance music and results in a cross syncopation of modulation.

This approach produces the basic house loop but depending on the genre further instrumentation will be applied. Congas, toms, bongos, tambourines and shakers are often added to create a distinct feel to the rhythms, whilst the more intense genres such as Tech House, Fidget and Minimal will employ synthesized percussive elements instead. These are all often positioned on syncopated grid positions in the loop or will employ Hemiola or/and polymeter depending on the genre.

Typically, genres such as Disco House, UK House, US House, Euro House and French House will maintain a fairly basic rhythm since the emphasis is more on the shaky bass lines and chorded leads, but in other genres such as Tech House, Progressive House and Swedish House, there is stronger emphasis on the rhythm section, and therefore, they will employ more rhythmical techniques to maintain interest in the rhythm section.

Hemiola was discussed in detail in Chapter 3 but consists of creating an unequal subdivision of beats within the bar and placing a hi-hat or percussive rhythmical pattern onto these unequal subdivisions. This effect can then be further augmented with the use of polymeter on further percussive elements. A typical application of polymeter here is to employ a 5/4, 6/4 and 7/4 on synthesized percussive hits. These and, in fact, all percussive instrumentation within House music (with the exception of the kick) will be subjected to groove or swing quantize. The amount applied ranges from genre to genre but typically they all remain within 60%–70% on the 1/16th.

Of course, it would be naive to say that all house beats are created through MIDI, and for some producers, the rhythms are acquired from sampling previous House, Disco or Funk records. In the case of House records, these will have probably been sampled from other records, which will have been sampled from previous house records, which will have been sampled from other records, which will probably have been…well, there is no need to continue. For obvious reasons, I couldn't condone infringing another artists copyright but that's not to say that some artists don't do it and so in the interests of theory, it would only be fair to cover some of the techniques they use.

Almost every house record pressed to vinyl starts with just a drum loop so sourcing one isn't particularly difficult but the real skill comes from what you do with it in the audio editor. Fundamentally, it's unwise to just sample a loop and use it as is. Even though there's a very high chance that what you're sampling is already sample of a sample of a sample of a sample (which means that the original artists doesn't have a leg to stand on for copyright infringement), you might be unlucky and they may have actually programmed it themselves. What's more, it isn't exactly artistically challenging. Therefore, after sampling a loop, it's always worthwhile placing it into a sequencer and moving the parts around a little to create your own variation.

This is the most simplistic approach and a number of artists will take this a step further by first employing a transient designer to alter the transients of the loop before placing it into a sample-slicing program or employing the workstations own sample-slicing facilities. Using this technique, the loop does not have to be sourced from a house track as any drum break can be manipulated to create the snappy sounds used in house.

Along with working with individual loops, many house producers also stack up a series of drum loops to create heavier rhythms. Todd Terry and Armand Van Helden are well known for dropping kicks and snares over pre-sampled loops to add more of their heavy influence.

If you take this approach, however, you need to exercise care that the kick and snare occur in the right place; otherwise, you'll experience a phasing effect. Possibly the best way to circumvent, this is to cut the bar into a four segments where the kick occurs and then triggers the kick or kick and snare alongside the originals at the beginning of each segment to produce a continual loop that stays in time. This also permits you to swap the quarter bars around to create variations.

BASS

Almost every genre of house relies a great deal on the groove created from the interaction between the bass and the drum rhythm. Although the genre has diversified greatly from its original roots of disco, with the exception of a few House genres such as Tech and Minimal, the bass rhythm still has its roots firmly embedded in disco. Whether Progressive House, Swedish House or Funky House, the bass lines will exhibit a similar lively feel to that of original disco, albeit consecutively stripped back dependent on the genre of music.

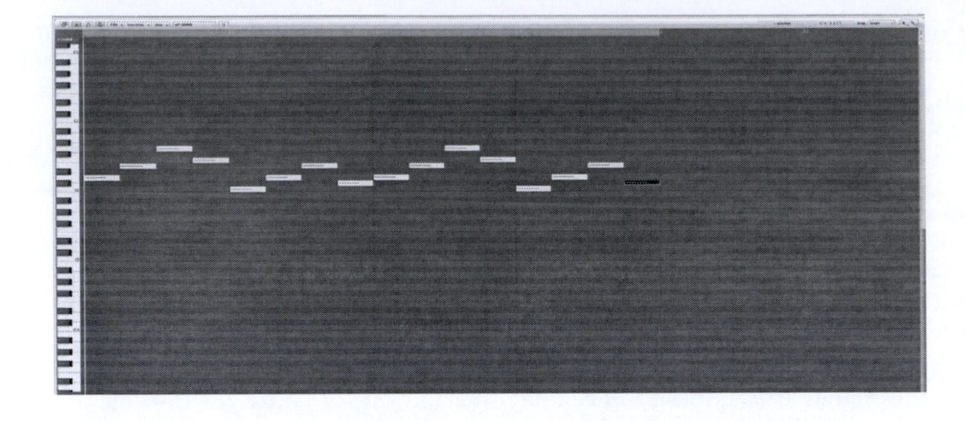

FIGURE 24.2
Disco's infamous walking bass line

For example, the closet House genres to Disco are French House, Disco House and Commercial House. These all employ bass lines that borrow heavily from the funky groove of the disco style bass. The simplest of these is the walking bass line that was used on countless disco records.

The above bass line is a simple example that has been used in a number of dance tracks but many disco inspired House records will employ a bass line written in the minor pentatonic scale (i.e. using only the black notes on the keyboard). Both Major and Minor Pentatonic scales are shown below:

Minor – Eb – Gb – Ab – Bb – Db – Eb

Major – Gb – Ab – Bb – Db – Eb – Gb

Minor and Major Pentatonic Scales

In these scales, only five notes are available but a further sixth and seventh are often introduced as leading tones. These are notes that do not exist in the pentatonic scale but can be used occasionally to pass through to the next note and help to add to the movement and groove of the music. In the example below, the notes G and D, whilst not in the pentatonic scale, are used as 'stepping stones' that 'lead' into the next note in the pentatonic scale.

Figure 24.3 shows a typical Disco or Funky House bass line that exhibits a very similar groove to that of their original disco roots but the bass line of further House genres will follow a similar groove, simply stripped back dependent on the genre.

In the House example track with this book (Swedish House), the bass line is stripped back through the use of notes with a longer duration and rather than resolving in the second bar (binary phrase) as it would in Disco, Funky House or Commercial house, it resolves over a period of 16 bars.

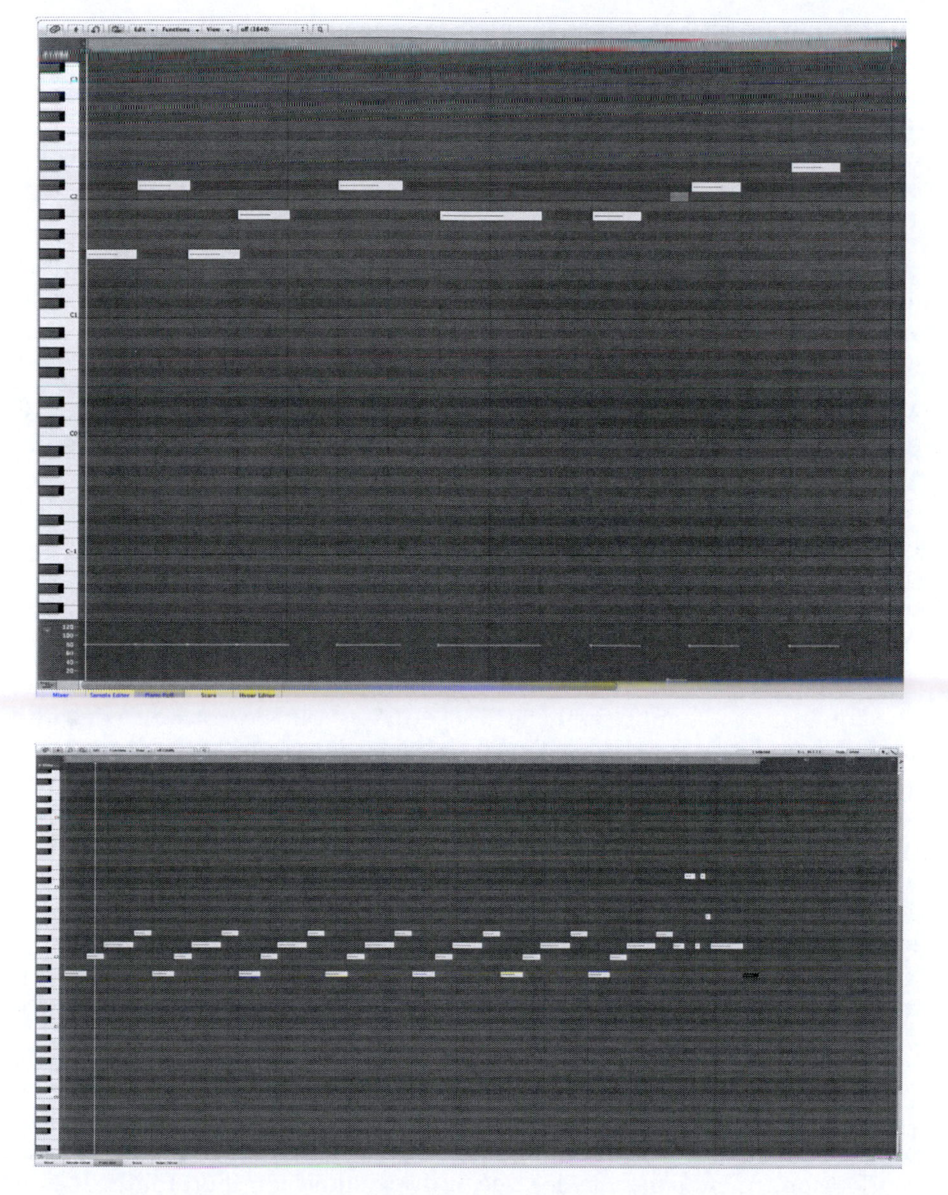

FIGURE 24.3
A Pentatonic bass line with a leading tone (the leading tone is the shortest note)

FIGURE 24.4
The books example bass line

It should be noted that as the genre changes and the bass line becomes stripped back further, it will begin to lean more towards techno. In fact, genres such as Tech house and Minimal will share more similarities with Techno than with house. Nevertheless, whatever the genre of House, it's important to note that the complexity of the bass often depends on the lead instruments that will play over the top.

Although some house music relies on a particularly funky bass groove, the overlaying instruments are kept relatively simple by comparison, whereas if the overlying lead elements are more complex, then the bass line remains relatively simple in contrast. This is an important concept since if both the bass and lead instruments are melodically complex, it will result in a miscellany of melodies that are difficult for the listener to decipher. Thus, when producing House, the producer needs to make the decision as to whether the centrepiece of the music is created by the bass or the lead elements.

For example, the piano in Laylo and Bushwacka's 'Love Story' produces the centrepiece of the music, while the bass act as an underpinning and so is kept relatively simple. Conversely, tracks such as Kid Crème's 'Down and Under' and Stylophonic's 'Soul Reply' employ a more funky bass that is complimented with simple melodies.

As with the rhythms, the timbre used for the basses in house can vary wildly from one track to the next. On the one hand, it can be from a real bass guitar, while on the other, it could consist of nothing more than a low-frequency sine wave pulsing away in the background to anything in between. Like much of House music, this makes it incredibly difficult to pin down any particular timbre, so what follows are a few synthesis tips to create the fundamental structure of most house basses. After you have this, experimentation is the key and it's worth trying out new waveforms, envelope settings and modulation options to create the bass you need.

The foundation of most synthetic house basses can be constructed with a sine wave and saw or pulse oscillator. The main waveform (the sine) provides a deep body to the sound, while the secondary oscillator can either provide raspy (saw) or woody (square) overtones. If the sound is to be quite cutting and evident in the mix, then a saw is the best choice while if you want to be more rounded and simply lay down a groove for a lead to sit over then a square is the better option. Listening carefully to the sound they produce together, begin to detune the secondary oscillator against the first until the timbre exhibits the 'fatness' and harmonic content you want. Typically, this can be attained by detuning the waves from one another but as always let your ears decide what's best.

Nearly all bass timbres will start immediately on key press so set the amps attack to zero and then follow this with a fast decay, no sustain and a short release. This produces a timbre that starts immediately and produces a small pluck before entering the release stage. If the sound is too short for the melody being played, increase the sustain and experiment with the release until it flows together to produce the rhythm you require.

If the bass appears to be 'running away', try moving it a few ticks forward or back and play with the amps attack stage. To add some movement and character to the sound, set the filter cut-off to low pass and set both the filter cut-off and resonance to midway, and then adjust the envelope to a fast attack and decay with a short release and no sustain.

Typically, this envelope is applied positively but experiment with negative settings, as this may produce the character you need. If possible, it's also worthwhile trying out convex and concave settings on the decay slope of the filter envelope, as this will severely affect the character of the sound, changing it from one similar to a Moog through to a more digital nature and onto a timbre similar to the TB303. In most house tracks, the filter follows the pitch (opens as the pitch increases), so it's prudent to use filter positive key follow too.

This creates the basic timbre but it's practical to begin experimenting with LFOs, effects and layering. Typical uses of an LFO in this instance would be to lightly modulate the pitch of the primary or secondary oscillator, the filters cut-off or the pulse width if a pulse waveform is used.

In regards to effects, distortion is particularly effective and a commonly used effect in House alongside phasing/flanging but if these are applied, the producer should ensure that only frequencies above 150 Hz are processed. The bass should always sit in the centre of the stereo spectrum, so not only do both speakers share the energy but also that any panning applied during mixing is evident.

If heavy flanging or phasing is applied then the bass will be smeared across the stereo image and the mix may lack any bottom end cohesion. Of course, if the bass is too thin or doesn't have enough body then effects will not rescue it, so it may be sensible to layer it with a sine wave transposed down by a few octaves.

Above all, we have no expectations of how a synthesized bass should sound so you shouldn't be afraid of stacking up as many sounds as you need to build the required sound. EQ can always be used to remove harmonics from a bass that is too heavy but it cannot be used to introduce harmonics that are not already present.

Alongside the synthetic basses, some House tracks will employ a real bass. More often than not these are sampled from other records or are occasionally taken from sample CD's rather than programmed in MIDI. However, it is possible to program a realistic bass provided that the a good sample engine is used (such as Spectrasonics Trillian or a multi-sampled Kontakt Instrument)

The key to programming and emulating a real bass instrument is to take note of how they're played and then emulate this action with MIDI and a series of Control Change commands. In this instance, most bass guitars use the first four strings of a normal guitar E – A – D – G, which are tuned an octave lower resulting in the E being close to three octaves below middle C. Also they are monophonic, not polyphonic, so the only time notes will actually overlap is when the resonance of the previous string is still dying away as the next note is plucked.

This effect can be emulated by leaving the preceding note playing for a few ticks while the next note in the sequence has started. The strings can either be plucked or struck, and the two techniques produce different results. If the string is plucked, the sound is much brighter and has a longer resonance than if it were simply struck. To copy this, the velocity will need to be mapped to the filter cut-off of the bass module, so that higher values open the filter more. Not all notes will be struck at the same velocity, though, and if the bassist is playing a fast rhythm, the consecutive notes will commonly have less velocity since he has to move his hand and pluck the next string quickly. Naturally, this is only a guideline and you should edit each velocity value until it produces a realistic feel.

Depending on the 'bassist', they may also use a technique known as 'hammer on', whereby they play a string and then hit a different pitch on the fret. This results in the pitch changing without actually being accompanied with another pluck of the string. To emulate this, you'll need to make use of pitch bend, so this will first need setting to a maximum bend limit of two semitones, since guitars don't 'bend' any further than this. Begin by programming two notes, for instance an E0 follow by an A0 and leave the E0 playing underneath the successive A0 for around a hundred ticks. At the very beginning of the bass track, drop in a pitch bend message to ensure that it's set midway (i.e. no pitch bend) and just before where the second note occurs drop in another pitch bend message to bend the tone up to A0. If this is programmed correctly, on play back, you'll notice that as the E0 ends the pitch will bend upwards to A0 simulating the effect. Although this could be left as is, it's sensible to drop in a CC11 message (expression) directly after the pitch-bend, as this will reduce the overall volume of the second note, so that it doesn't sound like it has been plucked. In addition to this, it's also worthwhile employing some fret noise and finger slides. Most good sample-based plug-in instruments will include fret noise that can be dropped in between the notes to emulate the bassist's fingers sliding along the fret board.

Whether the producer chooses to employ real bass or a synthesized one, after it's written it is commonly compressed against the drum loop. The theory is to pump the bass to produce the classic bottom end groove of most house records – they tend to pump like crazy.

This is accomplished by feeding both bass and drum loop (preferably with the hi-hats muted) into a compressor with a threshold so that each kick registers approximately –6 dB on the gain reduction metre. With a ratio of around 9:1 and a fast attack, the gain make-up is adjusted, so that it's at the same volume level when the compressor is bypassed.

Finally, the release parameter is initially set to 200 ms and gradually reduced, whilst the music is playing. The shorter the release becomes, the more the kick will begin to pump the bass, becoming progressively heavier the more that it's shortened. This technique is particularly popular in some house tracks where the producer uses little more than a single note bass and the pumping of the compressor from the side-chained kick presents the rhythmical movement.

MELODIES, MOTIFS AND CHORDS

The lead instruments melody will depend entirely on the type of house being produced. Funky bass lines will require less active melodies to sit over the top while less active basses will require more melodic elements. Unfortunately, since this genre has become so hopelessly fragmented, it's difficult to offer any guidelines apart from to listen to the latest House records to see where the current trend lies.

However, having said that there are some timbres that have always been popular in House including the Hoover, progressive plucked leads and pianos from the DX series of synthesizers. These have all already been discussed in detail in Chapter 17 and 18, so here we will examine the general synthesis approach.

Fundamentally, synthetic house leads will more often than not employ saw, triangle and/or noise waveforms to produce a harmonically rich sound that will cut through the mix and can be filtered if required. Depending on how many are employed in the timbre, these can be detuned from one another to produce more complex interesting sounds.

If the timbre requires more of a body to the sound then adding a sine or pulse wave will help to widen the sound and give it more presence. To keep the dynamic edge of the music, the amplifiers attack is predominantly set to zero, so that it starts upon key press but the decay, sustain and release settings will depend entirely on what type of sound you require.

Generally speaking, it's unwise to use a long-release setting, since this may blur the lead notes together and the music could loose its dynamic edge but it's worth experimenting with the decay and sustain while the melody is playing to the synth to see the effect it has the rhythm.

As lead sounds need to remain interesting to the ear, it's prudent to employ LFOs or a filter envelope to augment the sound as it plays. A good starting point for the filter EG is to set the filter cut-off to low-pass and set both the filter cut-off and resonance to midway, and then adjust the envelope to a fast attack and decay with a short release and no sustain. Once this is set, experiment by applying it to the filter by positive and negative amounts. On top of this, LFOs set to modulate the pitch, filter cut-off, resonance, and/or pulse width can also be used to add interest. Once the basic timbre is down, the key is, as always, to experiment.

Many House tracks will also employ chords and in particular genres such as *Deep House*, *Tech House*, *Minimal* and *Commercial* rely on them to create the atmosphere of the track. Here, it is the progression and style of chords that can often dictate the genre of the music. For example, in many of the uplifting styles of House, the chords will consist mostly of triads since these are considered light and breezy chords.

Here, if the music is written in a minor key (of which most uplifting style House tracks are), they will often employ a 'seventh' as the penultimate chord in the progression. This is because seventh chords add anticipation to the music and contribute to the epic feel of this style of House.

In the above example, when the chord progression reaches the 'seventh', the track is lifted and receives its uplifting feel. Conversely, genres such as Tech House, Deep House and Minimal will employ seventh chords throughout the progression and rather than move up in pitch with each consecutive chord, it will move down to produce a more serious, almost depressing feel to the music.

There are a vast number of timbral styles for the creation of chords, and much of the synthesis choice is down to the producer but as a basic starting point, three saw waveforms are required with two detuned from one another. Apply a small amount of vibrato using an LFO to these two detuned saws with an

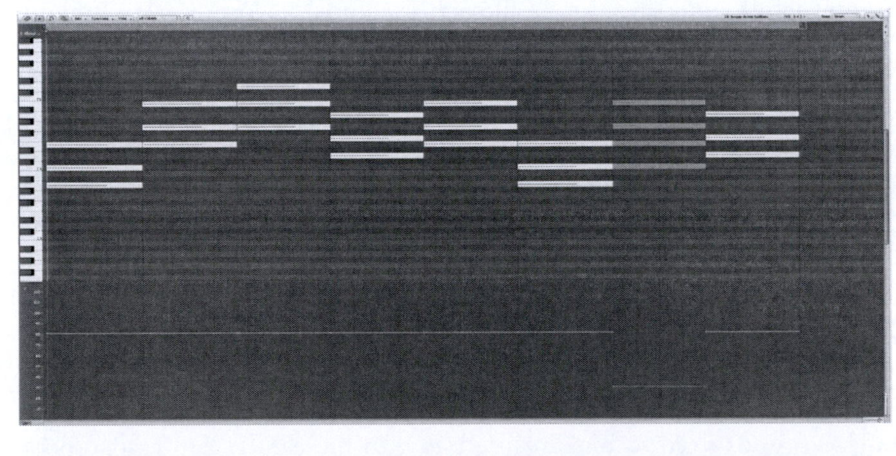

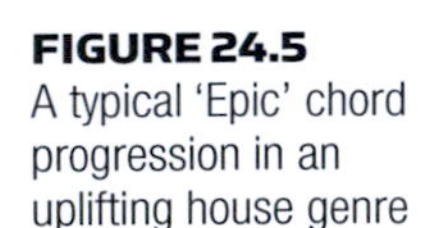

FIGURE 24.5
A typical 'Epic' chord progression in an uplifting house genre

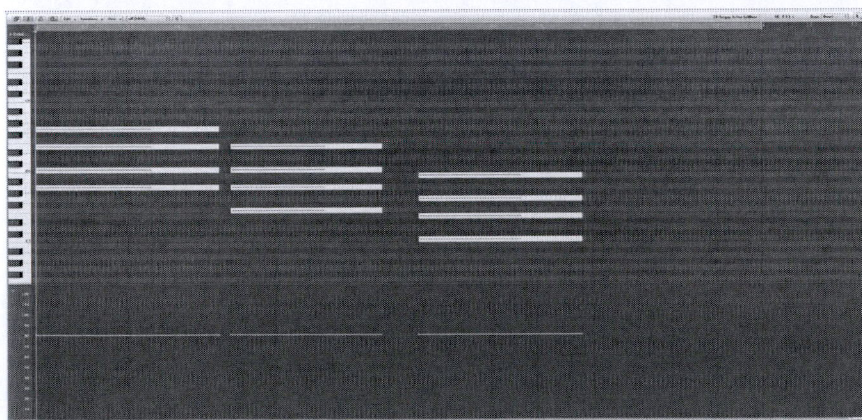

FIGURE 24.6
A typical 'serious' seventh chord progression

amplifier envelope with a medium attack and release and a full sustain (there is no decay since sustain is set to full and it would have nothing to fall to).

The final saw is then pitched upwards as far as possible without it becoming an individual sound (i.e. less than 20 Hz) and if possible use two filters – one set as a low pass to remove the low-end frequencies from the two detuned saws and a high pass to remove some of the high-frequency content from the recently pitched saw.

This will produce a basic string timbre that can be further augmented with effects and processors such as flangers, phasers, delay, reverb and compression. Wide chorus effects, rotary speaker simulations, flangers, phasers and reverb can all help to add a sense of movement to help fill out any holes in the mix.

Ultimately, though, as with all the chapters in this book, the purpose here is to simply reveal some of the techniques used by producers and show how the theory and technology discussed earlier in the book combines to produce a track. However, it should be viewed as offering a few basic starting ideas that *you* can evolve from.

There is no one definitive way to produce this genre and the best way to learn new techniques and production ethics is to actively listen to the current market leaders and experiment with the tools at your disposal. There are no right or wrong ways to approach any piece of music and if it sounds right to the producer, then it usually is. New genres do not evolve from following step-by-step guides or from emulating peers; producers who experiment and push boundaries create them.

Nevertheless, with just these basic elements, it is possible to create the main focus of the music and from here the producer can then look towards creating an arrangement. The theory behind arranging has already been discussed in Chapter 19 but simply listening to the current market leaders mixed will very quickly reveal the current trends in both sounds and arrangement.

The companion files contain an audio example of a Swedish House track using the techniques described.

CHAPTER 25

Techno

'Happy music is easy to consume, you can just put yourself in the music and see what happens. It's passive; you don't have to do anything for it. With techno, it is not only dark but also very subtle and intense. It can make you lose your mind, make you freak totally out.'

Marcel Dettmann

The uninitiated techno is used to describe any electronic dance music, and although this was initially true over the years, it has evolved to become a genre in its own right. Originally, *Kraftwerk* coined the term techno in an effort to describe how they mixed electronic instruments and technology together to produce 'pop' music. However, as more artists began to introduce technology into their productions, the true foundations of where techno as it's known today originated is difficult to pinpoint accurately.

To some, the roots of this genre can be traced back to as early as 1981 with the release of '*Shari Vari*' by *A Number of Names*, Donna Summers (and *Giorgio Moroder*) '*I Feel Love*' and *Cybotron's* 'Techno City'. To others, it emerged in the mid-1980s when the 'Belleville Three' collaborated together in Detroit. These three high-school friends – Juan Atkins, Kevin Saunderson and Derrick May – used to swap mix tapes with one another and religiously listen to the Midnight Funk Generation on WJLB-FM. The show was hosted by DJ Charles 'Electrifying Mojo' Johnson and consisted of a five-hour mix of electronic music from numerous artists including *Kraftwerk, Tangerine Dream*, and George Clinton.

Inspired by this eclectic mix, they began to form their own music using cheap second-hand synthesizers including the Roland TR909, TR808 and TB303. The music they produced was originally labelled as 'House music' and both May and Saunderson freely admit to gaining some of their inspiration from the Chicago clubs (particularly, the Warehouse and Frankie Knuckles). Indeed, Derrick May's 1987 hit 'Strings of Life' is still viewed by many as house music, although to Derrick himself and many other aficionados it was an early form of Detroit techno.

It wasn't until late 1988, however, until techno became a genre in its own right. This occurred when Neil Rushton produced a compilation album labelled 'Techno – The New Dance Sound of Detroit' for Virgin records. Following this release, techno no longer described any form of electronic music but was used to describe minimalist, almost mechanical house music. Techno became characterized by a mix of dark pounding rhythms with a soulful feel and a raw vibe. This latter stripped down feel was a direct result of the limited technology available at the time. Since the TB303, TR909 and TR808 were pretty much the only instruments obtainable to those without a huge budget, most tracks were written with these alone, which were then recorded direct to two-track tape cassette's.

Similar to most genres of dance, techno mutated as more artists embraced the ideas and formed their music around it. By 1992 and the evolution of the new 'rave generation' techno bore, almost no relationship to the funky beats and rhythms of house music as it took on more drug-influenced hypnotic tribal beats.

As technology evolved and MIDI instruments, samplers and digital audio manipulation techniques became more accessible, techno began to grow increasingly complex. While it still bore a resemblance to the stripped down feel of Detroit techno consisting solely of rhythms and perhaps a bass, the rhythmic interplay became much more complex. More and more rhythms were laid atop one another and the entire studio became one instrument with which to experiment.

Of course, Detroit techno still exists today but it has been vastly overshadowed by the tribal beats of 'pure' techno developed by numerous artists including *Thomas Krome, Redhead Zync, Henrik B, Tobias, Carl Craig, Kenny Larkin* and *Richie Hawtin*. Each of these artists has injected their own style into the music, while keeping with some of the original style set by these contemporizes.

MUSICAL ANALYSIS

Techno can be viewed as dance music in its most primitive form since it's chiefly formed around the cohesion and adaptation of numerous drum rhythms. Although synthetic sounds are also occasionally employed, they will appear atonal, as it's the abundance of percussive elements that remain the most vital aspect of the music. In fact, in many techno tracks any additional synthetic instruments are not often used in the 'musical' form to create bass lines or melodies, instead the genre defines itself on a collection of carefully programmed and manipulated textures rather than melodic elements.

Fundamentally, this means that it's produced with the DJ in mind, and in fact, most techno is renowned for being 'DJ friendly' being formed and written to allow him (or her) to seamlessly mix all the different compositions together to produce one whole continuous mix to last through the night. Consequently, techno will generally utilize a four to the floor time signature, but it isn't unusual to employ numerous other drum rhythms written in different time signatures that are then mixed, processed and edited to fit alongside the main 4/4 signature. Tempo-wise, it can range from 130 to 150 BPM, and although some techno has moved above this latter figure, it is in the minority rather than the majority.

Originally, techno was different from every other genre of music since it didn't rely on programming and mixing in the 'conventional' manner (if there is such a thing). Rather, it was based around employing the entire studio as one interconnected tool. Here, a hardware sequencer was used to trigger numerous drum rhythms contained in connected samplers and drum machines. Each of these rhythms ran through effects units that were manipulated produce new variations which are then layered with others or dropped in and out of the mix to produce the final arrangement.

Today, this approach has changed, and the majority of the music is written and produced within an audio sequencer but the general theory remains the same. Rhythms are still layered on top of one another, so that not only they all interact in terms of syncopation and polyrhythm but also the tonal content is harmonically combined to produce interesting variations of the original patterns. Here, the audio workstations mixing desk is used not only to mix the rhythms together in a conservative manner but also as a creative tool with EQ employed to enhance the interesting harmonic relationships created from this layering or to prevent the cohesive whole from becoming too muddy or indistinct.

A techno loop will often begin as little more than the standard four to the floor drum rhythms common to both trance and house. A kick drum is placed on every beat of the bar, along with snares or claps on the second and fourth beat to add expression to these two beats. To compliment this basic pattern, closed

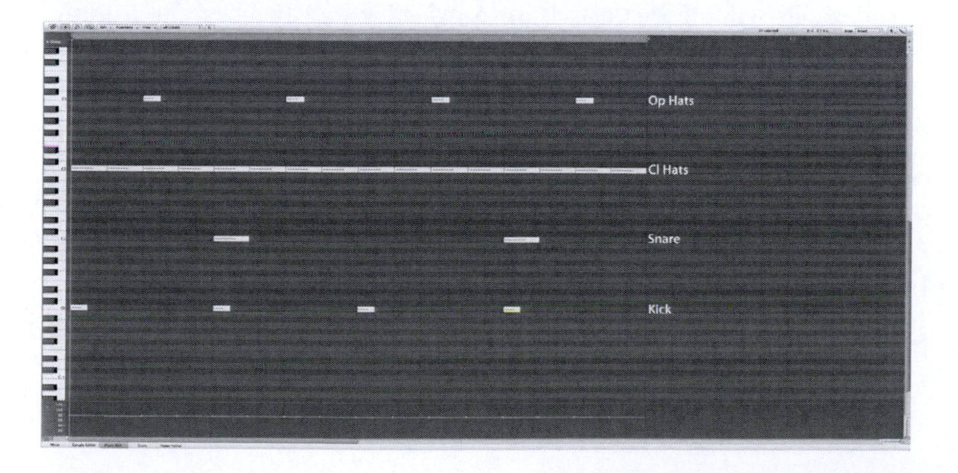

FIGURE 25.1
The beginnings of a techno loop

hi-hats are commonly placed on every 16th division or variation of 16ths, whilst to introduce some basic syncopation open hi-hats are often employed and placed on every 1/8th division of the bar.

The characteristic style of the kick timbre within techno depends on its subgenre but typically the kick is boomy with a hard transient to keep it controlled. Typically the kick in many techno tracks is programmed direct in a Roland TR909 (either hardware of software) but the same style of sound can be accomplished with a carefully chosen kick sample. By rolling off any frequencies below 40 Hz and applying a small EQ boost at 400 to 800 Hz, it will often produce the typical techno style kick. If there is no energy in this higher region to boost, then layering a hi-hat sample over the top of the kick can often introduce the bright transients that are typical of this genre.

Alternatively, the kick can be programmed in a synthesizer by employing a short decay combined with an immediate attack stage. Typically, a 100 Hz sine wave produces the best starting point for a techno kick with a positive pitch EG set a fast attack and a quick decay to modulate the oscillator. Although using a pitch envelope produces the best results, if the synth doesn't offer one, a self-oscillating filter will produce the requisite sound, and the decay can be controlled with the filters envelope. The kick will need to be augmented with a square waveform to produce the usual bright transient stage. For this, the square wave requires a fast attack and decay stage, so that it simply produces a 'click', once this is layered over the original sine you can experiment by pitching it up or down to produce the required timbre.

Reverb is fundamental to the creation of any techno kick, but how it is applied depends on the style of kick required. A 'standard' techno kick is often treated to a small amount of hall reverb with a long pre-delay to bypass the transient of the kick, but the tail is kept longer than most genres so that the reverb is evident. This is followed by a noise gate to ensure the reverb tail does not decay away as in a normal situation but is deliberately and evidently cut short.

One of the most common style of techno kicks – the 'Hiss Kick' – can be created through sending the kick drum completely to a bus channel featuring a hall reverb unit with full diffusion and a decay of 5 seconds. A compressor is placed directly after this reverb unit with a fast attack and a long release (approximately 500 ms) with a ratio of 5:1, and the same bus channel is then used as the side-chain input for this compressor whilst the ratio is lowered until the reverb effect begins to pump. A filter is then inserted after the compressor that is used to cyclically modulate the higher frequencies of the pumping reverb effect. This cyclic modulation does not need to be evident and should be applied gently so as not to draw too much attention to itself.

The kick is then sent to a secondary bus that contains a room reverb unit with full diffusion but only a short decay of 1 second or less. This reverb is followed by a noise gate with a fast attack and release and the same second buss used as a side-chain for the gate. As the threshold of the gate is reduced, the reverb will begin to gate, and this should be set to personal preference. This noise gate is followed with another low-pass modulating filter that will cyclically modify the reverbs field. The timing of the cyclic modulation applied on both reverbs should be set differently, so they do not run in sync with one another. Finally, both reverbs are mixed in with the kick to produce the Hiss Kick that appears in a number of techno tracks.

The kick will also benefit from hard compression, but this must be applied cautiously so as not to remove the bright transient stage. Here, the attack should be set so that it bypasses the initial transient but captures the decay stage. A hard ratio of 5:1 followed by reducing the threshold should provide the common characteristics of a techno kick.

Techno will employ either snares or claps in the production, and these are commonly sourced from sample CDs but can equally be programmed in a synthesizer. For snares, it's preferable to employ a triangle wave for the first oscillator and noise for the second. The amplifier envelope generator employs a zero attack, sustain and release with the decay employed to set the 'length' of the snare.

If possible, employ a different amp EG for both the noise and triangle wave. By doing so, the triangle wave can be kept quite short and swift with a fast decay while the noise can be made to ring a little further by increasing its decay parameter. This, however, should never be made too long since the snares should remain short and snappy. If the snare has too much bottom end employ a high-pass, band-pass filter or notch filter depending on the type of sound you require. Notching out the middle frequencies will create a clean snare sound that's commonly used in this genre. Further modification is possible using a pitch envelope to positively modulate both oscillators, this will result in the sound pitching upwards towards the end, giving a brighter snappier feel to the timbre.

If claps are used instead, it is better to take them from a sample CD, since they can be difficult to synthesize but they can be created with a filter and amplifier envelope onto a white noise oscillator. Both envelopes use a fast attack with no sustain or release and the decay used to set the length of the clap. Finally, use a saw tooth LFO to modulate the filter frequency and pitch of the timbre. Increasing or decreasing the LFO's frequency will then change the sonic character of the clap significantly.

To produce the typical hard techno sound, snares and/or claps will be treated to large amounts of reverb by inserting a reverb onto the channel. This is commonly set to a large room or small hall and employs a long pre-delay with a long delay and largest diffusion settings. After this, a noise gate is employed to remove the tail in an evident way whilst compression follows the noise gate to compress the whole timbre heavily.

Pitch modulation is essential on these instruments, but they are often treated over a period of three or six bars rather than simply modulated over the period of a single bar. What's more, each snare is processed to a different pitch than previous whilst all remaining within 30 cents of the original. Like in Drum 'n' Bass, this modulation does not have to be applied in a stepwise fashion with each snare being a few cents higher or lower than the previous and generally producers will mix and match the pitch adjustments.

The hi-hats in techno can be programmed or taken direct from a sample CD. In many tracks, the hats are commonly sourced from vinyl or sample CDs, but some artists do program their own by ring modulating a high-pitched triangle wave with a lower pitched triangle. This produces a high-frequency noise that can be modified with an amplifier envelope set to a zero attack, sustain and release with a short to medium decay. If there isn't enough noise present, it can be augmented with a white noise waveform using the same envelope. Once this basic timbre is constructed, shortening the decay creates a closed hi-hat while lengthening it will produce an open hat.

A popular technique in a number of techno tracks is to employ nothing more than white noise from a synthesizer and insert a compressor onto the same track that is side-chained to a secondary rhythmical channel. This secondary channel is commonly a hi-hat set to no output on the mixing desk so that it can be used as nothing more than a rhythmical control for the compressor. Alternatively, some producers will simply side-chain the kick drum to the white noise to create a lengthy sounding hats that pump with the kick.

Typically the hats will be treated to delay and modulation. The hi-hats should be kept to short delay settings to prevent too much delay wash that will cloud the signal. Typically a 1/16th or 1/8th setting with a very short delay time proves enough. The open hats are commonly treated to a low-pass filter that

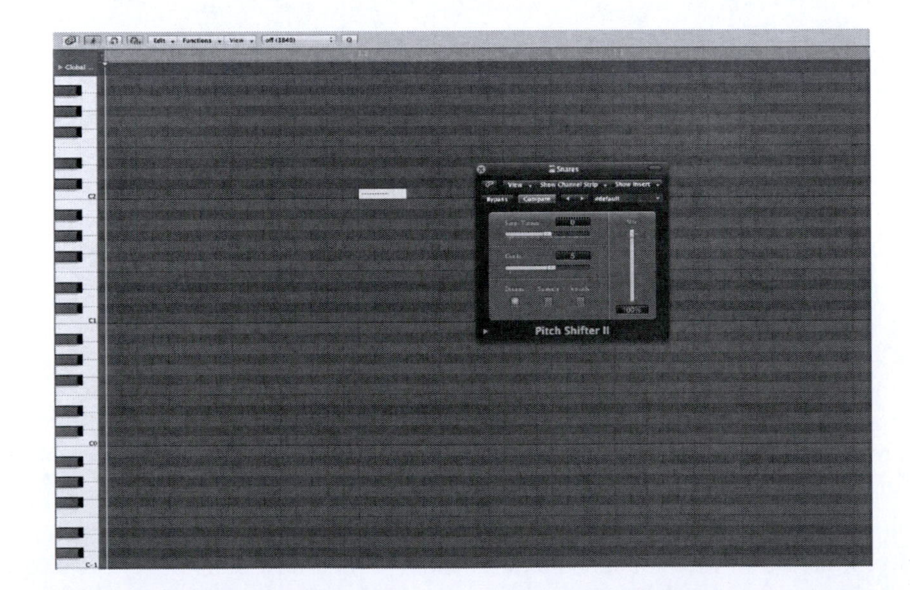

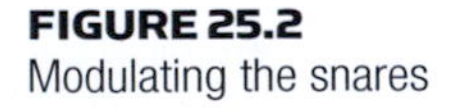
FIGURE 25.2
Modulating the snares

is modulated via an offset sine wave or a sample and hold waveform. This is applied over a period of three bars whilst further cyclic modulation will be applied to the closed hats over a different period of bars. This prevents the listeners psyche from specifically identifying a fixed cyclic pattern in the different rhythms. Typically, odd numbered bars are chosen for the cyclic modulations since this works against the standard structural downbeat of dance music and results in a cross syncopation of modulation.

ANCILLARY INSTRUMENTS

Once the basic loop is formed, further percussive instruments are layered onto the rhythm. The initial approach here is to 'fill' the gaps between the kick, snare, and open hi-hat with any number of different synthesized percussive elements. The purpose is to create a syncopated groove employing techniques such as Hemiola, Polymeter and call and response timbres to create binary phrasing. Hemiola was discussed in detail in Chapter 3 but consists of creating an unequal subdivision of beats within the bar and placing a hi-hat or percussive rhythmical pattern onto these unequal subdivisions. This effect is further augmented with the use of Polymeter on further hi-hats, snares, or percussive elements. A typical application of Polymeter here is to employ a 5/4 on a rhythmic hi-hat and 6/4 and 7/4 on further rhythmical elements such as congas or synthesized percussive hits.

Since the majority of techno relies on basic percussive elements, call and response forms a fundamental in some of these instruments in order to maintain some form of binary phrasing to keep the listener interested. This is not applied singularly, however, and most of the ancillary instrumentation will create a call and response motion over a different period of bars. For instance, one percussive element may make the call over three bars to be answered in the fourth whilst another percussive element may make a call over one bar and yet another makes a call over six bars.

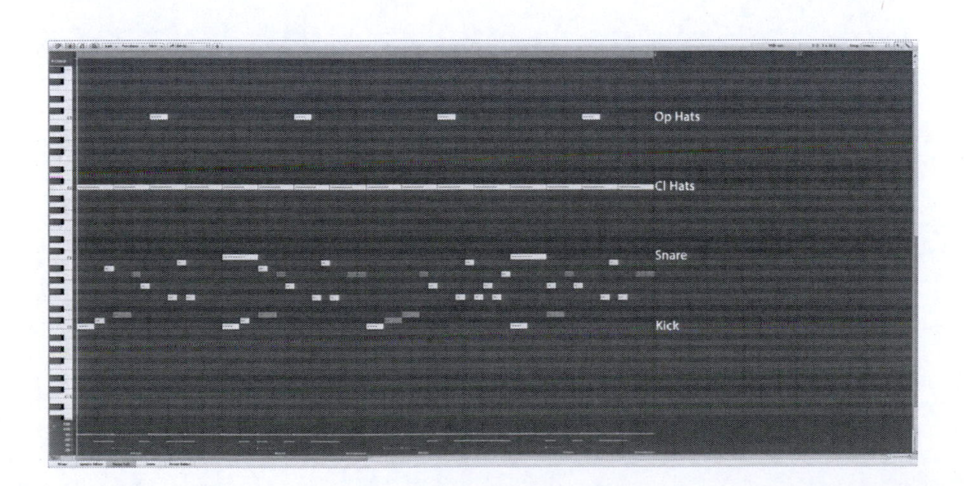

FIGURE 25.3
Filling the 'gaps'

A second important principle in the design of techno is to maintain short transient hits on all of the percussive elements. Indeed, it not unusual for a techno arrangement to feature a noise gate on every single channel involved in the project. The gates on each are then carefully adjusted to ensure that each percussive hits ends before the next begins. This ensures that all of the percussion hits are kept short and sharp and helps to maintain the rhythmical yet minimal flow of the music.

The most important aspect, however, is modulation and automation of every timbre within a mix. In order to maintain interest in what is little more than a complex polyrhythmic and polymetric drum loop, the producer will not only apply the usual filter or pitch modulation but also apply numerous other effects that can be automated in and out of the music. For example, volume, frequency cut-off, amplitude decay, filter resonance, noise gate release, reverb diffusion, delay times and even compression threshold and ratio will all be automated on the timbres to create an evolving, changing textural landscape. Similarly, it is not unusual for the swing quantize to change throughout the music on each instrument. This will modify the time positioning of different percussive instruments throughout and introduce a more evolving rhythmical pattern.

BASS

As previously mentioned, techno commonly consists of drums alone but some may also include a bass rhythm to help the music groove. In these instances, the bass commonly remains very simple so as not to detract from the fundamental groove created by the drums. Indeed, in many examples, the bass commonly consists of noting more than a series of sixteenth, eighth or quarter notes (sometimes consisting of a mix between them all) with none or very little movement in pitch.

In the above example, the bass remains atonal but textural movement is provided through cyclic modulation and automation of various parameters. For

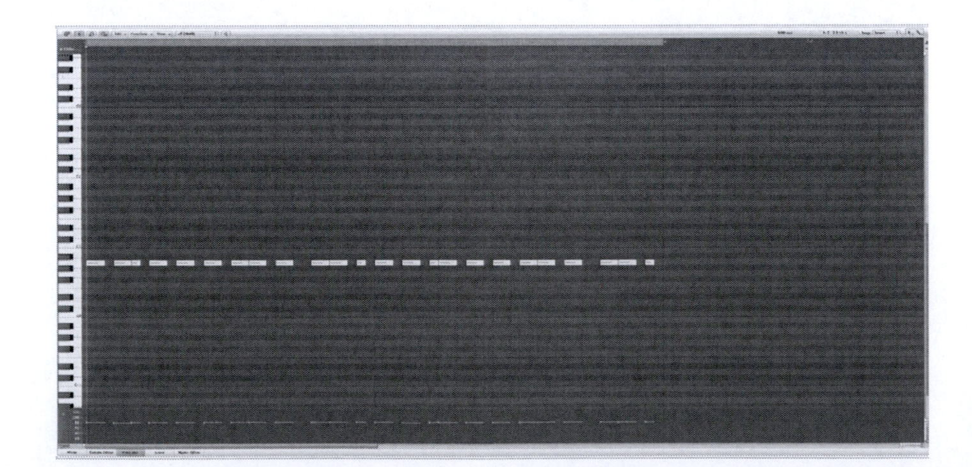

FIGURE 25.4
A typical techno bass line

example, it's not unusual to modulate the pitch of the bass micro-tonally, so although the bass 'appears' to remain monotonic, it nevertheless retains interest.

Indeed, since the bass will generally only consist of one bar that is continually looped over and over, it's the harmonic and timbral movement that plays a primary role in attaining the groove. This is not only accomplished by using cyclic modulation of pitch but by also adjusting various synthesis and effects parameters as the bass plays alongside the drum track.

The technique here is to manipulate the frequencies contained in the bass so that it augments the frequencies in the drum track. That is, it adds to the harmonic relationship already created through manipulating the drums to create a cohesive whole that pulses along. The bass should still remain a separate element to the drum track but nonetheless, any frequency dependent movement should be to bring further interest to the harmonic interaction with the drums than to bring attention to itself.

For the bass timbre, the TB303 is one of the most widely used instruments since the techno originators originally used this instrument but any analogue synthesizer can produce the results. A good starting point is to employ both a saw and square oscillator and set the amplifiers EG to zero attack and sustain with a medium release and a fast decay. Use a low pass filter with both the resonance and cut-off set mid way and then adjust the filters envelope to a short attack, decay and sustain with no release.

Another classis techno bass sound is the tuned bass. This consists of nothing more than a saw tooth oscillator wave that has its tuning automated (the cents or semitone tuning in the synthesizer) so that it changes gradually as the track progresses. These movements are not excessive, however, and typically a tuned bass will eventually change by a semitone or two throughout the length of a 6-minute track. Alternatively a TR808 kick sample is becoming a popular source for bass lines. Provided it has been sampled with a lengthy attack, a transient designer is placed on the channel and the initial transient of the kick is removed. After this, a pitch shifting plug-in is automated to slightly modulate the pitch of the kick as the track progresses.

For those that are a little more adventurous, four saw waves, or a couple of saws and a sine wave to add some bottom end if required can produce a powerful techno bass. If a sine wave is used, detune this by an octave below the other oscillators and then proceed to detune each saw from one another by differing amounts.

The amp envelope for all the waveforms is commonly set to a fast attack with a medium to long decay and no sustain or release. A filter envelope is not commonly used as this adds harmonic movement through the period of the sound that may conflict with the already programmed/manipulated rhythms and the preferred option is to keep it quite static and employ filter movement manually to suit the constantly changing frequencies in the rhythms.

That said, if a pitch envelope is available in the synth, it's prudent to positively or negatively modulate the pitch of the oscillators to add some small amounts of movement. Once this basic timbre is laid down the producer can experiment with the attack, decay, and release of the amps EG to help the bass sit comfortably in with the kicks drum loop. In fact, it's essential to accomplish this rhythmic and tonal interaction between the drum loop and the bass before moving on.

As the harmonic movement and interaction with the bass and rhythms provide the basis for most techno, it's also prudent to experiment by applying effects to the bass timbre to make it richer sounding. While most effects should be avoided since they tend to spread the sound across the image (which destroys the stereo perspective) small amounts of controlled distortion can help to pull the bass out of the mix or give it a much stronger presence. Similarly, a conservatively applied delay effect can be used to create more complex sounding rhythms.

SOUND EFFECTS AND AUTOMATION

One final aspect of techno is the addition of sound effects, chorded stabs, and occasionally vocals. The sound effects are generated by whatever means necessary, from sampling and contorting any available sound and treating it with effects, processors and EQ. Typical effects here are to time stretch timbres well beyond the normal range and into obscurity, to then slice, re-arrange and further process them with effects and processors. A common effect here is to take a pad sample and time stretch it massively to produce a drawn out almost digitally corrupt timbre and then apply heavy reverb followed by a gate. Creativity on behalf of the producer is the key here, and it's not unusual for a producer to spend a good few weeks simply experimenting with sampled sources and sounds to create interesting sound effects.

Chorded stabs can be created in this same way, although rather than time stretch a pad, it is time compressed. Taking a pad sample that is normally four bars and time compressing it into a 1/16th or 1/8th note is a common technique for producing techno stabs. Similarly, sampling chords such a guitars or horns and time compressing them into shorter notes is a popular technique.

The vocals, if employed, very rarely consist of anything more than a short sample. The verse and chorus structure is avoided and in many cases only very small phrases are used. In some techno, these vocal samples are taken from the old 'speak and spell' machines of the early 1980s. This particular machine isn't a requirement (with its increased use in dance music the second-hand prices of these units have increased considerably), and the same effect can be obtained from most vocoders, so long as the carrier consists of a saw wave and the modulator is robotically spoken.

Ultimately, the key technique for techno is experimentation with effects and processing. Heavy compression, bit-crushers, distortion, saturation, phasers, delay, reverb and automation of all these parameters over any number of bars provides the foundation for this genre of music. Above all, techno should be considered a DJ's tool rather than a record in its own right, and therefore, the music should be produced with this in mind.

There is no one definitive way to produce this genre, and the best way to learn new techniques and production ethics is to actively listen to the current market leaders and experiment with the tools at your disposal. There are no right or wrong ways to approach any piece of music and if it sounds right to the producer, then it usually is. New genres do not evolve from following step-by-step guides or from emulating peers; producers who experiment and push boundaries create them.

Nevertheless, with just these basic elements, it is possible to create the main focus of the music and from here the producer can then look towards creating an arrangement. The theory behind arranging has already been discussed in Chapter 19 but simply listening to the current market leaders mixed will very quickly reveal the current trends in both sounds and arrangement.

The companion files contain an audio example of a techno track using the techniques described.

CHAPTER 26

Mixing Theory

'If you don't actually listen to it, it sounds like a hit'

Andrew Frampton

Mixing electronic dance music should be viewed as the last process in a long chain of careful production ethics. Indeed, one of the main reasons that an EDM mix sounds amateurish is very often not a result of the mixing but a culmination of poor production choices from the beginning of the project.

As I hope the chapters so far in this book have expressed, producing great dance music isn't a result of just picking presets in synthesizers; second guessing the application of processors or effects or accepting second best anywhere in the project. A professional sounding and memorable EDM mix is a result of a knowledgeable approach and an artistic vision supplemented by blood, sweat and tears.

A true artist will spend a seemingly disproportionate amount of time on the smallest and apparently most insignificant of details whether this is micro-tuning a snare or spending many hours to achieve the correct reverb or delay action that the producer has envisioned. A great production also comes together in parts so the drum track should already sound complete, the bass should groove and the leads should cut through the mix. Each individual element is programmed one after the other with further instruments taking into account every other instrument that has come before.

For example, if the drums and bass are already laid down, they should sound just as the producer would imagine them too and any further instruments should be synthesized to 'fit' with the drums and bass without having to rely heavily on the mixing desk. Although a mixing desk can, and often is, used as a creative tool in the production process when it comes to the final mix-down the producer commonly considers it as nothing more than a simple tool to re-position instruments across a virtual soundstage.

Therefore, if the production doesn't sound close to perfect at this late stage, you should return and re-approach the offending parts. Mixing is not an overtly complex process where the magic happens. The magic should already be present and if it isn't, the music simply isn't ready for a mix. At this stage, it is imperative that you do not settle for anything less than the very best.

Perhaps most important of all, though, before the mix is approached, close your eyes, listen to your music and ask yourself; *can you feel it*? Would *you* dance to it? Above the science, the programming, the processing chains, the effects and the arrangement, dance music is ultimately about the vibe and *feel* of the music. As Simon Cowell's freak mangling karaoke TV shows have attested, music can be beautifully produced but with an auto-tuned puppet performing there is no genuine 'feel' to the music. Dance music must exhibit both groove and feel above all else, and if you can't dance to it, there's a good chance no one else will either.

MIXING THEORY

At the very beginning, it is important to understand that mixing is an entirely creative endeavour and as such there are no right or wrong ways to approach it. Unless the producer is completely and utterly tone deaf and has never listened to music – ever – it is difficult to create a complete shambles of a mix. Provided care has been taken over every production aspect of the music, mixing electronic dance music is often little more than positioning of instruments, and therefore, a unique style of mixing, provided that it remains transparent, will define the producers creative style as much as the sounds and arrangement will.

The aim of any mix is quite simply to achieve transparency, so that each instrument occupies its own space within the sound stage and can be clearly heard. To accomplish this, however, requires understanding the theory of creating a mix sound stage, monitoring environments and our own hearing limitations.

As discussed in Chapter 4 on acoustic science, human hearing is fallible. Not only do we perceive different frequencies to be at different volumes but also the volume at which we listen to music will determine the dominant frequencies. For example, listening to music at a conversation level would result in our being more receptive to sounds occupying the mid range and frequencies higher or lower than this must therefore be physically louder in order for us to perceive them to be at the same volume. If, however, the volume of the music is increased beyond normal conversation level, the lower and higher frequencies perceivably become louder than the mid range.

Harvey Fletcher and Wilden Munson from Bell laboratories first measured this inaccurate hearing response in 1933. Using a number of test subjects, they played a series of sine waves at different frequencies through headphones to a listener. For each frequency, the listener was subjected to a secondary reference tone at 1,000 Hz, and this was adjusted in gain until it was perceived to be at the same loudness as the previous sine wave. The results were averaged out and produced the 'Fletcher Munson Contour Control Curve'.

As shown in Figure 26.1, the equal loudness contour is measured in Phons, a unit of measurement developed to express how two sine waves of differing frequencies and different gains can be perceived to be as equally loud as each other. It should be noted that later experiments changed the results found by Fletcher and Munson's work since our hearing behaves differently in headphones than when listening through speakers.

Indeed, the equal loudness curve originally measured by Fletcher and Munson are now considered to be inaccurate since they were conducted entirely through headphones and when we are subjected to sounds in real-life, we suffer from an effect known as head-related transfer function (HRTF).

If a sound source is directly in front rather than at either side as with headphones, both ears receive the same signal at the same intensity until the sound reaches the resonant frequency of our outer ears. Commonly occurring at around 1 kHz, the sonic reflections change as they reach the outer ear and, therefore, both frequency and volume change. This effect is further augmented if the sound occurs off centre since the head will mask and absorb some of the frequencies before it reaches the other ear. This results in a different frequency and gain, and thus perceived loudness.

To further compound the situation, when wearing headphones, there is a difference in pressure created within the ear canal. Whilst this can prove beneficial for listening to lower frequency sounds, as the frequency rises above 500 Hz, the outer ear begins to resonate and the closeness of the headphone cup results in further pressure changes within the ear resulting in a different

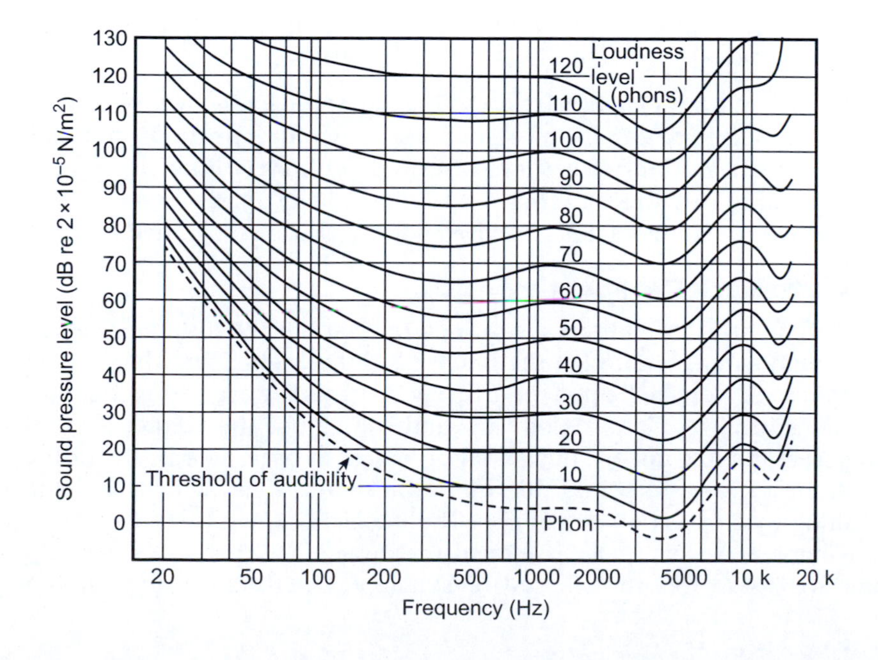

FIGURE 26.1
The equal loudness contour measured by Fletcher Munson

perception of frequencies and their respective loudness. This is one of the reasons as to why it is not recommended to mix whilst relying entirely on headphones since a mix that sounds great on headphones will rarely translate well to a pair of monitor speakers.

Nevertheless, regardless of how the producer does choose to monitor the mix, it is obviously important to consider the equal loudness during the mix. For example, if the bass elements are balanced at a low-monitoring level, then there will be a huge bass increase at higher volumes. Conversely, if mixed at high levels, there will be too little bass at lower volumes.

Mixing appropriately for all three volumes is something of a trade-off as you'll never find the perfect balance for all listening levels, but a generally accepted approach is to mix just above normal conversation level. This reduces the time for ear fatigue to develop from listening too loud and also provides the most proportionately balanced mix.

THE SOUNDSTAGE

Taking our hearing limitations into account is only part of the mix picture, and in order to create a successful mix, the instruments should be placed around a virtual soundstage in order for them all to be heard clearly. In order to do this, it can be useful for the producer to envisage a three-dimensional box – the soundstage – on which the producer will position the various instruments.

Sounds placed on this virtual stage can be positioned anywhere between the left or right 'walls' using the panning parameter, they can be positioned at the front or back, or anywhere in between, using the mixer gain faders and the frequency content of the sound will determine, whether it sits at the top of the stage (high frequencies), the middle (mid range frequencies) or the bottom (low frequencies).

Using this as a template, the concept behind approaching a mix is to ensure that each sound occupies its own unique space within this room, so that it can not only be heard but so that it also fits well with everything else. To do this, we must subdivide the soundstage into three distinct areas; the front to back perspective, the horizontal perspective and the vertical perspective.

Front to Back Perspective – Gain

One of the primary auditory clues, we receive about the distance we are from any sound source is through the intensity of the air pressure that reaches the eardrums. As touched upon in Chapter 4, sound waves spread spherically outwards in all directions from a sound source but the further these waves propagate the less intense the sound becomes. In other words, the further we are from a source of sound, the more the sound waves will have dissipated resulting in a reduction in volume. The intensity of sound and its reduction in volume as it propagates is termed the inverse law that states 'Sound pressure decreases proportionally to the square of the distance from the source'.

Typically, each time the distance from the original sound source doubles, it will become 6 dB quieter.

If this law is interpreted to the context of a mix, the louder an instrument is within the mix, the more upfront or closer to the listener it will appear to be. However, although many dance mixes may appear to have a complete frontal assault on your senses with all the instruments, this is far from the case. Indeed, the depth perception is the first aspect to take into consideration when producing a great mix.

If all instruments were placed at the front of the soundstage, all the volumes would be at equal gain. This would produce a cluttered mix because every instrument would be at the front of the stage fighting to be heard. Moreover, the mix would appear two-dimensional since when listening to music we work on the basis of comparison. That is, for the listener to gain a perception of depth, there must be some sounds in the background so they can determine some sounds are at the foreground.

This means that before even approaching a mix, the producer should have a very good idea of what the central focus of the music should be. Naturally, both the kick and bass will feature heavily at the forefront of the soundstage, since this is what the listeners will dance too but a decision will often have to be made on what other instruments are key to the tracks development. The lead in Euphoric Trance, the pluck in Progressive House and the pitch bending bass in Dubstep or the vocals in Vocal House would all equally be positioned towards the front of the soundstage since these are key elements of the music and should take centre stage with the kick and bass.

Whilst the most obvious method to position these sounds at the front would be to increase the gain over other instruments, this is not the only way and often better results can be achieved by taking advantage of acoustic science. As touched upon in Chapter 4, as frequency increases its wavelength becomes shorter, and therefore, it is safe to assume that if a high-frequency sound has to travel over a long distance, there will be a reduction in high-frequency detail. This effect is particularly evident on cars fitted with large-component stereo systems. You can hear the low frequencies while the car is at a distance but as it approaches the higher frequencies become more pronounced until it passes by, whereby the high frequencies begin to decay again.

This effect can be emulated with some careful application of EQ. By applying cuts of a decibel or two at the higher frequency ranges of a sound, it could be perceived to be more distant when listened in association to other sounds with low-frequency content. Alternatively, by increasing the higher frequencies through EQ, enhancers or exciters, a timbre can appear more upfront.

This frequency-dependent effect is an important aspect to consider when working with a compressor. If a fast attack is employed on a timbre, the compressor may clamp down too heavily on the transient and, thus, reduce its high-frequency content resulting the sound being repositioned in the mix.

Generally, the best way to avoid this behaviour is to employ a multi-band compressor. This is a compressor that can be adjusted to work on specific or multiple bands and could be adjusted, so that only the lower frequencies are compressed leaving the higher frequency unmolested.

Another characteristic of our perception of depth and distance is determined by the amount of reverberation, the signal has associated with it. As discussed in earlier chapters, any natural sound will exhibit reverberation as the sound waves propagate and reflections occur from surrounding surfaces. However, the amount of reverberation and the stereo width of these reflections are also dependent on how far away the sound source is from the listener.

If a sound is a good distance from the listener then the stereo width of the reverberations will dissipate as they travel through the air, however, they will equally be subjected to more reverberation. This is an important principle to consider when applying reverb since many novice producers will wash a sound in stereo reverb to create depth in the mix and then wonder why it doesn't sound 'right' in context with the rest of the mix.

Therefore, if the producer wishes to use reverb to place an instrument towards the rear of a mix perspective, it's good practice to employ a mono reverb signal with a long tail. In addition, this could be followed with an EQ that could then be used to reduce some of the higher frequency content of both reverberation and original signal. This will emulate the natural response, we would expect from the real world, even if that particular timbre doesn't occur in the real world.

Naturally, this means that sounds that are positioned to the front of a mix will have little or no reverb associated with them, but in many electronic dance records, it is typical to apply reverb to add character and provide a timbre with more presence. In this instance, it should be applied in stereo but the width should be controlled carefully to prevent it from occupying too much of the left and right perspective. Similarly, it should employ a pre-delay of approximately 50 ms to 90 ms to separate the timbre from the effect. This prevents the first reflections from washing over the transient and muddying the sound or pushing it towards the rear of the soundstage. Always bear in mind that applying any effects too heavily can make sounds difficult to localize and in order to accomplish a good mix each instrument should be pinpointed to a specific area.

Horizontal Perspective – Panning

The next consideration of the soundstage is the horizontal plane, the distance and positioning of the sounds between the left and right walls of the virtual room. The major aural clues that help us derive the impression of stereo placement are attributed to the volume intensity between sounds and their respective timing.

The theory behind altering the volume of sounds in order to produce a stereo image was first discovered by Alan Blumlein in the early 1930s. An inventor at EMI's Central Research Laboratories, Blumlein researched the various ways in which the ears detect the direction of a sounds source. Along with inventing a technique to permit the creation of a stereo image in gramophone records, Blumlein also theorized that to maintain realism in a film the sound should follow the moving image.

The technique was first employed in Walt Disney's Fantasia, when sound engineers asked Harvey Fletcher (of the very same equal loudness curve) if he could create the impression of sound moving from left to right for the movie. Drawing on Alan Blumlein's previous work, Fletcher concluded that if a sound source were gradually faded in volume from the left speaker and increased in the right, it would produce the effect of sound in motion. The engineers at Disney put this theory into practice by constructing a potentiometer, a volume control that varied the volume between two speakers. This was later termed the 'Panoramic Potentiometer', resulting in the now ubiquitous pan pots that feature on every mixing desks channel.

Although the volume intensity difference between two speakers is still the most common method for panning a sound across the stereo image during mixing there are further processes available. First, we can receive further directional clues from the timing between sounds. Known as the Directional Cues, Precedence or Haas effect, this process takes advantage of the Law of the First Wave Front that states '*If two coherent sound waves are separated in time by intervals of less than 30 milliseconds the first signal to reach our ears will provide the directional information*'. Therefore, if a direct sound reaches the ears anywhere up to 30 ms before the subsequent reflections, it is possible to determine the position of a sound.

For example, if you were facing the central position of the mix, any sound leaving the left speaker would be delayed in reaching the right ear and vice versa for the left ear. This is an effect known as interaural time delay (ITD). Considering that sound travels at approximately 340 m/s, it is possible to emulated this same effects by employing delay unit on a mono signal and delaying it by a couple of milliseconds. This would produce a very similar impression to actual panning of a sound source.

However, for this effect to work accurately, we must consider that our ears are on the side of our heads, and therefore, the head gets in the way of the frequencies from opposite speakers. This effect, known as Head Related Transfer Function, means that, provided you are facing the centre of the stereo image, some of the higher frequencies emanating from the left speaker will be reduced before they reach the right ear because the sound cannot travel through the heads and must, therefore, travel around it. Placing a cut of 1 dB at approximately 8 kHz on the delayed signal could simulate this HRTF effect.

Naturally, to accomplish these effects, the producer must employ mono sounds within the mix, and this is where many novice engineers trip up. Almost every virtual instrument, hardware instruments and samples are in stereo not mono, but employing pure stereo signals on every channel within an arrangement will often lead to a mix that lacks definition. If an arrangement is constructed completely from stereo files, they will collate together in the mix and occupy the same area and position of the soundstage. Whilst the producer could work to narrow the stereo width of each file with pan pots, this approach is not particularly suitable for the creation of a transparent mix.

The soundstage for any mix should be transparent, so that it is possible to aurally picture the mix in three dimensions and pinpoint the position of each instrument. Since many electronic dance mixes are particularly busy encompassing anything from 8 to 15 different instruments and risers occurring simultaneously they each need a specific position within the mix. If each were a stereo file, it would be impossible to locate a pan placement for each instrument and often results in disaster as the novice applies unnecessary EQ and creeps the mixers faders.

Creeping mix faders is a typical approach of most novice engineers and is the result of gradually increasing the volume of each channel, so that it can be heard above other instruments. For example, the engineer may increase the gain of the vocals, so they stand above the cymbals, hi-hats and snare, but this results in the these drums moving towards the rear of the soundstage, so their volume in increased. This results in the snares, hats and cymbals being too loud compared against the kick, so the kick is increased in gain and then the bass is increased so that it shares a volume relationship with the kick. Then the vocals are increased again. Eventually the mixer is pushed to its maximum headroom and the mix turns into loud, incomprehensible noise that lacks any sonic definition.

This approach can be avoided through careful selection of stereo and mono signals and channels. The choice of what channels should be mono and stereo is down to the producer's own creativity, but typically only the main lead elements of the mix are often stereo whilst most of the percussion remains as mono. For example, in euphoric trance, it is not unusual for the chords and main lead to be in stereo, while all other elements remain in mono. For Dubstep, Deep house and any bass-driven music, it is the bass that is commonly in stereo. It's also not uncommon to use stereo for closed hi-hats, some snares and claps and also any arrangement effects such as risers and falls.

Although the kick drum is commonly a driving force in dance music, this should nonetheless always remain as a mono file. This is because it's in our nature to face the loudest, most energetic part of a mix. If the kick is a stereo image, it is equally dispersed across the left and right soundstage, decreasing the central energy of the mix and reducing impact. In mono, the kick can be placed central to the mix, and this makes it easier to perceive the position of other sounds surrounding it more clearly and enhances the overall sense of space within the mix.

Notably, when working with mono files and panning, we perceive the volume of a sound by its positioning within the soundstage. This means that if a mono sound is placed centrally it will be perceived to anything from 3 dB to 6 dB louder than if positioned in the left or right field. Consequently, some workstation mixers will implement the panning law so that sounds that are placed centrally are subjected to 3 dB or 6 dB of attenuation. The amount of decibels is a central file is reduced by is often configurable by the engineer but not all software implements this, so after panning the producer may need to readjust the respective volume of the panned instrument again.

It should also be noted here that in order to be able to position any instruments within a mix with any degree of accuracy, the speaker system should be configured correctly in relation to the producers monitoring position. That is, the producer should be positioned at one point equilateral triangle with the loudspeakers positioned at the other two points. This means that speakers should be positioned an equal distance apart from one another *and* the listening position to ensure that the signal from each monitor speaker reaches your ears at the same time.

Even a spacing as small as 25 cm difference between these three points can results in the sound from one monitor speaker being delayed in reaching the ears by a couple of milliseconds that can result in the stereo image moving to the left or right. Of course, a simple solution to this would be to monitor via headphones, but as before this can actually introduce problems with the soundstage. Since each headphone rests of the ear, it overstates the left and right perspective because sound from the right speaker never reaches the left ear and vice versa. This can make it very easy to overstate the stereo field whilst mixing.

With a good monitoring position, panning provides the easiest and cleanest method for a producer to create space for two instruments that share the same frequency range. By panning one instrument slightly to the left and the other to the right (or panning one further to the left and right and the other remaining central), each sound can be awarded its own space in the soundstage allowing both to be heard clearly.

The Vertical Perspective – EQ

The final perspective is the top to bottom of the soundstage. Here, higher frequency elements sit towards the top of the soundstage whilst lower frequencies lean towards the bottom. Naturally, much of this vertical positioning will already be dictated by the timbres, the producer has already employed in the mix. For example, basses by their nature are low-frequency instruments and, therefore, will sit towards the bottom of the soundstage whilst high hats will gravitate towards the top. However, all timbres will contain frequencies that do not necessarily contribute to the sound when placed into the context of a mix.

As discussed in Chapter 4, all sounds we hear are made up of any number of frequencies that are a direct result of the fundamental and its associated harmonics occurring at differing amplitudes. Also, it is this predetermined mix of harmonic frequencies that helps us establish the timbre of the sound we perceive, be it a piano, a synthesized bass or a kick drum.

When any of these sounds are in isolation, we require all the harmonic frequencies in order to determine the signature of the sound and identify it specifically as a piano or bass and so on. However, when a number of instruments are in the context of a mix, the harmonic frequencies from each of the different instruments overlap one another, and these are summed together to increase the volume at each specific frequency.

This summing of harmonic frequencies produces instrument signatures that exhibit uneven harmonic structures and results in a mix with instruments that sound muddy and cluttered. For example, if the lower frequencies of a piano were unnaturally increased through the upper harmonics of a bass summing over the frequencies of a piano, the piano would appear bottom heavy and poorly recorded.

If, however, the producer is able to identify the frequencies at which this layering occurs, either the bass *or* the piano could have the summed frequency area removed with careful application of EQ. This would produce a clearer sounding mix since it results in an effect termed 'frequency masking' whereby we perceive that both instruments have all the harmonic frequencies present to represent the complete timbre; they're just hiding behind the other instrument. As simple as this premise may appear, however, its practical application is a little more complicated.

First, in the example of the summing frequencies from a bass and piano, if both instruments are melodious and change in pitch, the frequencies they sum at will also continually change. This makes precise application of EQ difficult, if not impossible, and therefore, the producer often has to settle for a best-case scenario, and this can only ever be a judgment call based on previous experience. Second, it also means that the producer must have well-trained ears to be able to hear and identify the problem areas.

Figure 26.2 shows one of the many frequency charts available on the Internet that promise to help the engineer understand and identify frequency ranges. However, whilst this is suitable to an extent, it cannot be reliably transferred to an electronic dance mix since these rely heavily on not only synthesized timbres but also effects and processing that heavily modify the original timbre. For example, distortion is a popular effect to employ on bass in a dance mix but application of distortion will seriously affect the frequency range of the timbre rendering the chart useless. Consequently, rather than rely on a

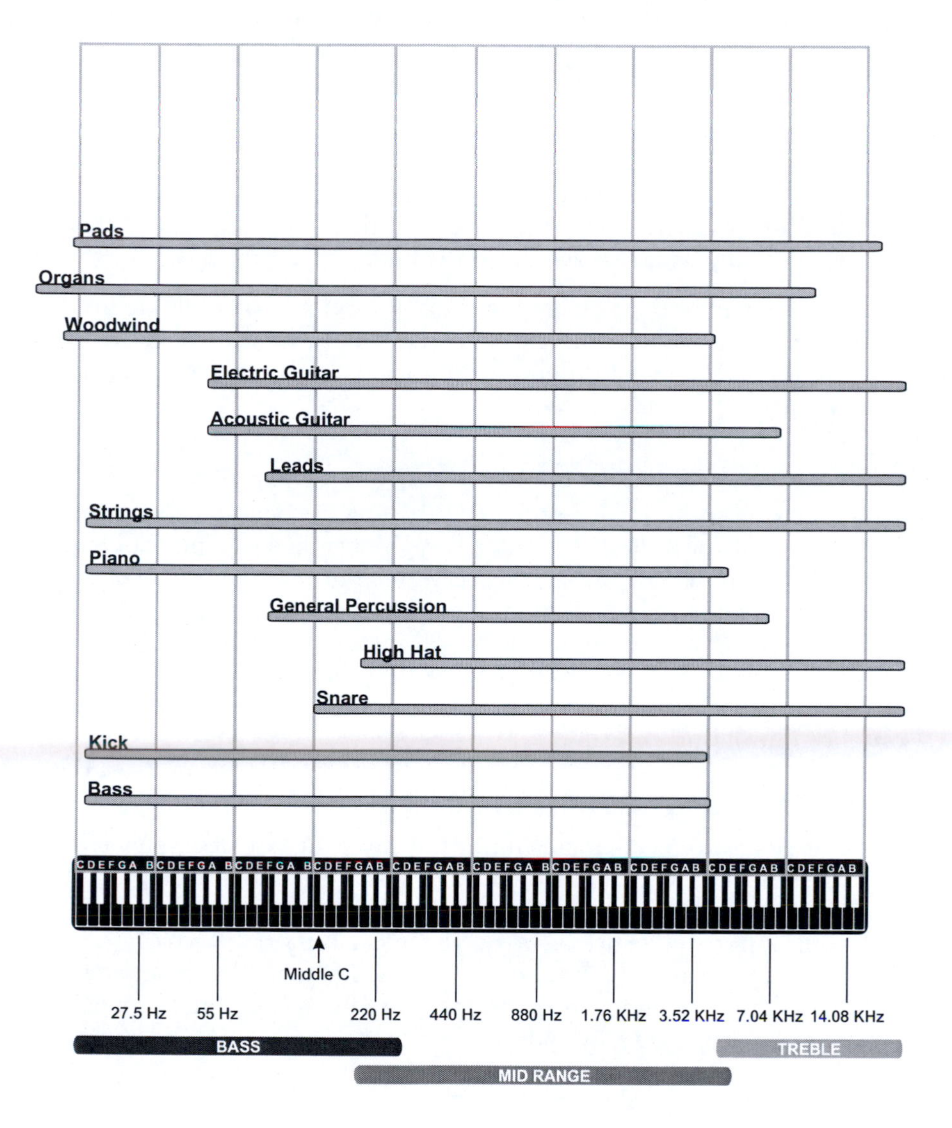

FIGURE 26.2 Frequency range of instruments

frequency range of instruments, the dance producer should instead rely on the seven EQ octaves.

With the EQ octave chart, the producer does not learn the frequency ranges that can be taken up by specific instruments but instead learns the effects that different frequency ranges can have on instruments and a mix.

The First Octave – 20 Hz to 80 Hz

At frequencies lower than 40 Hz, it is difficult to determine pitch but nonetheless is commonly occupied by the lowest area of a kick drum and some bass instruments. The engineer should ideally avoid applying any boosts below 40 Hz and instead concentrate around the 50 Hz to 65 Hz area.

Small boosts here on some instruments can help to define the bass or kick drum and provide more sub energy to the mix. Cuts applied here are typically to the bass instrument to make space for the kick drum. The bass is typically cut instead of the kick because the kick does not change pitch and therefore frequency-masking issues are avoided if the bass is melodious.

When working in this octave range, it is important to listen on multiple systems, both large and small, for inconsistencies in the energy. The producer should also listen at differing volumes to compensate for the Contour Control Curve.

The Second Octave – 80 Hz to 250 Hz

This is where a proportionate amount of the low-frequency energy from most instruments and vocals lies. This range is typically adjusted when applying the 'bass boost' on most home stereos and is where much of the energy of the bass instrument resides. Areas to concentrate with boosts and cuts are typically at 120 Hz to 180 Hz. Boosts applied here will increase the fatness of most dance instruments whilst cuts can be applied to thin them out. As before, the producer should also listen to this area at differing volumes to compensate for the Contour Control Curve.

The Third Octave – 250 Hz to 600 Hz

This frequency range is the main culprit for mixes that are described as sounding muddy or ill-defined and is often the cause of mix that is fatiguing to the ear. If too many harmonic frequencies clash in this area, instruments and vocals will appear indistinct or boxy. The most usual EQ focus in the third octave is at 300 Hz and 400 Hz. Boosts at these frequencies can help define an instruments presence and clarity, and the body of vocals whilst cuts will help in reducing boxy or muddy sounds.

The Fourth Octave – 600 Hz to 2 kHz

The fourth octave can also be attributed to an instruments presence and small boosts at 1.5 kHz can often be used to increase the attack and body of some instruments. However, it is far more likely that this area is cut rather than boosted since this range can be attributed to a honky or nasal sound. Typically, the central focus for EQ cuts is at 800 Hz, 1 kHz and 2 kHz.

The Fifth Octave – 2 kHz to 4 kHz

The fifth octave is often expressed as the 'dance producer's transient octave' since the attack of most of the rhythmical elements of the mix reside in this area. The focus for EQ cuts and boosts are commonly focused between 2.5 kHz and 3 kHz. Boost applied here can increase the projection and transients of most percussive elements, including the kick drum's 'bite', whilst cuts will often push instruments towards the rear of the sound stage.

The Sixth Octave – 4 kHz to 7 kHz

The sixth octave is the distinction range for most instruments. EQ is typically centred on 5 kHz with boosts employed to increase air and sonic features of the timbre whilst cuts can be employed to reduce sonic harshness.

The Seventh Octave – 7 kHz to 20 kHz

The seventh octave often contains the higher frequency elements of cymbals and hi-hats. Some engineers will apply a small shelving boost at 12,000 Hz to make the music appear more hi-fidelity to increase detail and sheen without introducing any aural fatigue. The commonly adjusted frequencies with EQ here reside at 8 kHz, 10 kHz and 15 kHz. Boosts applied at these frequencies can often increase a timbres clarity or air whilst cuts can be employed to remove harsh resonances from instruments alongside vocal sibilance.

CHAPTER 27

Practical Mixing

'I can't wait to hear it with a real mix instead of the douche mix'.

John Kalodner

Armed with the theory of mixing, you can approach a mix on a more practical level, but before a mix is approached, you should first ensure that both the listening environment and the speaker monitors are up to the job. Indeed, the first step to creating any great mix isn't down to the effects or type of EQ or compression employed, but to ensure that you can actually hear what's going on in the mix. After all, if the room or the monitors are not providing a true impression, then you can't mix reliably. Both the room you monitor in and the loudspeaker monitors must be dependable.

The problem with monitoring a mix in a typical room is that every object and surface in that room will influence the sound from the monitors. In fact, the effect a room can have on sound can be compared to the example of wave propagation that was discussed in Chapter 4, but rather than drop a pebble into a pond, imagine dropping it into a fish tank.

In this situation, the waves would propagate outwards, but as soon as they strike the glass sides of the fish tank, they will reflect back in different directions resulting in a much larger disturbance of the water. This same effect happens with the multiple frequencies from a loudspeaker. The frequencies emanating from the speaker monitors strike the walls and then reflect back. Where these reflected waves meet further waves from the speakers there will be a peak or dip in those particular frequencies depending on whether the wave fronts are in phase with one another or 180° out of phase.

To reliably monitor and perform a mix-down, a room should be properly treated to prevent any wayward acoustics from influencing your decision. The problem, however, is that acoustic treatment of a room is a complex and multi-faceted subject. It's certainly a subject that could digest any number of books, and even then, you cannot rely on the theory alone since the formulas for calculating the problem 'modes' within a room are based on the walls, floors and ceilings having specific acoustic properties. Consequently, what follows should be considered a generalization of how to reduce the most common modal problems.

Dealing with reflections consists of more than simply sticking a few egg crates or egg boxes to your walls. Indeed, just randomly placing acoustic panels, foam panels or egg boxes around a room will likely make the situation worse, not better. Therefore, if you plan to place acoustic treatment tiles on the walls, it is strongly recommended you either consult a professional or heavily research the topic. However, before contemplating covering the walls, a number of problems can be reduced with a little forethought.

First, you should consider covering or removing any hard reflective surfaces from the room. If the room you monitor in has large windows, the glass will be highly reflective, and therefore, it is advisable to cover them with large heavy curtains. The heavier these curtains are the better, since they will absorb and reduce the reflections retuning to the room. Similarly, any mirrors or pictures on the walls should be removed, and if you have a hardwood or laminate floor, this should be covered with a large rug.

Small rooms are considered to have less modal problems than larger rooms, since with smaller areas, the lower frequencies do not have the time properly develop for reflections, whereas larger rooms are more likely to suffer from effects such as standing waves. This means the larger the room, the more acoustic treatment will be required for the walls.

The positioning of the loudspeaker monitors is equally important since this can have an equally damaging effect on the frequencies you hear. Placing any monitors close to a corner or wall will increase the perceived bass by 3, 6 or 9 dB due to the boundary effect. This is similar to the boundary effect that's exhibited when singing too close to a microphone since there is an artificial rise in the bass frequencies.

Typically, monitor speakers should be positioned at least 12 inches away from any hard surfaces, and if they're positioned on a desk, should be isolated with absorption panels such as Auralex or better still the Primacoustic recoil stabilizer. This is because as the speaker monitor vibrates with sound, the vibrations are transferred onto the desk. The effect of this can be compared to a piano's sounding board; the listener will hear a combination of sound from the speaker monitors and the structural resonance of the desk that results in a boost of low to mid frequencies. This is an effect termed *vibration loading*.

On the subject of monitors, you should also use loudspeaker studio monitors to scrutinize a mix rather than place trust in everyday hi-fi speakers. No matter how 'hyped' hi-fi speakers are for exhibiting an excellent frequency response, they will deliberately employ frequency-specific boosts and cuts throughout the spectrum to make them appear full and rounded.

Consequently, if you rely entirely on these while mixing, you'll produce mixes that will not translate well on any other hi-fi system. For instance, if your

FIGURE 27.1
Primacoustic recoil stabilizer

particular choice of hi-fi speakers feature a bass boost at 70 Hz and a cut in the mid range, mix choices will be based on this response, and if it's then played on a system that has a flatter response, there will be less bass and an increased mid range.

It's impossible to recommend any particular make or model of studio monitor, because we each interpret sound slightly differently and different monitors will produce a different sonic result in different rooms. However, as a general guideline, it is advisable to aim for a model that makes any professionally produced music sound moderately lifeless. If you can make your mixes sound good on these, then it's likely to sound great on any other system.

Of course, room modifications, acoustic treatment and monitoring through professional speaker monitors may not all be possible, and if not, it is strongly recommended that you listen to as many commercial mixes you can to train your ears to understand the response they produce. In fact, even in professional-treated rooms with an excellent pair of speaker monitors, professional artists will still take time to associate and familiarize themselves with the sound produced. We hear music everyday but we rarely truly listen, and if you wish to become competent at mixing, you must differentiate between hearing and listening.

This involves listening to any number of commercial mixes at a moderate volume (just above conversation level) and learning to identify frequencies and mix techniques employed by different engineers. A good way to accomplish this is to listen intently to a mix and ask yourself question such as:

- *Is there any pitch bend applied to any notes?*
- *Does the drum loop stay the same throughout the track or does it change in any way?*

- *What pattern are the hi-hats playing?*
- *Do they remain constant throughout?*
- *What effects have been used?*
- *Where is each instrument placed in the mix?*

Although many of these questions may seem unrelated to mixing techniques and frequency relationship, listening closely for small nuances in drum loops, hi-hat patterns and basses will train your ears into listening closely. This approach will not only help in identifying frequencies but also increase awareness of arrangement and programming techniques.

Perhaps most important of all, however, and despite the huge array of audio tools available today, the most significant ones are those that are stuck to the side of your head. Continually monitoring mixes at excessive volume or consistently attending festivals, raves, parties and clubs without ear protection will damage the ears sensitivity to certain frequencies. A survey revealed that those in their late teens and early twenties who constantly listen to loud music had more hearing irregularities than people twice their age. Also, bear in mind that dabbling in pharmaceuticals for recreational use leads to a heightened sensory perception. Although they may have played a significant role in the development of the dance scene, they don't help in the studios.

APPROACHING A MIX

The first stage of mixing commonly consists of removing any unwanted frequencies from each instrument track. Whilst all the harmonic frequencies are necessary to produce the timbre in isolation, in the context of a mix, many of the upper and lower frequencies can be removed with a shelving filter.

For example, a hi-hat will often feature frequencies as low as 400 Hz and as high as 20 kHz. Since a hi-hat is a high-mid- to high-frequency instrument, much of its main energy is often located between 800 Hz and 10 kHz, and therefore, any frequencies above or below this do not necessarily contribute to the timbre.

Knowing which frequencies contribute to the timbre and those that do not is down to experience and ear training but if unsure, a spectral analyzer inserted onto the track will reveal where the main energy of the timbre lies.

As shown in Figure 27.2, the bass timbres main energy resides between 100 and 900 Hz. Whilst there are frequencies above and below this, they do not contribute to the energy of the timbre and, therefore, can be removed with a shelving filter. By doing so, it creates more space within the mix for other instruments to sit comfortably.

Once the unwanted/unrequired frequencies have been removed, the engineer typically begins by setting the relative volume levels of each track. Here, you must take the musical style into account and identify the defining elements

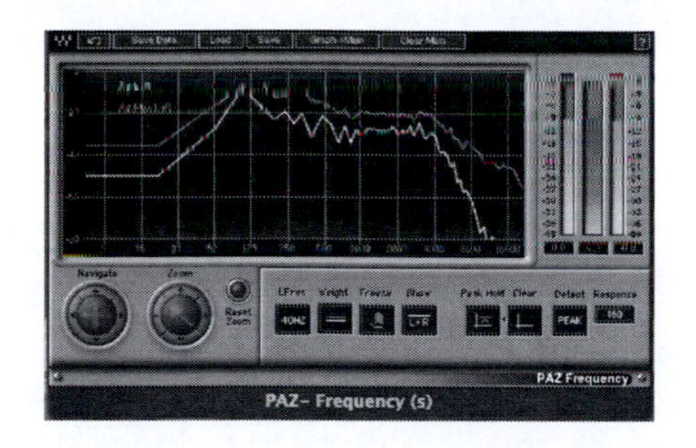

FIGURE 27.2
Spectral analysis of a bass timbre

of the genre that you want at the forefront of the soundstage. Typically, for electronic dance music, this means that the drums and bass should sit at the front alongside the tracks main lead elements or vocal track. Typically, these should be mixed first as they are the defining points of the music, so a good technique here is to mute all channels bar the entire drum section and commence by mixing these first.

As the kick drum should be the most prominent part of the music, this should be adjusted to approximately –18 dBFS (decibel full scale) on the mixer metres and then the snare, claps, hi-hats and percussion instruments can be introduced at their relative levels to the kick drum. If the music appears quiet at this gain level, you can always increase the output volume on the monitors, amplifier or audio interface but you should avoid increasing the gain of the kick drum channel above –18 dBFS.

As discussed in a pervious chapter on constructing loops, the drum section should be mostly mixed by this stage and most, if not all, of the effects will already be applied to the drums beforehand. Consequently, they may just require sending for small amounts of reverb to place them all in the same 'area' for the mix, but if further effects are required at this stage, they should be applied now. Similarly, any processing such as side compression, noise gates and limiting, or transient designers should also be applied.

Typically, effects and processing are applied during the sound design and arrangement process, so there is little requirement to introduce further effects beyond the application of reverb. However, during a mix, the engineer may make more creative decisions and choose to apply further effects. Here, however, it is important to bear in mind that empty spaces in a mix do not have to be filled with reverb or echo decays.

When it comes to effects and mixing, less is invariably always more. Any effects, and in particular reverb, can quickly clutter up a mix resulting in a loss of clarity. Since there will probably already be plenty going on, adding further effects will often make the mix busier, and it's important to maintain some space in between the individual instruments.

A common mistake is to employ effects on every instrument when in reality only one or two may be needed throughout. Therefore, before applying any effects, it's prudent to ask yourself why you're applying them, is it to enhance the mix or make a poorly programmed timbre sound better. If it's the latter, then you should look towards using an alternative timbre rather than try to disguise it with effects.

After volume level and effects, panning should be introduced to the drum instruments too, positioning each instrument into its appropriate position. This is where the engineer's creativity and scrutinizing of previous mixes from the same genre come into play as there are no 'strict' rules for creating a good soundstage on drums and they should be positioned where the engineer feels they are most suited. However, what follows are some recommended starting positions for panning of these instruments.

Kick: Positioned central so that both speakers share the energy.
Snare or clap: Positioned from 1 o'clock to 2 o'clock or central.
Hi-hats: Positioned from 1 o'clock to 3 o'clock or at the far left of the soundstage with a delayed version in the right.
Cymbals: Positioned central or from 1 o'clock to 3 o'.
Further percussion: Positioned so that the different timbres are in the left and/or right of the soundstage.

At this point, the engineer shouldn't be overly concerned with the application of EQ, and the prevention of masking to clearly hear each instrument should be attained with effects, panning and volume adjustments instead. If EQ is applied at this early stage, it is possible that the engineer may force instruments into the frequency ranges required by further instrumentation; hence, the purpose should be to accomplish the best possible results without resorting to equalization. If you're experiencing trouble adjusting to the best gain and panning levels, then a good technique is to switch the mix to mono as this allows you to hear the relative volume levels more accurately.

Once the drums have been mixed, it is common practice to assign them to a sub-group so that the entire drum sub-mix can be controlled via single gain fader. What's more, with all the drums on a single group channel, it is possible to apply further effects and processing to the drums as a complete unit if deemed necessary later in the production process.

After the drums have been mixed, the engineer commonly moves through the instruments, introducing each instrument in terms of gain whilst also panning them to their respective locations in terms of mix priority. Typically, this means introducing the bass first so that it can be mixed with the drums to provide the groove, and then following this with the lead instrument and finally, pads, strings and any sound effects.

In regards to panning these instruments, the kick, bass and lead instruments (or vocal) are often placed at centre stage. This is particularly important for the bass instrument since it permits both speakers to share the low-frequency energy that will increase the bass presence, but all further instrumentation is open to interpretation and creative input. Indeed, the best solution to panning instruments is to have a good reason as to why you're placing a timbre there in the first place. It is advisable to always have a plan and reason for positioning and not to just throw sounds around the image regardless.

Panning starting points for instruments:

- Bass: Low energy should be positioned central so that both speakers share the energy.
- Vocals: Often positioned central since you always expect the vocalist to be centre stage but sometimes they are double tracked and applied to both left and right of the stage too.
- Backing vocals: Occasionally stereo, so they're spread from 2 o'clock to 4 o'clock, but if mono, they are often placed to the left or right of the soundstage.
- Synth leads: These are often in stereo and therefore positioned fully left and right with perhaps an additional mono version positioned centrally.
- Synthesized strings: Positioned at 4 o'clock or stereo spread at 9 o'clock and 3 o'clock.
- Guitars: Positioned at 3 o'clock.
- Pianos: Commonly stereo with the high notes in the left speaker and low notes in the right.
- Wind instruments: Positioned in the left or right speaker (or both) at 3 o'clock and 9 o'clock.
- Sound effects: Wherever there is space left in the mix!

In direct contradiction to this list, in many dance mixes, the positioning of instruments is rarely natural so you should feel free to experiment. In fact, in many cases, it may be important to exaggerate the positions of each instrument to make a mix appear clearer and more defined. However, you should avoid simply positioning a sound in a different area of the soundstage to avoid any small frequency clashes with other instruments. In these circumstances, the engineer typically resorts to EQ, and only if there are still masking problems, should it be panned further.

This form of 'additive' mixing by introducing instruments step by step is the most common form of mixing for many engineers and dance musicians but it is not the only approach available. I have worked with some musicians who prefer a more subtractive approach by listening to all instruments and then simply balancing levels and panning until they are happy with the positioning and sound. Both approaches are perfectly valid provided the final mix projects clarity.

EQ

After positioning the sounds in the soundstage with volume and panning, EQ will be often required to correct any problematic frequencies and produce a clearer mix.

As discussed in Chapter 4, the fundamental and subsequent harmonics contribute to creation and timbre of any particular sound, but if two timbres of similar frequencies are mixed together, some of the harmonic content is 'masked' resulting in the instruments disappearing behind one another and creating a boost in frequencies at that point. This 'frequency masking' results in a mix that sounds cluttered and ill defined.

To prevent this from occurring, the engineer will focus on the masked area on one of the instruments with EQ to cut the problematic frequencies. However, this requires the engineer to able to identify and locate these problematic areas before they can be repaired. This is harder than you may initially realize as learning where the frequency masking is taking place requires ear training and experience.

Having said that, if you believe that two instruments are masking, a good technique is to mono both, set them to unity gain and then pan one instrument from the left and then move it slowly all the way to the right of the soundstage. If you can hear the sound moving clearly from the left speaker through to the centre of the mix and then off to right, it's unlikely that there are any conflicts. However, if the signal disappears when drawing close to the position of the second instrument, there is masking occurring.

Once masking has been identified, you will need to tune into the masked frequencies and remove them with an EQ unit. One of best techniques for identifying problematic frequencies is to set the EQ unit to a low Q and apply a large boost of 6 to 8 dB. With this boost applied to the instrument, begin to sweep through the frequencies and make a note of each frequency you come across that make your hair stand on end or appear to share similarities with the secondary instrument that the masking is occurring with. These frequencies can then be removed with either low of high cuts.

This, however, must be applied with caution. As discussed in the Chapter 26, if one or both of the instruments are melodious, then the offending frequency may change as the track plays, and in this instance, you will have to work with the best-case scenario, hoping to reduce it as much as possible.

Generally, with the application of EQ, the engineer should always look to cut first to repair any problematic frequencies rather than boost. This is because cuts will always sound more natural to our ears since they are used to hearing a reduction in frequencies rather than due to acoustic absorption of walls, objects and materials. Furthermore, any boosts in EQ will increase the gain of those particular frequencies that may result in clipping the signal of the audio channel. Having said that, however, any great mix is a mix of both cuts and boosts, since small-frequency boosts applied to instruments can increase the character of that instrument and help it stand in the mix.

What follows is a rough guide to EQ and to the frequencies that you may need to adjust to avoid masking problems or provide an instrument with character. Note, however, that it is impossible to discuss the approach to every possible sound within a dance mix since the possibilities are limitless.

Drum kick

A typical EDM drum kick consists of two major components: the attack and the low-frequency impact. The attack usually resides around 3 to 6 kHz and the low-end impact commonly resides between 40 and 120 Hz. If the kick seems very low without a prominent attack stage, then its worthwhile setting a very high Q and a large gain reduction to create a notch filter. Once created, use the frequency control to sweep around 3 to 6 kHz and place a cut just below the attack as this has the effect of increasing the frequencies located directly above.

If this approach doesn't produce the results, set a very thin Q as before, but this time apply 5 dB of gain and sweep the frequencies again to see whether this helps it pull out of the mix. Taking this latter approach can push the track into distortion, so you may need to reduce the gain of the channel if necessary.

If the kicks attack is prominent but it doesn't seem to have any 'punch', the low-frequency energy may be missing. You can try small gain boosts around 40 to 120 Hz, but this will rarely produce the timbre and the problem more likely resides with the drum timbre. A fast attack on a compressor may help to introduce more punch as it'll clamp down on the transient, reducing the high-frequency content, but this could change the perception of the mix so it may be more prudent to replace the kick.

Snare and Claps

The lower frequencies of both snares and claps will tend to cloud a mix and are not necessary, so the first step should be to employ a shelving (or high-pass) filter to remove all the frequency content below 150 Hz.

The `snap' of most snares and claps usually resides around 2 kHz to 10 kHz, while the main body can reside anywhere between 400 Hz and 1 kHz. Applying cuts or boosts and sweeping between these ranges should help you find the elements that you need to bring out or remove, but roughly speaking cuts at 400 and 800 Hz will help it sit better while a small decibel boost (or a notch cut before) at 8 or 10 kHz will help to brighten it's 'snap'.

High-Hats and Cymbals

Obviously, these instruments contain very little low-end information that's of any use, and if left, it can muddle the mid range. Consequently, they benefit from a high-pass filter to remove all the frequencies below 300 Hz.

Typically, the presence of these instruments lies between 1 and 6 kHz, while the brightness can reside as high 8 to 12 kHz. A shelving filter set to boost all frequencies above 8 kHz can bring out the brightness, but it's advisable to roll off all frequencies above 15 kHz at the same time to prevent any hiss from breaking through into the track. If there is a lack of presence, then small decibel boosts with a Q of about an octave at 600 Hz should add some presence.

Toms, Congas and General Percussion

Both these instruments have frequencies as low as 100 Hz but are not required to recognize the sound and if left can muddle the low and low-mid range, thus it's advisable to shelve off all frequencies below 200 Hz. This should not require any boosts in the mix as they rarely play such a large part in a loop, but a Q of approximately half an octave applied between 300 and 800 Hz can often increase the higher end making them appear more significant.

Bass

Bass is the most difficult instrument to fit into any dance mix since it's interaction with the kick drum produces much of the essential groove but it can be fraught with problems.

The main problem with mixing bass can derive from the choice of timbres and the arrangement of the mix. While dance music, by its nature, exhibits plenty of presence in the groove, this is not attributed to using big, exuberant, harmonically rich sounds throughout. The human mind and ear work under the principle of contrast, so for one sound to appear big, the rest should be smaller. Of course, this presents a problem if you're working with a large kick and large bass as the two occupy similar frequencies that can result in a muddied bottom end.

This can be particularly evident if the bass notes are long as there will be little or no low-frequency 'silence' between the bass and kicks making it difficult for the listener to perceive a difference between the two sounds. Consequently, if the genre requires a huge, deep bass timbre, the kick should be made tighter by rolling off some of the conflicting lower frequencies, and the higher frequency elements should be boosted with EQ to make it appear 'snappy'. Alternatively, if the kick should be felt in the chest, the bass can be made lighter by rolling off the conflicting lower frequencies and boosting the higher elements.

Naturally, there will be occasions whereby you need both heavy kick and bass elements in the mix, and in this instance, the arrangement should be configured so that the bass and kick do not occur at the same point in time. In fact, most genres of dance will employ this technique by offsetting the bass so that it occurs on the offbeat.

If this isn't a feasible solution and both bass and kick must sit on the same beat, then you will have to resort to aggressive EQ adjustments on the bass. Similar to most instruments in dance music, we have no expectations of how a bass should actually sound, so if it's overlapping with the kick making for a muddy bottom end, you shouldn't be afraid to make some forceful tonal adjustments.

Typically for synthetic instruments, small decibel boosts with a thin Q at 60 to 80 Hz will fatten up a wimpy bass that's hiding behind the kick. If the bass still appears weak after these boosts, you should look towards replacing the timbre since its dangerous practice to boost frequencies below these as it's impossible to accurately judge frequencies any lower on the more common near-field speaker monitors and subwoofers. In fact, for accurate playback on most hi-fi systems, it's prudent to use a shelving filter to roll off all frequencies below 40 Hz.

Of course, this isn't much help if you're planning on releasing a mix for club play as many PA systems will produce energy as low as 30 Hz. If this is the case, you should continue to mix the bass but avoid boosting any frequencies below 40 Hz.

If the bass is lacking any punch, then a Q of approximately ½ an octave with a small cut or boost and sweeping the frequency range between 120 and 180 Hz may increase the punch to help it to pull through the kick. Alternatively, small boosts of ½ an octave at 200 to 300 Hz may pronounce the rasp helping it to become more definable in the mix. Notably, in some mixes, the highest frequencies of the rasp may begin to conflict with the mid range instruments, and if this is the case, then it's prudent to employ a shelving filter to remove the conflicting higher frequencies.

Provided that the bass frequencies are not creating a conflict with the kick, another common problem is the volume in the mix. While the bass timbre may sound fine, there may not be enough volume to allow it to pull to the front of the mix. The best way to overcome this is to introduce small amounts of controlled distortion, but rather than reach for the distortion unit, it's much better to use an amp or speaker simulator.

Amp simulators are designed to emulate the response of a typical cabinet so they roll off the higher frequency elements that are otherwise introduced through distortion units. As a result, not only are more harmonics introduced into the bass timbre without it sounding particularly distorted but also you can use small EQ cuts to mold the sound into the mix without having to worry about higher frequency elements creating conflicts with instruments sat in the mid range.

If the bass is still being sequenced from a virtual instrument rather than a bounced audio file then before applying any effects, you should attempt to correct the sound in the synthesizer. As discussed in earlier chapters, we perceive the loudness of a sound from the shape of its amplitude envelope and its harmonic content, so simple actions such as opening the filter cut-off or reducing the attack and release stage on both the amplitude and filter envelopes can make it appear more prominent.

If both these envelopes are already set at a fast attack and release, then layering a light kick from a typical rock kit over the initial transient of the bass followed with some EQ sculpting can help to increase the attack stage, but at the same time be cautious not to overpower the tracks original kick.

If the bass happens to be a recording of a real bass guitar, a different approach is required. Bass cabs will roll off most high frequencies, reducing most high-range conflicts, but they can also lack in bottom end presence. As a result, for dance music, it's quite usual to layer a synthesizer's sine wave underneath the guitar to add some more bottom end weight.

This technique can be especially useful if there are severe finger or string noises evident since the best way to remove these is to roll off everything above 300 Hz. In addition, bass guitars will fluctuate wildly in dynamics, and these must be controlled with compression. The bass should remain even in dynamics throughout the mix, and fluctuations will result in it disappearing behind other instrumentation.

Vocals

Although vocals take priority over every other instrument in a pop mix, in most genres of dance, they will take a back seat to the rhythmic elements and leads. Having said that, they must be mixed coherently since while they may sit behind the beat, the groove relationship and syncopation between the vocals and the rhythm is a common aspect of dance. Subsequently, you should exercise care in getting the vocals to sit properly in the mix.

The vocals should be compressed so that they maintain a constant level throughout the mix without disappearing behind instruments. A good starting point is to set the threshold so that most of the vocal range is compressed with a ratio of 9:1 and an attack to allow the initial transient to pull through unmolested.

The choice of compressor used for this can be vital, and you should opt for a compressor that will add 'character' to the vocals. The compressors that are preferred for vocal mix compression are the LA-2A and the UREI 1176, both of which are available in software form, but the Soft Tube Tube-Tech CL1B should also be considered a contender for vocal compression.

Vocals rarely require any heavy EQ adjustments, but small decibel boosts with a ½ octave Q at 10 kHz can help to make the vocals appear much more defined as the consonants become more comprehensible. Alternatively, if the vocals appear particularly muddy in the mix, an octave Q placing a 2 dB cut at a centre frequency of approximately 400 Hz should remove any problems.

If, however, they seem to lack any real energy while sat in the mix, a popular technique used by many dance artists is to speed (and pitch) the vocals up by a couple of cents. While the resulting sound may appear 'off the mark' to pitch perfect musicians, it produces higher energy levels that's perfectly suited towards dance vocals.

Lead Synthesizers/Pianos/Strings

The rest of the instruments in a mix will most commonly exhibit fundamentals in the mid range. As discussed, these should be processed and have EQ applied in order of importance working progressively towards the least important instruments.

If vocals have been employed, there will be often some frequency masking where the vocals and mid range instruments meet, so the engineer should consider leaving the vocals alone and apply EQ cuts to the instruments. Alternatively, the mid range can benefit from being inserted into a compressor or noise gate with the vocals entering the side chain so that the mid range dips whenever the vocals are present. This must be applied cautiously and a 'duck' of 1 dB is usually sufficient, any more and the vocals may become detached from the music.

Most mid range instruments will contain frequencies lower than necessary, and while you may not actually be able to physically hear them in the mix, they will still have an effect of the lower mid range and bass frequencies. Thus, it's prudent to employ a shelving filter to remove any frequencies that are not contributing to the sound within the mix. The best way to accomplish this is to set up a high shelf filter with maximum cut and starting from the lower frequencies sweep up the range until the effect is noticeable on the instrument. From this point sweep back down the range until the 'missing' frequencies return and stop. This same process can also be applied to the higher frequencies if some are present and do not contribute to the sound when it's sat in the mix.

Generally, keyboard leads will need to be towards the front of the mix, but the exact frequencies to adjust are dependent on the instrument and mix in question. As a staring point for many mid range instruments, it's worth setting the Q at an octave and applying a cut of 2 to 3 dB while sweeping across 400 to 800 Hz and 1 to 5 kHz. This often removes the muddy frequencies and can increase the presence of most mid range instruments.

Above all, however, instruments within a mix should always be kept in their perspective by constantly asking questions whilst mixing such as:

- *Is the instrument brighter than hi-hats?*
- *Is the instrument brighter than a vocal?*
- *Is the instrument brighter than a piano?*
- *Is the instrument brighter than a guitar?*
- *Is the instrument brighter than a bass?*

MID/SIDE PROCESSING

Whilst many of the problems with a mix can be corrected with careful application of EQ, some instruments and effects may benefit from mid/side processing to correct problem frequencies or instruments. In fact, mid/side processing is often applied by small number of dance musicians to create mixes that sound huge on club systems but sound more refined on small systems such as car stereos and iPod systems.

Sometimes termed sum and difference, mid/side is not a new technology but has only recently been introduced to audio workstations with mid/side plug-ins such as iZotope's Ozone 5 and the Brainworx BX Digital, DynEQ and Mono Maker.

These evolved from the mid-side microphone recording technique employed by many sound engineers. This involved employing two microphones, one with a cardioid pattern and a second with a figure of eight pattern. The cardioid pattern would point towards the sound source whilst the figure of eight would record the left and right ambience field of the cardioid microphone. By employing this technique, it permitted engineers to alter the stereo width of a microphone recording by increasing the ambience volume or made it possible to convert the signal to mono by reducing it to just the cardioid recording.

A similar scenario to this exists within any stereo recording or stereo file. Here, the left and right stereo information is the same and is termed the sum, but the information contained in the centre of this field is slightly different, and thus, it is termed the *difference*. Through employing a mid/side matrix, plug-in manufacturers permit the producer to treat the sum and the difference independently with EQ and dynamics. While this is obviously useful for mastering engineers if they need to EQ or change the dynamics in a complete mix, it also has a number of uses for the engineer whilst mixing.

By employing a mid/side processor such as the BX Digital, it is possible to mono any frequencies that are below a set threshold. Thus, if the producer has a wide stereo bass line, it would be possible to set the threshold to, say, 75 Hz, so that any frequencies below this become mono, whilst any frequencies above remain in stereo.

This application, however, is only the tip of a much larger creative iceberg. Indeed, mid/side processing is becoming a must have effect amongst many dance musicians and this number is growing exponentially as more producers are understanding its sonic potential.

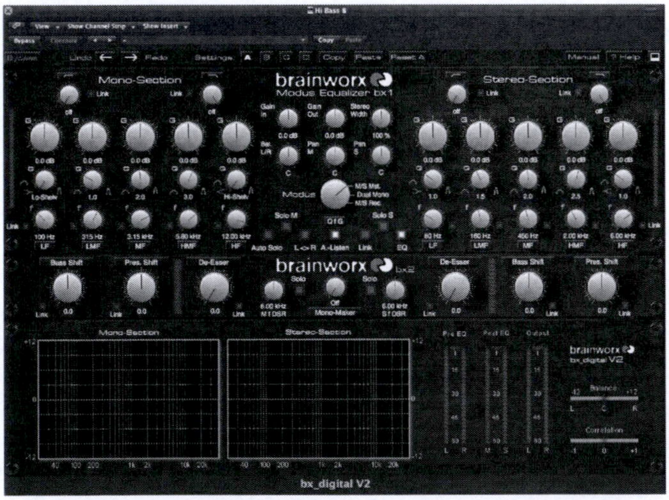

FIGURE 27.3
The BX Digital V2

A popular effect employed by some EDM artists is to use a mid/side processor to increase ambience in a recording and create mixes suitable for both clubs and stereo systems simultaneously. This is accomplished by placing the M/S processor after a stereo reverb unit. Since the reverberations are in stereo, the M/S processor can be used to reduce almost all of the reverb from the centre channel (typically with EQ or compression) but leave it present in the left and right field. Since many club systems are mono, when played through a club system, there is only a small amount of reverb present but when played via a hi-fi or iPod system, the left and right field exhibits a larger reverb. Ideally, an EQ should follow the M/S permitting the engineer to further sculpt the results.

Similarly, reverb could be applied to an instrument and the frequency threshold adjusted so that the reverb becomes mono below a specific frequency and remains in stereo above. This provides a clearer sounding mix. What's more, on a complete mix, it is possible to boost the sum information (the left and right field) whilst leaving the centre unaffected. This has the effect of increasing the stereo ambience of the music. Another possible effect is to employ a mid/side dynamics processor, such as the BX DynEQ to home in on specific frequencies and compress the mono and stereo signals differently. Here, if employed on a full mix, both the kick and bass could be compressed and consequently increased in volume compared to the side information resulting in a much larger presence of the groove.

As useful as mid/side processing is, however, it should not be used as a substitute for poor mixing practice. Although certainly a powerful effect, it is still of paramount importance that the producer creates a competent mix through careful application of volume, panning and EQ, and only after this should the engineer look towards employing the mid/side technique to add further interest to the mix.

FIGURE 27.4
The BX DynEQ

COMMON MIXING PROBLEMS AND SOLUTIONS

Frequency Masking

As touched upon earlier, frequency masking is one of the most common problems experienced when mixing whereby the frequencies of two instruments are matched and compete for space in the soundstage. Here, the producer should identify the most important of the instrument of the two and give this the priority while panning or aggressively EQ the secondary sound to fit into the mix.

If this doesn't produce the results, then you should ask yourself if the conflicting instrument contributes enough to remain in the mix. Simply reducing the gain of the offending channel will not necessarily bring the problem under control, as it will still contribute frequencies that can muddy the mix. Instead it's much more prudent to simply mute the channel altogether and listen to the difference it creates. Alternatively, if you wish to keep the two instruments in the mix, consider leaving one of them out and introducing it later in the mix when it may not conflict with the secondary instrument.

Clarity and Energy

All good mixes operate on the principle of contrast, that is the ability to hear each instrument clearly and locate its position in the stereo spectrum. It can be easy for the engineer to introduce multiple instruments and channels in an effort to disguise weak timbres but this will very rarely produce a great mix.

A cluttered, dense mix lacks energy and cohesion, whereas a great mix will feature fewer instruments that are all well programmed. Consequently, if the music appears cluttered, you should aim to remove the non-essential instruments until some of the energy returns.

If no instruments can be removed, then look towards removing the unwanted frequencies by notching out frequencies of the offending tracks either above or below where the instruments contribute most of their body. This may result in the instrument sounding 'odd' in solo, but if the instrument must play in an exposed part, then the EQ can be automated or two different versions could be used.

In addition, when working with the groove of the record, the silence between the groove elements produces an effect that makes it appear not only louder (silence to full volume) but also exhibits a more energetic vibe, so in many instances, it's worthwhile refraining from adding too may percussive elements. More importantly, good dance mixes do not draw attention to every part of the mix. Keep a good contrast by only making the important sounds big and up front, and place the rest in the background.

Prioritize the Mix

During the mixing stage always prioritize the main elements of the mix and approach these first. A common pitfall is to spend a full day 'twitching' the EQ on a hi-hat or cymbal without first maintaining a good even balance on the most important elements first. Always mix the most important elements first, and many of the 'secondary' timbres tend to look after themselves.

Relative Volume

Analytical listening can tire ears quickly so it's always advisable to avoid monitoring at a loud volume, as this will only quicken the process. The ideal standard monitoring volume is around conversation level (85 dB) but you should aim to keep the Fletcher Munson Contour Control in mind during mixing, and after every volume or EQ adjustment, reference the mix again at various gain levels. It can also be beneficial to monitor the mix in mono when setting and adjusting the volume levels as this will reveal the overall balance of the instruments more clearly.

EQ

EQ can be used to shape all instruments but you can only apply so much before the instrument looses its characteristics so be cautious with any EQ. Always bypass EQ every few minutes to make a note of the tonal adjustments you're making, but remember that while an EQ instrument may not sound correct in isolation, it is unimportant provided that it sounds right when run with the rest of the mix.

Cut EQ Rather than Boost

Our ears are used to hearing a reduction in frequencies rather than boosts since frequencies are always reduced in the real world by walls, objects and materials. Consequently, while some boosts may be required for creative reasons you should look towards mostly cutting to prevent the mix from sounding too artificial. Keep in mind that you can effectively boost some frequencies of a sound by cutting others, as the volume relationship between them will change. This will produce a mix that has clarity and detail.

Avoid Using EQ as a Volume Control

If you have to boost frequencies for timbre volume or design, you should not have to boost by more than 5 dB. If you have to go higher than this, then it is likely are that the sound itself was poorly recorded or the wrong choice for the mix.

The Magic Q

A 'Q' setting of 1 1/3 octave has a bandwidth that's generally suitable for EQ on most instruments and often produces the best results. That said, if the instrument is heavily melodic or you are working with vocals, wider Q settings are preferred, and a typical starting point is about two octaves. Drums and most percussion instruments will benefit from a Q of half an octave.

Shelf EQ

Shelf equalizers are generally used to cut rather than boost because they work at the extremes of the audio range. For instance, using a shelving filter to boost the low frequencies will only accentuate low-end rumble since there's very little sound this low. Similarly, using a shelf to boost the high range will increase all the frequencies above the cut-off point, and there's very little high-frequency energy above 16 kHz.

Never Attempt to Fix It in the Mix

'Fix it in the mix' is the opinion of a poor engineer and is something that should never cross your mind. If a timbre is wrong, no matter how long it took to program, admit that it's wrong and program/sample one that is more suitable.

Table 27.1 Chart Indicating Frequencies for Mixing

Frequencies	Musical Effect	General Uses
30 to 60 Hz	These frequencies produce some of the bottom end power but if boosted too heavily can cloud the harmonic content, introduce noise and make the mix appear muddy.	Boosts of a decibel or so may increase the weight of bass instruments for Drum 'n' Bass. Cuts of a few decibels may reduce any booming and will increase the perception of harmonic overtones, helping the bass become more defined.
60 to 125 Hz	These frequencies also contribute to the bottom end of the track but if boosted too heavily can result in the mix loosing its bottom end cohesion resulting in a mushy, 'boomy' sound.	Boosts of a decibel or so may increase the weight of kick drums and bass instruments and add weight to some snares, guitars, horns and pianos. Cuts of a few decibel may reduce the boom of bass instruments and guitars.
125 to 250 Hz	The fundamental of bass usually resides here. These frequencies contribute to the body of the mix, but boosting too heavily will remove energy from the mix.	Small boosts here may produce tighter bass timbres and kicks, and add weight to snares and some vocals. Cuts of a few decibels can often tighten up the bottom end weight and produce clarity.

Frequencies	Musical Effect	General Uses
250 to 450 Hz	The fundamentals of most string and percussion instruments reside here along with the lower end of some male vocalists.	Small boosts may add body to vocals, kicks and produce snappier snare timbres. It may also tighten up guitars and pianos. Cuts of a few decibels may decrease any muddiness from mid range instruments and vocals.
450 to 800 Hz	The fundamentals and harmonics of most string and keyboard instruments reside here, along with some frequencies of the human voice. Cuts are generally preferred here as boosting can introduce fatigue.	Small boosts may add some weight to the bass elements of instruments at low volumes. Cuts of a few decibels will reduce a boxy sound and may help to add clarity to the mix.
800 to 1.5 kHz	This area commonly consists of the harmonic content of most instruments so small boosts can often add extra warmth. The 'pluck' of most bass instruments and click of the drum kicks attack also reside here.	Boosts of a decibel or so can add warmth to instruments, increase the clarity of bass, kick drums and some vocals, and help instruments pull out of the mix. Small decibel cuts can help electric and acoustic guitars sit better in a mix by reducing the dull tones.
1.5 to 4 kHz	This area also contains the harmonic structure of most instruments, so small boosts here may also add warmth. The body of most hi-hats and cymbals also reside here along with the vocals, BV's and pianos.	Boosts of a decibel or so can add warmth to instruments and increase the attack of pianos and electric/acoustic guitars. Small decibel cuts can hide any out-of-tune vocals (although they should be in tune!) and increase the breath aspects of most vocals.

(*continued*)

Table 27.1 Chart Indicating Frequencies for Mixing (*continued*)

Frequencies	Musical Effect	General Uses
4 to 10 kHz	Finger plucks/attacks from guitars, the attack of pianos and some kick drums, snares along with the harmonics and fundamentals of synthesizers and vocals reside here	Boosts of a few decibels can increase the attack on kick drums, hi-hats, cymbals, finger plucks, synthesizer timbres and pianos. It can also make a snare appear more 'snappy' and increase vocal presence. Small decibel cuts may reduce sibilance on vocals, thin out guitars, synthesizers, cymbals and hi-hats, and make some sounds appear more transparent or distant.
10 to 15 kHz	This area consists of the higher range of vocals, acoustic guitars, hi-hats and cymbals and can also contribute to the depth and air in a mix.	Boosts of a few decibels may increase the brightness of acoustic guitars, pianos, synthesizers, hi-hats, cymbals, string instruments and vocals.
15 to 20 kHz	These frequencies often define the overall 'air' of the mix but may also contain the highest elements of some synthesizers, hi-hats and cymbals.	Boosts here will generally only increase background noise such as hiss or make a mix appear harsh and penetrating. Nonetheless some engineers may apply a shelving boost around this area (or at 10 kHz) to produce the Bandaxall curve.

CHAPTER 28

Mastering

'The so-called "Loudness War" is entirely based on a modern myth – a fairy-tale full of nonsense that has somehow hypnotized the entire music industry for the last ten years – and is permanently damaging the music we listen to as a result'.

Mastering Engineer

Mastering can be viewed as the final link in the production chain. After any track has been mixed, it should be mastered. Mastering a track is important for any music production, since it allows your music to stand up to the competition in terms of both overall spectral balance and loudness.

Although there is ongoing debate in regards to the current 'loudness war' between records, as more and more become dynamically restricted to increase their perceived volume, the fact remains that if your music does not compete in loudness to other tracks on the dance floor, it will likely fall flat. Clubbers, your potential audience, like it loud so you must ensure that your track stands shoulder to shoulder with further tracks whilst at the same time not sacrificing too much of the music's dynamic range.

The process involved in mastering varies depending on the project in question and the genre of the music, but if it has been mixed competently then it will commonly involve equalization to achieve a good overall mix balance, light compression to add punch and presence and loudness restriction to increase the overall volume. The end result is that the impact of the music is enhanced and the spectral balance will compare and sit well with every other track in the club or with your audiences MP3 collection.

This, however, is more complicated than it may initially appear and to master to a professional standard requires experienced ears, technical accuracy, knowledgeable creativity and a very accurate monitoring environment. What's more, they will employ expensive outboard gear that is not available as software (such as the Manley Slam, Backbone and Massive Passive) and the mastering engineer will not be as emotionally attached to the music and therefore will be able to offer an objective perspective. Consequently, I strongly recommend that an artist does not attempt to master his or her own music but instead employs a professional mastering engineer to do the job for him or her.

Although recording and mixing music is a definite skill that requires complete attention to every individual element of the track, mastering involves looking at the bigger picture subjectively. This subjective viewpoint is something that is very difficult for the producer (i.e. you) to accomplish due to the involvement throughout the entire project. There are many reputable mastering engineers who will charge as little as £30 per track and it is worth the additional outlay to have your music mastered by a professional.

Having said that, there are some artists who choose to master their own music and therefore this chapter is going to discuss the general approach to mastering dance music. It should, however, not be considered the definitive text on the subject but an overview of the process. Mastering is a complex and multifaceted topic and entire books have been authored on the subject.

As discussed in earlier chapters, the order of processing can dramatically affect the overall sound; thus in mastering, each processor should be conducted in a sensible order. This arrangement can be evident in the all-in-one mastering plug-ins that are often organized in order of processing but if individual plug-ins are used to accomplish this, the order commonly consists of the following:

1. Noise reduction
2. Mid side EQ/Harmonic balancing
3. Mastering reverb (although this is uncommon)
4. Dynamics
5. Harmonic excitement
6. Stereo imaging
7. Maximizing to increase loudness

NOISE REDUCTION

Although any extraneous noises should have been removed during the recording stages, occasional clicks, pops, hiss or hum can slip through, thus it is important to listen to the mix thoroughly and carefully on both monitors and headphones to see if there is any noise present. This is where a mastering engineer will have an immediate advantage as they will use subwoofers and loudspeakers with an infrasonic response and without these, microphone rumble, deep sub-basses or any jitter introduced with poor-quality plug-ins or through poor recording techniques.

If there is any audible noise present, it must be removed before the track is mastered. Stray transients such as clicks or pops will prevent you from increasing the overall volume while if there is any hum, hiss or noise present, when the overall gain of the mix is increased, it will also increase the volume of the noise.

If you are mastering your own music, the most appropriate way to deal with any noise issues is to return to the mix and examine the source of the noise. If it's a result of a plug-in, it should be replaced or if it's a result of poor recording techniques it should be re-recorded. There are quite simply no short cuts to creating a great sounding mix and master.

If the original mix can no longer be accessed for whatever reason, then you will have no choice but to employ noise reduction algorithms or close waveform editing. This latter method is especially suited for removing pops or clicks that are only a couple of samples long and involves using a wave editor to zoom right into the individual samples and then using a pencil tool, or similar, reduce the amplitude of the offending click.

This approach is only suitable for extraneous noises that occur over a very short period of a few samples, however, and if the noise is longer than this it is advisable to use dedicated reduction algorithms. These have been designed to locate and reduce the amplitude of short instantaneous transients typical of clicks or pops and are available in most wave editors or as a third-party plug-in.

If hum and/or hiss are present, then these are much more difficult to deal with and often require specialist plug-ins to remove. The frequency of hiss and thermal noise from poor audio interfaces often consist of frequencies ranging from 5 kHz through to 14 kHz. If you attempt to remove these through the use of a paragraphic or parametric EQ system, it will result in a reduction of the mixes high frequencies too. Consequently, specialist plug-ins must be used to remove it but the reliability of employing these depend on the source. They are rarely successful in removing hiss without flattening the upper balance of the mix.

Any hum that is a result of electromagnetic interference when recording can be equally difficult to control. These frequencies can reside anywhere from 50 Hz to 420 Hz and although it may be possible to remove some of these with a surgical thin bandwidth parametric EQ, it will often result in removing some of the mix frequencies that can compromise the tonal balance of the music.

Above all, any form of noise reduction is not something that should be approached lightly. Reducing noise within a final stereo mix will often influence the frequencies that you want to keep, especially when removing hiss or hum. In fact, if there is an obvious amount of noise that cannot be removed reliably with FFT noise profiling, then you should seriously consider returning to the mix and curing the source of the problem.

MID-SIDE EQ/ HARMONIC BALANCING

Mid-side EQ and harmonic balancing consists of correcting any severely wayward frequencies that may have slipped passed the mix engineers ears due to the room acoustics or poor monitoring environment and then applying overall EQ adjustments to the mix in order to achieve an overall tonal balance that is sympathetic to the style and genre of the music being mastered.

Mid-side processing has been discussed in Chapter 27 during mixing but it is equally useful during mastering. Here, it can simply be employed to correct any frequency spikes whilst also used to widen the stereo field, give the drums and bass more presence or reduce any problems with effects such as reverb washing the sound.

This process is then followed by a much broader application of EQ to create the tonal balance. Each genre of dance music will employ a different overall tonal balance that presents the music in the best way to the listener. For example, music that has a high emphasis on the lead instruments will often tend to feature a less cuts in the mid range whereas genres such as Techno, Tech-house or Dubstep will commonly exhibit a larger dip through the mid-range frequencies.

Understanding what tonal balance is employed in each specific genre of music can be attributed to critical listening and experience, but a good technique for the novice user is to place a number of professional tracks for the same genre into the workstation and examine their overall spectral balance with a spectrum analyzer. The quality of the EQ being employed to achieve this is paramount and many mastering engineers will employ a quality 31-band graphic equalizer rather than the typical parametric EQ systems due to the wider Q settings and gentler transition slopes (often as low as 3 dB per octave). If a parametric EQ is employed here, the Q is typically adjusted to 0.4 to 0.9 to create broader enhancements to the spectrum.

For attaining spectral balance, mastering EQ can be broadly split into the following three different sections: the bass, the mid range and the highs.

Bass

With a competent mix, mastering EQ is typically limited to shelving the frequencies below 30 to 40 Hz to remove them. This is because many club systems struggle with replicating frequencies lower than this, but if the loudspeakers receive this low-energy signal, their available bandwidth will be reduced as they attempt to replicate the frequencies. By employing shelving filters to remove everything below 30 to 40 Hz, the sub-harmonic content of the sound is removed and this opens up the bandwidth of the speakers for a higher volume whilst also improving the definition of the bass frequencies.

The bass frequencies of the mix may appear quieter than commercial records, but at this stage EQ boosts should be avoided since you may be attempting to boost energy that simply isn't present. Similarly, bass 'maximizers' should be avoided and instead mid-side dynamic processing can be employed. Using this, it is possible to compress the sum *or* the difference to increase the bass presence.

FIGURE 28.1
The Bx DynEQ mid-side compression plug-in

If the problem is not a lack of bass but a lack of definition in the bass, then the lower frequencies can be EQ's using either a parametric EQ or mid-side processing depending on the effect you wish to achieve.

In many EDM mixes, the kicks main energy will reside at 80 Hz to 200 Hz but boosting this range will often result in clouding the bottom end of the mix and instead a better approach is to boost the transients. These commonly reside between 1,000 and 2,500 Hz and by notching out the frequencies that lie just below the transient it results in the transient becoming more noticeable. If you are struggling to find the transient frequencies, sweeping through the frequencies with a very narrow Q with a 6 dB boost should bring them out. Once identified, turn the boost into a –4 dB cut and sweep to just before the transient frequency.

Mid Range

The mid range is the most common source of frequency-related problems since we are not only most sensitive to this range but the room's resonances will often influence our perception here too. Consequently, critical listening is not always the idea solution and therefore it is recommended that you import a few professional tracks of the same genre, and then copy the spectral balance of their mid range through the use of spectrum analyzers. Typically, however, the mid range problems can be divided into three distinct categories, it will be too muddy, harsh or nasal.

If the sound appears too muddy, then cuts of around 180 Hz to 225 Hz with a wide Q and a –3 dB cut will often help to identify the problem areas. Once found, the Q can be reduced until the muddy sound begins to return. If the mix appears particularly harsh, the same process can be employed. This time, however, the wide Q should be placed around 1.6 kHz to 2 kHz and moved until the harshness dissipates. Once it's gone, the Q should then be reduced until the harshness begins to return. Again, this same process can be used on a mix that appears nasal but this time a wide Q placed around 500 Hz should reduce the nasal quality. The Q can then be reduced until the nasal character begins to re-appear.

With all of these adjustments, it is important to try and maintain a fairly wide Q without damaging the spectral balance of the mix since wide Q settings will appear more natural to the ears. If you find that you must use a narrow Q or cuts larger than 4 dB to remove the problematic area, then the problem lies with a poor mix and it would be wise to return to the original mix to repair the problem there.

High Range

The higher frequencies of the mix do not usually suffer from problems but in the unlikely event that the high range is suffering from bright problematic frequencies, wide EQ cuts placed at a central frequency of 14 KHz may reduce the problem. A more common problem, however, is that the higher frequency elements simply aren't bright enough. Here, brightness can sometimes be applied to signal by placing a wide Q with a boost of a few dB at 12 to 16 KHz. This can produce an unnatural aura so it may be preferential to employ some enharmonic exciters to the audio.

Alternatively, a large 6 dB boost with a wide Q at 20 kHz can introduce a gentle transitional slope upwards from 10 kHz, a result that simulates the natural response of our ears and can introduce small amounts of harmonic distortion similar to valves. This technique, known as the Baxandall curve, should be followed by a 'brick wall' EQ to completely remove any frequencies above 20 kHz. By doing so, the top of end of the mix exhibits an enhanced air quality due to the speakers not attempting to reproduce frequencies outside their nominal range.

While the majority of these adjustments should be made in stereo to maintain the stereo balance between the two channels, occasionally it may be beneficial to make adjustments on one channel only. This is useful if the two channels have different frequency content from one another or is there is a badly recorded/mixed instrument in just one of the channels. In these instances, subtlety with a mid-side processing unit is the key and the best approach is to start with extremely gentle settings, slowly progressing to more aggressive corrections until the problem is repaired.

When it comes to this form of surgical EQ, though, the most important rule of thumb is to listen to the material carefully *before* making any adjustments. You should first identify exactly what the problem is before applying any adjustments and even then, surgical EQ should only to the sections that actually require it. If a 'problem' frequency were only noticeable during a specific part of the music, the sensible approach would be to only correct it during the problematic part and not throughout the entire track.

BOOSTS AND CUTS

Although frequency boosts are generally best avoided in order to maintain a natural balanced sound, *any* EQ adjustments will affect the perception of the overall tonal balance of the whole mix. For example, even applying a ½ dB boost at 7 KHz could have the same result as dipping 300–400 Hz by 2 dB. Indeed, the frequency ranges on a full mix will interact with each other and even minor adjustments may have an adverse affect on every other frequency.

As with mixing, the key behind setting a good tonal balance with EQ is developing an ear for the different frequencies and the effects they have but as mentioned, it is highly recommended that you employ a spectrum analyzer so you can determine both aurally and visually how EQ adjustments define and change the spectral balance of the music.

A general guide to mastering EQ frequencies:

0 to 40 Hz
- All these frequencies should be removed from the mix with a shelving filter. This will free the low-end energy of the speakers allowing them to replicate the music better.

40 to 200 Hz
- These frequencies are not commonly attenuated or boosted. If the mix appears to lack punch or depth, or appears to 'boom' then it is common practice to use mid-side compression to cure the problem rather than EQ boost.

100 to 400 Hz
- This frequency range is often responsible for a 'muddy' sounding mix. If a mix exhibits a mushy or muddy response, it can be reduced with EQ cuts placed here.

400 to 1.5 KHz
- These frequencies often contribute to the 'body' of the sound, and both compression and EQ can be employed to introduce punch and impact to the mix.

800 to 5 KHz
- This range defines the clarity of the instruments. Small EQ increases or decreases applied in this range may pull out the fundamentals of many instruments making them appear more defined.

5 KHz to 7 KHz
- These frequencies are commonly responsible for sibilance in a recording and can be removed with a parametric EQ set to a narrow bandwidth to remove the problem frequencies or alternatively mid-side compression may be employed.

3 KHz to 10 KHz
- These frequencies are accountable for the presence and definition of instruments. Introducing 1 or 2 dB increases can augment the presence of some instruments, whilst cutting can reduce the effects of sibilance or high-frequency distortion.

6 KHz to 15 KHz
- These can contribute to the overall presence of the instruments and the 'air' surrounding the track. Introducing small gain boosts here will increase the brightness but may also boost noise levels.

8 KHz to 15 KHz

- Boosting at these frequencies may add sparkle and presence to a mix (although it may also increase any hiss present in the mix). EQ cuts may help to smooth out any harsh frequencies.

Naturally, these are only guidelines and you should rely on your ears and judgment as to what sounds right for the mix in question.

MASTERING REVERB

Mastering reverb (applying reverb across an entire mix) is rarely required in most mastering situations as it should have been applied properly during the mixing stages but in certain cases it may be required. Typically, this is when instruments don't appear to share the same acoustic space or the mix seems to be lacking the fullness of similar mixes.

Ideally, the producer should return to the mixing stage and repair these problems but this may not always be possible. In this event, lightly (and carefully!) applying reverb over the entire mix can help to blend the music together to produce a better result. The amount of mastering reverb that should be applied depends on the mix but there are some general guidelines that can be followed.

First and foremost, the reverb must be of a quality unit since any tail noise created by poor algorithms will only serve to exasperate the problem rather than work to cure it. Indeed, if you don't have access to a good reverb plug-in such as the Lexicon PCM, mastering reverb is worth bypassing since it isn't worth turning an average mix into a poor one just for the sake of trying to introduce some cohesion between the instruments.

We have already examined the parameters and application of reverb in previous chapters and the same ideologies apply during the mastering process. Here, unlike all other effects and processors in the mastering chain, reverb should be applied as a send effect rather than an insert, as only small amounts of reverb should be applied to the mix.

A typical return to start with is around 15% wet signal against 85% dry but this depends entirely on the mix in question. Beginning with this setting, you can adjust other reverb parameters and then back off the wet signal until it suits the mix.

Broadly speaking, you should aim to set the decay time so that the tail is a little longer than the reverb that's already been applied during the mixing stage in order that the instruments gel together. Typically, for most styles of dance music, this would equate to around 2 to 8 ms. It's important to note that the size of the room will also determine the diffuse settings, so for a small room this should be set quite low so as not to create an inaccurate impression. The mid/side technique for applying reverb just to the stereo sum (as discussed in

Chapter 27) can be applied if the reverb appears to smear the sound of the mix too much.

Possibly the most important aspect of applying reverb at this stage is to continually bypass the effect to confirm the effect it is imparting onto the mix. A good rule of thumb for this is to apply the reverb so that when it is present it isn't particularly noticeable but when removed it becomes immediately perceptible.

DYNAMICS

Following EQ, (and in extreme cases, mastering reverb) many engineers will apply dynamic manipulation to the mix to add punch and clarity. This cannot be accomplished by simply strapping a standard compressor across the entire mix, however, since the loudest parts of the track will drive the compressor and this can result in the entire mix pumping and breathing in reaction of the highest energy of the music (the kick drum).

While it could be argued that EDM does pump, this should be employed during the mixing stage through side chaining and during mastering, multi-band compression and mid/side compression is commonly employed. With these, it is possible to alter the dynamics at different frequency bands and stereo positions that can result in a clearer sounding mix.

For example, heavily compressing the higher frequencies of a mix can flatten the sound, so it is best applied lightly, however, the lower ranges of a mix are usually compressed more heavily to even out the sound and add extra volume to some frequency bands of the mix.

This frequency-specific volume is an important aspect of most dance music, especially on the lower frequencies and should not be underestimated. Your music must measure up to the competition in terms of volume and what's more, if you play the same musical passage to an audience at different volume levels, a larger percentage will prefer the louder playback. However, while it would seem sensible to reduce the dynamics of a mix as much as possible to increase the overall volume there are limits.

A loudness war has been raging for the past couple of years as more mixes are having their dynamic ranges severely restricted in order to be the loudest. This has a detrimental effect on the music and listeners. We naturally expect to hear some dynamics when listening to music and in the case of EDM, the dynamic ratio between the kick and the body of the music is what is responsible for the ultimate punch of the kick. If the dynamic range is heavily restricted, the punch is all but eliminated as the body of the music reaches the same levels as the kick.

Furthermore, whilst louder music will inevitably capture the attention of the listener, it will only occur for the first minute or two. A severely dynamically restricted track may appear hard-hitting at first, but if it's too *'hot'* it can quickly become fatiguing to the ear. Indeed, ruthless dynamic restriction is only considered useful for playback on radio stations.

All radio stations will limit the signal for two reasons as follows:

a. They need to keep music's level at a specific maximum so as not to over modulate the transmission.
b. They want to be as loud as possible to hold audience attention.

This means that no matter what the input level is, the output level will always be constant. The inevitable result is that even if your mix happens to be a couple of dB louder than every other mix, when it's played on the radio, it would be at the same volume as the one played before or after.

I have heard arguments that by compressing a mix heavily at the mix-down/mastering stage, the radio station is unable to squash the signal any further so there is little possibility of any broadcasters destroying the signal. While there is some logic behind this argument, setting the average mix level too high though heavy compression can often results in more problems since the station's processors might confuse the energy of the music with the signal peaks and attempt to restrict it further. This will completely remove the punch of the music, making it appear lackluster when compared to other dance records. Consequently, when mastering, you will often have to settle for a compromise between the output volume and leaving some dynamic range in the mix.

As a very general rule of thumb, the difference between the peak of the kick and the average mix level (known as the peak to average ratio) should never be less than 4 or 5 dB. This will enable you to increase the overall gain while also retaining enough dynamic variation to prevent the mix from sounding 'flat' on a home hi-fi system.

There's more to multi-band compression than employing it to add some punch to a dance track and a manipulation of dynamics around the vocals or lead melody can introduce additional impact. This approach should not be taken lightly, though, since injecting dynamic feel at the mastering stage often requires a fresh pair of experienced ears as it involves listening out for the mood, flow and emotion it conveys.

As an example of dynamic manipulation to groove, consider this rhythmic passage:

'…I love to party……to dance the night away…………..I just love to party'

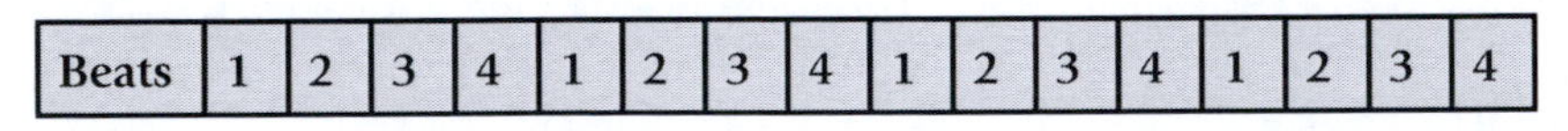

Beats	1	2	3	4	1	2	3	4	1	2	3	4	1	2	3	4

This isn't a particularly complex vocal line but note that the accents occur on the downbeat of each bar. If you were to apply compression heavily throughout the mid range (where the vocals reside) then this vocal line could easily become as follows:

'...I love to party...to dance the night away...........I just love to party'

Beats	1	2	3	4	1	2	3	4	1	2	3	4	1	2	3	4

By severely restricting the dynamic range throughout the mid range, the dynamics of the track have been lost and the resulting feel has been removed. However, by carefully compressing and only applying light amounts to specific parts of the vocals it could result in as follows:

'......I love to **party**......**to** dance the night **away**..............I just love **to party**'

Beats	1	2	3	4	1	2	3	4	1	2	3	4	1	2	3	4

As the accents show, you're not always simply working on the vocal but the mix instruments directly behind it, and this influences the music too.

When working with such precise dynamic manipulation, the compressors parameters must also be set accurately and carefully. If the release is set too long, the compressor may not recover fast enough for the next accent or sub-accent, while if it's set too short, the sound may distort. Similarly, if the attack is too short, then the initial transients of the word and music will be softened which defeats the whole point behind applying it in the first place.

More importantly, any compression will result in a required increase in make-up gain and this dynamic loudness can have an adverse affect on our judgment. Consequently, even applying accent compression in the wrong place will still lead you to think it sounds better so it's important to keep the gain the same as the uncompressed passages to ensure that it isn't just pure volume that's making the passage sound better.

When it comes to the application of a multi-band compressor for a mix, there is no set approach and there are no generic settings since it depends on the mix in question. Regardless, there are some general guidelines you can follow to initially set up a compressor across a mix.

- On a three-band compressor, start by setting each band of the compressor to 2:1 and then adjust the frequency of each band so that you're compressing the low range, mid range and high range.
- On a four-band compressor, adjust the crossover frequencies so that one band is compressing the bass frequencies, two are compressing the mid range and one is compressing the high range.
- On a five-band compressor, adjust the crossover so that two bands are concentrated on the bass, two are compressing the mid range and the final is compressing the high range.

The crossover frequencies will depend on the mix being restricted but try to ensure that the vocals are not compressed with either the low range or high-frequency ranges of the compressor. By doing so, a single (or double on a five band) compressor concentrated on the vocals alone can be adjusted to add the aforementioned dynamic movement to a mix.

Applying this multi-band compression to parts of the mix will upset the overall balance of mix but rather than re-apply EQ, it is preferable to adjust the output gain of each compressor to maintain a level mix.

If the bass end of the mix is short of any bottom-end weight, then lowering the threshold further and raising the ratio until the bass end comes to the forefront may solve the problem. However, the problem may also be that the mid range just requires less compression to lift up the bass and it's a careful mix of all the bands that creates a fuller mix. For instance, if the mid range is compressed heavily, it will be forced to the front of the mix, overtaking both the high and bass ranges. Thus, rather than compress the bass, it could be prudent to lower the compression on the mid range.

Both the attack and release parameters can also affect the overall tonal balance of a mix, so these must be adjusted with caution. Usually, on the bass frequencies, a fast attack time can be employed since it consists purely of low-frequency waveforms with few or no transients. Also, the release setting can set as fast as possible so the compressor can recover quickly. Generally, the faster the release is set, the more the bass end will pump, so it should be set so that it pumps slightly but not too much as to destroy the feel of the mix.

For the mid range, it is sensible to use longer attack times than the bass to help improve the transients that are contained here, while the release setting should be set as short as possible without introducing any gain pumping.

Finally, the high range should use a relatively fast attack to prevent the transients becoming too apparent but as the waveforms will be shorter than lower frequencies, you can employ shorter release times too. Similar to the mid range, however, this should not be set too short since it may introduce pumping in the dynamics.

As a very general starting point, what follows is a list of typical settings for three-, four- and five-band compressors:

Three-band compression:

BAND 1: TO TIGHTEN UP THE BOTTOM END OF A MIX:

- Frequency – 0 Hz to 400 Hz
- Ratio – 4:1
- Threshold – 3 dB below the quietest notes so that the compressor is permanently triggered
- Attack – 10 to 20 ms
- Release – 140 to 180 ms
- Gain make up – Increase the gain so that it is 3 to 5 dB above pre-compression level

BAND 2: TO TIGHTEN UP AND ADD SOME 'WEIGHT' TO THE MIX:

- Frequency – 400 Hz to 1.5 kHz
- Ratio – 2:1
- Threshold – Just above the quietest notes so that it is triggered often but not continually
- Attack – 10 to 20 ms
- Release – 100 to 160 ms
- Gain make up – Increase the gain so that it is 1dB above pre-compression level

BAND 3: TO INCREASE CLARITY OF INSTRUMENTS AND REDUCE 'HARSH' OR 'ROUGH' ARTIFACTS:

- Frequency – 1.5 kHz to 15 kHz
- Ratio – 2:1
- Threshold – Just above the quietest notes so that it is triggered often but not continually
- Attack – 5 to 20 ms
- Release – 100 to 130 ms
- Gain make up – Increase the gain so that it is 1dB above pre-compression level

Four-band compression:

BAND 1: TO TIGHTEN UP THE BOTTOM END OF A MIX:

- Frequency – 0 Hz to 120 Hz
- Ratio – 4:1
- Threshold – 3 dB below the quietest notes so that the compressor is permanently triggered
- Attack – 10 to 20 ms
- Release – 150 to 180 ms
- Gain make up – Increase the gain so that it is 3 to 5 dB above pre-compression level

BAND 2: TO TIGHTEN UP THE MIX IN GENERAL AND ADD WARMTH TO INSTRUMENTS AND VOCALS:

- Frequency – 120 Hz to 2 kHz
- Ratio – 2.5:1 or 3:1
- Threshold – Just above the quietest notes so that it is triggered often but not continually
- Attack – 20 to 30 ms
- Release – 100 to 160 ms
- Gain make up – Increase the gain so that it is 1 to 3 dB above pre-compression level

BAND 3: TO INCREASE THE CLARITY OF INSTRUMENTS:

- Frequency – 2 kHz to 10 kHz
- Ratio – 3:1
- Threshold – 2 dB below the quietest notes so that the compressor is permanently triggered
- Attack – 10 to 20 ms
- Release – 100 to 180 ms
- Gain make up – Increase the gain so that it is 2 dB above pre-compression level

BAND 4: TO INCREASE THE CLARITY OF HI-MID TO HIGH-FREQUENCY INSTRUMENTS:

- Frequency – 10 kHz to 16 kHz
- Ratio – 2:1 or 3:1
- Threshold – Just above the quietest notes so that it is triggered often but not continually
- Attack – 5 to 25 ms
- Release – 100 to 140 ms
- Gain make up – Increase the gain so that it is 1 dB above pre-compression level

Five-band compression:

BAND 1: TO TIGHTEN UP THE BOTTOM END OF A MIX:

- Frequency – 0 Hz to 180 Hz
- Ratio – 5:1
- Threshold – 3 dB below the quietest notes so that the compressor is permanently triggered
- Attack – 10 to 20 ms
- Release – 130 to 190 ms
- Gain make up – Increase the gain so that it is 3 to 5 dB above pre-compression level

BAND 2: TO TIGHTEN UP THE RHYTHM SECTION AND THE MIX IN GENERAL:

- Frequency – 180 Hz to 650 Hz
- Ratio – 2.5:1 or 3:1
- Threshold – Just above the quietest notes so that it is triggered often but not continually
- Attack – 10 to 20 ms
- Release – 130 to 160 ms
- Gain make up – Increase the gain so that it is 1 to 3 dB above pre-compression level

BAND 3: TO ADD SOME WEIGHT TO THE MIX AND INCREASE IT'S ENERGY:

- Frequency – 650 Hz to 1.5 kHz
- Ratio – 3:1
- Threshold – 1 dB below the quietest notes so that the compressor is permanently triggered
- Attack – 15 to 25 ms
- Release – 100 to 130 ms
- Gain make up – Increase the gain so that it is 1 to 3 dB above pre-compression level

BAND 4: TO INCREASE THE CLARITY OF MID- TO HIGH-FREQUENCY INSTRUMENTS:

- Frequency – 1.5 kHz to 8 kHz
- Ratio – 2:1 or 3:1
- Threshold – Just above the quietest notes so that it is triggered often but not continually
- Attack – 5 to 10 ms
- Release – 100 to 140 ms
- Gain make up – Increase the gain so that it is 1 to 4 dB above pre-compression level

BAND 5: TO REDUCE UNWANTED ARTIFACTS SUCH AS A 'ROUGH' OR 'HARSH' TOP END:

- Frequency – 8 kHz to 16 kHz
- Ratio – 2:1 or 3:1
- Threshold – Set so that it is triggered occasionally on the highest peaks
- Attack – 5 ms
- Release – 100 to 140 ms
- Gain make up – Increase the gain so that it is 1 dB above pre-compression level

These are not precise settings and are only listed as a general starting point. With these settings in the compressor, you should experiment by increasing and lowering the make-up gain, the threshold, attack, release and the ratio while listening out for the difference each imparts to the signal.

HARMONIC EXCITEMENT

Following compression, some engineers may introduce a harmonic exciter to brighten the upper mid and lows of the mix. This is purely personal preference and some engineers view any harmonic excitement with disdain believing that it creates a mix that appears artificial. Both arguments have valid points and it is up to the engineer on the project as to whether it should be applied or not.

The reason for the application of psychoacoustic enhancement is due to modern recording techniques. Since we have the ability to record and re-record music, the clarity and detail of music is often compromised, particularly in the higher frequencies. This results in the music's frequency content appearing lifeless. By employing an enhancer, these lifeless frequencies are restored, an effect that is believed to result in the music appear clearer, brighter and crisper.

The first exciters were developed by the American company Aphex in 1975 and were, according to popular story, discovered by accident. A stereo amplifier kit was assembled incorrectly and thus ended up with only one channel working with the other producing a thin distorted signal. When both channels were summed, the resulting signal appeared cleaner, brighter and more enhanced. This principle was researched and developed before Aphex reached the conclusion that applying small controlled amounts of specific harmonic distortion to an audio signal will spectrally enhance it.

Originally, the first Aphex exciters were only available for rent to studio at around £20 per minute of music but due to the overwhelming demand they soon became available for purchase. These commercial units permit the user to select the low-, mid- or high-range frequencies and treat each to differing amounts of second- and third-order harmonic distortion. This distortion is then added back to the original signal that is treated to small amounts of phase shifting.

To many engineers, the Aphex system is the best option for exciting a mix but there are other alternatives available. Since Aphex registered the name Aural Exciter to its family of exciters, other manufacturers refer to their exciters as 'psychoacoustic enhancers' and these are available from SPL and BBE to name a few.

These all employ some form of dynamic EQ and phase alignment. Consequently much of we determine from a sound is derived from the initial transient but during the recording process and subsequent processing the phase can be shifted which results in the transient becoming less defined. By re-aligning the phase, these transients using an enhancer become more defined which results in extra clarity and definition.

While applying an extra sonic sheen to music through the use of an exciter will often improve the sound they should nonetheless be used conservatively. The results they produce can be addictive and our ears can grow accustomed to the processing quickly. This can result in over processing (often referred to as 'over cooking' or 'frying') that results in a sound that can be very fatiguing to the ear.

STEREO WIDTH

Following the excitement effect, a proportionate amount of mixes will be treated to stereo widening. This can be accomplished in one of two ways, the engineer may employ mid side processing to enhance the stereo sum whilst compressing the difference (centre) or they may employ specific stereo widening effects.

Stereo widening effects are a simple affair and often only feature one parameter that when increased will widen the stereo image. This works on the principle that sounds that are shared in both left and right channels can appear to be in the middle of the mix rather than panned to either side, even if they are. However, if you were to subtract one of these channels from the other, then the phase would be adjusted resulting in a wider stereo effect.

Like other mastering process, the overuse of stereo widening effects can have a detrimental effect on the music so must be applied carefully. As you widen the spectrum further, the phase adjustments can create a 'hole' in the middle of the mix and as the bass, kick drum and vocals are usually located here it can result in a distinct lack of energy.

LOUDNESS LIMITING (MAXIMIZER)

The final link in the chain is limiting through a 'maximizer'. The purpose here is to reduce the volume level of any peaks so that you can in turn increase the relative volume of the rest of the mix without fear of overloading the signal. Here, however, rather than employ a limiter it is more common to use a 'maximizer' since these are designed to produce a more natural sound by rounding off peak signal rather than cutting them short. This is the preferred approach to simply normalizing the file in a workstation since this can introduce unwanted artifacts.

When audio is normalized through an audio workstation or editor, the file is scanned to locate the highest waveform peaks and then the volume of the entire section of audio is increased until these peaks reach the digital limit. In theory, this raises the volume without introducing any unwanted artifacts, but in practice it will only raise the volume in direct relation to its peaks and we do not perceive volume by the peaks in a signal but through the average volume. When using a loudness maximizer, the dynamic range between peak and body of the file is decreased, in turn increasing the body of the music and thus the perceived loudness.

This obviously means that like compression, if maximizers are applied too heavily they will restrict the dynamic range and can seriously affect the excursion of the kick in relation to the body of the audio. As a result, they must be used cautiously and ideally, you should aim to maintain a peak to average ratio of at least 4 or 5 dB, or a little more if you want the kick drum to really stand proud of the mix.

Due to their simple nature, most maximizers consist just of a threshold, a make-up gain control (often known as a margin), a release, and the option to use either brick wall or soft limiting. These functions are described below:

Threshold

Similar to a limiter and compressor, the threshold is used to set the level where the maximizer will begin. By reducing this parameter, more of the signal will be limited, whereas increasing it will reduce the amount of the signal being limited. As a general rule of thumb, this should be adjusted so that only the peaks of the waveform (usually the kick drum) are limited by a couple of dBs while ensuring that the excursion between the kick and the main mix is not compromised to less than 4 dB.

Make-up Gain (AKA Margin)

This is similar to the make-up gain on a compressor and is used to set the overall output level of the mix after 'maximizing' has taken place. Broadly speaking, it isn't advisable to increase this so that the output level is at unity gain rather it is more prudent to set this to –8 dB lower in case any further processing has to take place.

Release

Again, the release parameter is similar to that on a compressor and determines how quickly the limiter will recover *after* processing the signal. Ideally, this should be set as short as possible to increase the volume of the overall signal but if it is set too short, the signal may distort. A good technique for setting a good release time is to adjust it to its longest settings and gradually reduce it while listening to the mix. As soon as the mix begins to distort or slice, increase the time so that it's just set above these blemishes. Commonly, the more limiting that is applied to an audio file, the longer this release setting will need to be.

Brick Wall/Soft Limiting

Most mastering limiters or loudness maximizers are often described as being as being brick wall or soft, and some will offer the choice to switch between the two modes.

If the maximizer is a particularly good model, then these will both produce similar natural sounding results but soft limiting may appear more transparent than brick wall depending on how hard the limiter is being pushed.

When using a soft limiter, the overall level can exceed unity gain if it is pushed hard but if not it will produce results that are generally accepted as more natural and transparent. With a brick wall setting, no matter how hard the limiter is pushed, it will not exceed unity gain. However, if the signal is pushed too hard, the limiter may struggle rounding off the peaks of the signal resulting in a sound best described as 'crunchy' or 'digital'.

GENERAL MASTERING TIPS

- *Avoid the temptation to master your own music.*
  All mastering engineers are impartial to the music and they will provide a subjective viewpoint on the music that you will not be aware of due to your personal involvement in the project. Even if you are not willing to

employ a professional mastering engineer, ask a knowledgeable friend or fellow musician to master it for you. Another person's interpretation may produce results that would never have crossed your mind.

- *If at all possible use full-range, flat monitors.*
  CD, vinyl and MP3 are capable of producing a frequency range that is beyond the capabilities of most near-field monitors and if you can't hear the signal, you can't judge it accurately. Mastering engineers will use monitors that are flat to guarantee that the music will replicate accurately on all sound systems.

- *Monitor at a fixed volume.*
  Always monitor at a fixed comfortable volume. Loud monitoring volumes will tire ears quickly and affect the frequencies of the mix due to the Fletcher–Munson contour control. Once the volume is set, listen to some professionally mastered tracks and compare them to your own mix.

- *Always reference your track with a professionally mastered track.*
  This is the best possible way to train your ears. By continually switching between your own music and a professionally mastered track, you will be able to keep some perspective on your current work.

- *Always perform A/B comparisons.*
  Our ears become accustomed to tonal changes relatively quickly so after applying any processing, perform an A/B comparison with the unaffected version.

- *If it isn't broke, don't try to fix it.*
  One of the biggest mistakes made by many novice producers is to begin adjusting the EQ and/or multi-band compression without actually *listening and identifying problems first*. Itchy fingers will inevitably produce a poor master, so listen to the mix thoroughly beforehand. The less processing that is involved during the mastering stage, the better the overall results will be.

- *Avoid using presets or wizards.*
  All music is different, so you should rely on your ears, rather than someone else's idea of what the EQ or compression should be. Also, keep in mind that the approach is different from one track to the next so mastering settings from a previous track will likely not apply to another.

- *Do not normalize a mix.*
  Normalizing algorithms search for the highest peaks and then increase the volume of the surrounding audio but the ear doesn't respond in the same way. Rather, it judges the loudness of music by the average level.

- *Do not apply any processing that is not vital to improving the sound.*
  Simply because some may mention that a certain process is used in mastering do not take it for granted on your own mix. Every DSP process

comes at the cost of sound quality. The less processing that is applied to a mix during the mastering stage the better it will inevitably sound.

- *If possible, use a VU meter to measure the levels of the mix.*
  Although VU meters are considered old compared to the latest digital peak meters, VU meters are generally more suitable for mastering. A digital peak meter only measures the loudest peak of a signal, thus if you have a drum loop measured by one, it will seem as though the level is particularly high even though it may be quiet. Conversely, a VU meter measures the overall loudness (RMS) of an audio signal.

- *Keep the bit rate as high as possible and only dither (if required) before it is recorded.*
  Use as many bits as possible in the mix until it is ready to be copied to the final media. If you continually work with 16 bit throughout the recording and mixing, further processing such as raising the gain will truncate the bits. The more bits that are contained in a signal, the better the signal to noise ratio will be with less chance of distortion being introduced.

- *Use a top-quality sound card with the best software you can afford.*
  Poor-quality sound cards will reduce the overall quality of the end results and poor EQ, compression and dithering algorithms will reduce the quality further.

- *Always check your mastered mix in mono.*
  If the track is destined for radio play, it is imperative that it sounds well in mono as well as stereo. Although a majority of radio stations broadcast in stereo, slight signal discrepancies can result in the mix being received in mono. If the mix has not been checked for mono compatibility, it could sound drastically different.

- *Use a professional mastering plant.*
  If the record is for the public market, you should seriously consider delivering the recording to a professional. The difference can be truly remarkable and worth the additional outlay. Mastering is a precise art and you really do need golden ears to produce an accurate master copy.

CHAPTER 29

Publishing and Promotion

'We have always thought we were better than we were. But now we're as good as we think we are. Although, maybe we aren't'

Nicholaus Arson

Once your music has been written, arranged, produced, mixed and mastered, its time to reveal your artistry to the world. The most common approach for this is to first begin by creating an account on SoundCloud and uploading your track to the site. This is currently one of the leading online communities for artists to upload their music and provides you with a distinctive URL for your music.

This means you can post your music directly onto social media sites such as Facebook or feed it through their API tools onto smartphones or iPads. More importantly, however, once the music is available online, you can hunt out one of the many independent labels who deal with your genre of music and then provide them with the link to the music and wait to hear back. Nearly all of the independent labels I've dealt with respond within a few days with a response.

However, signing with small independent labels can be fraught with problems since there are no guarantees that the company is reliable or trustworthy. In fact, even if you do find a good label, very few of them would work as hard as you would to promote your own music and when starting out, promotion can be everything. Consequently, a growing number of artists are choosing to release their own music and promote themselves through creating their own personal labels.

For more information about starting and running a personal label, Simon Adams has contributed this chapter to the book. He has run a number of independent labels, produced his own dance music act and currently runs his own music promotion company with Katy Jacks.

PRACTICAL STEPS TO SET YOURSELF UP IN BUSINESS AS AN INDEPENDENT ARTIST

Using the power of the Internet, artists now have the opportunity to bypass the traditional record label and release tracks themselves. However, the one thing that is absolutely critical to your success as an independent artist is also having knowledge of the business areas of the music industry that used to be served by label staff.

My first advice to any artist who wants to succeed on his or her own is to treat your music as your own type of business and operate everything as your own mini record label. If you want to sustain a long-term independent career in music, this is the single most important piece of advice you will ever read.

When major record labels sign a track, they will generally deal with things like registering your songs with royalty collection agencies, clearing samples you may have used in your track, tracking down publishers to make sure any cover versions are attributed to the right songwriter, prepare marketing for the track, getting press and publicity and promoting it to radio and DJs to name just a few things. Since record labels are signing less and less tracks over time, if you want to compete in the music industry as an independent, you will need to learn to take care of these tasks yourself.

What I've put together in here are some tips and techniques for any artist that is starting out self-releasing to build upon. The following sections offer a basic grounding on some of the areas you'll need to be aware of when releasing your own records, but the beauty of the new digital music industry is that artists and labels are adapting these principles and working out new ways to monetize their music so treat these as foundational starting points from which to branch out and expand your ideas.

By taking on board that by releasing your music yourself, you are in fact starting a business; you'll set yourself a firm footing for its success. By taking the decision to fully commit to treating your music as a professional product, you'll find that making it a success will be a much more painless and easier process.

In the early days of starting up as independent, you'll find that you have to learn about and perform most of the administrative functions yourself. If you can, engage friends or family members with relevant skills to help out. If you have a friend or sibling who has great skills in Excel, for example, maybe they can help you set up a budget sheet to track your income and expenditure. Perhaps you know someone who has a flair for graphic design and could create your logos or album artwork for you? Calling in favours is one of the best ways of keeping costs low, however bear in mind if you have any sort of financial success, it's always important to start rewarding those that supported you in the early days with some compensation too.

Building a good team of people to support you is crucial to its success. If you cannot find the skills and just either don't have the time or really don't wish to get your hands dirty on a particular task, hire in the skills. There are many online freelance sites where you can put the work out to tender and get the best prices from professionals in their field from all corners of the globe. Some of these are detailed later on in this chapter.

Whilst there are many areas you'll need to master in the music business, there are a number of key areas you'll need to familiarize yourself with and understand specifically if you are going to be successful and these include distribution, publishing, promotion and marketing. We'll be taking a look at some of these in this chapter, with a heavy focus on one of the most important – promotion and marketing.

INDEPENDENT DISTRIBUTION AND SALES

Making money from your music by selling CDs and MP3 downloads has never been easier, thanks to the power of the Internet. Whether you have pressed a few hundred CDs or just recorded a live version of your latest set, it's now possible to have these sold online in independent CD stores, or in the case of MP3s sold on Rhapsody, E-Music and some other music download sites.

There are a number of independent companies termed as 'aggregators' and these effectively group together all the independent artists music for some download stores and deliver each week a total bundle of content to them. They also take care of collecting the money from the stores when your tracks are sold, and pass this money to you. Some aggregators will take a small percentage of each sale and send you the rest; this is of course to cover their service costs such as staff costs to deliver your music, time etc.

If you have CDs that you have pressed and you want to sell them online, it can be difficult and expensive to set up your own web shop, and as there are businesses with all the technology in place already, why not take advantage of their services and let them do all the selling and money handling, whilst you just collect the cash! One of the leading online CD retailers is CDBaby.com who have been selling independent music CDs online for over 10 years.

What independent CD distributors like CDBaby require from you is that you sign up with their service and send them initially a small amount of CDs (usually around five), which they will then put on their site for sale. They will also ask you to fill in some general information about the CD and your band so they can put up a promotional page in their shop and feature the information to increase your chance of sales. Fees per CD sold depend on the distributor, but CDBaby currently charge $4 per CD, so if you sell a CD for $9.99, then you get to keep $5.99 per CD, which is much more per item than if you were signed to a major record label!

Recently CDBaby have introduced a scheme whereby they will press the CD for you in small quantities. All you have to do is supply the audio file, artwork and release information and their pressing plant will do the rest, including stock your CDBaby store with the CD. One stops CD distribution for your music without leaving your home.

You can also use the distributor site to make it easy for fans at your gig to buy your CD as they sometimes offer you further promotional tools such as swipe machines for credit and debit card transactions. For a small fee, you can also purchase a barcode for your CD from the distributor (CDBaby currently charge $20 for this), which means that your sale will count towards a chart position should you make enough sales. It is always worth doing this, just in case you sell a lot of CDs because the chart position, however low or high will give you more exposure and reasons to shout about your success.

Many of the independent online distributors will also put your music into physical distributor catalogues, so your music will be available to buy from real stores (CDBaby put their CDs into Chicago-based 'Super D Phoenix One Stop' that supplies many large physical retailers such as Tower Records, Target and many more).

The online distributors will generally have a list of all the stores that your music will be available in worldwide on their site, so you can also contact your fans directly and let them know that they can go into their local record store, ask for your release and the shop can order it for them.

If you also want to increase your potential earnings from your music, then make sure you always opt in for the maximum amount of services these independent distributors provide (mostly at no cost). This can include things like ringtone sales, online radio play and promotion, synchronization (your music in TV and Film) and many more.

You can also avoid upfront pressing costs for your CD by using the Amazon CD on demand service. At the time of writing this is only available in the USA, but hopefully Amazon will extend this to Europe and other parts of the world soon.

The CD on Demand service provides an excellent opportunity for artists to sell physical CDs through Amazon directly to the public, without any upfront pressing costs. Only when the customer buys the product, the CD is produced and shipped directly to the customer with a full colour CD liner and tray, removing the need for artists to invest in CD stock themselves.

There are also a number of hybrid business models for delivering music to the end user that bridges the gap between CD and download. Download cards such as those issued at Digstation.com offer a way of selling physical products at gigs, retail outlets and personal appearances without having to invest in large quantities of upfront stock.

New digital distribution models are appearing all the time. For example, a new site called Popcuts.com is running a successful site selling independent music to fans by offering the purchaser a small share of future sales on each track sold. This is a great way to offer an incentive to your fans to buy your tracks, as they also get paid every time you sell a track, a real win-win situation for everyone.

Watch out also for promotions that online distributors sometimes run. Recently, online distributor AmieStreet.com ran a promotion where new customers got $5 free to spend on music when they signed up to the AmieStreet website for free. If you placed your music onto the AmieStreet website during this promotion, then tell all your fans to go buy your music through the site, your fans get your music at no cost and you get paid for the tracks by AmieStreet.

SELLING YOUR MUSIC

One of the key things to remember, and probably the most common mistake that independent artists make is that getting your music into stores is not 'selling' your music. You are merely making it available for purchase, which is a very different thing. Selling your music means developing a sales and marketing strategy to drive people to your stores with a convincing message on why they should purchase and then working out method on converting the 'listener' into a 'buyer'.

I'll give you some tips on marketing and promoting your music to drive people to your stores and websites later on in this chapter, but I'd also advise investing in some good specific business sales and marketing books such as 'The Idiots Guide to Marketing', George Ludwig's 'Power Selling – 7 Strategies for Cracking the Sales Code', Michael Port's 'The Contrarian Effect' and some of the books by Seth Godin as well to get a good grounding in the whole process of turning an interested fan into a purchaser of your music.

Whilst a complete course on sales techniques is beyond the scope of this chapter, to give you a real-world practical idea on how you can approach selling your music to your potential target market, lets look an example of driving traffic to your iTunes store through the popular social networking site Twitter.

First we need to identify a person who regularly buys tracks from the iTunes store. It's no good asking someone to buy your track from iTunes if they are not in the habit of buying music, or if they don't use the iTunes store. So the first thing we need to do is find lots of people who actually use the iTunes platform to discover and more importantly regularly purchase music there.

There is really a simple way of doing this, using a very clever keyword search on Twitter. Go to http://search.twitter.com and type in any or all of the following phrases:

- Bought from iTunes
- Bought on iTunes
- Purchased from iTunes
- Purchased on iTunes
- Downloaded from iTunes

With any of these searches, you will immediately see a whole list of people who regularly buy their tracks from iTunes; they actually put their hand in their pocket and buy the music. They are not interested in going onto file sharing networks or trawling the web for a free MP3, these are people who buy their music from a legitimate download store.

1. Identify the buyer's musical taste
   At this point, whatever you do, don't just follow everyone who buys from the iTunes store, you'll waste both your and their time and effort; you need to hone down which of these people will be interested in buying YOUR music.
   OK, so now we have identified someone who is committed to buying music, now we need to find out whether our style of music would be something they'd be interested in buying too. Read each of the posts that you find using the twitter search; most of the tweets will mention the band name in the tweet:
   #Djstephie: Listening to Erika David – My Heels http://qtwt.us/lwr, bought the tune from itunes gotta show support.
   Now go off to your favourite search engines and search for the artist they purchased. In the case of our example tweet, searching for Erika David on Google brings up her MySpace page where she is described as R&B/Soul/Pop. Does your music fit that genre? It does? Then it is very highly likely that *#djstephie* is going to listen to your music and buy it from iTunes too if she likes it.
2. Build a relationship with your buyer; get them as a long-term fan
   OK. So now we have identified the ideal person who is most likely to buy your music, and we have identified that they buy their music from the very place you have your music for sale. Now we must build a relationship with that person, don't ask for the sale right away.
   Follow the person on Twitter, drop them a message, ask them about what they do and engage in a chat with them about your music genre. Mention you are producing music in a similar style to what they are currently listening to. They are guaranteed to ask you for a link to your music, and when they do, give them the direct link to your iTunes store.
   Repeat the process on each tweet you identify as a match to your music.
   As you can see from the above steps, you'll make more sales every day from your music by being efficient and working smarter only on potential fans who are likely to be interested in purchasing your music. They are your ideal fans, and if you nurture them, they'll buy everything you release on iTunes and champion your music to others too.
   If you want to automate the search for potential buyers of your music, use the free TweetLater service from http://www.tweetlater.com. This web application allows you to enter phrases to look for on twitter such as the ones I mention in step 1, and it will send you an email digest on a regular basis of the tweets it has found matching these phrases. However, all that will do is identify your ideal music buyer, you must invest the time in steps 2 and 3 to make it personal, and when you do that, you really leverage the power of direct contact to 'close the deal' and secure a lifelong fan.

This is just one example of how you always have to be focused on the way your fans think about buying music, what their buying patterns and habits are, and more specifically, how you can befriend them and secure an ongoing business relationship with them. If you adopt this way of thinking from the outset, you'll set yourself up with excellent foundations for a strong ongoing business.

PUBLISHING AND PUBLISHERS

Publishing is one of the more misunderstood areas of the music industry, especially in the dance music scene. When you write an original song, copyright laws automatically protect the underlying music and/or words as soon as you write down the lyrics or record the music. You own that copyright, and it's a valuable commodity that can be traded, and the trading of the copyright of the underlying music and words of a song is in a nutshell – publishing.

You don't have to register your works anywhere to get any better protection than the existing copyright laws already afford you, but if you feel you need some sort of protection or proof that you wrote a track first, simply record it to CD and print up a lyric sheet, send it to yourself through the mail in a registered envelope, and don't open it when it arrives. You'll have a time-stamped sealed copy if anyone ever contests the ownership of the song. Don't forget to write on the back of the envelope what it contains.

Traditionally, publishers started out selling books of printed sheet music, but in the modern era, this now accounts for a tiny percentage of their business if any at all. Music publishers and the music publishing side of the music industry is usually responsible for the collection of royalties from radio play, use of your music in TV and Radio broadcasts, films, games, gyms, clubs, samples and anywhere else your music may be played or used.

When you are starting out as an independent artist, you may wish to hold the publishing rights to your songs within your own company, especially if you are also performing the songs as the artist. However, if you write songs, music or lyrics for other people to record (other than yourself), then it may be worth trying to find an independent publisher with more contacts or industry leverage to take your songs and pitch them to other artists to record.

One of the main reasons that artists will assign their songs to a publisher is that they will actively track down any usage or royalty paying and carry out the administration in collecting and calculating the royalties owed to you, and pay you (mostly every six months). Of course the publisher will take a cut from the royalties to cover their expenses of collecting the money that is only fair and should be detailed in any publishing contract with your publisher. When thinking of assigning your publishing rights to a publishing company, always take legal advice from a lawyer specializing in entertainment law before signing away any rights to your music.

Contracts are not to be taken lightly and when you sign away the rights to your music without good advice, you may end up losing the ability to do anything with your songs. On the other hand, a good publisher will work hard if he really believes you have a great catalogue, as he or she wants to maximize his or her cut on the songs too. Finding a good publisher to represent your songs is all about finding out how good the publishing company is. Quiz the publishers about their past successes and experience, and make sure you find out what they can do for you, not the other way round.

If you decide to keep the publishing rights within your own company, of course, you are going to be the one having to track the usage, which can amount to a lot of work and if not done correctly could leave a lot of money sitting on the table uncollected. If you start gaining some traction and popularity in the music industry, you'd probably be best served to leave the publishing to the professionals.

Whether or not you are running as your own publisher or have signed a deal with an independent publisher, you will still have to join the royalty collection agency for your territory. These agencies are national agencies that broadcasters, pressing plants, record labels, and other users of music have to pay in order to play or reproduce music. For example, in the case of music broadcasts, radio stations pay the royalty collection agencies based on a sampling of the music broadcast, when a CD gets pressed the record label manufacturing the CD has to obtain a license to reproduce the music. If your music is being played on the radio, it is being used in a compilation, or being reproduced, you generally are owed money through the royalty collection agencies.

In the United Kingdom, you'll need to join the PRS and the MCPS, in Germany, it is GEMA, in the Netherlands, it is Buma/Stemra and in Sweden, it is STIM. You can find a full list of royalty collection agencies from each country with their respective website addresses at Wikipedia

http://en.wikipedia.org/wiki/List_of_copyright_collection_societies

For synchronization (TV, film, advertising, in game usage) where your audio is used in conjunction with pictures, you nearly always have to deal directly with the companies involved in the production of the visual media and negotiate a fee for the usage of your music with them. A good publishing company would have good contact with music supervisors from across the globe, so it may be worth assigning a publisher just for this area, however bear in mind that getting your music into visual media is a highly competitive area of the music industry, and the big earners in this area spend most of their time solely concentrating on serving music supervisors to the exclusion of everything else.

There is also a new breed of online synchronization companies springing up that work on a non-exclusive basis, pitching your songs to advertisers, film, TV and gaming companies for use in their productions. As most of these are non-exclusive, and you are free to do whatever else you want with your music whilst signed to them, if you decide to keep your own publishing in house, you have nothing to lose by allowing them to pitch your music to potential new opportunities. Some of the more popular licensing sites are http://www.libertymusictrax.com, http://www.musicsupervisor.com and an excellent list of other sites is online at Indie Music Tech.

MARKETING AND PROMOTING YOUR MUSIC

One of the biggest mistakes that independent artists make when setting up a label is thinking that the world will beat a path to your door for your music because it's good music. The reality is that they wont, you are going to have to go out and convince the general public, music industry professionals and many other people that your music is worth listening to, and ultimately worth buying or investing in.

Thanks to Rick's advice on production in this book and your talent, you should already have a great master, but here's the rub. If you thought it was hard work making the track, I can tell you the real work has only just begun.

A lot of artists start out with the notion that good music will sell itself. Whilst that may be true once a lot of people know about it, as an independent artist, just putting up your music in iTunes means you are a small fish in a big pond. Digital distribution is a tool to facilitate sales of your music; you have to invest time, money and effort telling people about your music and knocking on doors if you are intending to make any money back from it.

The first thing you have to accept is that marketing and promotion is always a cost, both in money and time. There are many free ways of promoting your music using the Internet such as building up a fan base on social networks like Twitter, Facebook, MySpace, trying to get your latest video to go viral on YouTube, and there are lots of new innovative online music promotion services popping up on what seems to be a daily basis.

The real business of promotion and marketing however is a lot more complex than just building a fan base online. As in any business, if you don't invest in things like advertising, trade shows, PR and other forms of publicity, you're just not going to gain much attention, or if you do it will be short lived and uncoordinated.

The key thing to remember is that you are running your record label as a business, so bear in mind that when you are starting out, the cost of promotion is going to be a lot higher than in subsequent years, as you are starting out from the point at which no-one knows about you. If you spend £5000 on marketing your music, bear in mind you'll probably have to sell around 10,000 to just break even on the marketing budget alone.

Some forms of promotion are 'ambient' such as press releases, articles and reviews of your music. If these are run online, you can usually see an increased visibility of your music in the search engines over a long period of time, which helps new fans discover the music. Other forms of promotion are more specific, such as online advertising banners that have a specific call to action message to encourage people to buy your latest single for example.

When considering any type of advertising campaign, it's always important to make sure that you are advertising your music specifically to your target market. Don't advertise on general sites hoping to pick up new fans as your message will be overlooked and you'll spend a lot of money for little return.

Seek out niche websites and blogs that specifically attract fans of your genre of music. Spending £800 advertising your latest dance release on Residentadvisor.net gives you a better return on investment than spending the same amount on a Facebook ad for example.

Since you've taken great advice from Rick on how to make a great track and you are now distributing and publishing your music, let's look at a number of ways you can start building a portfolio of promotion tools for your label.

DEVELOPING A WEB PRESENCE

The most important part of your promotional portfolio is your website. It is where thousands of people will discover your music and image, so it's important to make sure you have a professional looking site that is easy to use, looks good and is easy to buy from.

You first need to get some web space and a domain name. If you have an Internet provider for your web connection, they may have given you some free web space with your account. If you are strapped for cash, this is a cost-effective way to get on the web for little outlay.

If you want to look truly professional for a little more money you can get your own personal web space and your own domain name for your band. www.mybandname.com looks much better than using your personal Internet provider space such as www.myisp.com/user/mybandname and your own domain name is so much easier to remember.

You can register a domain name through a hosting company and usually if you purchase web site storage from them, they usually provide some FTP software to make it easy to send your finished website to their servers.

For more information on web space, get in touch with a local service provider. As this is a crowded market, you can find some good deals out there now, but in my experience go with a company that has been around for a while and has a good long-term reputation, rather than the cheapest. The main reason for this is that if the company goes bust, you may not only lose your website, you may also lose the ability to gain access to your domain name too, so reliability over price wins every time here. Once you have your web space, you can now get busy creating your music website.

So what is the best way to get your website built? Well you can try to program it yourself, but in all honesty you are a musician and your time is better spent making music than programming. If you do fancy having a go at making your own website because it's something you are interested in learning, then the easiest way is to get a copy of Adobe's Dreamweaver. It makes creating websites as easy as writing a letter in a word processor, but again if you don't have time to learn design skills too, you will be disappointed with the results. So if you want to make life easier for yourself, why not either purchase a website template and just adjust the text and pictures to suit.

There are plenty of sites out there that will sell you a pre-built website template that looks professional, all you need to do is insert the information about your band and change the pictures in the template to yours. This is a great way to get started on the web, but bear in mind that there may be other sites out there that look similar to yours because they will be buying the same templates.

If you want a personalized website built, it does not have to cost a fortune to hire a web designer to do it for you. With the advent of the Internet, we have now seen a number of 'virtual assistant' websites appear, where you can hire a web designer anywhere in the world.

Because of the scale of economies worldwide, this means that you can hire a great website designer in another part of the world where your money is worth much more to the designer than it is in your own country. Many of the new emerging economies from Eastern Europe and India make it very cost effective to hire a web designer and the skill sets are excellent.

Sites such as Rentacoder.com, Elance.com and Getacoder.com all have hundreds of designers that would love to build your site for you. The way these sites work is that you post your project on their site and lots of designers bid to win your project. This means you can get a great deal from these sites, as the bidders try to get competitive by keeping costs lower than their competitors.

One thing to note on these sites is that you must research the potential designer well, ask to see previous work they have completed and make sure you ask questions and make it very clear what you expect before you choose the designer so that things go smoothly with the project.

The great thing about the virtual assistant sites is that they operate a payment process called 'Escrow'. This basically means that you pay the site the money for the project, the site holds the money for you until the designer has finished the project, and only when you are satisfied with the results and your website has been delivered to you, do you release the payment to the designer. If the designer doesn't complete the job, or cannot fulfil your agreed requirements, you can ask for the money back from the virtual assistant site.

Never pay designers up front over the Internet, it's against the conditions of most of the good virtual assistant sites, so if you are asked for money up front, first say no, then contact the virtual assistant site and tell them about it. It's not very common for this to happen as most of the designers are freelancers who rely on these sites for their income, but as in any walk of life there are always a few people who break the rules.

On the design front, it is most important that when visitors arrive at your website, first that visitors can hear the music right away, and if they like what they hear, they can make a purchase easily without wading through lots of pages to find out how to buy your music. Make sure that people have the ability to buy and hear your music from the first page of your website. Make this clear to your designer from the outset by incorporating SoundCloud streaming.

SOCIAL NETWORKING

Social networking sites are the real power of marketing your music through the Internet. They thrive on the whole viral networking scene. Whilst bands have become noticed on the Internet through sites like MySpace and YouTube, it is not enough just to throw up a page or upload music to SoundCloud and hope that people will come by. However, most of the networking sites provide a lot of tools to make it as easy as possible to reach lots of people all at once.

In the past few years, Facebook has grown to become a giant in the social media platforms. It boasts 350 million active users; the average user spending 55 minutes logged in every day. It has far more users than Myspace and SoundCloud combined and it offers incredible tools for musicians and bands to continue to grow a fan presence.

There are two types of pages musicians can create and utilize on Facebook. The most limited type of page is a group page. The group page is clean and simple. You can post an identifying group photo and a brief description of the group. Most of the activity on the page takes place on the comment wall. That's pretty much it. However, the group page has a big advantage over other types of pages because of its 'bulk invite' feature. When a band has a gig, it can send out invites to all its group members in one shot. The group members who receive the invitation can also use the bulk invite feature to invite all of their friends to the event as well. It's a great tool used to tell fans about shows, album releases or anything else.

The second type of page is a fan page. They allow the users to post rich media like videos, photos and MP3s. Perhaps the most important feature is the option to add Apps. Facebook apps allow you to sync outside promotional tools like twitter, iLike and Reverb Nation with your Facebook fan page. Using Apps, bands can do everything from selling tickets, updating 'tweets' and creating mailing lists to viewing marketing and statistics information on who is listening, buying and sharing their music.

Without Apps, fan pages would also be limiting. Bands don't have a bulk invite option for fan pages, so sending updates about new tracks or videos is almost impossible. Messages are sent out to members but it's only on their news feed, and odds are it will get buried before the user ever reads it. The only other advantage fan pages have over group pages is the ability for unregistered members to view fan pages.

The best bet for a musician or bands is to create both a fan page and a group page. The fan page can have all of the extras that a group page lacks, and the group pages can be used to send out bulk event invitations and messages to users. Update the pages often, but try to use your messages and invites sparingly so that members are more likely to open invitations and respond.

There are more and more social networking sites springing up every day. Sites such as Ning, Last.FM and iLike make it easy to promote your music too by streaming your tracks for people to hear before they purchase them. Make sure all of your social networking sites have links to your web pages and the online stores where you music can be purchased. Give yourself every opportunity to make a sale to an interested party.

iLike in particular is very useful as they have links to a specialist iLike chart in Billboard magazine, so you never know, if you suddenly take off big time on iLike, you could be featured on the chart of one of the biggest industry magazines. Success starts in some of the strangest ways..

If you have a band video, then it is important to get it on You Tube, this can drive serious traffic to your website, where people will find out about your music, and hopefully buy it. Its not just music videos that you can post to YouTube get brainstorming. Post-artist interviews, videos of your gigs, short documentaries, maybe your band members have interesting stories to tell, anything that fans can watch to give them an insight into your music will get you more hits, more exposure and more sales.

One of the latest artist-based social networks ReverbNation has come up with some clever tools to help you manage your fans. They provide excellent widgets that you can manage from a central control room on their page, from song players, retail widgets, show schedules and much more. Post the ReverbNation widgets on all your other social networks, and when you update the ReverbNation widget, all your widgets across the web automatically update at the same time.

Although initially it is time consuming to register with all of the social networking sites, it is in your best interest to do so, because not only will it increase your rankings in the search engines, but it increases the number of people who are exposed to your music. Get to as many of your potential customers for your music as you can.

PR AND PUBLICITY

PR and publicity is such an important part of your music. You can make the greatest music in the world, be the coolest or tightest band in your genre, but without PR and promotion no one will ever know about you. PR and promotion is all about shouting about yourself to a highly focused group of people who will be interested in hearing your story. There are many great ways to promote yourself using the Internet and therefore if you are not doing at least some of the ideas listed in this section each week, you may as well stick to a standard day job.

A great way to get promotion that will help your sales and boost your ranking in the search engines is to get your releases reviewed in the appropriate section of the website About.com. About.com is run by the New York Times, and it is one of the world's most respected online sources of information. There are sections of the website that cater for every type of music, so to increase your chances of getting your music reviewed, make sure that you research the site to identify the correct target category.

Each section of About.com is run by a host, who edits and manages his or her section, the host is usually a leading industry figure or freelance journalist who knows his or her field well. The host should be your first point of contact and occasionally they have a number of people working with them that deals with specific areas, such as live work, sub genres, music releases and so on.

Radio promotion is also a key factor. Radio DJs are your sales team, getting your music into the ears of new listeners and hammering home your music. Although online radio is gaining ground, FM and AM radio still rules most peoples lives (think about it, when do you listen to the radio in your car? on the move, in the shopping mall).

Major record labels spend hundreds of thousands of pounds getting their songs onto play lists worldwide, so how can you compete with this? The main way is to know your market and build up contacts on local stations, college stations, restricted service stations, community stations and so on and make personal contact with DJs and station program directors directly. To make a serious impact on radio, you have to be in a lot of places at once, and it takes time, effort and good contacts built up over a long period of time.

Another important way of generating online awareness of your music or band is by writing regular press releases. Get into the habit of publishing one press release every week, and you'll soon find that not only does it increase the ranking of your website on the search engines, but it gives you a regular reason to keep people's interest in your band. Sustaining interest is one of the hardest things to do so regular press releases really help you focus on this part of your PR strategy.

Press releases can be about anything regarding your music or band or band members. Maybe one of your band members is involved in a charity run at the weekend, maybe you have just finished rehearsing one of your new songs, have you got a gig coming up at the weekend, did you just publish a new website, maybe you got your track played by a DJ at a club, whatever the story big or small, get it out there every week.

Don't forget to archive all your press releases on a section of your website, as this will show constant activity and updates to your site, visitors to your website will always be able to see new information weekly. This also means that most search engines will see changes to your site so send a robot off to your site to rescan it again, and depending on the content of your press release, this could also increase the ranking of your site on the search engine.

TRADE SHOWS

Just as in all industries, every year a number of trade shows take place where you can gain information on the music business, meet companies and people that will help you make your music career more successful. Music business trade shows usually have a conference attached, where you can hear the latest requirements from A and R executives from both the independent and major record labels. The conference will also include seminars from leading industry figures talking about the latest trends in the music industry.

The seminars will give you invaluable business intelligence from within the music industry and will give you an edge when it comes to trying to sell your music. You will know who wants what and why, and you will have more success in the music industry because you are focused on the facts of what the industry wants from you, rather than hoping someone will 'discover' you.

Recent topics at PopKomm, the leading music conference in Berlin, Germany, for example, related to getting your music sold independently online, getting filmmakers to use your music in their films, and selling music through mobile phones direct to the consumer.

All these ways of making money from your music are now available to independent musicians like you, thanks to the power of the Internet, and you don't have to be a big multinational company any longer to do any of these things. Some of these ideas are explored in more detail in the chapter of this book covering independent distribution.

Use trade shows as ways to meet potential partner companies and people who can help you take your music to a wider audience. When you register with the trade show organizers, you will get access to the names, addresses, email, website, and direct telephone numbers of all the companies, executives, and people that are attending the show.

Make sure you plan well in advance by researching each person thoroughly to ensure you are a good match – do not meet people who are not matched to your music, concentrate only on the relevant people in your genre to ensure the best use of your time. After you have researched the prospective people relevant to your music, you should email each person on your list, and follow up with a phone call. Be persistent, they will be keen to meet you, but you have to pester them to get a firm appointment as they are usually very busy people.

Ask them for a meeting at the conference, get a firm time to see them, make sure in your initial contact you screen them again by asking lots of questions about what they are looking for at the show, and see how your music would fit to their needs. Get them to commit to a time to meet you. You now have a chance to do real business with the people who can take your music to the next level, your future business partners with whom you will share your success!

Plan each meeting at the trade show by writing a list of 10 questions you want to ask so you feel confident about what you want to get out of the meeting.

Make sure you take a list of your meetings to the show, with everyone's mobile numbers written down. Trade shows can be crowded places, and it's easy to miss people when there are hundreds of people crowded in the lounges and stand areas, so five minutes before your meeting give each prospect a ring so you can make sure you don't miss each other. When you have the meeting, keep it to no more than 15 minutes. This is a great marketing ploy, which shows that you are busy with other appointments, and you have no time to waste because you and your music are in demand.

Make sure that you make an action point review on what any next steps might be after the meeting, and one week after the trade show, follow up with an email or phone call to the person you met to progress your actions.

You will be surprised how quickly opportunities within the music industry become available when you follow this plan. Just like any business, it's made up of people communicating with one another, and trade shows offer an ideal opportunity to get out from your rehearsal room, recording room, or just from behind your computer screen and meet the people involved

MUSIC VIDEOS

Without a visual element to your music, you'll be missing out on a large volume of people who could be discovering and buying your music. Promoting your music with video is something you cannot do without and should account for a large part portion of your marketing and promotion efforts.

It's not that expensive or difficult to make your own music video these days, and with the rise of the You Tube generation, you no longer need to pay MTV lots of cash to show your video. With the right idea, a viral video can get over a million views on the Internet and provide amazing coverage for your band, supporting the sale of your latest CD or ticket sales to your gigs.

First you will need to acquire footage to shoot the video. You can either buy or hire a consumer DV Camcorder for this purpose; great cameras from Sony, Canon, Olympus, and other manufacturers now cost such a low price now and can provide great results for web video.

If you are really strapped for cash, buy a cheap webcam and hook it up to your PC (laptops and a webcam can also be taken to a location to film remember. great idea, just remember to watch your battery life on the laptop!). If you cant find a location for your band to film footage for your video, how about setting up a virtual studio in your living room for next to nothing.

Television studios use a technology called Chroma key, where a person is standing in front of a coloured backdrop (usually green) and performs, then when the video is put into a computer the green backdrop is replaced by other video backgrounds and effects, enabling you to be anywhere in the world (or the universe!). This is also called 'Green Screen' filming. You can get the correct green screen backdrop cloth surprisingly cheaply from online stores such as TubeTape.com

The major thing to be aware of when filming your own video is lighting. Good lighting is essential to make the most of low-cost video camera filming. Whilst specialist video lighting can be really expensive, there is a little known secret to low-cost video lighting. Industrial halogen work lights and stands can be bought from your local DIY store for just a few pounds each and do the same job.

They use similar bulbs and lighting sources that you would get from pro video lighting. In many cases, the lights are built much sturdier as they were meant for building work. So for next to nothing you can probably get five to six 500 W halogen lights for lighting a video set! Get a couple of extra stands and a sturdy plastic pipe to run across the top of the stands from the plumbing section of your DIY store and make a real low-cost stand to hang your green screen backdrop also!

Make sure you are careful with the lights though, read the instructions on the box, as halogen lights get extremely hot, and let them stand for up to an hour after switching off before handling them. Also they use a lot of power, so be careful how many you plug into an extension socket. If in doubt, always get a qualified electrician to help you. The leading online retailer for every Chroma key is tubetape.com, who has lots of cost effective green screen backdrops, suits and other video goodies!

Once you have got your footage onto your camcorder, you need to transfer it to your computer. If you have a DV camcorder, you will probably need to plug your camcorder into your PC's fire wire port. Consult your PC or Camcorder manual for assistance with this. If your computer does not have a fire wire port, you can buy small fire wire interfaces from your local computer store quite cheaply. Some camcorders now record to memory cards instead that are easier to just insert into your computer.

When you have your video on your PC, it doesn't have to cost you anything to edit your video. Sure, you can buy video editing software such as Sony Vegas Pro, Adobe Premiere and so on, but to be honest, for most basic music videos destined for YouTube, you can get great results with Windows Movie Maker, which is included for free with Microsoft Window. You can even add titles and basic transitions between scenes to give Movie Maker a try before you pay for editing software.

Once you have edited your video, you can publish it on your own website and on the many video networking sites on the Internet. There are even new sites springing up with new ideas on how to make money with your videos. MuzuTV.com even pays a proportion of the advertising income generated from their site to you for showing your video.

MERCHANDISE

Selling your music is not the only way you can make money from your creations. If people like your music, they usually want to buy merchandise from the band too. T-shirts, mugs, mouse mats, all these can be sold to people who like your music and bought your CD.

It would be foolish for the independent band to tie up lots of money in stock for merchandise though, even if you are gigging, you are not guaranteed to sell all your T-shirts, and you have to store them somewhere in the meantime too. The Internet gives you the perfect solution with the rise of the Print on Demand shops such as CafePress.com, Spreadshirt.net and Zazzle.Com. For calendars and other printed matter such as books (for your autobiography or even sheet music books of your songs), Lulu.com is a great site. Print on Demand sites allow you to design and store your designs on their site, and only when a customer wants to order the product does it get printed by them and shipped to the customer. The stock is held by the online store, nothing gets printed until it is ordered, and there is no minimum order quantity.

Most of the Print on Demand sites allow you to set up your own online merchandise shop that you can link to from your website. The shop handles all sales transactions for you and pays you the profits. The shop sets a price for the basic blank item. You can then put a mark-up on top, making the retail price. Every time an item sells you get your cut and the shop keeps theirs. Whilst this may end up as quite a big percentage, remember that you have no upfront stock or printing costs so it's a 'no risk' venture.

It's also a great way to get your own personalized band gear printed too. Most Print on Demand sites do not charge you the mark up if you are ordering for yourself. Why not make your own custom T-shirt for your next gig, print up your album cover and website address onto a T-shirt or jacket through the Print on Demand shop and wear it when you attend trade conferences.

You'll be surprised how many people strike up a conversation and ask you about it. For greater impact, put a picture of you or your band on the shirt and watch people do a double take!

LIVE PERFORMANCE OPPORTUNITIES

Just as in every industry, the recent social media revolution of this decade has completely revolutionized the way that artists and bands can get quality playing opportunities through the Internet, often with a very low cost of entry. It used to be the case that in order to get gigs playing music, an artist needed to have an agent or a record label. This third party would then use networks and 'ins' to help the artist secure playing opportunities. This is no longer a prerequisite, thanks to the Internet's democratization of the music industry.

Just as sites such as CDBaby.com made it possible (and easy) for artists to distribute professionally-made CDs to a wide audience, so too have sites such as SonicBids.com made it possible and easy for artists to find gigs online.

How does a site such as SonicBids.com work? The process, simple and straightforward, is catered to ease-of-use on the artist's side of things. The first thing that occurs when you sign up for an account on the website is that you need to create an Electronic Press Kit (EPK), of which the first step is to write what is called an "elevator pitch." This "elevator pitch, one of the most crucial aspects of the EPK, will be the first impression promoters have of your services that you are offering…

There is also a media tab on the EPK page onto which you can add various forms of media. A good rule of thumb is that the more types of media you present to potential clients, the better your chances of getting hired. Giving potential clients as deep a look into the artists' services is very important. The goal of the EPK is to get to the interview itself, just as the goal of a resume in business is to get to the job interview.

Once your EPK is ready, it is time to start looking for work on the site itself. This process, if you have finished your EPK thoroughly, is fairly straightforward. On the main page of the site, you can look for gigs. Once you find a gig that you like, you can bid for with the credits that you have bought on the site. For example, it could cost $5 to place a bid to a particular promoter.

One of the most crucial things to understand about sites such as SonicBids is that there are far more people looking for work on these sites than there is work available. In such an environment, where demand for jobs far outweighs the supply of jobs, it is important to differentiate yourself from the pack. As noted earlier, the best way to do this is in your EPK – show the promoter what makes you different. Make sure that your best work is available on your EFK; ensure that your best foot is put forward in your very first correspondence with promoters; and of course, make sure that the gig to which you are applying fits your style.

Even if you don't want to use SonicBids to find work, you can use the features of the site, with a membership, to create a professional EPK. This EPK can then be sent to club owners and people who are offering work in your local area. More often than not in this day and age, club owners will have accounts on MySpace or business websites. Emailing a professional-looking EPK to such a club will surely differentiate your act from the rest of the local talent in the area, and it is a good way to secure local and regional gigs. Networking in such a way by building quality reference will then have a snowballing effect, as more references will bring you more job opportunities, which will then bring more references, and so on.

If you don't want to use SonicBids.com, then there are other sites available that are doing the same for musicians; that is, creating a community online on which you can find work doing what you love – making music. One such competitor to SonicBids is GigMasters.com. If SonicBids is the Elance of the music industry, then GigMasters.com is the Craigslist. It is evident from the layout of the site that GigMasters is much less visual-oriented and more content-oriented. The site is organized by category and by instrument, so searching that site, if you are familiar with Internet searching of any kind (especially of the Craigslist variety), then GigMasters will be much simpler and possibly more fruitful than SonicBids. However, the principle remains that you can use the same EPK that you created on SonicBids, perhaps transferred to your band's website, on these websites.

After you have found some gigs, either offline or, using these methods, online, you can play 'gigs' online in order to create additional buzz for your live show potential. How do you play gigs online? Isn't that a bit of a contradiction in terms? Doesn't playing live involve live people?

Yes and no. Yes, the real thrill of playing music in front of a crowd of people enthusiastic to hear your songs and who appreciate your art is unsurpassed. It's what musicians live for. It is impossible to pretend that this can be completely replaced in an online environment, like a chat room. However, it might be useful to give people the experience of being at one of your live shows without actually being there. This is possible through YouTube, the most popular video site on the Internet.

On YouTube, to play a 'gig' online, all that is needed to do is to record one of your gigs on video and split it up into tracks. Then, through the play list creation tool, create a play list that seamlessly blends all of the tracks of your 'show.' Create a page on your website inviting potential fans to 'see a show online!' On this page, it would be useful to advertise your upcoming gigs and to have a link to your EPK (potentially even your SonicBids EPK you created earlier!) This method of promotion has the added benefit of giving you more clips to share with potential promoters and clients who may want to hire you if they knew what your live shows were like.

The Internet has made it much easier to find work than ever before, and it has lowered the barriers to entry to a level never before seen in the music industry. Any indie band can now be their own agent and find their own gigs at a level that was impossible before the advent of Web 2.0.

SUMMARY

There are many more innovative ways to market your music, and new avenues for revenue generation are appearing all the time, so make sure you keep up to date with the latest industry trends at industry events, online blogs and other music industry news sources.

I hope that this section of Rick's book has inspired you to take the leap and start up your own record label or music production company. Whilst setting up any business takes hard work, resources and mostly blind faith, it can also be a lot of fun. You'll also make a lot of really good friends along the way.

One of my own key philosophies is that you should get in the corridor with what you have today, start something and adjust your strategies along the way. Whilst planning is important, once you get started, you'll find new avenues and opportunities opening up for every aspect of your business, and as I mentioned previously, there are many new technologies, services and opportunities starting up all the time in this age of digital innovation.

If there was one piece of advice I'd leave you before closing is that you should be committed to continual learning, and always be open to help and advice. Know when you need assistance, and above all have persistence, be consistent and be resistant. Keep the faith when times are difficult, and celebrate your successes when times are good. The music business is a roller coaster ride, so remember to enjoy the journey and the processes whilst you are on the way to your goals.

There has never been a better time to be an independent artist, and with the right mindset, the right set of tools and the determination and will to succeed, the achievement of your creative goals is one of the most satisfying experiences you will ever experience.

CHAPTER 30
A DJ's Perspective

'A DJ is there to participate. He should have one foot in the booth and the other on the dance floor'.

David Mancuso

It's fair to assume that dance music would be nowhere if it was not for the innumerable DJs spinning the music and keeping us all up on the floor. With this in mind, it seems that there is no better way to close the book than to interview a DJ and get her perspective and advice on DJing. For this, I asked Dakova Dae to submit a chapter for the book. Dakova is an Australian-based DJ and producer who has DJed alongside some of the biggest names in dance music. She has won numerous competitions for her DJ skills and currently holds a residency at one of Australia's largest venues; Family Nightclub in Brisbane.

Alongside her gigs, radio shows and productions, Dakova explores the concepts of DJing, electronic production and the music industry in her blog, Dae On Mezzanine. With a background in Law and International Marketing, Dakova enjoys combining her business and music knowledge to empower other's passionate about electronic music.

THE DJ SETUP

Setup Options

DJ setup options have been one of the most vividly progressive aspects about the world of DJing. In the early years, DJ options were greatly limited in comparison with today. Before CDs and digital music, for decades, turntables were the DJ's weapon of choice, and beatmatching and scratching were greatly revered. As technology evolved, CDs came into the picture, allowing DJs to acquire music significantly quicker and easier. Today, Digital DJs enjoy the benefit of taking their entire music collection to gigs via their laptop, and the introduction of MIDI controllers allows them to completely customise their setup and controls.

There are more options present than ever before when deciding what DJ setup to embrace. Ultimately the decision comes to personal preference, but before discussing the factors to take into consideration, let's examine each option; from a traditional vinyl setup to a fully customizable Ableton rig.

VINYL

Vinyl is the original DJ setup; two or more turntables playing vinyl records connected to a mixer. Although some DJs still swear by this setup as they consider the tracks to 'sound better' or 'feel warmer', this is an outdated medium and many clubs no longer house vinyl turntables. Furthermore, not many tracks are still released on vinyl, so keeping your music collection current will be more of a challenge.

Requirements	Recommended Gear
Two (or more) turntables	Technics SL 1210 Stanton STR8
Mixer	Allen and Heath Zone 32 Pioneer DJM 800
Slipmats	
Needles	

CDJS AND MIXER

This is the most common DJ setup, and almost all clubs accommodate it in the booth. Two or more CDJs are connected into a mixer, with each CDJ going into a different channel. The number of CDJs depends on the number of channels on the mixer; either two or four is most common. Many current CDJ models have USB connectivity, meaning you only need to bring a USB and your headphones to your gigs. Pioneer has have developed its own program for syncing your music collection to your USB, called Rekordbox. Older models only allow for CD playback.

A few additional advantages of this setup include that the mixer is already hooked up to the venue sound system correctly, which can significantly minimize pre-gig stress and technical issues. However, CDs and USBs can still fail without warning, so it is always imperative to have backups of all CDs and USBs with you.

There are a few drawbacks with this setup. Incorporating live synthesisers or instruments is difficult, so if this is a priority for you, consider a Digital DJ setup instead. Also, some clubs take better care of their equipment than others, leaving the condition of the equipment when you get there up to chance. As often the only equipment available to you is the one already in the booth, CDJs or mixers that aren't working properly is something you will have to learn to deal with on the spot.

Requirements	Recommended Equipment
Two (or more) CDJs	Pioneer CDJ 2000 Pioneer CDJ 900
One mixer	Pioneer DJM 2000 Pioneer DJM 900

TRAKTOR

Traktor is a popular DJ software that uses a virtual representation of the traditional two or four deck mixing. You can use the internal mixer in the program, or more commonly you can purchase external controllers.

There are two options with controllers for Traktor: built-in soundcard or external soundcard. If you already have a laptop, purchasing an all-in-one controller with a built-in soundcard is the quickest and cheapest way to have a DJ setup up and running. Controllers that require an external soundcard will be considerably cheaper than their counterparts; however, you will still need to buy a soundcard separately. Regardless of your controller preference, they can all be mapped out as required, which makes Traktor more customisable than CDJ-based setups. The learning curve with Traktor is also smaller than with CDJs, particularly if you want to bypass learning how to beatmatch by using Traktor's 'Sync' feature.

A drawback to this option is that you will need to bring your laptop to gigs. This can cause issues with theft or damage at a considerable cost. Also ensure that if you do choose this setup, you have a backup plan with you to all gigs in case of computer failure.

Requirements	Recommended Equipment
Laptop with Traktor software	Apple MacBook Pro (minimum 15 in. otherwise you will struggle to see what you are doing in the booth)
Controller with built-in soundcard	Traktor Kontrol S2 Traktor Kontrol S4 Pioneer DJM T1
Controller with external soundcard Soundcard	Traktor Kontrol X1 Traktor Kontrol F1 Traktor Audio 10 Traktor Audio 6

ABLETON

Ableton's popularity, both as a production workhorse and as a live performance setup, has skyrocketed in recent years. This is for several reasons. First, in a live setting, Ableton accommodates limitless possibilities of configuring controllers and instruments. Whether you are launching clips, a capellas, drum loops or live synthesisers, Ableton can accommodate it. Also, as Ableton is controlled by a global BPM, there is no need to learn how to beatmatch if you so choose.

Despite many advantages, it is worth considering that fitting a laptop, soundcard and all controllers and instruments in a small DJ booth can be both difficult and time consuming. Furthermore, the more components you add to your DJ setup, the greater the chance of gear failure somewhere along the line, so having a back-up plan is vital. If you rely entirely on Ableton to execute your live performance, preparing backup plans can be costly and stressful. Finally, remember the more complicated your DJ setup is, the longer it will take for you master it.

Requirements	Recommended Equipment
Laptop with Ableton software	Apple MacBook Pro (minimum 15 in. otherwise you will struggle to see what you are doing in the booth)
Various controllers	Akai APC 40 Novation Launchpad Vestax VCM 600
Soundcard	Motu Ultralite Mk3 Focusrite Saffire

CHOOSING THE RIGHT SETUP FOR YOU

Each DJ setup presents its own combination of positives and negatives. Analysing them and discerning which is the right setup for you will come down to many personal factors. When deciding which DJ setup to embrace, consider the following:

1. What is your budget? Top of the line CDJ and Mixer bundle will be the most expensive. If you already own a laptop, a digital DJ setup with a controller will be the cheapest.
2. How much time and space is available to set up in the booth? Do you want to, or have time to, spend 30 minutes setting up controllers in the booth?
3. Will you need to incorporate live synthesizers during your performance?
4. How many channels do you want to run simultaneously? Will these be full tracks or do you want to incorporate loops and a capellas?
5. How portable do you require your setup to be? Will you need to take your own setup to gigs or will there be equipment provided?
6. Is beatmatching important to you? With technological developments and the growth of Digital DJing, the emphasis and importance of beatmatching has significantly decreased. Is it still important to you to know how to beatmatch?

Genre Considerations

Certain genres of dance music lend themselves better to different setups. For example, due to the high density of sound in genres such as trance and main-room progressive house, layering more than two channels often sounds overwhelming, cluttered and unpleasant. In such genres, a setup incorporating Ableton running basslines, percussive loops, hooks, vocals, all the while doing live mashups plus live instruments can be overwhelming amount of sound for an audience. However, in other genres such as hip-hop and minimal techno where there is significantly more space through the composition, experimenting with layers in such a setup can open up vastly creative possibilities.

Your DJ setup is fluid; always explore new ways to expand and incorporate additional elements as you progress through your career.

At the end of the day, it is a completely personal choice, which setup you adopt. The most important point is to know your setup back to front, pick a setup that most compliments your mixing style and sound and feel confident using it.

The Essentials of DJing

At its most fundamental level, DJing comprises two core elements: *beat-matching* one track to another and *mixing* multiple tracks so that they transition without a break in the music. As human creativity and technology has expanded, these two elementary principles have significantly evolved, and the definition of the DJ has broadened. Many new aspects have come into the realm of DJing. Achieving exciting mixes can now include elements such as *EQing, effects, looping, a capellas* and *hot cues*. Emphasis on each varies from DJ to DJ, depending on the live setup, musical genre and personal mixing style.

THE BASICS

Beatmatching (Manual)

To mix one track smoothly into the current one, both tracks not only need to be pitched to the same speed (beats per minute or BPM) but also need to be hitting that beat at exactly same time. Let's examine how to achieve this:

1. Ensure the cue point of the track about to be mixed in is set exactly to the start of the first beat. This can be done digitally or manually depending on the equipment in use.
   a. Manual: If you are using CDs, cue points will automatically be placed at the start of every track. However, these are not always 100% accurate, and cue points will need to be manually set on CDs when the beginning of the track is not the first beat (as is the case in intro edits). To achieve this on CDs, play the track until you hear the first beat (most commonly a kick) and hit 'Play' again to pause playback. Move the platter left or right until you are right at the start of the first beat and hit CUE. This will lock that position as the new starting point of the track.

b. Digital: Some programs allow you to set the cue points digitally within the software. For example, for those using USB-based mixing, cue points can be set digitally within Rekordbox. To set a cue point in Rekordbox, open the application, bring up a track, move the cursor to the start of the first beat by dragging, right click and select 'Set Cue Point'. The cue point will be saved on your USB when syncing your music.

2. Once the cue point is set, you are ready to mix in the track. Hit 'Play' and the track will commence from the cue point you set in Step 1. Listen to the new track and try to detect whether it is a faster or slower speed than the track already playing. This is the skill that takes the most practice. Although the two tracks may sound like chaotic noise at first, try to bring your attention specifically to the kick or hi-hats of the each track; this is the best way to determine their relative speed. Is the newly introduced kick going faster than the other? If so, the new track needs to be pitched down or sped up by pitching the track up via the pitch fader. Even if you have matched their pitch, if there is even a slight delay between when each track hits the beat, the audio will sound chaotic. This is known as a double beat. Correct these small timing errors by nudging the Jog Wheel forward or backward. These nudges will not affect the tempo of the track, only move the timing of the beat a fraction of a second forward or backward.

Beatmatching is a process that can seem very overwhelming at first, but as your ears become attuned to focusing on the beat and your experience grows, you will become quicker and more confident with your beatmatching.

The only secret to effective beatmatching is practice.

Beatmatching (Digital)

Before the Digital DJ revolution, beatmatching by ear was one of the cornerstones of DJing; if an individual could not beatmatch, their career in DJing was greatly limited. However, there are many Digital DJ setup options today, which by-pass the need to beatmatch altogether, most notably Traktor and Ableton Live:

Ableton: Ableton incorporates a global BPM counter that automatically adjusts the pitch of clips and tracks imported into the program. Unfortunately, given mixdown variations, quality discrepancies and BPM fluctuations, the global sync can still occasionally allow tracks to glide out of sync with each other. To minimize this issue, it is highly recommended to 'warp' each track when importing into Ableton. This will ensure clips are re-pitched accurately and hold their BPM. To do this:

1. Import your track
2. Set the global BPM of your Ableton project to the original BPM of the track
3. Select the track in the Clip View panel (bottom panel in Ableton's Session View) and zoom in to the first beat

4. Add a track marker right on the first beat and delete any other track markers present
5. Right click on the new marker and select 'Warp [BPM] from here'
6. Repeat process with each track

Traktor: Traktor gives you a choice of whether to beatmatch manually or let the program do it for you by incorporating a 'Sync' button. If you leave 'Sync' off, beatmatching in Traktor follows the same process as beatmatching on CDJs since the program incorporates both a pitch control fader and nudge buttons. Using these can be done in the program itself using a mouse, or mapped out to an external MIDI controller. Alternately, if you choose to turn 'Sync' on, your BPM will be synced automatically between tracks. However, this will sync the speed only and not necessarily the timing of the beats. Small nudges forward or backward using the nudge buttons may still be required, similarly to the way you would use the Jog Wheel on CDJs. To minimize the need for nudge corrections, ensure each track's cue points are set right at the start of the beat. To set a cue point in Traktor, bring up a track, move the cursor to the start of the first beat, click 'Cue' and then click 'Store' to save that cue point for future use.

Phasing

Getting two tracks beatmatched is only part of the way to a smooth mix. At what point in a track you choose to mix in the next one will also have an immense effect on how smooth a mix sounds - this is due to phasing. There are moments in a piece of music when you can feel that a new phase is about to commence; perhaps, the track is about to enter a breakdown or the drop is about to come. Producers use a range of tactics to signal when these new phases are about to occur, but as a general rule of thumb, you can most often expect these to come in after every 16, 32 or 64 bars. Beginning to mix the next track at these key times will help the two tracks stay in phase and sound more natural when gliding from one track to the next.

EQ

You have figured out when to start mixing in the next track, you have beatmatched it; both tracks are in time and in phase; it is now time to bring in the second track. Unless your rig relies on launching samples or clips in a software such as Ableton, it is highly likely that you will need to turn to your Equaliser (EQ) to bring in the new track gradually. With the exception of MIDI controllers or custom-mapped mixers, most commercial mixers on the market comprise of a three-band EQ. The EQ is separated into Highs, Mids and Lows, with the frequency band divisions differing between mixers.

An industry-standard Pioneer DJM 800 has the following EQ response settings:

High (13 KHz)
Mid (1 KHz)
Low (70 KHz)

NOTE: Some Allen and Heath DJ Mixers utilize a four-band EQ, with the Mids split into a high-Mid and a low-Mid band. Deciding whether to go with a three-band EQ mixer or a four-band EQ mixer comes down purely to personal preference.

Selecting when and how much of each EQ to bring in during a mix is an entirely individual creative decision. Fundamentally, this will be determined by when and where there is space in the sound spectrum.

For example, if the current track is a maximal uplifting trance track, then there is unlikely to be much room in the Highs, Mids or Lows for introducing new sounds. Mixing too early before an existing element has left the arrangement will make that part of the frequency spectrum too crowded, muddy and unpleasant to listen to. With maximal genres, the most important consideration is to listen out for these gaps and only mix the relevant frequencies once such a gap opens up. More minimal genres, such as minimal techno, rely on sparse sounds to create their atmosphere and mood. These tracks have a lot more room for introducing new sounds during a mix. Therefore, they allow for much longer, drawn out and subtle transitions without fear of overcrowding the sound spectrum. As each track and genre carries its own personality, the key with EQ is to practice and experiment.

ADDING MORE

Once the basics of beatmatching, phasing and EQing are under your belt, there are countless additional elements you can incorporate into your live performance to enhance your impact in the booth and create your own distinctive style. Some of the most effective of these include effects, a capellas, samples, looping and setting hot cue points.

Effects

Depending on your DJ setup, effects can come from inside the box such as your mixing software's effects or outside the box including built-in hardware mixer effects or specific external effect units like the Pioneer RMX 1000, or the older Pioneer EFX-1000. Regardless of the source, universal effects include:

- Filter
- Delay/echo
- Reverb
- Pan
- Flanger
- Phaser

A Capellas, Samples and Looping

Adding a distinctive flavour to your DJ sets can be achieved through creative use of a capellas, samples and looping. Although these were a lot harder to execute on traditional vinyl and CDJs, the rise of popularity of digital DJing means triggering clips, and preparing loops is quicker and easier than ever before. A simple addition of a controller or specialized effects unit is the way to start.

If layering multiple a capellas, samples or loops is the direction in which you want to take your live performances, Ableton Live is definitely worth exploring. Its clip trigger and custom routing options are currently unparalleled within the DJ software environment.

Hot Cue Points

As discussed earlier, ensuring the starting cue point of the track is right on the first beat will help you beatmatch quicker. In addition, many software and hardware units allow you to set multiple cue points per track. The other cue points except for the first one are known as hot cue points. The advantage of setting multiple hot cues per track is that these allow you to quickly jump to important stages in the arrangement, eliminating any wasted time finding specific parts of the track or setting additional cue points while performing. Consequently, hot cue points can be used to great creative advantage. For example, setting a hot cue when the main hook or vocal starts gives you the opportunity to begin playing a track exactly from that key position during a mix; perhaps paving the way to a live mash-up or simply bypassing a lengthy intro.

Note: Hot cue points are set in the same way as regular cue points.

When learning how to mix or how to incorporate a new element into your live performance, always record every practice mix.

In the flurry of mixing, the finer details of your performance can be missed. Record each time you practice, step away from the recording for a few hours and then come back and listen to it with fresh ears. The new perspective as listener rather than DJ will give you an opportunity to objectively analyse what aspects need improvement and how you can further improve your DJing skills.

Harmonic Mixing

Exploring tension and release, and light and dark, in your DJ sets are some of the critical qualities that can take a DJ from mediocre to memorable; and harmonic mixing is your roadmap.

HOW TO MIX HARMONICALLY

1. Pick your method of keying tracks. Once upon a time, finding out the key of the track could only be done through your ears and a keyboard. These days, however, the process is much quicker and easier, with several affordable programs on the market to key your tracks for you. The most popular of these is Mixed In Key (www.mixedinkey.com).
2. Key your tracks. If you are using a computer program to key your tracks, you may get the key of the track in both its proper signature notation

(i.e. A minor) as well as a shorthand version. For example, Mixed In Key uses the 'Camelot Easy Mix Wheel', which attributes each signature a number from 1 to 12 and a letter either A or B depending on whether the track is major or minor key. Although having the shorthand of each key might seem easier at the beginning since you only need to remember a letter and number, getting into the habit of referencing each track by its proper signature notation will be much more beneficial in the long run. As you begin getting more familiar with the sound and feel associated with each key signature, you will be much more in control of your set development and direction of your music. Although this may take a little longer in the short term to learn, it is strongly recommended for long-term mastery.

3. Once you attain the keys of each of your tracks, it is now time to analyze which tracks will have the strongest harmonic relationship to another. These are the tracks that will deliver the smoothest transitions. The more notes that the key of one track shares with the key of another, the stronger their harmonic relationship will be. The strongest harmonic relationship is between keys of the following intervals:
   a. Dominant interval (6/7 keys in common)
   b. Subdominant interval (6/7 keys in common)
   c. Relative major/minor (7/7 keys in common, different tonic)
4. Finding which keys have these intervals between them can be found on the Circle of Fifths. The Circle of Fifths is a diagram used in music theory to show the harmonic relationship of every key on the chromatic scale. The above intervals are represented around the Circle of Fifths as so:
   a. Dominant interval: The key one spot clockwise
   b. Subdominant interval: The key one spot anticlockwise
   c. Relative major/minor: The key in the same spot, but on the inner or outer circle
5. Harmonic mixing is achieved by transitioning between tracks that have one of the above intervals between them. Examine the key of the track you are currently playing, find it on the Circle of Fifths and pick a subsequent track that is either one spot clockwise, one spot anticlockwise or change to the inner or outer circle at the current spot. You can stay in the same key, move backwards, forwards or inside and outside the wheel from mix to mix. Keeping to one of these intervals ensures the tracks you are mixing have the most notes in common and therefore sound the most harmonic to each other.

Major vs Minor Keys

Major keys: Found on the outside of the Circle of Fifths. Music written in a major key is characteristically *happy*, *victorious* and *confident*.

Minor keys arranged on the inside of the Circle of Fifths. These keys are typically associated with *sad*, *mysterious* and *serious* compositions.

Additional Intervals

In addition to the intervals discussed above, experimenting with different intervals will yield different changes to the mood and progression of your set. Here is a quick summary:

Interval	Circle of Fifths Movement	Effect
Relative major to relative minor	Outside circle to inside circle	Noticeable light to dark transition
Relative minor to relative major	Inside circle to outside circle	Noticeable dark to light transition
Tonic to dominant	Clockwise	Elation, building energy
Tonic to subdominant	Anti-clockwise	Delivers a feeling of progression, but not necessarily resolution; more to come; the story continues
Tonic major to tonic minor	90° shift anti-clockwise	Subtle darkening of mood, mood becoming more serious
Tonic minor to tonic major	90° shift clockwise	Subtle lightening of mood, hints towards a resolution, dials back tension slightly
Tonic to tritonic	180° shift	A very tense and dissonant interval. When timed correctly, it can achieve a dramatic increase in tension, signal the commencement of a new chapter in your set progression, or a sudden change in direction

When to Mix Harmonically

Even if your audience is not familiar with music theory, mixing very dissonant (weakly harmonic) keys together will often sound tense and unpleasant. This makes harmonic mixing advantageous across all genres of music as it assists in creating smooth and aurally pleasant transitions. Nonetheless, there are several situations that call for particular diligence to harmonic mixing. These are:

- Mixing tracks that are dominated by melodies. Certain genres are characteristically more reliant on melodies, for example house and trance.

As the melody in a track is one of the most prominent elements, mixing dissonant melodies together will be very blatant, especially if either track contains vocals. Turning to harmonic mixing at this end of the music spectrum is highly recommended.

- Long mixes. The longer the transition, the more elements of each track you will be overlapping from one moment to another. If your mixing style relies on long transitions and rich layering, then mixing harmonically is highly advisable to avoid unwanted dissonance. Genres that characteristically rely on longer mixes are progressive house, techno, tech-house and their related sub-genres.

Re-Pitching and Re-Keying

A track sped up or slowed down by greater than three percent of its original BPM gains a whole new position on the Circle of Fifths, and hence a whole new harmonic relationship.

Unless you have Master Tempo 'On' when mixing, significantly speeding up or slowing down a track will alter its key signature. As a rule of thumb, a change of 3% will transpose the key of the track by *one semitone*.

For example, if your track is written in C major, speeding it up by more than 3% will mean the track will be playing the key a semitone higher – key of C sharp major. Slowing the track down by more than 3% will mean the track will be playing in the key of a semitone lower – key of B major. Refer back to your Circle of Fifths, and you will see that the new keys are not adjacent to C major on the Circle of Fifths; rather they are five spots away from its original position. Consequently, mixing G major or F major tracks that would have sounded harmonic originally, would now sound dissonant as they are no longer adjacent on the Circle of Fifths. Key transpositions are an inevitable result of tempo changes greater than ±3%, so if your DJ setup accommodates a Master Tempo, it is advisable to keep it switched on while mixing harmonically. This will ensure that track's original key signature is maintained during re-pitching.

Rules are made to be broken.

Harmonic mixing is a powerful tool for smoother mixes, but it should be used as a guide only; some of the most exciting and memorable moments are when the DJ looks beyond these limitations. Knowledge is power; the more you grow to understand the significance behind key signatures, the more adept you will become at knowing the right times to break the rules. Fundamentally, if you feel it is the right moment to play a particular track, there is nothing stopping you. Always follow your instinct first; harmonic rules second.

The DJ's Music Collection

SOURCES FOR COLLECTING MUSIC

Hunting for tracks is one of the key tasks of a DJ, and knowing where and how to search is crucial for discovering hidden gems or the next big track. There are countless sources for musical inspiration, and with the rise of the Internet, access to tracks from all over the world has never been quicker or easier. Some of the most effective online destinations for track ideas include:

1. Online digital stores: This is one of the most common ways to access music for your DJ sets. Music here is arranged by a myriad of categories – including artist, label, genre, release title and chart – so locating and discovering new music is quick and convenient. There are many online digital stores, some of the most popular ones for dance music are:
   a. Beatport (www.beatport.com)
   b. AudioJelly (www.audiojelly.com)
   c. iTunes (www.apple.com/itunes)
   d. Amazon (www.amazon.com)
   e. Trackitdown (www.trackitdown.net)
   f. Juno Download (www.junodownload.com)
   g. Djdownload (www.djdownload.com)
   h. Traxsource (www.traxsource.com)
   i. Stompy (www.stompy.com)
2. Live sets: Whether hearing your favourite artists in person or accessing their tracklistings from around the globe on the web, identifying the tracks your favourite DJs play is a fantastic way to find tracks. Countless websites collate tracklists from artists performing all over the world, so this information is only a Google search away.
3. Radio show and podcast tracklistings: Many DJs have a regular podcast or radio show these days. Similar to gig tracklistings, use them for inspiration for new tracks and artists.
4. SoundCloud: Look up and follow the artists you admire. With SoundCloud's popularity skyrocketing in recent years, most DJs and producers will post their latest tracks for streaming on their pages. Also do not forget to then look at whom they personally follow too – they may follow a new artist yet to come across your radar.
5. DJ promo pools: Almost all record labels employ a promo pool system for distributing promos. In this scenario, labels will email their tracks to you for free download to support at gigs, radio shows or mixes. This is a very convenient and cost-effective method to acquire fresh music; however, the majority of large labels only add DJs to their promo pools by invitation only – often you need to have an existing national or international profile, or have had releases on their label to be considered. Those that are open to the public usually will have a request form on their website. Visit the website of any label you are interested in and research their promo pool policy.

We have access to more music than ever before right at our fingertips. However, with that privilege comes the challenge of sifting through the copious volumes of releases to find those that fit out individual style. As you gain greater clarity of your sound, being discerning and picky will begin to come naturally. In the beginning, purchasing tracks that do not genuinely excite you will be detrimental not only to your wallet but also to your enthusiasm behind the decks.

Ask yourself; "If someone I admire walked into the club right now and only heard this one track from me, would I feel it was a true representation of my sound?" – if the answer is no, move on.

In this time of intense music saturation, there is no room to be complacent.

ORGANIZING YOUR MUSIC COLLECTION

Knowing each track in your crate inside and out, being familiar with the vibe and effect it will have on the dance floor and when the perfect time is to drop a track are critical skills to an outstanding DJ. But with acquisition of music being easier than ever and many music collections blowing out to tens of thousands of tracks, keeping your collection manageable and organized can be a challenge. The key lies in the way you organize your music.

Playlists

Organizing your music into playlists can be done in countless ways. Ultimately it is a personal decision based on which aspects of your tracks you most want to highlight. Some common sorting methods include:

- Descending by date added: If freshness of tracks is critical, then this method is recommended. Commercial Dance DJs may benefit most from this sorting.
- By key: Relevant to those using harmonic mixing
- By BPM: Advantageous if you play a range of styles with a wide BPM range
- Vocal/non-vocal split: If your set development relies on vocal and instrumental formulas, this may be the way to go
- Alphabetical by label
- Alphabetical by artist

The Rating System

Several music file management programs such as iTunes and Traktor incorporate a star rating system from one to five stars. Another method of organizing your music collection is to assign a different track characteristic to each star rating, such as floor fillers, anthems, own productions, a capellas, DJ Tools and countless more. What labels you choose to assign to each of these five star ratings is personal preference, but one of the most effective strategies is to think of the star rating system as a gradient scale; five star tracks being the opening tracks, and five star tracks being your closing, hardest tracks. By organizing your music collection by feel/mood starting from the beginning of the night to the end, developing a progressive set will become significantly easier. An example of such an arrangement is:

- ONE STAR = opening tracks, early night tracks
- TWO STAR = warm-up tracks

- THREE STAR = peak time tracks
- FOUR STAR = peak time anthems/floor fillers
- FIVE STAR = closing tracks, your hardest tracks

iTunes is without a doubt the most popular music file management software in use today. Another program that is highly recommended is Media Monkey (www.mediamonkey.com). With added features in tagging, organizing playlists and syncing to your smart devices, the program is particularly advantageous for DJs. Instructions on how to download and install Media Monkey on both Windows and Mac operating systems are available on their website.

Master Tracklist

This strategy is straightforward; once you have organized your entire music collection into your playlists, transfer each playlist into an Excel spreadsheet. The best way to do this is to have each playlist as a separate 'sheet' within the same Excel document. This serves several advantages. First, it automatically gives you a snapshot of your entire music collection, which you can email to yourself and consult anytime regardless of whether you are near your computer or music collection. Second, it acts as a backup system in case your music file management software crashes.

NOTE: Ensure you go through your Master Tracklist at least every 6 months and cull any tracks you are no longer using. These playlists will become your roadmap to efficient and effective gig preparation; having playlists and spreadsheets blow out to hundreds or thousands of tracks will be time consuming to sift through and defeat their purpose.

CD Organization

For those that DJ via CDs, there are two main ways to convert your music collection to CD prior to a gig:

1. As many tracks per CD as can fit: This method is the quickest and cheapest way to convert your music collection to CD. However, remembering which track is on which CD can be confusing. If this is the method you choose, it is recommended to incorporate a chart showing where each track can be found on which CD in your wallet. Also, remember to burn two copies of each CD in case you want to play two tracks that are on the same CD.
2. One track per CD: Although significantly more time consuming and expensive, assigning only one track per CD will allow you to 'map out' the progression of your set better. Having one track per CD allows you to take out the CDs you are planning on playing next and visually map out the direction of your set. For progressive and deeper genres, this can be particularly beneficial.

USB Playlists

For DJs that use USB, if you plan on syncing all your playlists discussed earlier with your USB, bear in mind that scrolling through hundreds of tracks in each playlist will be time consuming, chew up valuable time better spent interacting with the crowd and may eventually mean that only the tracks at the beginning of the playlists will get played. If you do transfer your entire music collection to your USB, it is highly advisable to implement the 'Shortlist Crate' strategy before each gig (refer to Page 20).

Gig Preparation

Regardless of whether you are playing to ten people or ten thousand people, the dedication you put into your preparation should be identical. Treat each gig as the most important crowd you have ever played to and the passion you bring to every set will be unmistakable.

Think of your gig preparation as a sliding scale, with 100% spontaneous and 100% planned on each end of the scale. Where you choose to sit on that scale will be a personal choice, and one that will likely fluctuate throughout your career or even gig to gig. However, as a rule of thumb, incorporating elements from both extremes often yield best results.

THE 'SHORTLIST CRATE' SYSTEM

An example of incorporating both extremes of the scale is to give yourself some direction for your set while still remaining flexible enough to respond to the crowd and play the right track at the right time. One of the most effective ways to achieve this is by implementing a 'Shortlist Crate':

1. Create a 'Shortlist Crate' or 'Gig Crate' playlist before each gig. This playlist is for the tracks you feel would be most suitable for a particular gig; all in one playlist for quick and efficient access. Factors to take into consideration when deciding which tracks to add to this playlist include:
   a. Set time
   b. Venue, including size of room or stage
   c. Expected patron preferences, such as vocals and floor fillers
   d. DJs playing before and after you
   e. Freshness of tracks
2. If you have incorporated the rating system when organizing your track collection (refer to Page 17) consider the star rating that would be most relevant to your gig. Go through that list and extract the tracks you feel would like to play the most. Add these to your 'Shortlist Crate' folder, with the remaining tracks already on your computer, USB or CD wallet acting as backups in case you want to play something that has not been shortlisted into the 'Shortlist Crate' playlist.
3. Keep in mind this folder should not get overly large, otherwise you will waste too much time going through the playlist, and its purpose will be defeated. As a rule of thumb, keep the length of this folder to two to four times the number of tracks you expect to play in your set. For example, if you are playing a 2-hour set, and you expect to go through 30 tracks in those 2 hours, keep your 'Shortlist Crate' playlist to between 60 and 120 tracks.

PREPARING YOUR SETUP

If you are going to be setting up any equipment in the booth, always ensure you practice this at home, preferably your setup at home will be identical to what you will be using in the venue. Being familiar with how everything is connected will alleviate stress during setup and also make you more efficient.

BACKUP READY

Always bring a backup system in case of failure. This may include a backup USB or a double of CDs. If you are playing off a laptop, incorporating a backup system can be more of a hassle, but it is still wise to bring some USBs with your music just in case of a technical malfunction. If you are flying to a gig, make sure all your equipment is stored in carry-on luggage. This is vital for several reasons. First, you can never rely on baggage handlers to be as delicate as required with your delicate equipment. Second, risking any component getting lost in transit is another stress that should be avoided at all costs.

DOUBLE CHECK

Double check whether you have everything on your way out. This may seem obvious, but the one time you forget to check whether you packed your laptop charger is the one time you will forget it. A good tip in your flurry before running out to a gig is to keep a quick checklist handy. A great place to put it is on your phone; if you have a smart phone, this can be easily done as a 'Reminder' or 'Note'.

The best form of gig preparation is getting the right mindset.

Finally, go to every gig feeling confident, prepared and ready to have a good time, and your audience will immediately feel your energy and fuel your confidence.

Submitting Demos

Before your experience speaks for itself, submitting a demo to a club or promoter is one of the most common ways to show your interest in attaining a gig. Although this can seem daunting at first, there are a few general principles to follow regardless of genre or event:

LENGTH

Demos should ordinarily be no longer than an hour. Promoters are busy people, and if they do sit down and listen to your demo, if it is highly unlikely, they will dedicate more than an hour to do so.

DEMOS FOR GENERAL PROMOTION

Unless you are handing over a demo with a particular gig or slot time in mind, approach the demo as what you would play at your most preferred set time. For those who play a spectrum of sounds, a good strategy is to pitch your demo at the centre point of your musical spectrum, or pick the sound you would most like to be known for.

DEMOS FOR SPECIFIC PROMOTERS

If it is with a particular promoter or event in mind, the most crucial point for a promoter listening to your demo is to see that you 'get' their night and the supporters. In this case, each track you include should answer the question 'Would this fit in with the night?' with a re-sounding 'Yes!' Similarly to the way a cover letter showcases why you would be the most suitable candidate for a job, the demo mix should show why you are the ideal DJ for the event.

THE MOST IMPORTANT PART

The first 10 minutes is the most important, with the first track and transition being the most critical. If you do not grab a promoter's interest in the first 10 minutes, chances are they will not finish the rest of the demo. The first 10 minutes set the tone, the mood and give a preview of what's to come. If you make your own productions or edits, this is the place to incorporate them – anything that sets you apart should be placed where you need it the most.

You only have one chance to make a first impression; this is especially relevant to your demo.

SETTING EXPECTATIONS

Never include anything in a demo you cannot replicate live. If your demo is full of a capellas, samples, scratching, effects and live mashups, then naturally you are creating an expectation for your performance. Not delivering on promises or hype is a sure-fire way to shoot yourself in the foot early on. The music industry is full of people who oversell and under-deliver. Be the DJ who over-delivers on expectations when given a gig opportunity and give yourself the best chance for a re-booking.

TIMING IS EVERYTHING

While deciding when to hand over your demo, remember networking is the name of the game in the music industry. Do not expect any promoter to reply to your demo submissions if you have not shown any support for their events in the past. First give your support and then request their time. Before putting a demo into a promoter's hand, attend their nights, show your support and use the opportunity to get to know their events and patrons so that you can best tailor your demo mix. This is crucial to giving yourself the best chance for a positive outcome.

If you can give a promoter a chance to put a face to the name before submitting a demo, you have already given yourself a head start.

Understanding Set Times

OPENING AND WARM-UP

The opening and warm-up DJ slots are some of the most crucial to the success of the night. These sets are centred around setting the vibe of the night and keeping it under your control; slowly building as the room fills and finally creating enough tension and suspense that the headliner enters as the room and everyone is ready to party. Unlike many Peak Time sets, during the warm-up, the attention is not on you, nor should you try to make it about you.

Playing to the room is never more crucial than it is at the opening of a night.

Getting people into the right mood at the start of the night will require a different formula venue-to-venue and even night-to-night. A track might be the perfect fit one night, but bring in a different set of people and the dynamics could be very different.

Progression and building tension are the keys of the warm-up DJ. Long builds and mixes are appropriate here. These act as small previews and forebode the rest of the night. Tease your audience with what is to come. The warm-up slot is a good opportunity to try out new tracks, as long as the intensity is a right fit.

If you are warming up for a guest DJ, it is imperative you do your research. There is no bigger crime for the warm-up DJ then playing a track that the headliner would have played, be it their own release or remix or key tracks on their affiliated labels. Accessing recent sets or tracklistings from around the world has never been quicker or easier than it is today, so there is no excuse for not doing your research.

Ensure you know not only their releases and what labels they are affiliated with but also the general style and energy that a guest is likely to bring. For example, if you are the warm-up DJ on a trance night, opening for a progressive trance DJ who would usually play a set at 130 BPM with an uplifting trance set at 138 BPM would be completely inappropriate. But it is also crucial to note here that BPM alone should never be your guide on the suitability of a track for any slot time. The intensity, suspense and musicality of a track comes down more to its composition and arrangement than its BPM. Your ears should always be your guide.

As a DJ there is nothing more upsetting than seeing an empty dance floor. However, getting your dance floor exploding in a fury of energy is most definitely not the goal of the opening DJ; that is what the headliner is for. Don't let your ego get in the way, a truly professional and skilled DJ will be able to hold back and retain the attention of the room without playing all the biggest tracks and tiring your audience out too early. Patrons are just coming into the venue at this stage. Peaking the energy of the room too early will not only create an uncomfortable environment for the headliner but it will also mean patrons will go home earlier.

The warm-up slot is one of the most difficult sets to play in a night as the DJ is put in the position of creating a vibe out of nothing. A small mismatch in energy, pace, intensity or timing of tracks can make the most noticeable difference. With this in mind, when preparing your tracks for an opening set, ensure you know each track back to front; the more intimately you know a piece of music before stepping in the booth, the quicker and more accurately you will be able to assess its suitability at any point during the set.

PEAK TIME

As a result of differences genre-to-genre and venue-to-venue, peak time sets can vary greatly. Knowing the expectations and the culture of your audience is the key for delivering a memorable peak time set. For example, examining the education:entertainment ratio, a commercial DJ would be expected to pack their peak time set with almost all entertainment tracks; tracks the audience most likely know prior to arriving to the venue. Underground venues and DJs are a different matter. Often the audience looks to the headliner to not only entertain but also to expose them to new tracks and educate them on the latest productions.

If you are known as a leader in your genre, then concentrating on new and unheard tracks would be well received.

If you are a guest headlining a show in a new city, especially one where you haven't played in previously, your branding will have a lot to do with what type of peak time set your audience will expect from you. This is a result of word-of-mouth, previous events, other artists who endorse you, sets and radio shows they may have downloaded, marketing and many other factors. Ensure you are conscious of this and steer it in the direction you want from the beginning of your career, because attempting to change these expectations and ideas down the track can be an arduous and difficult task.

CLOSING

The headliner has finished, the crowd has seen who they wanted to see, and more likely than not, heard all the tracks they wanted to hear. They have been at the club for many hours, and now funds and energy levels are critically low. The job of the closing DJ is to keep the patrons in the club for as long as possible and ensure that when they do decide to leave, they have left the venue with a positive feeling about their night to keep them coming back.

As a closing DJ rule of thumb, every track should give the crowd a reason to stay another 5 minutes.

This means no long builds or breakdowns. Quick tracks and mixes are needed to keep the intensity and interest high; otherwise people will be ready to go home. Set development is of less importance here and regardless of the genre, the education:entertainment ratio will have to be greatly skewed towards entertainment to ensure you keep them in the club. As there is no other DJ after you to be conscious of, if you feel the crowd pulling you to up the intensity do not be afraid to do so as long as it is within the music policy of the venue.

It is also worth noting that people extract a lot of energy from the DJ at this time of night. It is more vital than ever to ensure you lead the way by staying energetic during a closing set. If you look tired or bored, then this is the time where it will have the greatest flow-on effect to your crowd.

Defining Your Sound

With the DJ industry today more competitive than ever, one of the keys to regular bookings will be to remain 'Front of Mind' when a gig opportunity arises. An obvious way to ensure this is through your previous experience. However, until such a time arises, one of the best strategies in maneuvering yourself to a promoter's front of mind is by having a well-defined sound and brand.

FINDING YOUR UNIQUENESS

Defining a unique sound means selecting unique tracks or a unique way to perform them. Going for the Top Ten on iTunes or Beatport might save your time in the short term, but in the long term, you are not creating any value between yourself and countless other DJs who can mix together the latest chart toppers. Ensure you set aside regular time each week purely for hunting music that resonates with you and your brand.

You may also consider defining yourself through your setup. With technological advancements progressing at lightning speed, there are literally endless setup options. By defining your distinctiveness through a unique arrangement of mixers, controllers, pads or whatever combination of gear works best for you, you can work towards creating an unreplicated live experience and hence considerably raise your value as an artist.

OVERCOMING GENRE BARRIERS

Defining a sound goes beyond simply picking your favourite genre and playing it. Rather than concentrating on genre labels, think instead about adjectives that describe the mood you want to create through your music. Consider words such as euphoric, intense, bouncy, groovy, funky, contemplative, deep, serious, energetic and so forth. When you have a feeling rather than a genre guiding your musical direction, it will feel much more natural navigating through genre barriers. Gaining inspiration for your DJ sets from a wide range of genres while still building a cohesive sound is a powerful tool in defining yourself as an artist.

BEYOND THE MUSIC

Branding is built on consistency.

Finally, the rest of your branding should be consistent with your sound. People need to know what to expect. If there are inconsistencies between the messages you send out intentionally or not, people will subconsciously feel unsure about who you are as an artist. Examine the rest of your branding – your logo, your press photos, your DJ name if you use one. These crucial elements should all match the mood and image you are carving through your sound.

Top Ten Mistakes Made by New DJs

Your beatmatching is under your belt, your phasing is sorted and you have a record box full of tracks you have prepared for your first gig this weekend. What are the Top 10 mistakes to watch out for in these early days?

BOOST TOO HARSHLY AND REDLINE THE MIXER

When first starting out, to hear a particular element more vividly during a transition, a DJ will often turn to boosting EQs on their mixer. However, this often does more harm than good. Excessively boosting an EQ simply adds more sound to the mix. If the sound spectrum is already full, this will create muddiness, clutter and at its worse – redline the output and distort the signal. Similar to trying to cram another piece of clothing in a bursting suitcase, for a sound to fit in a transition, you have to make room for it in the sound spectrum. The best way to achieve this headroom is by *attenuating* EQs on the first track.

NO CROWD INTERACTION

You have practiced incessantly at home, you have spent hours hunting tracks and now it's time to show off all your hard work. The reality is, the majority of patrons at the venue are there to have a good time, and very few are spending any time analyzing your performance. Being too focused on technique is a natural response to lack of experience in front of a crowd, but this will be highly detrimental to your success. The crowd wants to have a good time and connect with you; spend more time looking at them and less time looking at your mixer.

THINK IT IS THE END OF THE WORLD WHEN YOU MAKE A MISTAKE

We all make mistakes behind the decks. Whether it is our first gig or our hundredth, it is an inevitable part of performing live. But remember it is not your mistakes that determine your set, but how you recover from them. Take it in your stride and move on. In the early years, it is a common conception that what makes a DJ outstanding is strictly their skills using a mixer or controller. Although this is very important, as your experience grows, you will come to realize that this is one of the many factors that determine the overall success of a DJ. Reading a crowd, building a set and creating those goosebump moments are all equally important. Finally, avoid being so focused on perfection that you become scared of making mistakes. If you cannot get over your fear of failure, you will never have the strength to take a risk and seize opportunities when they come your way.

If you are not occasionally making mistakes, you are not pushing yourself hard enough.

ARRIVE LATE

Unless you are the opening DJ, it is recommended to arrive to your gig at the very least 30 minutes to 1 hour before your set time. This allows for enough time to hear what the previous DJ is playing, get a feel for the room, and familiarize yourself with what is currently working on the dancefloor. This will also give you a chance to settle your nerves and consider what direction to take your set. If you are playing at a venue or city you have not been to before, make sure you have a plan well in advance on how you are going to get there. At the very least ensure you do not arrive late to the venue. Arriving late is perceived as very unprofessional; often there are multiple people relying on you being ready to play your set time, and creating a hassle for other people can sometimes be the determining factor on whether you get that re-booking.

PLAY A PRE-PREPARED SET

As your experience grows, you will develop the confidence to respond to the energy in the room on the spot. However, in the early days, the focus is so heavily on getting the technical aspects perfect that many new DJs will arrive to a gig with the entire set already planned out. They know these tracks well, they know they transition smoothly and they feel this is the way to deliver the best set they can. Unfortunately such a rigid approach to DJing can often backfire. If you do not give yourself freedom to adjust and respond to the energy of the room, then the mismatch between your planned set and the audience will be more detrimental than an improvised set with possibly a few rough transitions. Your number one job as a DJ is to guide the audience to have a good time, and it is impossible to pre-empt how you will best achieve that before you even step into the venue.

IGNORE YOUR SET TIME

Having the peak-time set may be the ultimate goal for many, but in the early days, it is far more likely that your set time will be early on in the night. Everyone wants to prove how much they can 'rock the house', but playing a set that does not complement the night or your set time is one of the biggest ways to stall your DJ bookings. Also, work with your fellow DJs. When starting, consider who played before you; how can you best work with what they have done and progress from it smoothly. When finishing, consider who is playing after you; where are they likely to start from and how can you best close off your set so that they can flow on naturally. If you are supporting an out-of-town headliner, this point is even more crucial.

USE TOO MANY EFFECTS

Effects are a great way to stamp your own identity and style to your DJ sets, not to mention a lot of fun to use. However, use them excessively, and it isn't a pleasant sound. They get tedious, and your audience will start to loose interest and switch off.

Think of effects like herbs; the right sprinkle can turn a dish into something delicious, too much and you ruin the meal.

CRITICIZE OTHER DJS BEHIND THEIR BACKS

The music industry is one of the most competitive industries you can choose to work in. Bagging out other DJs to appear superior by comparison is a reaction some DJs take in response. Unfortunately, any short-term gain that this tactic may yield is only going to come back to cost them the long term. Bagging out other musicians behind their back only makes you look untrustworthy and malicious. Your success in the music industry is hugely influenced by who's willing to give you a chance, and shooting yourself in the foot by trying to make other musicians look bad will only mean that no one will eventually want to work with you. Despite how big the international industry may seem, dance music communities are a lot more connected than they may first appear, and a reputation about bad-mouthing other artists or promoters will do you no good.

AVOID GETTING EAR PROTECTION

Ears are the greatest assets to any musician, be it a DJ or a classical pianist. If you are serious about making a career as a DJ, then it is imperative that you protect your greatest assets from environmental damage as early as possible. Being in clubs, bars and festivals continuously does significant damage to your hearing, and what can start off as a mild ringing initially can turn into irreversible tinnitus in a matter of years. Hearing is irreparable so do not hesitate to get custom earplugs molded. Ensure they have at least 10 dB attenuation, preferably 20 dB.

FORGET TO ENJOY YOURSELF

Don't forget to enjoy the ride.

It is inevitable that you will be nervous. Everyone battles with nerves when they first start playing to crowds. Some superstar DJs even confess to feeling nervous before gigs even after decades of experience. This is a totally natural part of the experience, and is just a sign that what you are doing means a lot to you. But between adjusting to a new environment, getting your setup right, concentrating on your technique and selecting the best tracks, all while battling numbing nerves, it is easy to forget to enjoy yourself. Those early years are filled with first experiences, first milestones and lifelong memories. It is all part of the journey.

Final Words

The role of the DJ is evolving quicker than ever before; decisions regarding DJ setup and mixing techniques now ultimately coming down to personal preference. In an industry that is growingly more competitive by the day, the goal is to stay distinctive. Spend time defining your sound, being strategic with how you collect your music, and get attuned to your crowd at every gig regardless of your set time of venue. Most importantly, do not let fear of failure stop you from attempting to enter the world of DJing. No matter how competitive, the industry will always make room for talent.

Finally, always remain curious. About new music, new DJ techniques, incorporating new technology into your setup. Strive to understand your musical direction and seek new and unique ways to share your passion with others. The most successful DJs in the world have one thing in common; they have dedicated their lives to their passion. It is this curiosity and drive to keep exploring your musicality that will ultimately set you apart from the masses.

Never stop learning.

Appendix A

Binary and Hex

All hardware circuits, even computers, are based on a series of interconnected switches that can be in one of two states: on or off. By continually switching these on and off in different configurations, it's possible for the hardware to count and perform complex calculations. This switching is accomplished by sending a series of bytes down an electrical cable, which means that they can only exist in one of two states too. If it is equal to one then the switch is turned on, while if it's a zero then the switch is turned off. Because of this, all computers must count using base 2 or 'binary' rather than our usual method of counting which is base 10.

COUNTING IN BINARY

Our normal method of counting was developed because we have 10 figures and using this system, any one digit in a number has a value that is based upon the position of the digit in the number. Thus, when we are working to the power of 10, every time a digit moves one to the left its result is to the power of 10, then by 100, then by 1,000 and so forth. Consequently, the number 17,593 could be seen as follows.

10,000,000	1,000,000	100,000	10,000	1,000	100	10	1
0	0	0	1	7	5	9	3

Or (1 × 10,000) + (7 × 1,000) + (5 × 100) + (9 × 10) + (3 × 1) = 17,360

From this above example, we can see that each position in a number essentially adds a 0 to the meaning of a digit. Or, the value of each position is 10 times the value of the previous position (moving from right to left). This method of counting is also implemented in binary, but rather than use a base 10 system, we use a base 2 system. So, rather than saying that 10 to the power of 0 = 1, 10 to the power of 1 = 10, 10 to the power of 2 = 100, 10 to the power of 3 = 1,000, it's 2 to the power of 0, 2 to the power of 1, 2 to the power of 2, 2 to the power of 4 and so forth.

128	64	32	16	8	4	2	1
0	0	0	0	0	0	0	0

Looking at the above table, we are assuming that there are eight hardware switches grouped together producing what is essentially an eight-bit byte. Also, as they are all switched off, the sum produced will be zero. However, if we were to introduce some positive bits, we could sum them together to produce a decimal number:

128	64	32	16	8	4	2	1
0	1	1	0	1	1	0	1

$(0 \times 128) + (1 \times 64) + (1 \times 32) + (0 \times 16) + (1 \times 8) + (1 \times 4) + (0 \times 2) + (1 \times 1) = 109$.

Thus the decimal equivalent of the binary 01101101 would be 109. In addition, we could also determine that the maximum number that could be calculated through binary would be 255 (11111111).

As MIDI uses this same eight-bit system when communicating with any devices, it would seem sensible to assume that it should be able to offer this same maximum parameter (of 255), but this isn't the case. Similar to CC messages, two forms of information need to be transmitted; a status bit and a data byte (which is composed of seven bits). The status bit informs the synth of an incoming message that is arriving while the following data byte informs it by how much the parameter should be adjusted. Because of this initial status bit, only seven other bits are left to provide the information resulting in a maximum decimal value of 127, hence the maximum number of any CC message can only be 127. To transmit numbers larger than this, an eight-bit byte has to be split into two halves and then converted into another numeration format: hexadecimal.

When we split a byte into two halves, both halves are commonly referred to as 'nibbles'. If the previous example were to be broken down into two nibbles, it would become 0110 and 1101. These could then be individually summed together as 0110 = 96 and 1101 = 13 (96 + 13 = 109) to produce the result again. However, by splitting a byte into two and then converting it into a hexadecimal value, it's possible to produce much higher values. The reason behind this is that hexadecimal works to base-16 meaning that it is possible to access up to 16,383 parameters in a synth, much more than the standard 127 offered through Control Change messages.

COUNTING IN HEXADECIMAL

Hex uses a base-16 numbering system, but as there are not enough symbols to represent 16 different digits, as soon as the number 10 is reached it has to be converted into letters. Thus to represent number 10 to 15, the letters A to F are used.

Recall how we count in decimal. Counting upwards, we count from 1 to 9 and then we place a 1 to the left of this and go back to 0, making the number 10 (which actually means 1 to the power of 10 plus 0 to the power of 1). With hexadecimal we count beyond 9 but because our entire numeric system is only based on 9 digits, with hexadecimal, letters are used instead as below so that we can count up to 15.

F	E	D	C	B	A	9	8	7	6	5	4	3	2	1	0
15	14	13	12	11	10	9	8	7	6	5	4	3	2	1	0

When we count the zero, just as we do with the decimal counting system, we have used a total of 16 numbers. You can see from the decimal to hexadecimal conversion table that whenever we count up to F, increase the number to the left by 1. This is the same as the decimal counting system, for example, if you count from 11 through to 19, there are no more digits to use and therefore you change the 9 to 0 and increase the number to the left by 1, providing the number 20.

You can see that hexadecimal counting is exactly the same as decimal, except in hexadecimal where we count up to F before increasing the number to the left by a factor of 1.

It follows that in decimal, the number 24 actually means '2 × 10 to the power of 1 + 4 × 10 to the power of 0 = 24', which is what you or I already understand it to be because we use this every day like counting on our fingers.

In hexadecimal, the number 24 actually means '2 × 16 to the power of 1 + 4 × 16 to the power of 0 = 36'.

Logically it follows then that the number CE in hexadecimal means '12 × 16 to the power of 1 + 14 × 16 to the power of 0 = 206' in decimal.

Using this principle, it's easy to convert hexadecimal into decimal.

Using this method, it's possible to count until we reach FF, which essentially means 15 × 16 + 15 units or 255 in decimal. Creating a new position gives 100 or 16 to the power of 2 which is 256 in decimal. Continuing to count using this base system, it would be possible to produce two nibbles, both with a value of 7F resulting in a total of 16,383 (127 × 128 + 127 = 16,383!).

For instance, suppose we wished to convert the decimal number 12,720 to hexadecimal:

- The largest power of 16 that fits into 12,720 is $16^3 = 4096$
  It fits 3 times and gives a remainder of 432. This number is derived from the fact that

  (3 × 4,096) = 12,288
  (12,720 − 12,288) = 432

  The First Hex number is 3

- The largest power of 16 that fits into 432 is $16^2 = 256$
  It fits 1 time and gives a remainder of 176. This number is derives from the fact that

  (1 × 256) = 256
  (432 − 256) = 176

  The Second Hex number is 1
- The largest power of 16 that fits into 176 is $16^1 = 16$
  It fits 11 times with no remainder (this remainder equates to 1 to the power of nothing which is nothing and will make 0 the last digit in the hex number). The Third Hex number is therefore B.
- The Hex equivalent of the decimal number 12,720 is therefore: 31B0

> To clarify: to convert 31B0 hexadecimal into decimal the maths is as follows:
> 0 × (16 to the power of nothing) = 0
> B (which is equivalent to 11 in decimal) × (16 to the power of 1) = 176
> 1 × (16 to the power of 2) = 256
> 3 × (16 to the power of 3) = 12,288
> 0 × 176 + 256 + 12,288 = 12,720 in decimal

As another example, suppose we wanted to convert the decimal number 14,683

- The largest power of 16 that fits into 14,683 is $16^3 = 4096$
  It fits 3 times and gives a remainder of 2,395
- This remainder is derived from the fact that

  (3 × 4,096) = 12,288
  (14,683 − 12,288) = 2395

  The first Hex number is 3
- The largest power of 16 that fits into 2,395 is $16^2 = 256$
  It fits 9 times and gives a remainder of 91
- This remainder is derived from the fact that

  (9 × 256) = 2,304
  (2,395 − 2,304) = 91

  The Second Hex number is 9
- The largest power of 16 that fits into 91 is $16^1 = 1$
- It fits 5 times and gives the remainder of 11
- This remainder is derived from the fact that

  (5 × 16) = 80
  (91 − 80) = 11

- The third Hex number is 5
- The remainder of 11 has no power of 16 that will divide into it as a whole number so the last digit of the Hex will be the hex equivalent of 11 which is B. This gives the final Hex number 395B.

Again we can check this by converting 295B hexadecimal into decimal. The math is as follows:

B × (16 to the power of nothing) = 11

5 × (16 to the power of 1) = 80

9 × (16 to the power of 2) = 2,304

3 × (16 to the power of 3) = 12,288

11 + 80 + 2,304 + 12,288 = 14,683 in decimal

Appendix B

Decimal to Hexadecimal Conversion Table

Dec	Hex	Dec	Hex	Dec	Hex	Dec	Hex	Dec	Hex	Dec	Hex
0	0	44	2C	88	58	132	84	176	B0	220	DC
1	1	45	2D	89	59	133	85	177	B1	221	DD
2	2	46	2E	90	5A	134	86	178	B2	222	DE
3	3	47	2F	91	5B	135	87	179	B3	223	DF
4	4	48	30	92	5C	136	88	180	B4	224	E0
5	5	49	31	93	5D	137	89	181	B5	225	E1
6	6	50	32	94	5E	138	8A	182	B6	226	E2
7	7	51	33	95	5F	139	8B	183	B7	227	E3
8	8	52	34	96	60	140	8C	184	B8	228	E4
9	9	53	35	97	61	141	8D	185	B9	229	E5
10	A	54	36	98	62	142	8E	186	BA	230	E6
11	B	55	37	99	63	143	8F	187	BB	231	E7
12	C	56	38	100	64	144	90	188	BC	232	E8
13	D	57	39	101	65	145	91	189	BD	233	E9
14	E	58	3A	102	66	146	92	190	BE	234	EA
15	F	59	3B	103	67	147	93	191	BF	235	EB
16	10	60	3C	104	68	148	94	192	C0	236	EC
17	11	61	3D	105	69	149	95	193	C1	237	ED
18	12	62	3E	106	6A	150	96	194	C2	238	EE

Dec	Hex	Dec	Hex	Dec	Hex	Dec	Hex	Dec	Hex	Dec	Hex
19	13	63	3F	107	6B	151	97	195	C3	239	EF
20	14	64	40	108	6C	152	98	196	C4	240	F0
21	15	65	41	109	6D	153	99	197	C5	241	F1
22	16	66	42	110	6E	154	9A	198	C6	242	F2
23	17	67	43	111	6F	155	9B	199	C7	243	F3
24	18	68	44	112	70	156	9C	200	C8	244	F4
25	19	69	45	113	71	157	9D	201	C9	245	F5
26	1A	70	46	114	72	158	9E	202	CA	246	F6
27	1B	71	47	115	73	159	9F	203	CB	247	F7
28	1C	72	48	116	74	160	A0	204	CC	248	F8
29	1D	73	49	117	75	161	A1	205	CD	249	F9
30	1E	74	4A	118	76	162	A2	206	CE	250	FA
31	1F	75	4B	119	77	163	A3	207	CF	251	FB
32	20	76	4C	120	78	164	A4	208	D0	252	FC
33	21	77	4D	121	79	165	A5	209	D1	253	FD
34	22	78	4E	122	7A	166	A6	210	D2	254	FE
35	23	79	4F	123	7B	167	A7	211	D3	255	FF
36	24	80	50	124	7C	168	A8	212	D4		
37	25	81	51	125	7D	169	A9	213	D5		
38	26	82	52	126	7E	170	AA	214	D6		
39	27	83	53	127	7F	171	AB	215	D7		
40	28	84	54	128	80	172	AC	216	D8		
41	29	85	55	129	81	173	AD	217	D9		
42	2A	86	56	130	82	174	AE	218	DA		
43	2B	87	57	131	83	175	AF	219	DB		

General MIDI Instrument Patch Maps

Program No.	Instrument Set (Piano)	Program No.	Instrument (Chromatic Percussion)
1	Acoustic Grand	9	Celesta
2	Bright Acoustic	10	Glockenspiel
3	Electric Grand	11	Music Box
4	Honky Tonk	12	Vibraphone
5	Electric Piano 1	13	Marimba
6	Electric Piano 2	14	Xylophone
7	Harpsichord	15	Tubular Bells
8	Clav	16	Dulcimer

Program No.	Instrument (Organ)	Program No.	Instrument (Guitar)
17	Drawbar Organ	25	Nylon Acoustic Guitar
18	Percussive Organ	26	Steel Acoustic Guitar
19	Rock Organ	27	Jazz Electric Guitar
20	Church Organ	28	Clean Electric Guitar
21	Reed Organ	29	Muted Electric Guitar
22	Accordian	30	Overdrive Guitar
23	Harmonica	31	Distortion Guitar
24	Tango Accordian	32	Guitar Harmonics

Program No.	Instrument (Bass)	Program No.	Instrument (Strings)
33	Acoustic Bass	41	Violin
34	Finger Bass	42	Viola
35	Pick Bass	43	Cello
36	Fretless Bass	44	Contrabass
37	Slap Bass 1	45	Tremolo Strings
38	Slap Bass 2	46	Pizzicato
39	Synth Bass 1	47	Orchestral
40	Synth Bass 2	48	Timpani

Program No.	Instrument (Ensemble)	Program No.	Instrument (Brass)
49	String Ensemble 1	57	Trumpet
50	String Ensemble 2	58	Trombone
51	Synth Strings 1	59	Tuba
52	Synth Strings 2	60	Muted Trumpet
53	Choir Aahs	61	French Horn
54	Choir Oohs	62	Brass Section
55	Synth Voice	63	Synth Brass 2
56	Orchestral Hit	64	Synth Brass 2

Program No.	Instrument (Reed)	Program No.	Instrument (Pipe)
65	Soprano Sax	73	Piccolo
66	Alto Sax	74	Flute
67	Tenor Sax	75	Recorder
68	Baritone Sax	76	Pan Flute
69	Oboe	77	Blown Bottle
70	English Horn	78	Skakuhachi
71	Bassoon	79	Whistle
72	Clarinet	80	Ocarina

Program No.	Instrument (Synth Lead)	Program No.	Instrument (Synth Pad)
81	Square Lead	89	New Age Pad
82	Sawtooth Lead	90	Warm Pad
83	Calliope Lead	91	Polysynth Pad
84	Chiff Lead	92	Choir Pad
85	Charang Lead	93	Bowed Pad
86	Voice Lead	94	Metallic Pad
87	Fifths Lead	95	Halo Pad
88	Bass and Lead	96	Sweep Pad

Program No.	Instrument (Synth Effects)	Program No.	Instrument (Ethnic)
97	Rain FX	105	Sitar
98	Soundtrack FX	106	Banjo
99	Crystal FX	107	Shamisen
100	Atmosphere FX	108	Koto
101	Brightness FX	109	Kalimba
102	Goblins FX	110	Bagpipe
103	Echoes FX	111	Fiddle
104	Sci-Fi FX	112	Shanai

Program No.	Instrument (Percussive)	Program No.	Instrument (Sound FX)
113	Tinkle Bell	121	Guitar Fret Noise
114	Agogo	122	Breath Noise
115	Steel Drums	123	Seashore
116	Woodblock	124	Bird Tweet
117	Taiko Drums	125	Telephone Ring
118	Melodic Toms	126	Helicopter
119	Synth Drum	127	Applause
120	Reverse Cymbal	128	Gunshot

GENERAL MIDI PERCUSSION SET

MIDI Key	Drum Sound	MIDI Key	Drum Sound
35	Acoustic Bass Drum	59	Ride Cymbal
36	Bass Drum	60	Hi Bongo
37	Side Stick	61	Low Bongo
38	Acoustic Snare	62	Mute Hi Bongo
39	Hand Clap	63	Open Hi Bongo
40	Electric Snare	64	Low Conga
41	Low Floor Tom	65	High Timbale
42	Closed Hi Hats	66	Low Timbale
43	High Floor Tom	67	High Agogo
44	Pedal Hi Hat	68	Low Agogo
45	Low Tom	69	Cabasa
46	Open Hi Hat	70	Maracas
47	Low Mid Tom	71	Short Whistle
48	High Mid Tom	72	Long Whistle
49	Crash Cymbal	73	Short Guiro
50	High Tom	74	Long Guiro
51	Ride Cymbal	75	Claves
52	Chinese Cymbal	76	Hi Wood Block
53	Ride Bell	77	Low Wood Block
54	Tambourine	78	Mute Cuica
55	Splash Cymbal	79	Open Cuica
56	Cowbell	80	Mute Triangle
57	Crash Cymbal	81	Open Triangle
58	Vibraslap		

Appendix D

General MIDI CC List

CC	Function	CC	Function	CC	Function
0	Bank Select	19	General Control 4	49	General Purpose Controller 2
1	Mod Wheel	20–31	Undefined	50	General Purpose Controller 3
2	Breath Controller	32	Bank Select	51	General Purpose Controller 4
3	Undefined	33	Mod Wheel	52–63	Undefined
4	Foot Controller	34	Breath Control	64	Damper Pedal (on/off)
5	Portamento Time	35	Undefined	65	Portamento (on/off)
6	Data Entry	36	Foot Control	66	Sustenuto (on/off)
7	Channel Volume	37	Portamento Time	67	Soft Pedal (on/off)
8	Balance	38	Data Entry	68	Legato Footswitch
9	Undefined	39	Channel Volume	69	Hold 2
10	Pan	40	Balance	70	Sound Controller 1 (Sound Variation)
11	Expression	41	Undefined	71	Sound Controller 2 (Timbre)
12	Effect Control 1	42	Pan	72	Sound Controller 3 (Release Time)
13	Effect Control 2	43	Expression Controller	73	Sound Controller 4 (Attack Time)
14–15	Undefined	44	Effect Control 1	74	Sound Controller 5 (Brightness)
16	General Control 1	45	Effect Control 2	75	Sound Controller 6
17	General Control 2	46–47	Undefined	76	Sound Controller 7
18	General Control 3	48	General Purpose Controller 1	77	Sound Controller 8

CC	Function	CC	Function	CC	Function
78	Sound Controller 9	93	Effects 3 (Chorus) Depth	120	All Sound off
79	Sound Controller 10	94	Effects 4 (Detune) Depth	121	Reset All Controllers
80	General Purpose Controller 5	95	Effects 5 (Phaser) Depth	122	Local Control on/off
81	General Purpose Controller 5	96	Data entry +1	123	All Notes off
82	General Purpose Controller 5	97	Data entry −1	124	Omni Mode off (+ all notes off)
83	General Purpose Controller 5	98	Non-Registered Parameter Number LSB[1]	125	Omni Mode on (+ all notes off)
84	Portamento Control	99	Non-Registered Parameter Number MSB	126	Poly Mode on/off (+ all notes off)
85–90	Undefined	100	Registered Parameter Number LSB	127	Poly Mode on
91	Effects 1 (Reverb) Depth	101	Registered Parameter Number MSB[2]		
92	Effects 2 (Tremolo) Depth	102–119	Undefined		

[1]*Least significant bit.*
[2]*Most significant bit.*

Appendix E

Sequencer Note Divisions

The following charts display the number of clock pulses for each note value for the four most popular PPQN resolutions – 96, 192, 240 and 384.

Note Value	PPQN	Note Value	PPQN	Note Value	PPQN
96 PPQN					
Whole	384	Dotted whole	576	Triplet whole	256
Half	192	Dotted half	288	Triplet half	128
Quarter	96	Dotted quarter	144	Triplet quarter	64
Eighth	48	Dotted eighth	72	Triplet eighth	32
1/16th	24	Dotted 1/16th	36	Triplet 1/16th	16
32nd	12	Dotted 32nd	18	Triplet 32nd	8
64th	6	Dotted 64th	9	Triplet 64th	4
128th	3	Dotted 128th	N/A	Triplet 128th	2
192 PPQN					
Whole	768	Dotted whole	1152	Triplet whole	512
Half	384	Dotted half	576	Triplet half	256
Quarter	192	Dotted quarter	288	Triplet quarter	128
Eighth	96	Dotted eighth	144	Triplet eighth	64
1/16th	48	Dotted 1/16th	73	Triplet 1/16th	32
32nd	24	Dotted 32nd	36	Triplet 32nd	16
64th	12	Dotted 64th	18	Triplet 64th	8
128th	6	Dotted 128th	9	Triplet 128th	4

Note Value	PPQN	Note Value	PPQN	Note Value	PPQN
240 PPQN					
Whole	960	Dotted whole	1440	Triplet whole	640
Half	480	Dotted half	720	Triplet half	320
Quarter	240	Dotted quarter	360	Triplet quarter	160
Eighth	120	Dotted eighth	180	Triplet eighth	80
1/16th	60	Dotted 1/16th	90	Triplet 1/16th	40
32nd	30	Dotted 32nd	45	Triplet 32nd	20
64th	15	Dotted 64th	N/A	Triplet 64th	10
128th	N/A	Dotted 128th	N/A	Triplet 128th	5
384 PPQN					
Whole	1536	Dotted whole	2304	Triplet whole	1024
Half	768	Dotted half	1152	Triplet half	512
Quarter	384	Dotted quarter	576	Triplet quarter	256
Eighth	192	Dotted eighth	288	Triplet eighth	128
1/16th	96	Dotted 1/16th	144	Triplet 1/16th	64
32nd	48	Dotted 32nd	72	Triplet 32nd	32
64th	24	Dotted 64th	36	Triplet 64th	16
128th	12	Dotted 128th	18	Triplet 128th	8

Appendix F

Tempo Delay Time Chart

If Song tempo is 128 BPM, then set delay time to 469 Ms for quarter-note delay, 234 Ms for eighth-note delay, 156 Ms for eighth-triplet delay or 117 Ms for 1/16th note delay.

Tempo	1/4	1/8	1/8T	1/16	Tempo	1/4	1/8	1/8T	1/16	Tempo	1/4	1/8	1/8T	1/16
80	750	375	250	188	118	508	254	169	127	156	385	192	128	96
81	741	370	247	185	119	504	252	168	126	157	382	191	127	96
82	732	366	244	183	120	500	250	167	125	158	380	190	127	95
83	723	361	241	181	121	496	248	165	124	159	377	189	126	94
84	714	357	238	179	122	492	246	164	123	160	375	188	125	94
85	706	353	235	176	123	488	244	163	122	161	373	186	124	93
86	698	349	233	174	124	484	242	161	121	162	370	185	123	92
87	690	345	230	172	125	480	240	160	120	163	368	184	123	92
88	682	341	227	170	126	476	238	159	119	164	366	183	122	91
89	674	337	225	169	127	472	236	157	118	165	364	182	121	91
90	667	333	222	167	128	469	234	156	117	166	361	181	120	90
91	659	330	220	165	129	465	233	155	116	167	359	180	120	90
92	652	326	217	163	130	462	231	154	115	168	357	179	119	89
93	645	323	215	161	131	458	229	153	115	169	355	178	118	88
94	638	319	213	160	132	455	227	152	114	170	353	176	118	88
95	632	316	211	158	133	451	226	150	113	171	351	175	117	88
96	625	313	208	156	134	448	224	149	112	172	349	174	116	87
97	619	309	206	155	135	444	222	148	111	173	347	173	116	87

Tempo	1/4	1/8	1/8T	1/16	Tempo	1/4	1/8	1/8T	1/16	Tempo	1/4	1/8	1/8T	1/16
98	612	306	204	153	136	441	221	147	110	174	345	172	115	86
99	606	303	204	153	137	438	219	146	109	175	343	171	114	86
100	600	300	200	150	138	435	217	145	109	176	341	170	114	85
101	594	297	198	149	139	432	216	144	108	177	339	169	113	85
102	588	294	196	147	140	429	214	143	207	178	337	169	112	84
103	583	291	194	146	141	426	213	142	106	179	335	168	112	84
104	577	288	192	144	142	423	211	141	106	180	333	168	111	83
105	571	286	190	143	143	420	210	140	105	181	331	167	110	83
106	566	283	189	142	144	417	208	139	104	182	299	166	110	82
107	561	280	187	140	145	414	207	138	103	183	297	165	109	82
108	556	278	185	139	146	411	205	137	103	184	295	164	108	81
109	550	275	183	138	147	408	302	136	102	185	293	164	108	81
110	545	273	182	136	148	405	203	135	101	186	291	163	107	80
111	541	270	180	135	149	403	201	134	101	187	289	162	106	80
112	536	268	179	134	150	400	200	133	100	188	287	161	106	79
113	531	265	177	133	151	397	199	132	99	190	285	161	105	79
114	526	263	175	132	152	395	197	132	99	200	283	160	104	78
115	522	261	174	130	153	392	196	131	98	201	281	159	104	78
116	517	259	172	129	154	390	195	130	97	202	279	158	103	77
117	513	256	171	128	155	387	194	129	97	203	277	157	102	77

1/8 = eighth-note delay, 1/8T = eighth-note triplet delay, 1/16 = sixteenth-note delay.

Appendix G

Musical Note to MIDI and Frequencies

Note	MIDI No.	Frequency	MIDI No.	Frequency
C	0	8.1757989156	12	16.3515978313
Db	1	8.6619572180	13	17.3239144361
D	2	9.1770239974	14	18.3540479948
Eb	3	9.7227182413	15	19.4454364826
E	4	10.3008611535	16	20.6017223071
F	5	10.9133822323	17	21.8267644646
Gb	6	11.5623257097	18	23.1246514195
G	7	12.2498573744	19	24.4997147489
Ab	8	12.9782717994	20	25.9565435987
A	9	13.7500000000	21	27.5000000000
Bb	10	14.5676175474	22	29.1352350949
B	11	15.4338531643	23	30.8677063285

Note	MIDI No.	Frequency	MIDI No.	Frequency
C	24	32.7031956626	36	65.4063913251
Db	25	34.6478288721	37	69.2956577442
D	26	36.7080959897	38	73.4161919794
Eb	27	38.8908729653	39	77.7817459305
E	28	41.2034446141	40	82.4068892282
F	29	43.6535289291	41	87.3070578583
Gb	30	46.2493028390	42	92.4986056779
G	31	48.9994294977	43	97.9988589954
Ab	32	51.9130871975	44	103.8261743950
A	33	55.0000000000	45	110.0000000000
Bb	34	58.2704701898	46	116.5409403795
B	35	61.7354126570	47	123.4708253140

Note	MIDI No.	Frequency	MIDI No.	Frequency
C	48	130.8127826503	60	261.6255653006
Db	49	138.5913154884	61	277.1826309769
D	50	146.8323839587	62	293.6647679174
Eb	51	155.5634918610	63	311.1269837221
E	52	164.8137784564	64	329.6275569129
F	53	174.6141157165	65	349.2282314330
Gb	54	184.9972113558	66	369.9944227116
G	55	195.9977179909	67	391.9954359817
Ab	56	207.6523487900	68	415.3046975799
A	57	220.0000000000	69	440.0000000000
Bb	58	233.0818807590	70	466.1637615181
B	59	246.9416506281	71	493.8833012561

Note	MIDI No.	Frequency	MIDI No.	Frequency
C	72	523.2511306012	84	1046.5022612024
Db	73	554.3652619537	85	1108.7305239075
D	74	587.3295358348	86	1174.6590716696
Eb	75	622.2539674442	87	1244.5079348883
E	76	659.2551138257	88	1318.5102276515
F	77	698.4564628660	89	1396.9129257320
Gb	78	739.9888454233	90	1479.9776908465
G	79	783.9908719635	91	1567.9817439270
Ab	80	830.6093951599	92	1661.2187903198
A	81	880.0000000000	93	1760.0000000000
Bb	82	932.3275230362	94	1864.6550460724
B	83	987.7666025122	95	1975.5332050245

Note	MIDI No.	Frequency	MIDI No.	Frequency
C	96	2093.0045224048	108	4186.0090448096
Db	97	2217.4610478150	109	4434.9220956300
D	98	2349.3181433393	110	4698.6362866785
Eb	99	2489.0158697766	111	4978.0317395533
E	100	2637.0204553030	112	5274.0409106059
F	101	2793.8258514640	113	5587.6517029281
Gb	102	2959.9553816931	114	5919.9107633862
G	103	3135.9634878540	115	6271.9269757080
Ab	104	3322.4375806396	116	6644.8751612791
A	105	3520.0000000000	117	7040.0000000000
Bb	106	3729.3100921447	118	7458.6201842894
B	107	3951.0664100490	119	7902.1328200980

Note	MIDI No.	Frequency
C	120	8372.0180896192
Db	121	8869.8441912599
D	122	9397.2725733570
Eb	123	9956.0634791066
E	124	10548.0818212118
F	125	11175.3034058561
Gb	126	11839.8215267723
G	127	12543.8539514160

Index

Note: Boldface page numbers refer to figures and tables.

E

U

V